Statistics for Industry and Technology

Series Editor

N. Balakrishnan
McMaster University
Department of Mathematics and Statistics
1280 Main Street West
Hamilton, Ontario L8S 4K1
Canada

Editorial Advisory Board

Max Engelhardt
EG&G Idaho, Inc.
Idaho Falls, ID 83415

Harry F. Martz
Group A-1 MS F600
Los Alamos National Laboratory
Los Alamos, NM 87545

Gary C. McDonald
NAO Research & Development Center
30500 Mound Road
Box 9055
Warren, MI 48090-9055

Peter R. Nelson
Department of Mathematical Sciences
Clemson University
Martin Hall
Box 341907
Clemson, SC 29634-1907

Kazuyuki Suzuki
Communication & Systems Engineering Department
University of Electro Communications
1-5-1 Chofugaoka
Chofu-shi
Tokyo 182
Japan

Advances in Stochastic Simulation Methods

N. Balakrishnan
V.B. Melas
S. Ermakov

Editors

Birkhäuser
Boston • Basel • Berlin

N. Balakrishnan
Department of Mathematics
  and Statistics
McMaster University
Hamilton, Ontario L8S 4K1
Canada
bala@mcmail.cis.mcmaster.ca

S. Ermakov
Faculty of Mathematics and Mechanics
St. Petersburg State University
St. Petersburg 198904
Russia
admin@niimm.spb.su

V.B. Melas
Faculty of Mathematics and Mechanics
St. Petersburg State University
St. Petersburg 198904
Russia
Viatcheslav.Melas@pobox.spbu.ru

Library of Congress Cataloging-in-Publication Data
Advances in stochastic simulation methods / N. Balakrishnan, V.B. Melas, S. Ermakov, editors.
    p.   cm.—(Statistics for industry and technology)
   Includes bibliographical references and index.
   ISBN 0-8176-4107-6 (alk. paper)
   1. Mathematical statistics—Data processing. 2. Mathematical models. 3. Computer simulation. I. Balakrishnan, N., 1956– II. Melas, V.B. (Viacheslav Borisovich)
III. Ermakov, S.M. (Sergei Mikhailovich) IV. Series.
   QA276.4 .A29   2000
   519.5—dc21
                                                                                                                                   00-039791
                                                                                                                                                            CIP

AMS Subject Classifications: 60H35, 82B31, 91B70

Printed on acid-free paper.
© 2000 Birkhäuser Boston   *Birkhäuser*

All rights reserved. This work may not be translated or copied in whole or in part without the written permission of the publisher (Birkhäuser Boston, c/o Springer-Verlag New York, Inc., 175 Fifth Avenue, New York, NY 10010, USA), except for brief excerpts in connection with reviews or scholarly analysis. Use in connection with any form of information storage and retrieval, electronic adaptation, computer software, or by similar or dissimilar methodology now known or hereafter developed is forbidden.
The use of general descriptive names, trade names, trademarks, etc., in this publication, even if the former are not especially identified, is not to be taken as a sign that such names, as understood by the Trade Marks and Merchandise Marks Act, may accordingly be used freely by anyone.

ISBN 0-8176-4107-6
ISBN 3-7643-4107-6   SPIN 10772170

Typeset by the editors in TeX.
Printed and bound by Edwards Brothers, Inc., Ann Arbor, MI.
Printed in the United States of America.

9 8 7 6 5 4 3 2 1

Contents

Preface — xiii
Contributors — xv
List of Tables — xxi
List of Figures — xxv

PART I: SIMULATION MODELS

1 Solving the Nonlinear Algebraic Equations with Monte Carlo Method — 3
S. Ermakov and I. Kaloshin

    Introduction 4
  1.1 Neumann-Ulam Scheme 4
  1.2 Simples Nonlinear Problems 7
    References 14

2 Monte Carlo Algorithms For Neumann Boundary Value Problem Using Fredholm Representation — 17
Y. N. Kashtanov and I. N. Kuchkova

  2.1 Introduction 17
  2.2 Integral Representation 18
  2.3 Monte Carlo Estimators 19
  2.4 The Two-Dimensional Case 23
  2.5 An Application to Navier-Stokes Equations 26
    References 28

3 Estimation Errors for Functionals on Measure Spaces — 29
N. Golyandina and V. Nekrutkin

  3.1 Introduction 29
  3.2 Strong Weakly-Continuous Derivatives 33
  3.3 General Results 34

    3.4 Stratification   40
        3.4.1 General stratification scheme   40
        3.4.2 Examples   41
        References   45

4 The Multilevel Method of Dependent Tests   **47**
Stefan Heinrich

    4.1 Introduction   47
    4.2 The Standard Method of Dependent Tests   48
    4.3 The Multilevel Approach   50
    4.4 Integrals Depending on a Parameter   52
        References   60

5 Algebraic Modelling and Performance Evaluation of Acyclic Fork-Join Queueing Networks   **63**
Nikolai K. Krivulin

    5.1 Introduction   63
    5.2 Preliminary Algebraic Definitions and Results   65
    5.3 Further Algebraic Results   67
    5.4 An Algebraic Model of Queueing Networks   68
        5.4.1 Fork-Join queueing networks   69
        5.4.2 Examples of network models   71
    5.5 A Monotonicity Property   72
    5.6 Bounds on the Service Cycle Completion Time   74
    5.7 Stochastic Extension of the Network Model   75
        5.7.1 Some properties of expectation   76
        5.7.2 Existence of the cycle time   77
        5.7.3 Calculating bounds on the cycle time   78
    5.8 Discussion and Examples   78
        References   81

PART II: EXPERIMENTAL DESIGNS

6 Analytical Theory of E-Optimal Designs for Polynomial Regression   **85**
V. B. Melas

    Introduction   85
    6.1 Statement of the Problem   86
    6.2 Duality Theorem   86
    6.3 The Number of Design Points   87
    6.4 Tchebysheff Designs   90
    6.5 Boundary Equation   91
    6.6 An Extremal Property of Positive Polynomial Representations   93

- 6.7 Differential Equation   95
- 6.8 Limiting Design   103
- 6.9 Taylor Expansion   109
- 6.10 Particular Cases   110
    - 6.10.1 Boundary equation   110
    - 6.10.2 Matrices $J_{1_{(0)}}$ and vectors $J_{z_{(0)}^2}$ 111
    - 6.10.3 Tables of coefficients   112
    - 6.10.4 Studying of convergence radius   113
    - References   114

7 Bias Constrained Minimax Robust Designs for Misspecified Regression Models   117
Douglas P. Wiens

- 7.1 Introduction   117
- 7.2 General Theory   118
- 7.3 Fitting a Second Order Response in Several Regressors   122
    - 7.3.1 S an ellipsoid   122
    - 7.3.2 S a q-dimensional rectangle   123
- 7.4 Fitting a Polynomial Response   123
- 7.5 Wavelet Regression   125
- 7.6 Extrapolation Designs   126
    - 7.6.1 Extrapolation of a polynomial fit   127
    - 7.6.2 Extrapolation of a first order response in several variables   128
- 7.7 Lack of Fit Testing   128
- 7.8 Generalized M-Estimation   130
    - References   130

8 A Comparative Study of MV- and SMV-Optimal Designs for Binary Response Models   135
J. López-Fidalgo and W. K. Wong

- 8.1 Introduction   135
- 8.2 MV- and SMV-Optimal Designs   139
    - 8.2.1 Logistic model   139
    - 8.2.2 Double exponential model   143
- 8.3 Robustness Properties of MV- and SMV-Optimal Designs   144
- 8.4 Conclusions   146
    - References   150

9 On the Criteria for Experimental Design in Nonlinear Error-In-Variables Models — 153
Silvelyn Zwansig

9.1 Introduction   153
9.2 Error-In-Variables Model   156
9.3 The Total Least Squares Estimator   157
    9.3.1 Asymptotic normality and the Hajek bound   159
9.4 The Alternative Estimator   160
9.5 Conclusions   162
    References   163

10 On Generating and Classifying all q^{n-m-1} Regularly Blocked Factional Designs — 165
F. J. Laycock and P. J. Rowley

10.1 Introduction   165
10.2 The Algorithm   168
10.3 Some Specimens   170
    10.3.1 $n = 10$, $k = 7$, $l = 2$, $q = 2$ 170
    10.3.2 $n = 8$, $k = 6$, $l = 3$, $q = 2$ 171
    10.3.3 $n = 7$, $k = 4$, various l, $q = 3$ 172
    10.3.4 Some $q = 4$ examples   174
    References   175

11 Locally Optimal Designs in Non-Linear Regression: A Case Study of the Michaelis-Menten Function — 177
E. P. J. Boer, D. A. M. K. Rasch and E. M. T. Hendrix

11.1 Introduction   177
11.2 Calculation of Optimal Design   178
    11.2.1 Replicationfree designs   178
    11.2.2 Unrestricted designs   179
11.3 Results   180
    11.3.1 Optimal replicationfree designs   181
    11.3.2 Optimal unrestricted designs   182
    11.3.3 Non-convexity of the continuous NLP formulation   185
11.4 Conclusions   185
    References   187

12 D-Optimal Designs for Quadratic Regression Models — 189
E. E. M. van Berkum, B. Pauwels and P. M. Upperman

12.1 Introduction   189
12.2 D-Optimal Designs   190
12.3 Optimality of the Designs   191
12.4 Conclusion   194
    References   194

13 On the Use of Symmetry in Optimal Design of Experiments — 197
Vladimir Soloviov

13.1 Symmetry in Convex Optimization Problems   197
13.2 Optimal Design of Experiments   198
13.3 Optimal Designs for Polynomial Regression   200
    References   203

PART III: STATISTICAL INFERENCE

14 Higher Order Moments of Order Statistics from the Pareto Distribution and Edgeworth Approximate Inference — 207
Aaron Childs, K. S. Sultan and N. Balakrishnan

14.1 Introduction   207
14.2 BLUE's of θ and σ 209
14.3 Exact Expressions for the Moments of Order Statistics   212
    14.3.1 Single moments of order statistics   213
    14.3.2 Double moments of order statistics   213
    14.3.3 Triple moments of order statistics   214
    14.3.4 Quadruple moments of order statistics   215
14.4 Recurrence Relations for Moments of Order Statistics   215
    14.4.1 Relations for single moments   215
    14.4.2 Relations for double moments   216
    14.4.3 Relations for triple moments   217
    14.4.4 Relations for quadruple moments   218
14.5 Approximate Inference   221
14.6 Numerical Illustration   222
14.7 Recurrence Relations for Moments of Order Statistics in the Doubly Truncated Case   224
    14.7.1 Relations for single moments   225
    14.7.2 Relations for double moments   225
    14.7.3 Relations for triple moments   226
    14.7.4 Relations for quadruple moments   228
    References   231

15 Higher Order Moments of Order Statistics from the Power Function Distribution and Edgeworth Approximate Inference — 245
K. S. Sultan, Aaron Childs and N. Balakrishnan

15.1 Introduction   245
15.2 BLUE's of θ and σ 247

15.3 Exact Expressions for the Moments of Order Statistics   250
    15.3.1 Single moments of order statistics   251
    15.3.2 Double moments of order statistics   251
    15.3.3 Triple moments of order statistics   252
    15.3.4 Quadruple moments of order statistics   253
15.4 Recurrence Relations for Moments of Order Statistics   253
    15.4.1 Relations for single moments   253
    15.4.2 Relations for double moments   254
    15.4.3 Relations for triple moments   255
    15.4.4 Relations for quadruple moments   257
15.5 Approximate Inference   259
15.6 Numerical Illustration   261
15.7 Recurrence Relations for Moments of Order Statistics in the Doubly Truncated Case   263
    15.7.1 Relations for single moments   263
    15.7.2 Relations for double moments   264
    15.7.3 Relations for triple moments   265
    15.7.4 Relations for quadruple moments   267
References   270

16 Selecting from Normal Populations the One with the Largest Absolute Mean: Comon Unknown Variance Case   283
S. Jeyaratnam and S. Panchapakesan

16.1 Introduction   283
16.2 Proposed Procedure R_t 285
16.3 Approximation for h 288
16.4 Expected Sample Size   288
16.5 Concluding Remarks   289
References   292

17 Conditional Inference for the Parameters of Pareto Distributions when Observed Samples are Progressively Censored   293
Rita Aggarwala and Aaron Childs

17.1 Introduction   293
17.2 Best Linear Unbiased Estimation   296
17.3 Conditional Confidence Intervals   297
17.4 Conditional Tolerance Intervals   299
17.5 An Example   300
17.6 Sensitivity Analysis   300
References   301

PART IV: APPLIED STATISTICS AND RELATED TOPICS

18 On Randomizing Estimators in Linear Regression Models   305
S. Ermakov and R. Schwabe

    18.1 Some Properties of the Δ^2 Distribution   306
    18.2 Least Squares Estimators and Their Randomization   307
    18.3 Δ^2-Distribution in Experimental Design   311
    References 313

19 Nonstationary Generalized Automata with Periodically Variable Parameters and Their Optimization   315
A. Yu. Ponomareva and M. K. Tchirkov

    19.1   Introduction   315
    19.2   Base Definitions and Problem Setting   316
    19.3   The Basic Matrices of the Automaton $\mathcal{A}_{gv}$ 317
    19.4   Algorithms for Construction of the Families of Basic Matrices   319
    19.5   Two Properties of Basic Matrices   320
    19.6   Reduces and Minimal Forms of the Automaton   321
    19.7   Theorems on Reduced Forms   322
    19.8   Theorems on Minimal Forms   329
    19.9   The Algorithm of Optimization   331
    19.10 Example 331
    References 335

20 Power of Some Asymptotic Tests for Maximum Entropy   337
M. Salicrú, S. Vives and J. Ocaña

    20.1 Introduction   337
    20.2 Taylor Series Approximations   339
    20.3 A Simulation Study   341
    20.4 Results and Discussion   343
    20.5 Expectation and Variance of the Havrda-Charvat Entropies   345
    20.6 Expectation and Variance of the Functional $=(h,\phi)$-Entropies   348
    20.7 Tables   350
        20.7.1 Tables of simulation results   350
    References 351

21 Partially Inversion of Functions for Statistical Modelling of Regulatory Systems 355
A. G. Bart, N. P. Alexeyeff (Klochkova) and N. Botchkina

Introduction 355
21.1 A Method for Partial Inversion of Functions 356
    21.1.1 The parametrical description of the partial inverse functions 356
    21.1.2 Double inversion 358
21.2 Generalised Binomial Distributions 359
21.3 Applications of Fiducial Distributions to Neurophysiology 362
21.4 Advertisement for Sales Marketing 363
    21.4.1 Sanogenesis (compensation) curve 363
    21.4.2 Example 367
    Appendix 368
    References 370

22 Simple Efficient Estimation for Three-Parameter Lognormal Distributions with Applications to Emissions Data and State Traffic Rate Data 373
N. Balakrishnan and Jun Wang

22.1 Introduction 373
22.2 Explicit Estimators 374
22.3 Simulation Results 375
22.4 Illustrative Examples 377
22.5 Concluding Remarks 380
    References 380

Subject Index **385**

Preface

This is a volume consisting of selected papers that were presented at the 3rd St. Petersburg Workshop on Simulation held at St. Petersburg, Russia, during June 28–July 3, 1998.

The Workshop is a regular international event devoted to mathematical problems of simulation and applied statistics organized by the Department of Stochastic Simulation at St. Petersburg State University in cooperation with INFORMS College on Simulation (USA). Its main purpose is to exchange ideas between researchers from Russia and from the West as well as from other countries throughout the World. The 1st Workshop was held during May 24–28, 1994, and the 2nd workshop was held during June 18-21, 1996. The selected proceedings of the 2nd Workshop was published as a special issue of the Journal of Statistical Planning and Inference.

Russian mathematical tradition has been formed by such genius as Tchebysheff, Markov and Kolmogorov whose ideas have formed the basis for contemporary probabilistic models. However, for many decades now, Russian scholars have been isolated from their colleagues in the West and as a result their mathematical contributions have not been widely known. One of the primary reasons for these workshops is to bring the contributions of Russian scholars into limelight and we sincerely hope that this volume helps in this specific purpose.

This volume contains 22 articles by experts from different countries. Based on the technical contents, the volume has been divided into four parts with

Part I – Simulation Models
Part II – Experimental Designs
Part III – Statistical Inference
Part IV – Applied Statistics and Related Topics

We thank all the authors for submitting their contributions to this volume in good shape and in a timely fashion. We thank Mr. Wayne Yuhasz and Ms. Lauren Lavery for taking a keen interest in this project and helping us in shaping the volume into its present form. Our final thanks go to Mrs. Debbie Iscoe for the fine typesetting of this entire volume.

Hamilton, Canada	*N. Balakrishnan*
St. Petersburg, Russia	*V. B. Melas*
St. Petersburg, Russia	*S. Ermakov*

January 2000

Contributors

Aggarwala, Rita
Department of Mathematics and Statistics, University of Calgary, Calgary, AB, Canada T2N 1N4
e-mail: *rita@math.ucalgary.ca*

Alexeyeff (Klochkova), N. P.
St. Petersburg State University, Bibliotechnaya sq. 2, 198904, St. Petersburg, Russia

Balakrishnan, N.
Department of Mathematics and Statistics, McMaster University, Hamilton, ON, Canada L8S 4K1
e-mail: *bala@mcmail.cis.mcmaster.ca*

Bart, A. G.
St. Petersburg State University, Bibliotechnaya sq. 2, 198904, St. Petersburg, Russia
e-mail: *agb@agb.usr.pu.ru*

Boer, E. P. J.
Department of Agricultural, Environmental and Systems Technology, Subdepartment of Mathematics, Dreijenlaan 4, 6703 HA Wageningen, The Netherlands
e-mail: *eric.boer@wts.wk.wau.nl*

Botchkina, H.
St. Petersburg State University, Bibliotechnaya sq. 2, 198904, St. Petersburg, Russia
e-mail: *nataly@LB2355.sbp.edu*

Childs, Aaron
  Department of Mathematics and Statistics, McMaster University, Hamilton, ON, Canada L8S 4K1
  e-mail: *childsa@mcmaster.ca*

Ermakov, S.
  St. Petersburg State University, Bibliotechnaya sq. 2, 198904, St. Petersburg, Russia
  e-mail: *sergej.ermakov@pobox.spbu.ru*

Golyandina, N.
  St. Petersburg State University, Bibliotechnaya sq. 2, 198904, St. Petersburg, Russia
  e-mail: *nina@neg.usr.pu.ru*

Heinrich, Stefan
  Fachbereich Informatik, Universität Kaiserslautern, Postfach 3049, D-67653 Kaiserslautern, Germany
  e-mail: *heinrich@informatik.uni-kl.de*

Hendrix, E. M. T.
  Department of Agricultural, Environmental and Systems Technology, Sub-department of Mathematics, Dreijenlaan 4, 6703 HA Wageningen, The Netherlands
  e-mail: *Eligius.hendrix@oa.wk.wau.nl*

Jeyaratnam, S.
  Department of Mathematics, Southern Illinois University, Carbondale, IL, USA 62901-4408
  e-mail: *sjeyarat@math.siu.edu*

Kaloshin, I.
  St. Petersburg State University, Bibliotechnaya sq. 2, 198904, St. Petersburg, Russia
  e-mail: *igor@vega.math.spbu.ru*

Kashtanov, Y. N.
  St. Petersburg State University, Bibliotechnaya sq. 2, 198904, St. Petersburg, Russia
  e-mail: *yuri.kashtanov@paloma.spbu.ru*

Krivulin, Nikolai, K.
  St. Petersburg State University, Bibliotechnaya sq. 2, 198904, St. Petersburg, Russia
  e-mail: *nikolai@knk.usr.pu.ru*

Contributors

Kuchkova, I. N.
St. Petersburg State University, Bibliotechnaya sq. 2, 198904, St. Petersburg, Russia
e-mail: *kin@vega.math.spbu.ru*

López-Fidalgo, J.
Department of Statistics, University of Salamanca, 37008-Salamanca, Spain
e-mail: *fidalgo@gugu.usal.es*

Laycock, P. J.
Department of Mathematics, University of Manchester Institute of Science and Technology, Sackville St. , Manchester M60 1QD, U.K.
e-mail: *pjlaycock@fs1.ma.umist.ac.uk*

Melas, V. B.
St. Petersburg State University, Mathematical-Mechanical Faculty, Bibliotechnaya sq. 2, St. Petersburg, Petrodvoretz, 198904, Russia
e-mail: *Viatcheslav.melas@pobox.spbu.ru.*

Nekrutkin, V.
St. Petersburg State University, Bibliotechnaya sq. 2, 198904, St. Petersburg, Russia
e-mail: *nekr@stat2.math.lgu.spb.su*

Ocaña, J.
Department of Statistics, Barcelona University, Spain
e-mail: *ocana@porthos.bio.ub.es*

Panchapakesan, S.
Department of Mathematics, Southern Illinois University, Carbondale, IL, USA 62901-4408
e-mail: *kesan@math.siu.edu*

Pauwels, B.
UFSIA, University of Antwerp, Antwerpen, Belgium
e-mail: *fte.pauwels.b@alpha.ufsia.ac.be*

Ponomareva, A. Yu.
Research Institute for Mathematics and Mechanics, Petrodvorets, St. Petersburg State University, Bibliotechnaya sq. 2, 198904, St. Petersburg, Russia

Rasch, D. A. M. K.
Department of Agricultural, Environmental and Systems Technology, Subdepartment of Mathematics, Dreijenlaan 4, 6703 HA Wageningen, The Netherlands
e-mail: *Dieter.Rasch@WTS.WK.WAU.NL*

Rowley, P. J.
> Department of Mathematics, University of Manchester Institute of Science and Technology, Sackville St. , Manchester M60 1QD, U.K.
> e-mail: *peter.rowley@umist.ac.uk*

Salicrú, M.
> Department of Statistics, Barcelona University, Spain
> e-mail: *miquel@bio.ub.es*

Schwabe, R.
> Free University, Berlin, Institute for Mathematics I, Arnimallee 2–6, D-14 195 Berlin, Germany
> e-mail: *schwabe@mathematik.th-darmstadt.de*

Soloviov, Vladimir
> Moscow State Social University
> e-mail: *solovio@orc.ru*

Sultan, K. S.
> Al-Azhar University, Nasr City, Cairo, Egypt
> e-mail: *OrObada@frcu.eun.eg*

Tchirkov, M. K.
> Research Institute for Mathematics and Mechanics, Petrodvorets, St. Petersburg State University, Bibliotechnaya sq. 2, 198904, St. Petersburg, Russia
> e-mail: *tchirkov@niimm.spb.su*

Upperman, P. M.
> Quality Engineering Consultancy, Oosterhesselen, The Netherlands

van Berkum, E. E. M.
> Eindhoven University of Technology, Eindhoven, The Netherlands
> e-mail: *wsevb@win.tue.nl*

Vives, S.
> Department of Statistics, Barcelona University, Spain
> e-mail: *sergi@porthos.bio.ub.es*

Wang, Jun
> Department of Mathematics and Statistics, McMaster University, Hamilton, ON, Canada
> e-mail: *junwang@loyalty.com*

Contributors

Wiens, Douglas, P.
    Department of Mathematical Sciences, University of Alberta, 429 Central Academic Building, Edmonton, AB, Canada T6G 2G1
    e-mail: *wiens@stat.uablerta.ca*

Wong, W. K.
    Department of Biostatistics, University of California, Los Angeles, CA 90095-1772
    e-mail: *wkwong@sun.sunlab.ph.ucla.edu*

Zwanzig, Silvelyn
    University of Hamburg, Hamburg, Germany
    e-mail: *Zwanzig@math.uni-hamburg.de*

Tables

Table 1.1:		12
Table 1.2:		13
Table 1.3:	$x = 0.1x^2 + 0.7$ $p^{direct}_{absorption} = 0.1$, $p^{conjugate}_{absorption} = 0.1$ $x_{exact} = 0.7573593$	13
Table 1.4:	$x = 0.7x^2 + 0.1$ $p^{direct}_{absorption} = 0.1$, $p^{conjugate}_{absorption} = 0.1$ $x_{exact} = 0.1081942$	13
Table 1.5:	$x = 0.15x^2 + 0.3$ $p^{direct}_{absorption} = 0.15$, $p^{conjugate}_{absorption} = 0.3$ $x_{exact} = 0.3148716$	14
Table 1.6:	$x = 0.15x^2 + 0.3$ $p^{direct}_{absorption} = 0.3$, $p^{conjugate}_{absorption} = 0.15$ $x_{exact} = 0.3148716$	14
Table 5.1:	Numerical results for a network with dependent service times.	79
Table 5.2:	Results for a network with a dominating service time.	80
Table 5.3:	Results for tandem queues at changing variance.	80
Table 6.1:	Table of coefficients ($m = 4$).	112
Table 6.2:	Table of coefficients ($m = 5$).	113
Table 6.3:	Table of coefficients ($m = 6$).	113
Table 8.1:	Sensitivities of MV- and SMV-optimal designs to the nominal values of the parameters in the logistic model. The value of $\sqrt{c}$ is 1.33379.	148
Table 8.2:	Sensitivities of MV- and SMV-optimal designs to the nominal values of the parameters in the double exponential model. The values of $\sqrt{c}$ and $\sqrt{v_0}$ are 1.35699 and 1.26239 respectively.	149
Table 11.1:	Measurements of Michaelis and Menten (1913) of the change of rotation y depending on time x.	181
Table 11.2:	Optimal replicationfree designs for the three different criteria, with the criterion value of the optimal design and the criterion value of the original design of Michaelis and Menten.	182

Table 12.1:	Values determining the D-optimal designs of Corollary 12.2.1.	**192**
Table 14.1:	Mean, variance and coefficients of skewness and kurtosis of $R1^*$ and $R2^*$ when $v = 5$.	**233**
Table 14.2:	Mean, variance and coefficients of skewness and kurtosis of $R1^*$ and $R2^*$ when $v = 15$.	**234**
Table 14.3:	Mean, variance and coefficients of skewness and kurtosis of $R1^*$ and $R2^*$ when $v = 25$.	**235**
Table 14.4:	Percentage points of the distribution of R_1 when $v = 5$.	**236**
Table 14.5:	Percentage points of the distribution of R_1 when $v = 15$.	**237**
Table 14.6:	Percentage points of the distribution of R_1 when $v = 25$.	**238**
Table 14.7:	Percentage points of the distribution of R_2 when $v = 5$.	**239**
Table 14.8:	Percentage points of the distribution of R_2 when $v = 15$.	**240**
Table 14.9:	Percentage points of the distribution of R_2 when $v = 25$.	**241**
Table 14.10:	Simulated percentage points of the distribution of R_3 when $v = 5$.	**242**
Table 14.11:	Simulated percentage points of the distribution of R_3 when $v = 15$.	**243**
Table 14.12:	Simulated percentage points of the distribution of R_3 when $v = 25$.	**244**
Table 15.1:	Mean, variance and coefficients of skewness and kurtosis of $R1^*$ and $R2^*$ when $v = 3/2$.	**271**
Table 15.2:	Mean, variance and coefficients of skewness and kurtosis of $R1^*$ and $R2^*$ when $v = 3$.	**272**
Table 15.3:	Mean, variance and coefficients of skewness and kurtosis of $R1^*$ and $R2^*$ when $v = 6$.	**273**
Table 15.4:	Percentage points of the distribution of R_1 when $v = 3/2$.	**274**
Table 15.5:	Percentage points of the distribution of R_1 when $v = 3$.	**275**
Table 15.6:	Percentage points of the distribution of R_1 when $v = 6$.	**276**
Table 15.7:	Percentage points of the distribution of R_2 when $v = 3/2$.	**277**

Table 15.8:	Percentage points of the distribution of R_2 when $v = 3$.	**278**
Table 15.9:	Percentage points of the distribution of R_2 when $v = 6$.	**279**
Table 15.10:	Simulated percentage points of the distribution of R_3 when $v = 3/2$.	**280**
Table 15.11:	Simulated percentage points of the distribution of R_3 when $v = 3$.	**281**
Table 15.12:	Simulated percentage points of the distribution of R_3 when $v = 6$.	**282**
Table 16.1:	Value of h satisfying equation (16.8) $P^* = 90$.	**287**
Table 16.2:	Value of h satisfying equation (16.8) $P^* = 95$.	**287**
Table 16.3:	Exact and approximate value of h $P^* = 90$ (top entry) and 0.95 (bottom entry).	**288**
Table 16.4:	Values of N_S and $E(N)/N_S$ with $n_0 = N_S$ $P^* = 0.90$.	**290**
Table 16.5:	Values of N_S and $E(N)/N_S$ with $n_0 = N_S$ $P^* = 0.95$.	**291**
Table 18.1:		**312**
Table 18.2:		**313**
Table 20.1:	Havrda-Charvat ($\alpha = 3D3$) $M = 3D\ 20$, $n = 3D\ 40$ $\varepsilon = 3D0.05$.	**344**
Table 20.2:	Asymptotic behaviour of Havrda-Charvat and Renyi entropies versus parameter alpha $M = 3D\ 20$, $n = 3D\ 40\ \varepsilon = 3D0.05$.	**345**

Figures

Figure 2.1:		27
Figure 2.2:		27
Figure 2.3:		27
Figure 7.1:	MVU design densities over the square $S = [-1,1] \times [-1,1]$, robust against heteroscedasticity as well as response uncertainty. (a) Design for estimating a second order response. (b) Design for extrapolation of a first order fit to $\mathcal{T} = \{[-1,2] \times [-1,2]\} \setminus \mathcal{S}$.	122
Figure 7.2:	MVU design densities on $[-1,1]$, robust against heteroscedasticity as well as response uncertainty. (a) Polynomial regression. 1: $k_*(x, q=3)$; 2: $k_*(x, \infty)$. (b) Wavelet regression. 1: $k_{3,0}(x)$; 2: $k_{3,2}(x)$. (c) Extrapolation of a cubic fit. 1: $k_0(x, t_0 = 1.5)$, symmetric extrapolation; 2: $k_0(x, t_1 = 1.5)$, asymmetric extrapolation; 3: $k_0(x, \infty)$.	124
Figure 8.1:		138
Figure 8.2:		141
Figure 8.3:		142
Figure 8.4:		146
Figure 8.5:		147
Figure 9.1:		155
Figure 9.2:		155
Figure 9.3:		162
Figure 11.1:	Measurements and the fitted Michaelis-Menten function.	181
Figure 11.2:	Contour plot of the minisation problems of x_1^* and x_2^* for d-optimality, $r_1 = r_2 = 5$. The plotted criterion values are equal to $K_1 \times 10^7$.	183

Figure 11.3: Criterion K_2 for different combinations of (r_1, r_2): ○ *corresponds to* $(7,3)$, ▲ *to* $(8,2)$ and ■ *to* $(9,1)$. Criterion value: $K_2 \cdot 10^3$. **184**

Figure 11.4: Non-convexity of the continuous NLP formulation of the optimal unrestricted design problem. Criterion value: $K_1 \cdot 10^7$. **186**

Figure 21.1: Graphs of the function $H_\gamma(t) = S_{\gamma 1}^{-2}$ with $\gamma = 1$ and $\gamma = 0.5$. **364**

Figure 21.2: Crucial points of sanogenesis curve. **364**

Figure 21.3: Dynamics of advertisement $(Y(t))$ and sales percentage $(X(t))$. **367**

PART I
SIMULATION MODELS

1

Solving the Nonlinear Algebraic Equations with Monte Carlo Method

S. Ermakov and I. Kaloshin
St. Petersburg State University, St. Petersburg, Russia

Abstract: Many problems of rarefied gas dynamics can be solved by imitation of the collision molecules process [see, for example, Bird (1994), and Illner and Neunzert (1987)]. It is known that this process is described with a good precision by the equation with quadratic nonlinearity (the Boltzman equation). The relation between the Boltzman equation and branching and collision processes was studied in the book [Ermakov, Nekrutkin and Sipin (1989)] and a number of following papers [for example, Goliandina (1996), and Nekrutkin and Tur (1997)]. It worth to note that a great number of papers were devoted to this relation. However, the problem of the relation between general type equations with quadratic nonlinearity and random processes was studied relatively small. Especially, numerical techniques for the related processes simulation were not enough investigated. At the same time many physical processes are described by equations with quadratic, or, in general, polynomial nonlinearity. The important example is the Navier-Stokes equations.

It is known also that many simulation techniques for the solution of equations are based on the Neumann-Ulam scheme (N. U. scheme). And it is of a great interest to study the relation of this scheme with known simulation techniques for the solution of problems of rarefied gas dynamic.

This work continues the research of the certain kind of Monte Carlo algorithms for solving the equations with polynomial nonlinearity. The authors have limited themselves with a detailed considering of the systems of algebraic quadratic equations.

Keywords and phrases: Monte Carlo method, Neumann-Ulam scheme, system of nonlinear equations, estimator by collision, estimator by absorption

Introduction

Neumann-Ulam scheme (N.U.scheme) is a well-known instrument for constructing the Monte Carlo algorithms in the linear case. This scheme shows the way to construct Markov chain that is characteristic for this problem as well as a large number of unbiased estimators for functionals from the solving problem.

The method of equivalent substitution of the equation was described in [Nekrutkin (1974)] with polynomial nonlinearity with infinite system of linear equations under some additional "majorant" conditions. Further this method is applied to the system of square equations. After that we consider the algorithms for solving the received system with the help of Neumann-Ulam scheme.

The formal application of N. U. scheme to the mentioned system leads to random processes that are certain modifications of the processes described earlier [Ermakov (1972), Nekrutkin (1974)]. In some cases these modifications can be more convenient for calculations than the well-known estimators in nonlinear case.

1.1 Neuman-Ulam Scheme

Let us recall some basic facts concerning N. U. scheme in the case of the linear algebraic equations systems. Proofs of these facts can be found in standard handbooks [Ermakov (1971)].

Let a system of equations

$$X = AX + F, \quad X = (x_1, \ldots, x_n)^T, \quad F = (f_1, \ldots, f_n)^T, \quad A = \|a_{ij}\|_{i,j=1}^n \quad (1.1)$$

be given.

It is supposed (majorant conditions) that the finite limits $\overline{X}_m = (\overline{x}_1^{(m)}, \ldots, \overline{x}_n^{(m)})$ for the vector exists

$$\overline{X}_m = |A|\overline{X}_{m-1} + |F|, \quad \overline{X}_0 = |F| \quad (1.2)$$

as m tends to ∞. The homogenous Markov chain $(p^0, \mathcal{P})$, $p^0 = (p_1^0, \ldots, p_n^0)$, $\mathcal{P} = \|p_{ij}\|_{i,j=1}^n$ is associated with the system (1.1) moreover it has n states and $(n+1)$-th absorbing state.

$$\sum_{j=1}^n p_{ij} = 1 - g_i, \quad 0 \leq g_i \leq 1, \quad i = 1, 2, \ldots, n \quad (1.3)$$

p^0 - is an initial distribution, $\mathcal{P}$ - is a transition matrix and g_i - is an absorption probability in states with number i. It is supposed that all trajectories $i_0 \to i_1 \to \ldots \to i_\tau$ of Markov chain are finite with probability 1.

Let us take a vector $H = (h_1, \ldots, h_n)$ and simulate a trajectory $i_0 \to i_1 \to \ldots \to i_\tau$ in correspondence with given p^0 $\mathcal{P}$.

Simultaneously with calculation i_t, $t = 0, 1, \ldots, \tau$ (τ is a stop point) we will calculate values Q_t (statistical weights).

$$Q_0 = \frac{h_{i_0}}{p_{i_0}^0}, \qquad Q_{t+1} = Q_t \frac{a_{i_t, i_{t+1}}}{p_{i_t, i_{t+1}}} \tag{1.4}$$

Then if concordance conditions (see below) are fulfilled, the random values

$$\xi = Q_\tau \frac{f_{i_\tau}}{g_{i_\tau}} \qquad \text{absorption estimator} \tag{1.5}$$

$$\eta = \sum_{t=0}^{\tau} Q_t f_{i_t} \qquad \text{collision estimator} \tag{1.6}$$

are unbiased estimators for $(H, \hat{X})$ where $\hat{X}$ - is the solution of system of equations (1.1):

$$E\xi = E\eta = (H, \hat{X}) \tag{1.7}$$

The concordance conditions play an important role because they determine a structure of process $(p^0, \mathcal{P})$ connecting them with a structure of system (1.1).

1^0 (Conditions for $\mathcal{P}$) $p_{ij} \neq 0$ if and only if $a_{ij} \neq 0$ (1.8)

2^0 (Conditions for p_i^0 and g_i in the case of estimator (1.5))

$$p_i^0 > 0, \text{ if and only if } h_i \neq 0 \tag{1.9}$$
$$g_i^0 > 0, \text{ if and only if } f_i \neq 0$$

3^0 In the case of estimator (1.6) only condition for p_i^0 is needed

$$p_i^0 > 0, \text{ if and only if } h_i \neq 0 \tag{1.10}$$

Generally speaking the equality takes place for more general conditions than (8), (9), (10). See [Ermakov (1971)]. Our conditions give simple computation algorithms. A dual (conjugate) scheme for the described scheme exists. It is constructed on the base of the well-known equality.

$$(H, \hat{X}) = (\hat{Y}, F), \tag{1.11}$$

where $\hat{Y}$ is solution of (conjugated) equation

$$Y = A^T Y + H, \quad Y = (y_1, \ldots, y_n) \tag{1.12}$$

After respective modification of the concordance conditions ($\mathcal{P}$ must be concorded with A^T in 1^0; h_i and f_i change there place in 2^0 and 3^0) we can construct estimators ξ' and η' on the trajectories of another, generally speaking, process $(p_0', \mathcal{P}')$.

$$\xi' = Q_\tau' \frac{h_{i_\tau}}{g_{i_\tau}'}, \qquad \eta' = \sum_{t=0}^{\tau} Q_t' h_{i_t}, \tag{1.13}$$

These estimators are dual to ξ and η and the following equalities hold

$$E\eta' = E\xi' = (\hat{Y}, F) = (H, \hat{X}). \qquad (1.14)$$

Let us consider a simple example to illustrate peculiarities of direct and conjugate N. U. scheme.

The simplest difference scheme for Heat equation

$$\frac{\partial U}{\partial t} = \frac{\partial^2 U}{\partial x^2}, \quad t \in [0, T], \quad x \in [0, 1] \quad \text{is}$$

$$U((n+1)\Delta t, kh) = \frac{1}{2}\left(U(n\Delta t, (k+1)h) + U(n\Delta t, (k-1)h)\right).$$

If $U(t, x)$ is given in initial point of the time and in each moment on the boundaries of the interval then $U(0, kh) = \phi_k$, $U(n\Delta t, 0) = \psi_1^n$ and $U(n\Delta t, 1) = \psi_2^n$ are known.

ϕ_k and ψ are components of vector F in the direct scheme and may be nonzeros.

The matrix A has a block structure

$$A = \begin{pmatrix} 0 & A_1 & 0 & . & 0 \\ 0 & 0 & A_1 & . & 0 \\ . & . & . & . & . \\ 0 & 0 & 0 & . & A_1 \\ 0 & 0 & 0 & . & 0 \end{pmatrix}, \text{ where} \qquad (1.15)$$

$$A_1 = \begin{pmatrix} 0 & 1/2 & & & \\ 1/2 & 0 & . & & \\ & . & . & . & \\ & & . & . & . \\ & & & . & 1/2 \\ & & & 1/2 & 0 \end{pmatrix}$$

Moreover A is substochastic and may be chosen as $\mathcal{P}$. If H has an unique nonzero component for example in the point $(n_0 \Delta t, k_0 h)$ of the net then in accordance with 2^0 a correspondent p_i^0 must be 1 (i is a number that corresponds to the point n_0, k_0 of the net).

All concordance conditions in this case are fulfilled and we have a random walk on the net. (n_0, k_0) is an initial point, "a particle" will pass with probability $1/2$ on each step from the point (t, k) in one of the two points $(t-1, k-1)$ or $(t-1, k+1)$. The walk is finished with probability one in the first of the points where U is known.

The dual scheme gives

$$A^T = \begin{pmatrix} 0 & 0 & . & 0 & 0 \\ A_1 & 0 & . & 0 & 0 \\ 0 & A_1 & . & 0 & 0 \\ . & . & . & . & . \\ 0 & 0 & . & A_1 & 0 \end{pmatrix}$$

In this case the initial distribution p^0 must be agreed with F and the vector of absorption probabilities with H.

If $\mathcal{P} = A^T$ then the walk has opposite direction in the time. "A particle" passes from the point (t, k) in one of the two points $(t+1, k+1)$ or $(t+1, k-1)$ with equal probabilities.

A particle is born in one of boundary points of the net where a correspondent component of F is nonzero and the walk is finished in one of the points where the correspondent component of H is nonzero. We should calculate a number of trajectories that are finished in this point to estimate solution in a point (n_0, k_0).

Under the described choice $\mathcal{P} = A$ or $\mathcal{P} = A^T$ statistical weights Q_t and Q'_t are constant on all trajectories of processes. As it is well known [Ermakov (1971)] some problems, especially problems of high precision estimation of separate inner products (H, X) demand a special choice of $\mathcal{P}$ and $\mathcal{P}'$ ($\mathcal{P} \neq A$, $\mathcal{P}' \neq A^T$). It demands calculation of Q_t (Q'_t) in simulation process.

1.2 Simplest Nonlinear Problems

We will consider the systems of algebraic equations with polynomial nonlinearity. The detailed considering of the simplest choice allows to receive some new results in comparison with [Nekrutkin (1974), Sizova (1976)] and in particularly we can describe structure of collision estimators.

The system of the algebraic equations of an aspect

$$x_i = f_i + \sum_{i=1}^{n} a_{ij} x_j + \sum_{j=1}^{n} \sum_{k=1}^{n} b_{ijk} x_j x_k, \quad i = 1, \ldots, n \qquad (1.16)$$

is considered in the supposition of existence of the final limit

$$\lim_{n \to \infty} \bar{x}_i^{(n)} = \bar{x}_i, \text{ where}$$

$$\bar{x}_i^{(n)} = |f_i| + \sum_{i=1}^{n} |a_{ij}| \bar{x}_i^{(n-1)} + \sum_{j=1}^{n} \sum_{k=1}^{n} |b_{i,j,k}| \bar{x}_j^{(n-1)} \bar{x}_k^{(n-1)}, \quad i = 1, 2, \ldots, n \qquad (1.17)$$

Construction of branching random process for a solution of system (1.16) with condition (1.17) is circumscribed in [Ermakov (1971) p.327]. In [Nekrutkin (1974)] the infinite system of linear equations is indicated (equivalent for (1.16)), for which the N. U. scheme is formal applicable.

The absorption estimators are constructed in both cases. As to collision estimator, its analog in the case of branching process was considered in [Sizova (1976)]. The realization of this estimator, however, is connected with handling of all subtrees (subsets of trajectories) of branching process and it appears to be rather difficult. Analogues of collision estimators are considered below, they appear when the N. U. scheme is applied to the mentioned system of linear equations. The algorithms of simulation are described and numerical examples are given.

Formal multiplication of (1.16) on a product $x_{i_1} x_{i_2} \cdots x_{i_l}$ $i_1, i_2, \ldots, i_l = 1, \ldots, n$ and denotation $x_{i_1}, \ldots, x_{i_l} = U_l(i_1, \ldots, i_l)$ leads to an infinite system of linear equations after some modifications of variables indexing:

$$U_l(i_1, \ldots, i_l) = f_{i_1} U_{l-1}(i_2, \ldots, i_l) + \sum_{j=1}^{n} a_{i_1,j} U_l(j, i_2, \ldots, i_l) +$$
$$+ \sum_{j=1}^{n} \sum_{k=1}^{n} b_{i_1,j,k} U_{l+1}(j, k, i_2, \ldots, i_l), \qquad (1.18)$$
$$i_1, \ldots, i_l = 1, 2, \ldots, n, \quad l = 1, 2, \ldots$$

Thus $U \equiv 1$ and $U_l(i_1, \ldots, i_l) = U_l(i'_1, \ldots, i'_l)$, where $i'_1, \ldots, i'_l$ any permutation of indexes $i_1, \ldots, i_l$.

Let us apply the N.U.Scheme to the system (1.18). The conditions (1.17) are equivalent to (1.2) and in order to construct the algorithm and corresponding estimators we must chose a homogeneous Markov chain in accordance with condition (1.8)–(1.10).

It is a simple problem if we can write the matrix U of system (1.18). In Table 1.1 we can see first lines of U for the case $n = 2$, and in Table 1.2 the first lines of U^T.

The authors suppose that a reader can easily continue each of these matrixes and rewrite them for the case of any n.

It is sufficient to use the formulae (1.16). If we have U then the Markov Homogeneous chain with a countable number of states is constructed as follows. The structure $\mathcal{P}$ coincides with the structure of a matrix $\mathcal{A}$ systems (1.18), i.e. $\mathcal{P}$ has zeros on the same places as $\mathcal{A}$. Instead of f_i in $\mathcal{A}$ we have p_i^0 in $\mathcal{P}$, instead of $a_{i,j}$ we have $p_{i,j}^1$ ($p_{i,j}^1 > 0$, if $a_{i,j} \neq 0$), and instead of $b_{i,j,k}$ we have $p_{i,j,k}^2$ ($p_{i,j,k}^2 > 0$, if $b_{i,j,k} \neq 0$). Thus

$$p_i^0 + \sum_{i=1}^{n} p_{i,j}^1 + \sum_{j=1}^{n} \sum_{k=1}^{n} p_{i,j,k}^2 = 1, \quad i = 1, \ldots, n.$$

It is interesting to mention that f_i is absent (enters into the right side) in the first (appropriate $l = 1$) row of matrixes $\mathcal{A}$ and in this line

$$\sum_{i=1}^{n} p_{i,j}^1 + \sum_{j=1}^{n} \sum_{k=1}^{n} p_{i,j,k}^2 = 1 - q_i, \quad q_i > 0.$$

Generally speaking we could suppose that the probabilities $p_i^0, p_{i,j}^1, p_{i,j,k}^2$ are different for the different l, but we assume further that they are identical.

Markov chain with discrete time $t = 0, 1, \ldots, n$ and transitional matrix arranged in this way allows the following natural interpretation (in a phase space of l particles). We have 1 particle with $t = 0$, and the distribution of the states $\pi = (\pi_1, \ldots, \pi_n)$ should be given.

If in time $t > 0$ $l > 1$ of particles is present, one of them is selected with equal probabilities and the following possibilities can be carried out for it. If i_t is number of the state, in which it is located (in a time t), then in a time $t + 1$ we have the next possibilities:

1. If the probability is $p_{i_t}^0$, it is lost $(l \to l - 1)$

2. If the probability is $p_{i_t, i_{t+1}}^1$, it passes to the state with number i_{t+1} $(l \to l)$

3. If the probability is $p_{i_t, i_{t+1}, i'_{t+1}}^2$, two particles in states with numbers i_{t+1} and i'_{t+1} correspondently appear. $(l \to l + 1)$

The remaining particles do not change their conditions.

The "statistical weight" Q_t is calculated simultaneously with modeling according to the following formulae

$$Q_0 = \frac{h_{i_0}}{\pi_{i_0}}; \qquad Q_{t+1} = Q_t \frac{f_{i_t}}{p_{i_t}^0} \quad \text{in case of 1,} \quad l > 1$$

$$Q_{t+1} = Q_t \frac{a_{i_t, i_{t+1}}}{p_{i_t, i_{t+1}}^1} \quad \text{in case of 2} \qquad (1.19)$$

$$Q_{t+1} = Q_t \frac{b_{i_t, i_{t+1}, i'_{t+1}}}{p_{i_t, i_{t+1}, i'_{t+1}}^2} \quad \text{in case of 3}$$

If $l = 1$ in case 1, trajectory is broken $(t = \tau)$ with probability q_{i_τ} and the quantity $\xi_\tau^{(1)} = Q_\tau \frac{f_{i_\tau}}{q_{i_\tau}}$ is a unbiased estimator for the scalar product (H, X), where $H = h_1, \ldots, h_n$, and $X = x_1, \ldots, x_n$ is an iterative solution of system (1.16). ξ_τ is named absorption estimator.

The construction of collisions estimator in our case has some peculiarities in connection with the special form of the right side of the system (1.18). The absorption can take place only with $l = 1$ and if $t_1, \ldots, t_s$ are time moments, when only one particle $(l = 1)$ was left, the collision estimator should look like

$$\xi_\tau^{(2)} = \sum_{i=1}^{s} Q_{t_i} f_{t_i} \qquad (1.20)$$

It obviously differs from the estimator considered in [Sizova (1976)].

To derive a "conjugate" estimator $\mathcal{A}$ is transposed and the countable Markov chain is created, and the structure of transitional matrix is coherence with $\mathcal{A}^T$ (see Table 1.2).

Thus the right side of a conjugate system has a form $(h_1, \ldots, h_n, 0, 0, \cdots)$.

Exposition of chain modeling and algorithm of the collision estimator construction follows. In terms of an ensemble l of particles - the process appears collisionly. It is designed by initial distribution, matrix of transition probabilities $\mathcal{P}$ and probability of a breakaway of a trajectory.

$\mathcal{P}$, as well as $\mathcal{A}$ has a block structure with rectangular and square blocks, which dimensionality will increase as n^l with growth l. Each line with $l > 1$ contains sequentially probabilities $p^2_{i,j,k}$, $i = 1, 2, \ldots, n$, $p^1_{i,j}$, $i = 1, \ldots, n$ and p^0_i, $i = 1, \ldots, n$. The first n lines do not contain $p^2_{i,j,k}$.

The conditions of the concordance and normalization are fulfilled: $p^2_{i,j,k} > 0$, if $b_{i,j,k} \neq 0$; $p^1_{i,j} > 0$, if $a_{i,j} \neq 0$ and $p^0_i > 0$, if $f_i \neq 0$,

$$\sum_{i=1}^{n}(p^2_{i,j,k} + p^1_{i,j} + p^0_i) = 1. \tag{1.21}$$

Thus

$$\sum_{i=1}^{n} p^2_{i,j,k} = p^2_j \text{ - should not depend on } k,$$

$$\sum_{i=1}^{n} p^1_{i,j} = p^1_j \text{ and } p^2_j + p^1_j \text{ should not depend on } j.$$

The probabilities are interpreted as follows

- p^0_i - is a probability of birth of a new particle in state with number i

- $p^1_{i,j}$ - is a probability of passage of one particle from state j to state i

- $p^2_{i,j,k}$ - with $l > 1$ - is a probability of the two particles in states j and k correspondently becoming one particle with number of state i. If $l = 1$, p^2_j is probability of a particle being lost in a state with number j.

According to this, algorithm of the process simulation and evaluation statistical weights looks as follows.

If $t = 0$ according to the distribution $\{p^0_j\}_{j=1}^n$ 1 particle in a state i_0 is born. Statistical weight

$$Q_0 = \frac{f_{i_0}}{p^0_{i_0}} \tag{1.22}$$

is calculated in this initial state.

With $t > 0, l = 1$, if the particle state number is i_t, we have

1^a. if the probability is $p^0_{i_t}$, one more particle appears in state i_t and

$$Q_{t+1} = Q_t \frac{f_{i_t}}{p^0_{i_t}}. \tag{1.23}$$

2^a. If the probability is $p^1_{i_{t+1},i_t}$, the particle state changes from i_t to i_{t+1} and for $l=1$ we have

$$l = 1, \quad Q_{t+1} = Q_t \frac{a_{i_{t+1},i_t}}{p^1_{i_{t+1},i_t}}. \tag{1.24}$$

3^a. If the probability is $p^0_{i_t}$, the process breaks $\tau = t+1$ and the absorption estimator is calculated

$$\xi_\tau = Q_i \frac{h_{i_\tau}}{p^2_{i_\tau}}. \tag{1.25}$$

If $t > 0, l > 1$ in contrast to $l = 1$ case 3^a is impossible. Either 1^a is carried out, with $l \to l+1$, or accidentally (with equal probability) one of particles is selected and 2^a is carried out for it, independently with equal probability from l particles 2 various are selected and if numbers of their states are i_t and i'_t correspondently, then with probability $p^2_{i_{t+1},i_t,i'_t}$ instead of them one particle in a state i_{t+1} will appear. Thus

$$Q_{t+1} = Q_t \frac{b_{i_{t+1},i_t,i'_t}}{p^2_{i_{t+1},i_t,i'_t}} \tag{1.26}$$

If t_i is a time moment, when there was $l = 1$, a collision estimator is

$$\xi^{(2)}_t = \sum_{i=1}^{\tau} Q_{t_i} h_{t_i} \tag{1.27}$$

The following circumstance must be pointed out.

With a solution of the evolutionary problems the physical time can differ from time t, appearing in our exposition of processes.

Some numerical results were received with modeling processes described above. There were used different estimators.

Results for the simplest case of one equation with the form $x = ax^2 + b$ are given in Tables 1.3–1.6.

NOTATIONS—

X^d_a - results in direct scheme, an estimator by absorption;
X^d_c - direct scheme, an estimator by collision;
X^c_c - conjugate scheme, an estimator by collision.
E - an average error,
D - variance of an estimator,
$5; 50; 500$ - a corresponding number of groups, 100 independent trials in each group.

Furthermore we consider a difference scheme for the next equation

$$\frac{\partial^2 U}{\partial t^2} = \frac{\partial^2 U}{\partial x^2} + U^2 \tag{1.28}$$

under conditions
$$U(t,0) = U(t,1) = 0 \quad \text{and}$$
$$U(0,x) = \sin^2(\pi x), \quad U(\Delta t, x) = \exp((-\Delta t)x)\sin^2(\pi x).$$

In result of simplest square net approximation (1.28) we have system of equations with square nonlinearity:

$$\begin{aligned}U_i^{n+1} &= \frac{\Delta t^2}{2\Delta t^2 + h^2}\left(U_{i+1}^{n+1} + U_{i-1}^{n+1}\right) + \frac{\Delta t^2 h^2}{2\Delta t^2 + h^2}(U^{n+1})^2 \\ &+ \frac{h^2}{2\Delta t^2 + h^2}\left(2U_i^n - U_i^{n-1}\right).\end{aligned}$$

Solution of given system using an estimator by "collision" (direct scheme) is given on Fig.1. Calculating parameters: number of points - 200 and hence $h = 0.005$, $\Delta t = 0.005$, number of time steps - 100, number of repetitions at each point - 100.

To determine a correct relation between h and Δt was used some results of [Ermakov and Wagner (1999)].

Table 1.1

	1	2	11	12	21	22	111	112	121	122	211	212	221	222
1	a_{11}	a_{12}	b_{111}	b_{112}	b_{121}	b_{122}	0	0	0	0	0	0	0	0
2	a_{21}	a_{22}	b_{211}	b_{212}	b_{221}	b_{222}	0	0	0	0	0	0	0	0
11	f_1	0	a_{11}	0	a_{12}	0	b_{111}	0	b_{112}	0	b_{121}	0	b_{122}	0
12	0	f_1	0	a_{11}	0	a_{12}	0	b_{111}	0	b_{112}	0	b_{121}	0	b_{122}
21	f_2	0	a_{21}	0	a_{22}	0	b_{211}	0	b_{212}	0	b_{221}	0	b_{222}	0
22	0	f_2	0	a_{21}	0	a_{22}	0	b_{211}	0	b_{212}	0	b_{221}	0	b_{222}
111	0	0	f_1	0	0	0	a_{11}	0	0	0	a_{12}	0	0	0
112	0	0	0	f_1	0	0	0	a_{11}	0	0	0	a_{12}	0	0
121	0	0	0	0	f_1	0	0	0	a_{11}	0	0	0	a_{12}	0
122	0	0	0	0	0	f_1	0	0	0	a_{11}	0	0	0	a_{12}
211	0	0	f_2	0	0	0	a_{21}	0	0	0	a_{22}	0	0	0
212	0	0	0	f_2	0	0	0	a_{21}	0	0	0	a_{22}	0	0
221	0	0	0	0	f_2	0	0	0	a_{21}	0	0	0	a_{22}	0
222	0	0	0	0	0	f_2	0	0	0	a_{21}	0	0	0	a_{22}

Table 1.2

	1	2	11	12	21	22	111	112	121	122	211	212	221	222
1	a_{11}	a_{21}	f_1	0	f_2	0	0	0	0	0	0	0	0	0
2	a_{12}	a_{22}	0	f_1	0	f_2	0	0	0	0	0	0	0	0
11	b_{111}	b_{211}	a_{11}	0	a_{21}	0	f_1	0	0	0	f_2	0	0	0
12	b_{112}	b_{212}	0	a_{11}	0	a_{21}	0	f_1	0	0	0	f_2	0	0
21	b_{121}	b_{221}	a_{12}	0	a_{22}	0	0	0	f_1	0	0	0	f_2	0
22	b_{122}	b_{222}	0	a_{12}	0	a_{22}	0	0	0	f_1	0	0	0	f_2
111	0	0	b_{111}	0	b_{211}	0	a_{11}	0	0	0	a_{21}	0	0	0
112	0	0	0	b_{111}	0	b_{211}	0	a_{11}	0	0	0	a_{21}	0	0
121	0	0	b_{112}	0	b_{212}	0	0	0	a_{11}	0	0	0	a_{21}	0
122	0	0	0	b_{112}	0	b_{212}	0	0	0	a_{11}	0	0	0	a_{21}
211	0	0	b_{121}	0	b_{221}	0	a_{12}	0	0	0	a_{22}	0	0	0
212	0	0	0	b_{121}	0	b_{221}	0	a_{12}	0	0	0	a_{22}	0	0
221	0	0	b_{122}	0	b_{222}	0	0	0	a_{12}	0	0	0	a_{22}	0
222	0	0	0	b_{122}	0	b_{222}	0	0	0	a_{12}	0	0	0	a_{22}

Table 1.3

$x = 0.1x^2 + 0.7 \qquad p_{absorption}^{direct} = 0.1, \quad p_{absorption}^{conjugate} = 0.1 \quad x_{exact} = 0.7573593$

		X_a^d	X_c^d	X_c^c
5	E	-0.000933	0.005568	-0.002619
	D	0.000031	0.000552	0.000653
50	E	0.000039	-0.002408	0.001796
	D	0.000033	0.000260	0.000419
500	E	0.000172	-0.001218	0.000336
	D	0.000040	0.000331	0.000339

Table 1.4

$x = 0.7x^2 + 0.1 \qquad p_{absorption}^{direct} = 0.1, \quad p_{absorption}^{conjugate} = 0.1 \quad x_{exact} = 0.1081942$

		X_a^d	X_c^d	X_c^c
5	E	-0.000244	-0.000145	0.000139
	D	0.000001	0.000006	0.000005
50	E	0.000130	0.000269	0.000093
	D	0.000001	0.000006	0.000006
500	E	-0.000020	-0.000068	0.000138
	D	0.000001	0.000007	0.000007

Table 1.5

$x = 0.15x^2 + 0.3$ $p_{absorption}^{direct} = 0.15,$ $p_{absorption}^{conjugate} = 0.3$ $x_{exact} = 0.3148716$

		X_a^d	X_c^d	X_c^c
5	E	0.003426	0.003119	0.001414
	D	0.000083	0.000011	0.000014
50	E	0.003383	0.000295	-0.000109
	D	0.000053	0.000017	0.000005
500	E	-0.000164	0.000134	-0.000010
	D	0.000087	0.000013	0.000008

Table 1.6

$x = 0.15x^2 + 0.3$ $p_{absorption}^{direct} = 0.3,$ $p_{absorption}^{conjugate} = 0.15$ $x_{exact} = 0.3148716$

		X_a^d	X_c^d	X_c^c
5	E	-0.009255	-0.000477	-0.000027
	D	0.000350	0.000005	0.000004
50	E	0.000101	0.000042	-0.000475
	D	0.000348	0.000007	0.000012
500	E	-0.001011	0.000157	-0.000110
	D	0.000310	0.000007	0.000015

References

1. Bird G.A. (1994). *Molecular Gas Dynamics and the Direct Simulation of Gas Flows*, Oxford: Clarendon Press.

2. Ermakov, S. M. (1971). *Monte Carlo method and related questions*, Moskow: Nauka (in Russian).

3. Ermakov, S. M. (1972). Monte Carlo method for nonlinear operators iterating, *Doklady Academii Nauk SSSR*, **2**, 271–274 (in Russian).

4. Ermakov, S. M., Nekrutkin, V. V. and Sipin, A. S. (1989). *Random Processes for Classical Equations of Mathametical Physics*, Dordrecht: Kluwer.

5. Ermakov, S. M. Wagner W. (1999). *Monte Carlo difference schemes for the wave equation*, to be published.

6. Golyandina, N. (1996). Markov processes for solving differential equations in measure spaces, In *Mathematical Methods in Stochastic Simulation and Experimental Design*, 2nd St. Petersburg Workshop on simulation, St. Petersburg, June 18–21, 1996 (Eds., S. M. Ermakov and V. B. Melas) pp. 75–80, St. Petersburg University Press.

7. Nekrutkin, V. V. (1974). Direct and conjugate Neumann-Ulam schemes for solving nonlinear integral equations, *Journal of Computation Math. and Math.-Phys.*, **6**, 1409–1415 (in Russian).

8. Nekrutkin V. V. and Tur N. I. (1997). Asymptotic expansions and estimators with small bias for Nanbu processes, *Monte Carlo Methods and Applications*, **3**, No. 1, 1–35.

9. Sizova, A. F. (1976). About a variance of an evaluation on collisions for a solution of the nonlinear equations by the Motne Carlo method, *Vestnik Saint Petersburg State University, Ser. Math., Mech., Astron.*, **1**, 152–155 (in Russian).

2

Monte Carlo Algorithms For Neumann Boundary Value Problem Using Fredholm Representation

Y. N. Kashtanov and I. N. Kuchkova
St. Petersburg State University, St. Petersburg, Russia

Abstract: The paper deals with Monte Carlo algorithms for the calculation of the solution of Neumann boundary value problem. Estimators, which have finite variance up to the boundary, are pointed out. The developed estimators are applied to the solution of Navier-Stokes equations by method of vortex simulation.

Keywords and phrases: Monte Carlo estimators, Neumann boundary value problem, Navier-Stokes equation

2.1 Introduction

Some problems, for example, vortex simulation [Chorin (1973)] need the calculation of derivatives of solution of Neumann boundary value problem at many points of the domain (of order 1000), the points of particular interest being situated on the boundary. In the case of boundaries with simple geometry conformal mapping may be used [Ghoniem and Gagnon (1987)]. As to Monte Carlo method, one can find some algorithms in the monographs Ermakov, Nekrutkin and Sipin (1989), Sabelfeld (1992). But the technique developed in the first monograph assumes the convexity of the domain, whereas the estimators variance in the second one are not bounded when the systematic error tends to zero. The Monte Carlo estimators, that we propose, are based on potential theory and Fredholm determinants and have bounded variance right up to the domain boundary.

2.2 Integral Representation

First, consider the general case of a bounded domain X in R^n with the boundary Γ of class $C^{(1,\lambda)}$. The derivatives of the solution of Neumann boundary value problem

$$\Delta\varphi(x) = 0, \quad \frac{\partial\varphi}{\partial n} = f \text{ on } \Gamma$$

can be represented as the derivatives of the single layer potential in the direction l:

$$\frac{\partial\varphi(y)}{\partial l} = \int_\Gamma dS_x \mu(x) g(x,y), \qquad (2.1)$$

where $g(x,y) = \partial_y h(x,y)/\partial l$,

$$h(x,y) = \begin{cases} \frac{1}{\pi}\ln\frac{1}{|x-y|}, & n = 2, \\ \frac{2}{\sigma_n(n-2)|x-y|^{n-2}}, & n > 2, \end{cases}$$

and μ satisfies the integral equation

$$\mu(y) = \int_\Gamma dS_x \mu(x) k(x,y) + f(y), \qquad (2.2)$$

$k(x,y) = -\dfrac{\partial}{\partial n_y} h(x,y)$, $y \in \Gamma$, n_y being the outer normal.

Denote by V and M the integral operators with kernels

$$v(x,y) = k(x,y)\chi_{(|x-y|>\delta)}(x,y),$$

$$m(x,y) = k(x,y)\chi_{(|x-y|<\delta)}(x,y)$$

respectively, where δ is small enough, then

$$\|v(\cdot,\cdot)\| = \sup_{x,y\in\Gamma}|v(x,y)| \leq v_1 < \infty \text{ and } \sup_{x\in\Gamma}\int_\Gamma dS_y |m(x,y)| \leq m_1 < 1. \qquad (2.3)$$

Using operators M and V, equation (2.2) can be represented in the form

$$\mu(I-M) = \mu(I-M)(I-M)^{-1}V + f. \qquad (2.4)$$

If we denote $\tilde{\mu} = \mu(I-M)$ and $\tilde{K} = (I-M)^{-1}V$ then (2.4) gives the equation for $\tilde{\mu}$:

$$\tilde{\mu} = \tilde{\mu}\tilde{K} + f. \qquad (2.5)$$

Further, assuming $\tilde{g} = (I-M)^{-1}g$, representation (2.1) is equal to the following:

$$\frac{\partial\varphi}{\partial l}(y) = \int_\Gamma dS_x \tilde{\mu}(x)\tilde{g}(x,y). \qquad (2.6)$$

Note that $\int_\Gamma dS_y \tilde{k}(x,y) = 1$ because

$$\left| \sum_{t=0}^{n} \int_{\Gamma^2} dS_z dS_y m^{(t)}(x,z) v(z,y) - 1 \right| = \left| \int_\Gamma dS_z m^{(n+1)}(x,z) \right| \le m_1^{n+1}.$$

So, $1(x)$ is the right eigenfunction for $\tilde{k}(x,y)$, corresponding to the eigenvalue equal to 1, and is unique, as is easy to see. Denote

$$\tilde{b}_t(x,y) = \frac{(-1)^t}{t!} \int_{\Gamma^t} dS_{x_1} \ldots dS_{x_t} \tilde{K}\begin{pmatrix} x, x_1, \ldots, x_t \\ y, x_1, \ldots, x_t \end{pmatrix},$$

$$\tilde{c}_t(x,y,x_0,y_0) = \frac{(-1)^t}{t!} \int_{\Gamma^t} dS_{x_1} \ldots dS_{x_t} \tilde{K}\begin{pmatrix} x, x_0, x_1, \ldots, x_t \\ y, y_0, x_1, \ldots, x_t \end{pmatrix}, \quad (2.7)$$

$$\tilde{b}(x,y) = \sum_{t=0}^{\infty} \tilde{b}_t(x,y), \quad \tilde{c}(x,y,x_0,y_0) = \sum_{t=0}^{\infty} \tilde{c}_t(x,y,x_0,y_0),$$

where $\tilde{K}\begin{pmatrix} \xi_1, \ldots, \xi_t \\ \eta_1, \ldots, \eta_t \end{pmatrix} = \det \{\tilde{k}(\xi_i, \eta_j)\}_{i,j=1}^{t}$ are Fredholm determinants. It follows from the general theory, that $\tilde{b}(x,y)$ is the right eigenfunction by the first argument for the kernel $\tilde{k}$ and the left eigenfunction by the second argument. Thus, $\tilde{b}$ has the form $\tilde{b}(x,y) = \tilde{\pi}(y)$ and

$$\int_\Gamma dS_x \tilde{\pi}(x) \tilde{k}(x,y) = \tilde{\pi}(y).$$

If we choose x_0 so that $\tilde{\pi}(x_0) \ne 0$ and assume $r(x,z) = \tilde{c}(x,z,x_0,x_0)/\tilde{\pi}(x_0)$, then Fredholm representation for $\tilde{\mu}$ may be written in the form:

$$\tilde{\mu} = f(I + R), \quad (2.8)$$

where R is the operator with kernel $r(x,z)$.

2.3 Monte Carlo Estimators

Considering (2.6) and (2.8), the task is to estimate $b_t(x_0, x_0)$ and J_t:

$$J_t = \int_\Gamma dS_x \tilde{\mu}_t(z) \tilde{g}(z,y),$$

where $\tilde{\mu}_t(z) = \int_\Gamma dS_x f(x) c_t(x,z,x_0,x_0)$. Represent $g(x,y)$ as a sum:

$$g(x,y) = g(x,y)\chi(|x-y|<\delta) + g(x,y)\chi(|x-y|>\delta) = g_1(x,y) + g_2(x,y).$$

This representation divides functional J_t into two functionals $J_t = J_{1t} + J_{2t}$.

First we construct the random sequence $x_{-1}, x_{-2}, \ldots, x_{-r}$, with random weights $G_1(i)$, $-r \leq i \leq -1$, so that for every smooth ψ, defined on the boundary, the sum $\sum_{i=-r}^{-1} \psi(x_i) G_1(i)$ is an unbiased estimator with finite variance for $\int_\Gamma dS_z \psi(z) \tilde{g}_1(x,y)$. For the sake of definiteness, we suppose that $y \in \Gamma$ and τ_y is a tangent direction at point y. Introduce a random point x' on the plane Γ', which is tangent to Γ at point y. Suppose, that $T'(x',y) = C|x'-y|^{-n+1+\lambda/2} \chi_{\{|x'-y|\leq\delta\}}$ is the substochastic density of the point x'. Let x be the projection of x' on Γ along the direction n_y. We assume that x is realized if $|x-y| \leq \delta$, then the random point x has a substochastic density $T(x,y) = (n_x, n_y) T'(x',y) \chi_{\{|x-y|\leq\delta\}}$. We denote this correspondence by the expression $x \sim T(x,y)$. A random sequence $\{x_i\}_{i=-1}^{-r}$ is constructed as follows:

$$x_{-1} = y, \; x_{-2} \sim T(x_{-2}, y), \; G_1(-1) = -\frac{g(x'_{-2}, y)}{T'(x'_{-2}, y)}, \; G_1(-2) = \frac{g(x_{-2}, y)}{T(x_{-2}, y)},$$

$$x_{-3} \sim T_3(x_{-3}, x_{-2}) = \frac{1}{2}[T(x_{-3}, x_{-2}) + T(x_{-3}, y)],$$

$$G_1(-3) = \frac{m(x_{-3}, x_{-2})}{T_3(x_{-3}, x_{-2})} \frac{g(x_{-2}, y)}{T(x_{-2}, y)} - \frac{m(x_{-3}, x_{-1})}{T_3(x_{-3}, x_{-2})} \frac{g(x'_{-2}, y)}{T'(x'_{-2}, y)},$$

$$x_i \sim T(x_i, x_{i+1}), \; G_1(i) = \frac{m(x_i, x_{i+1})}{T(x_i, x_{i+1})} g(x_{i+1}, y), \; i < -3.$$

If τ is the stopping time, then $r = \tau - 1$.

Now fix some n, x_0 and construct Monte Carlo estimators for $\tilde{b}_t(x_0, x_0)$, $0 \leq t \leq n$, and $\tilde{c}_t(x_i, x_j, x_0, x_0)$, $0 \leq t \leq n-2$. For that purpose, we simulate independent points $x_1, \ldots, x_n$ with some density $p(x)$ on Γ, $p(x) \geq p_0 > 0$. Then, for every x_i we simulate a Markov chain $z_0^{(i)}, \ldots, z_{\tau-1}^{(i)}$ in accordance with substochastic transitional densities $\bar{m}(z_j^{(i)}, z_{j+1}^{(i)})$, $\bar{m} = |m|$, $z_0^{(i)} = x_i$, τ being the stopping time. It is clear that under fixed x_i and x_j, the random values

$$\check{k}(x_i, x_j) = \sum_{t=0}^{\tau-1} v(z_t^{(i)}, x_j) \prod_{s=1}^{t} \operatorname{sign} m(z_{s-1}^{(i)}, z_s^{(i)}),$$

$$\check{g}_2(x_i, y) = \sum_{t=0}^{\tau-1} g_2(z_t^{(i)}, y) \prod_{s=1}^{t} \operatorname{sign} m(z_{s-1}^{(i)}, z_s^{(i)}),$$

are unbiased estimators for $\tilde{k}(x_i, x_j)$ and $\tilde{g}_2(x_i, y)$ respectively with uniformly (with respect to x_i, x_j) bounded variance. We form a matrix

$$K = \left\{ \frac{\check{k}(x_i, x_j)}{\sqrt{p_i p_j}} \right\}_{i=0, j=-r}^{n, n}, \; p_i = \begin{cases} p(x_i), \; i \geq 1 \\ 1, \; i \leq 0 \end{cases}$$

and vectors F and G_2:

$$F(i) = \frac{f(x_i)}{\sqrt{p(x_i)}}, \; G_2(i) = \frac{\check{g}_2(x_i, y)}{\sqrt{p(x_i)}}, \; 1 \leq i \leq n,$$

and define sequences A_t, B_t, C_t, $t = 0, \ldots, n$ by recurrent relations:

$$A_0 = 1, \quad B_0 = \{K(i,j)\}_{i,j=0}^n, \quad C_0 = \left\{K\begin{pmatrix} i,0 \\ j,0 \end{pmatrix}\right\}_{i=1,j=-r}^{n,n},$$

$$A_{t+1} = -\frac{1}{t+1}\mathrm{sp}B_t, \tag{2.9}$$

$$B_{t+1} = (B_t - A_{t+1}I)K, \tag{2.10}$$

$$C_{t+1} = [C_t + B_{t+1}(0,0)I]K - B_{t+1}(\cdot,0)K(0,\cdot),$$

here we assumed $\mathrm{sp}A = \sum_{i=1}^n A(i,i)$, $(A_1A_2)(i,j) = \sum_{s=1}^n A_1(i,s)A_2(s,j)$, $K\begin{pmatrix} i_1,\ldots,i_s \\ j_i,\ldots,j_s \end{pmatrix} = \det\{K(i_p,j_q)\}_{p,q=1}^s$.

Define random variables

$$\check{b}_t = \frac{(n-t)!}{n!}B_t(0,0),$$

$$\check{J}_{1t} = \frac{(n-1-t)!}{n!}\sum_{i=1}^n \sum_{j=-r}^{-1} F(i)C_t(i,j)G_1(j),$$

$$\check{J}_{2t} = \frac{(n-2-t)!}{n!}\sum_{i=1}^n \sum_{j=1, j\neq i}^n F(i)C_t(i,j)G_2(j),$$

Theorem $\check{b}_t$, $\check{J}_{1t}$ and $\check{J}_{2t}$ are unbiased estimators with finite variance respectively for $\tilde{b}_t(x_0,x_0)$, J_{1t} and J_{2t}.

Proof: Introduce matrices

$$\tilde{B}_t = \left\{\sum_{i_1=1}^n \cdots \sum_{i_t=1}^n K\begin{pmatrix} i,i_1,\ldots,i_t \\ j,i_1,\ldots,i_t \end{pmatrix}\right\}_{i=1,j=-r}^n, \tag{2.11}$$

where $K\begin{pmatrix} i_1,\ldots,i_s \\ j_1,\ldots,j_s \end{pmatrix} = \det\{K(i_p,j_q)\}_{p,q=1}^s$ and show that $B_t = \frac{(-1)^t}{t!}\tilde{B}_t$.

For $t = 0$ it is evident. Developing determinant in (2.11) by elements of the first column, we obtain

$$\tilde{B}_{t+1} = \mathrm{sp}(\tilde{B}_t)K - (t+1)\tilde{B}_tK. \tag{2.12}$$

Multiplying (2.12) by $\frac{(-1)^{t+1}}{(t+1)!}$, we receive that matrices $\frac{(-1)^t}{t!}\tilde{B}_t$ satisfy relations (2.10) and therefore coincide with B_t. Further

$$K\begin{pmatrix} i,i_1,\ldots,i_t \\ j,i_1,\ldots,i_t \end{pmatrix} = \begin{cases} 0, & \text{if } i_p = i_q \text{ for some } p \text{ and } q, \\ \sum_{(j_0,\ldots,j_t)} \pm K(i,j_0)K(i_1,j_1),\ldots,K(i_t,j_t), & \text{if } i_p \neq i_q, \end{cases}$$

Here $(j_0,\ldots,j_t)$ is a transposition of indices $(j,i_1,\ldots,i_t)$. In the case when $i_p \neq i_q$ and under fixed x_i, x_j, we have

$$\mathbf{E}K\begin{pmatrix} i,i_1,\ldots,i_t \\ j,i_1,\ldots,i_t \end{pmatrix}\bigg|x_i,x_j\bigg) = \frac{c_{ij}}{\sqrt{p_i p_j}}\int_{\Gamma^t} dS_{x_1}\ldots dS_{x_t} K\begin{pmatrix} x_i,x_1,\ldots,x_t \\ x_j,x_1,\ldots,x_t \end{pmatrix},$$

where
$$c_{ij} = \begin{cases} (n-2)!/(n-t-2)!, & \text{if } j > 0,\ i \neq j, \\ (n-1)!/(n-t-1)!, & \text{if } j \leq 0 \text{ or } j > 0,\ i = j, \end{cases}$$

and therefore we obtain:

$$\begin{aligned}
\mathbf{E}\check{b}_t &= \frac{(n-t)!}{n!}\mathbf{E}B_t(0,0) \\
&= \frac{(n-t)!}{n!}\frac{(-1)^t}{t!}\mathbf{E}\sum_{i_1=1}^{n}\cdots\sum_{i_t=1}^{n} K\begin{pmatrix} 0,i_1,\ldots,i_t \\ 0,i_1,\ldots,i_t \end{pmatrix} \\
&= \frac{(-1)^t}{t!}\int_{\Gamma^t} dS_{x_1},\ldots,dS_{x_t} K\begin{pmatrix} x_0,x_1,\ldots,x_t \\ x_0,x_1,\ldots,x_t \end{pmatrix} = \tilde{b}_t(x_0,x_0).
\end{aligned}$$

Now estimate the variance of $\check{b}_t$:

$$\begin{aligned}
\mathbf{E}\check{b}_t^2 &= \left[\frac{(n-t)!}{n!\,t!}\right]^2 \mathbf{E}\left[\sum_{\substack{i_1,\ldots,i_t=1 \\ i_l \neq i_k}}^{n} K\begin{pmatrix} 0,i_1,\ldots,i_t \\ 0,i_1,\ldots,i_t \end{pmatrix}\right]^2 \\
&\leq \frac{(n-t)!}{n!(t!)^2} \sum_{\substack{i_1,\ldots,i_t=1 \\ i_l \neq i_k}}^{n} \mathbf{E}K^2\begin{pmatrix} 0,i_1,\ldots,i_t \\ 0,i_1,\ldots,i_t \end{pmatrix} \\
&\leq \frac{(n-t)!}{n!(t!)^2} \sum_{\substack{i_1,\ldots,i_t=1 \\ i_l \neq i_k}}^{n} \mathbf{E}\prod_{i\in I}\sum_{j\in I} K^2(i,j), \quad\quad (2.13)
\end{aligned}$$

where $I = \{0,i_1,\ldots,i_t\}$. Since $K(i,j)$ are independent for different i and $\mathbf{E}K^2(i,j) \leq C$ then

$$\mathbf{E}\prod_{i\in I}\sum_{j\in I} K^2(i,j) \leq \prod_{i\in I}\sum_{j\in I}\mathbf{E}K^2(i,j) \leq (C(t+1))^{t+1}.$$

Applying this estimate and Stirling formula to 2.13, obtain $\mathbf{E}\check{b}_t^2 \leq (C/t)^t$. In particular, it means that the variance of $\check{b} = \sum_{i=0}^{n}\check{b}_t$ is bounded when n tends to infinity (or systematic error tends to zero).

Similarly, one can prove that the random values $\check{J}_{1t}$ and $\check{J}_{2t}$ are unbiased estimators respectively for J_{1t} and J_{2t} with bounded variance.

2.4 The Two-Dimensional Case

In this case we can obtain more simple integral representation for derivatives of the solution and consider some domains with non-smooth boundary. Let X be a bounded domain in R^2 with boundary Γ, which consists of finite number of curves of class C^2. Let φ be the harmonic function, $\varphi \in C^2(\bar{X})$, n_x be the external normal to the surface Γ at the point x, $f(x) = \partial \varphi(x)/\partial n_x$.

First point out an integral representation for derivative $\partial \varphi/\partial l$ in direction l. The function $\varphi(x)$, $x \in X$, has the integral representation:

$$\varphi(x) = -\int_\Gamma dS_y \frac{\partial H(x,y)}{\partial n_y} \varphi(y) + \int_\Gamma dS_y H(x,y) \frac{\partial \varphi}{\partial n_y}, \qquad (2.14)$$

where $H(x,y) = \frac{1}{2\pi} \ln \frac{1}{|x-y|}$. Taking derivative of φ at point $x \in X$ in the direction l and after some transformation obtain:

$$\frac{\partial \varphi(x)}{\partial l} = -\int_\Gamma dS_y \varphi(y) \frac{\partial_x}{\partial l} \frac{\partial_y H(x,y)}{\partial n_y} + \int_\Gamma dS_y \frac{\partial_x H(x,y)}{\partial l} f(y). \qquad (2.15)$$

Let R be the rotation on the right angle, then for arbitrary vectors n, m and $\tau = Rn$, $l = Rm$ the next equality is fulfilled:

$$\frac{\partial_x}{\partial l} \frac{\partial_y}{\partial n} H(x,y) = \frac{\partial_x}{\partial m} \frac{\partial_y}{\partial \tau} H(x,y). \qquad (2.16)$$

If we apply (2.16) to (2.15) and integrate by parts we obtain:

$$\frac{\partial \varphi(x)}{\partial l} = -\int_\Gamma dS_y \varphi(y) \frac{\partial_x}{\partial m} \frac{\partial_y H(x,y)}{\partial \tau_y} + \int_\Gamma dS_y \frac{\partial_x H(x,y)}{\partial l} f(y). \qquad (2.17)$$

where $\tau_y = Rn_y$ is the tangent direction for Γ at point y. Integrating by parts gives

$$\frac{\partial \varphi(x)}{\partial l} = \int_\Gamma dS_y \frac{\partial_x H(x,y)}{\partial m} \frac{\partial \varphi(y)}{\partial \tau_y} + \int_\Gamma dS_y \frac{\partial_x H(x,y)}{\partial l} f(y). \qquad (2.18)$$

Thus, to calculate $\partial \varphi(x)/\partial l$, $x \in X$, it is sufficient to find $\partial \varphi(y)/\partial \tau_y$ on Γ. Denote $\varphi_\tau(x) = \partial \varphi(x)/\partial \tau_x$ and deduce an integral equation for φ_τ. Let $x' \in \Gamma$, $n_{x'}$ be the normal at point x', then we can write $\tau_{x'} = Rn_{x'}$. Assume in (2.18) that $l = \tau_{x'}$ and turn x to x'. Passing to the limit in formula (2.18) and substituting x' by x, obtain:

$$\frac{\partial \varphi(x')}{\partial \tau_{x'}} = \int_\Gamma dS_y \frac{\partial_{x'} H(x',y)}{\partial n_{x'}} \frac{\partial \varphi(y)}{\partial \tau_y} + \frac{1}{2} \frac{\partial \varphi(x')}{\partial \tau_{x'}} + v.p. \int_\Gamma dS_y \frac{\partial_{x'} H(x',y)}{\partial \tau_{x'}} f(y),$$

the last integral being considered in the sense of principal value. If we denote

$$\tilde{f}(x) = v.p. 2 \int_\Gamma dS_y \frac{\partial_x H(x,y)}{\partial \tau_x} f(y)$$

and substitute x' by x, then the last equation may be rewritten in the form:

$$\varphi_\tau(x) = -\int_\Gamma dS_y \varphi_\tau(y) k(y,x) + \tilde{f}(x), \; x \in \Gamma, \quad (2.19)$$

where $k(y,x) = -2\partial H(y,x)/\partial n_x$. Note that the kernel $-k(y,x)$ is bounded if $\Gamma \in C^{(1,\lambda)}$ and Fredholm representation can be applied directly to (2.19). Since 1 is not an eigenvalue of $-k(y,x)$, the Fredholm representation for the solution looks like follows:

$$\varphi_\tau(x) = \tilde{f}(x) + \frac{1}{a}\int_\Gamma \tilde{f}(y) b(y,x),$$

where

$$a = \sum_{t=0}^{\infty} \frac{(-1)^t}{t!} \int_{\Gamma^t} dS_{x_1} \ldots dS_{x_n} \det\{-k(x_i,x_j)\}_{i,j=1}^{n},$$

$$b(x_0,y_0) = \sum_{t=0}^{\infty} \frac{(-1)^t}{t!} \int_{\Gamma^t} dS_{x_1} \ldots dS_{x_n} \det\{-k(x_i,x_j)\}_{i,j=0}^{n}.$$

Monte Carlo estimators for the terms a and b were already pointed out in formulas (2.9), (2.10). Note that in this representation we don't use the term c.

If Γ has a corner point, say, z, then $k(x,y)$ is not bounded when x and y tend to z. But the gradient $\nabla \varphi(z)$ is known and we approximate $\nabla \varphi(x) \approx \nabla \varphi(z)$ in δ-neighborhood Γ^δ of point z. In this case we substitute equation (2.19) by the equation:

$$\varphi_\tau(x) = -\int_{\Gamma \backslash \Gamma^\delta} dS_y \varphi_\tau(y) k(y,x) - \int_{\Gamma^\delta} (\nabla \varphi(z), \tau_y) k(y,x) + \tilde{f}(x), \; x \in \Gamma \backslash \Gamma^\delta.$$

Note that the kernel in the last equation is bounded and Fredholm representation of solution may be used.

If the distance from a point x to the boundary is not less than some fixed value δ, then the expression (2.18) allows us to construct simple statistical estimators for the derivatives which have uniformly bounded variance. Now let a point x be situated close to the boundary, i.e. $|x-x'| < \delta$, x' being the point on Γ nearest to x. Rewrite the expression (2.18) in the form $\int_\Gamma f(x,y) P(x,y) dS_y$, where $|f(x,y)| \leq C$ and $P(x,y)$ is a non-negative function on Γ, which satisfies the condition $\int_\Gamma dS_y P(x,y) < C$. Let $S_{x'}^\delta$ be the sphere with center at point x' and with radius δ, Γ' be the tangent plane to Γ at the point x', $\tau_x = \tau_{x'} = \tau$, $n_x = n_{x'} = n$. Let y' be the projection of y on Γ' along the direction $\tau_{x'}$. Taking into account that $\int_{\Gamma' \cap S^\delta_{x'}} H_{\tau_x}(y',x)dS'_y = 0$, we can obtain the expression for the tangent component in the form

$$\varphi'_\tau(x) = \int_{\Gamma \cap S^\delta_{x'}} \varphi'_\tau(y)\ H_{n_x}(y,x)dS_y + \int_{\Gamma \setminus S^\delta_{x'}} \varphi'_\tau(y) H_{n_x}(y,x)\ dS_y \qquad (2.20)$$

$$+ \int_{\Gamma \setminus S^\delta_{x'}} \varphi'_n(y) H_{\tau_x}(y,x)\ dS_y + \int_{\Gamma \cap S^\delta_{x'}} (\varphi'_n(y) - \varphi'_n(x')) H_{\tau_x}(y,x)\ dS_y$$

$$+ \int_{\Gamma' \cap S^\delta_{x'}} \varphi'_n(x') \left(\frac{H_{\tau_x}(y,x)}{(n_y, n_x)} - H_{\tau_x}(y',x) \right) |y' - x|^\alpha \chi_{S^\delta_{x'}}(y)\ \frac{1}{|y' - x|^\alpha} dS_{y'}.$$

In order to point out the expression for the normal component, first obtain expression for $U(x',y) = (\varphi'_\tau(y) - \varphi'_\tau(x'))/|y - x'|^\alpha$. After some simple transformation obtain:

$$U(x',y) = \int_\Gamma \varphi_\tau(u) \frac{H_{n_y}(u,y) - H_{n_{x'}}(u,x')}{|y - x'|^\alpha}\ dS_u$$

$$+ \int_{\Gamma \setminus S^\delta_{x'}} \varphi'_n(u) \frac{H_\tau(u,y) - H_\tau(u,y)}{|y - x'|^\alpha}\ dS_u$$

$$+ \varphi'_n(y) \frac{H(x' + \tau\delta, y) - H(x' - \tau\delta, y)}{|y - x'|^\alpha}$$

$$+ \int_{\Gamma \cap S^\delta_{x'}} \frac{(\varphi'_n(u) - \varphi'_n(y)) H_\tau(u,y) - (\varphi'_n(u) - \varphi'_n(x')) H_\tau(u,x')}{|y - x'|^\alpha (|u - x'|^{-\alpha} + |u - y|^{-\alpha})}$$

$$\left(\frac{1}{|u - x'|^\alpha} + \frac{1}{|u - y|^\alpha} \right) dS_u.$$

After all these substitutions we obtain expression for the normal component

$$\varphi'_n(x) = \int_{\Gamma \cap S^\delta_{x'}} U(x',y)\ |y - x'|^\alpha H_{\tau_x}(y,x)dS_y + \int_{\Gamma \setminus S^\delta_{x'}} \varphi'_\tau(y) H_{\tau_x}(y,x)\ dS_y$$

$$+ \int_{\Gamma \cap S^\delta_{x'}} \varphi'_n(y)\ H_{n_x}(y,x)dS_y + \int_{\Gamma \setminus S^\delta_{x'}} \varphi'_n(y) H_{n_x}(y,x)\ dS_y +$$

$$+ \int_{\Gamma' \cap S^\delta_{x'}} \varphi'_\tau(x') \left(\frac{H_{\tau_x}(y,x)}{(n_y, n_x)} - H_{\tau_x}(y',x) \right) |y' - x|^\alpha \chi_{S^\delta_{x'}}(y)\ \frac{1}{|y' - x|^\alpha} dS_{y'}.$$

Using above representations we can easily construct the estimators for $\varphi'_n(x)$ and $\varphi'_\tau(x)$ with uniformly limited variance.

2.5 An Application to Navier-Stokes Equations

Consider two-dimensional Navier-Stokes equations in the form of vortex diffusion:

$$\frac{\partial \omega(x,t)}{\partial t} + (u \cdot \nabla)\omega = \nu\omega, \quad \omega(x,t) = \text{rot } u(x,t),$$
$$\nabla u(x,t) = 0, \quad u(x,t) = u_0(x,t), \, x \in \Gamma,$$

where u_0 is the velocity of the solid boundary Γ.

For solution of these equations we use the method of vortex simulation. The algorithm of this method is the following.

The distribution of vortices $\omega(x,t)$ is approximated by the discrete set of vortex blobs with centers at points $x_i(t)$

$$\omega_\delta(x,t) = \sum \Omega_i f_\delta(|x - x_i(t)|),$$

where

$$\Omega_i = \int_{|x-x_i|<\delta} \omega(x,t)dx, \quad f_\delta(r) = \frac{1}{2\pi\delta r}\chi\{r < \delta\}.$$

At the first step, the velocity field is computed generated by this distribution:

$$u_\delta(x,t) = \sum \Omega_i K_\delta(x - x_i(t)), \quad K_\delta(x) = \frac{\min(|x|,\delta)}{2\pi|x|^2\delta}\begin{pmatrix} -x_2 \\ x_1 \end{pmatrix}.$$

At the second step, the potential velocity field u_p is calculated which on the boundary satisfies the condition $(u_p \cdot n) = -(u_\delta \cdot n)$, i.e. the Neumann boundary value problem is being solved

$$\Delta\varphi = 0, \quad \frac{\partial\varphi}{\partial n} = -(u_\delta, n), \quad u_p = \nabla\varphi.$$

At the third step, new vortex blobs are produced on the solid boundary Γ by the velocity field $u = u_\delta + u_p$

$$\Omega_i = u_0(x,t) - \int_{\Gamma_i} (u(x,t) \cdot dS_x),$$

where Γ_i is a part of boundary with the "center" at point x_i and length $\pi\delta$.

At the forth step, vortex blobs move along the velocity field with diffusion

$$x_i(t + \Delta t) = x_i(t) + u(x_i(t),t)\Delta t + \xi_i,$$

where ξ — is a two-dimensional vector of independent random normal variables with variance $\sigma^2 = 2\nu\Delta t$.

The foregoing algorithms for the solution of Neumann boundary value problem may be applied at the second step of vortex simulation. This method was used for numerical solution of some problems of viscous fluid flow. The first one was the problem of flow around the circle (Fig. 2.1). The appropriate Neumann boundary problem was solved for the ring $X = \{(x_1, x_2) : R_1 \leq r = \sqrt{x_1^2 + x_2^2} \leq R_2\}$. It was assumed, that $\nabla \varphi = (1, 0)$ on the exterior boundary ($r = R_2$), that produces the boundary condition of Neumann type $\frac{\partial \varphi}{\partial n} = \frac{x_1}{R_2}$. The second problem was the fluid flow in the rectangle cavity with moving wall (Fig. 2.2). The third problem was the movement of viscous fluid in a channel with a rearward-facing step (Fig. 2.3). Results of simulation are shown on Fig. 2.1–2.3 and give qualitatively proper pattern.

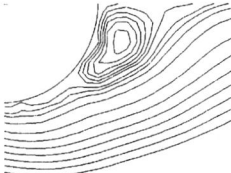

Figure 2.1

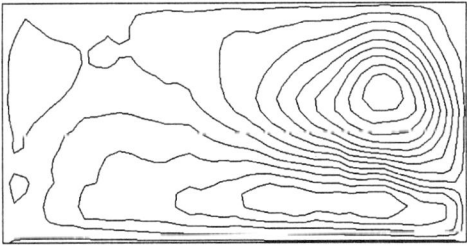

Figure 2.2

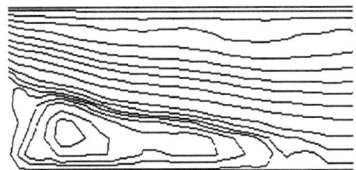

Figure 2.3

References

1. Chorin, A. J. (1973). Numerical study of slightly viscous flow, *Journal of Fluid Mechanics*, **57**, 785–796.

2. Ghoniem, A. F. and Gagnon, Y. (1987). Vortex Simulation of Laminar Recirculating Flow, *Journal of Computational Physics*, **68**, 346–376.

3. Ermakov, S. M., Nekrutkin, V. V. and Sipin, A. S. (1989). *Random Processes for the Solution of Classical Equations of Mathematical Physics*, Moscow: Nauka.

4. Sabelfeld, K. K. (1992). *Monte Carlo Methods in Boundary Value Problems*, Novosibirsk: Nauka.

5. Ermakov, S. M. and Kashtanov, Y. N. (1996). Monte Carlo Neumann function, *Proceedings of the 2nd St. Petersburg Workshop on Simulation*, pp. 69–74, Saint Petersburg: Saint Petersburg University Press.

3

Estimation Errors for Functionals on Measure Spaces

N. Golyandina and V. Nekrutkin

St. Petersburg State University, St. Petersburg, Russia

Abstract: The article is devoted to an investigation of bias and mean-square deviation when estimating smooth functionals on measure spaces. Such problems appear when one uses a Monte-Carlo procedure to solve certain linear or nonlinear equations for measures (e.g., nonlinear Boltzmann-like equations), and wants to estimate and diminish the part of an error corresponding to simulation of an initial distribution. It is shown that by means of a special variant of a stratification technique both bias and variance of this part of error may be reduced to $o(1/n)$ instead of the usual $O(n^{-1})$. Several simple examples of stratification for linear functionals (i.e., integrals) are presented.

Keywords and phrases: Non-linear functionals, measure-valued random variables, stratification technique

3.1 Introduction

We start with an example elucidating the problem under consideration. Let (D, ρ) be some metric space with Borel σ-algebra $\mathcal{B}$. Consider the following linear evolution equation for measures

$$\mu_t = \int_0^t e^{-(t-s)} \int_D T(\,\cdot\,; u)\mu_s(du)ds + e^{-t}\mu, \qquad (3.1)$$

where $\mu \in H$, H is the set of all probability measures on $(D, \mathcal{B})$, $T(\,\cdot\,; u) \in H$ for any u and $T(A;\,\cdot\,)$ is $\mathcal{B}$-measurable for any $A \in \mathcal{B}$. Surely, the solution $\mu_t = \mu_t(\mu)$ is a probability measure too.

The usual Monte-Carlo procedure for estimating the integral $\psi(\mu_t) = \int f d\mu_t$ may be described as follows. One simulates n independent copies $\zeta_i(s)$ of jump Markov process with the initial distribution $\mathcal{L}(\zeta_i(0)) = \mu$, jump law T and independent exponentially distributed (with the unit scale parameter) time intervals between jumps, and takes

$$\eta_n = \frac{1}{n} \sum_{i=1}^{n} f(\zeta_i(t))$$

as the estimator of $\psi(\mu_t)$.

Let us consider this procedure from the other viewpoint. If we gather up the processes $\zeta_i(s)$ into one n-particle process $\eta^{(n)}(s) = \big(\zeta_1(s), \ldots, \zeta_n(s)\big) \in D^n$, then $\eta^{(n)}(s)$ is also a jump Markov process. Its exponential scale parameter is equal to n, and the jump law may be expressed as follows. If $\mathbf{u} = (u_1, \ldots, u_n)$ is the phase coordinate of the n-particle process before a jump, we randomly choose one of the particles, and supposing this particle has number i, it jumps to a new position due to the distribution $T(\,\cdot\,; u_i)$. The initial distribution of the process is equal to $\mu^{\otimes n}$ (here $\mu_1 \otimes \mu_2$ stands for the outer product of measures μ_1 and μ_2, $\mu^{\otimes n} = \mu \otimes \ldots \otimes \mu$ (n times)).

On the other hand, any n-particle process $\eta^{(n)}(s)$ produces the corresponding measure-valued process

$$\xi_s^{(n)} = \frac{1}{n} \sum_{i=1}^{n} \delta_{\zeta_i(s)},$$

where δ_x is the Dirac measure, concentrated at the point x. Such processes may be called *empirical*. Surely, $\xi_s^{(n)}$ is also a jump Markov process.

In terms of the empirical process we have $\eta_n = \psi(\xi_t^{(n)})$, and the estimation error $\psi(\xi_t^{(n)}) - \psi(\mu_t)$ may be represented in the form

$$\psi(\xi_t^{(n)}) - \psi(\mu_t(\mu))$$
$$= \Big(\psi(\xi_t^{(n)}) - \psi(\mu_t(\xi_0^{(n)}))\Big) + \Big(\psi(\mu_t(\xi_0^{(n)})) - \psi(\mu_t(\mu))\Big), \qquad (3.2)$$

where $\mu_t(\xi_0^{(n)})$ stands for the solution of (3.1) with the initial value $\xi_0^{(n)}$.

As the starting point of the process $\xi_s^{(n)}$ is also $\xi_0^{(n)}$, the first bracketed term on the right-hand side of (3.2) describes the difference between transition mechanisms of the evolution equation (3.1) and the Markov process $\xi_t^{(n)}$, while the second may be interpreted as the part of the error, produced by the initial value $\xi_0^{(n)}$. Our aim is to investigate statistical properties of the second term.

If we fix time as t and define $F: H \mapsto \mathbf{R}$ by $F = \psi \circ \mu_t$ (that is $F(\mu) = \psi(\mu_t(\mu))$), then it occurs that we are interested in the difference $F(\xi_0^{(n)}) - F(\mu)$. When we deal with the linear functional ψ and with the linear equation (3.1),

F may be considered as linear. But if we want to estimate some nonlinear functional (such as $\psi(\mu_t) = (\int f d\mu_t)^2$) or if we consider a nonlinear evolution equation (e.g., a Boltzmann-like equation, where similar n-particle processes may be applied to), F is nonlinear.

Thus the first problem under consideration is to investigate the bias and variance for the estimator $F(\xi_n)$ of $F(\mu)$, where F is some smooth functional (the exact definitions will be given later) and ξ_n is the random empirical measure

$$\xi_n = \frac{1}{n} \sum_{i=1}^{n} \delta_{\zeta_i}$$

with independent ζ_i, $\mathcal{L}(\zeta_i) = \mu$.

The second problem is to understand the role of independence for ζ_i. Surely, if the ζ_i are 'almost independent', the result must be the same as if they are independent. Can one choose dependent $\zeta_i = \zeta_i^{(n)}$ in such a way, that for any smooth F both bias and variance would be less than in the simplest case of independence? It occurs that generally the answer is positive — if one manages to apply the proper stratification technique, both error characteristics would have the form $o(1/n)$, while in the case of independence they are of order $O(1/n)$.

To demonstrate such propositions we must deal with functions defined on the appropriate measure spaces, their derivative mappings etc. Definitions and some simple properties of such objects are collected in Section 3.2.

Section 3.3 contains the main theoretical results of the paper. Theorem 3.3.1 tells us that for independent $\zeta_i^{(n)}$ and under some smoothness conditions on F, the bias $m_0(F) = \mathbf{E} F(\xi_n) - F(\mu)$ has the form $A(F)/n + o(1/n)$. As for $A(F)$, it is a function of the second derivative $d^2 F(\mu)/d\mu^2$.

The same result takes place if the $\zeta_i^{(n)}$ are 'almost independent'. More precisely, these random variables must be equidistributed with the distribution μ, their mutual distribution must be invariant under their permutation, and the condition $\mathcal{P}_2^{(n)} = \mathcal{L}(\zeta_1^{(n)}, \zeta_2^{(n)}) = \mu^{\otimes 2} + o(1/n)$ must hold.

Corollary 3.3.1 deals with $s_0^2(F) = \mathbf{E}(F(\xi_n) - F(\mu))^2$. It is proved that under the same conditions, $s_0^2(F) = B(F)/n + o(1/n)$ with $B(F)$ depending on the first derivative $dF(\mu)/d\mu$.

The second proposition of Theorem 3.3.1 claims that if

$$\mathcal{P}_2^{(n)} = \frac{n+1}{n} \mu^{\otimes 2} - \frac{1}{n} \nu_n + o(1/n), \tag{3.3}$$

where ν_n is a probability measure on Borel subsets of D^2, ν_n converges weakly to ν, and ν is the 'diagonal measure' $\mathcal{L}(\zeta,\zeta)$, $\mathcal{L}(\zeta) = \mu$, then $m_0(F) = o(1/n)$ for any sufficiently smooth F. Moreover, $s_0^2(F)$ has the same form here. The proof of these propositions is based on the technical Lemma 3.3.1, which gives the appropriate variant of the Taylor expansion applied to the spaces under consideration.

The last Section 3.4 is devoted to investigation of the condition (3.3) and to several examples. It occurs that the proper variant of the stratification technique leads to (3.3). Roughly speaking (see Corollary 3.4.1 for precise results), this stratification may be described as follows:

For each n let us consider n subdomains $D_{i,n} \subset D$ with (up to some small number of subdomains) $\mathrm{diam}(D_{i,n}) \to 0$ uniformly in i. Let $\mu_i^{(n)} \in H$, $\mu = \sum_{i=1}^n \mu_i^{(n)}/n$ and $\mu_i^{(n)}(D_{i,n}) = 1$. Suppose $\widehat{\zeta}_i^{(n)}$ are independent and $\mathcal{L}(\widehat{\zeta}_i^{(n)}) = \mu_i^{(n)}$. Denote $(\zeta_1^{(n)}, \ldots, \zeta_n^{(n)})$ the result of random permutation for random variables $(\widehat{\zeta}_1^{(n)}, \ldots, \widehat{\zeta}_n^{(n)})$.

Then, by the assertion of Corollary 3.4.1, the conditions of the second proposition of Theorem 3.3.1 are fulfilled.

The linear functional $F(\mu) = \int f d\mu$, $f \in \mathbf{C}(D)$, is the simplest example of smooth functional on measure space. Surely, the case of independent ζ_i with $\mathcal{L}(\zeta_i) = \mu$ is trivial here, but integrals are good area for stratification, as one can try to get more precise results for them. Several examples of such a kind may be found in Subsection 3.4.2.

If $D = [0,1]$ and μ is the Lebesgue measure on D, the simplest stratification with $D_{i,n} = [(i-1)/n, i/n]$ and $\mu_i^{(n)}$ taken as the uniform distribution on $D_{i,n}$ leads to error $o(n^{-1})$ for any continuous f (this is just the proposition of Corollary 3.4.1) and to $O(n^{-3})$ if $f \in \mathbf{C}^1(D)$ (or even if f is Lipschitz continuous). This corresponds to the old result due to Bahvalov (1964). Moreover, in the case $f \in \mathbf{C}^1(D)$ a version of the Central Limit Theorem may be proved. Other similar examples deal with discrete measures, general one-dimensional integrals and the multidimensional case.

The last example is based on Theorem 3.3.1 itself but not on Corollary 3.4.1. The equality (3.12) shows that to diminish the variance, random variables $\zeta_i^{(n)}$ must 'avoid each other' with the probability equivalent to n^{-1}. Due to Corollary 3.4.1 this effect is provided with the help of small supports $D_{i,n}$ of distributions $\mu_i^{(n)}$. Here we produce such a property in the more direct way.

Denote $D = (0,1]$ and $\rho(x,y) = \min\{|x-y|, 1-|x-y|\}$. This means that we consider D not as the interval, but as the circle. Let B_n be a subset of D^n such that $\rho(x_i, x_j) > 1/2n$ if $i \neq j$ and $(x_1, \ldots, x_n) \in B_n$.

For fixed $n > 2$ consider random vector $\eta^{(n)} = (\zeta_1^{(n)}, \ldots, \zeta_n^{(n)})$ which is uniformly distributed on B_n. It may be easily seen that all conditions of Theorem 3.3.1 (second proposition) are fulfilled, in particular (3.3) is valid. Moreover, exact calculations show that such 'stratification without stratification' may give better results than the usual division of $[0,1]$ into n equal intervals.

3.2 Strong Weakly-Continuous Derivatives

Let us start with some general terms and notation.

For Banach spaces E_1 and E_2 with norms $\|\cdot\|_1$ and $\|\cdot\|_2$ we denote by $\mathbf{L}(E_1, E_2)$ the Banach space of bounded linear operators acting from E_1 to E_2. Let U be a certain open subset of E_1. The mapping $g' : U \mapsto \mathbf{L}(E_1, E_2)$ is called the *derivative mapping* (see Cartan (1967) for details) or briefly the *derivative* of the mapping $g : U \mapsto E_2$, if for any $\mu, \tilde{\mu} \in U$ with $\|\tilde{\mu} - \mu\|_1 \to 0$

$$\|g(\tilde{\mu}) - g(\mu) - g'(\mu)(\tilde{\mu} - \mu)\|_2 = o(\|\tilde{\mu} - \mu\|_1). \tag{3.4}$$

Note, that the derivative g' may be considered as the mapping $U \times E_1 \mapsto E_2$, which is linear in the second argument.

Let us introduce notation for derivatives of the k-th order, using the second derivative as an example. By definition, the second derivative mapping is a mapping

$$g'' : U \mapsto \mathbf{L}(E_1 \times E_1, E_2),$$

which satisfies the relation analogous to (3.4). Therefore we use the notation $g''(\mu)$ or $d^2 g(\mu)/d\mu^2$ for the value of this mapping at the point $\mu \in U$.

On the other hand one may deal with the second derivative as a mapping $U \times E_1 \times E_1 \mapsto E_2$, which is linear in the second and third arguments and is invariant under their permutation. Then we denote the second derivative by

$$\partial^2 g : U \times E_1 \times E_1 \mapsto E_2.$$

Thus $g''(\mu)(\nu_1, \nu_2)$ and $\partial^2 g(\mu; \nu_1, \nu_2)$ mean the same for $\mu \in U$ and $\nu_1, \nu_2 \in E_1$.

Now we suppose that (D, ρ) is some complete metric separable compact space, $\mathcal{F}^\rho$ is the topology induced by ρ, $\mathcal{B}$ is the corresponding Borel σ-algebra and E is the linear space of Radon measures on $(D, \mathcal{B})$. It is well known that the space E with the norm $\|\cdot\| = \mathrm{Var}(\cdot)$ is a Banach space (not separable in general).

Let $\mathcal{F}_w$ be the topology induced by week convergence of measures on E (namely, μ_n weakly converges to μ, $\mu_n \xrightarrow{w} \mu$, if $\int f d\mu_n \to \int f d\mu$ for any bounded continuous function f on D).

Definition 3.2.1 The function $\psi : E \mapsto \mathbf{R}$ is called *w-continuously differentiable*, if there exists a derivative mapping ψ' which is continuous as a mapping $(E, \mathcal{F}_w) \times (E, \mathcal{F}_w) \mapsto \mathbf{R}$.

Consider the following functional spaces:

1. $\mathbf{C}(E)$ is the space of continuous on $(E, \mathcal{F}_w)$ functions;

2. $\widetilde{\mathbf{C}}^1(E)$ is a set of all w-continuously differentiable functions;

3. $\mathbf{C}^1(E) = \mathbf{C}(E) \cap \widetilde{\mathbf{C}}^1(E)$.

The analogous definition may be done for $\mathbf{C}^k(E)$, $k \geq 2$. The following lemma gives a useful representation for derivative mappings induced by functions from $\mathbf{C}^1(E)$.

Lemma 3.2.1 *Let $\psi \in \mathbf{C}^1(E)$. Then there exists an $\mathcal{F}_w \times \mathcal{F}^\rho$-continuous function $\psi' : E \times D \mapsto \mathbf{R}$ such that for any $\nu \in E$*

$$\partial^1 \psi(\mu; \nu) = \int_D \psi'(\mu, v) \nu(dv).$$

PROOF. Let $u \in D$. Denote $\psi'(\mu, u) = \partial^1 \psi(\mu; \delta_u) = \int_D \psi'(\mu, v) \delta_u(dv)$. As $\psi \in \mathbf{C}^1(E)$, the function ψ' is continuous in its arguments. Besides, as D is a compact set, $\psi'(\mu, \cdot)$ is bounded for fixed μ. Evidently,

$$\partial^1 \psi(\mu; \nu_n) = \int_D \psi'(\mu, v) \nu_n(dv) \tag{3.5}$$

for any discrete measure $\nu_n = \sum_{i=1}^{n} \alpha_i^{(n)} \delta_{u_i^{(n)}}$. Let $\nu \in E$. As the linear span of the set of Dirac measures is dense in $(E, \mathcal{F}_w)$, there exists a sequence $\{\nu_n\}$ such that $\nu_n \xrightarrow{w} \nu$. Taking $n \to \infty$ in (3.5) we get the result. ∎

Note that these definitions and features may be extended to higher derivatives. For example, the analogue of Lemma 3.2.1 for second derivatives may be stated as follows:

Lemma 3.2.2 *Let $\psi \in \mathbf{C}^2(E)$. Then there exists an $\mathcal{F}_w \times \mathcal{F}^\rho \times \mathcal{F}^\rho$-continuous function $\psi'' : E \times D \times D \mapsto \mathbf{R}$ such that for any $\nu_1, \nu_2 \in E$*

$$\partial^2 \psi(\mu; \nu_1, \nu_2) = \int_{D^2} \psi''(\mu, v_1, v_2) \nu_1(dv_1) \nu_2(dv_2).$$

3.3 General Results

Let $\mu \in H$. For $n \geq 1$ and $1 \leq i \leq n$ consider random variables $\zeta_i^{(n)}, \widetilde{\zeta}_i^{(n)} \in D$, such that $\mathcal{L}(\zeta_1^{(n)}, \ldots, \zeta_n^{(n)}) = \mathcal{L}(\widetilde{\zeta}_1^{(n)}, \ldots, \widetilde{\zeta}_n^{(n)})$ and $\zeta_i^{(n)}, \widetilde{\zeta}_j^{(n)}$ are independent for any i and j. Let

$$\xi_n = \frac{1}{n} \sum_{i=1}^{n} \delta_{\zeta_i^{(n)}}, \quad \widetilde{\xi}_n = \frac{1}{n} \sum_{i=1}^{n} \delta_{\widetilde{\zeta}_i^{(n)}}$$

Estimation Errors for Functionals on Measure Spaces 35

and define $\mu_n = \mathcal{E}(\xi_n)$ as the distribution such that

$$\int_D f d\mu_n = \mathbf{E} \int_D f d\xi_n,$$

$f \in \mathbf{C}(D)$. The following technical proposition is the base of further estimates.

Lemma 3.3.1 *If uniformly on $1 \leq i, j \leq n$, $i \neq j$,*

$$\|\mathcal{L}(\zeta_i^{(n)}, \zeta_j^{(n)}) - \mathcal{L}(\zeta_i^{(n)}) \otimes \mathcal{L}(\zeta_j^{(n)})\| \xrightarrow[n \to \infty]{} 0 \tag{3.6}$$

and $\|\mu_n - \mu\| \xrightarrow[n \to \infty]{} 0$, then for any $F \in \mathbf{C}^3(E)$

$$m_0(F) \stackrel{def}{=} \mathbf{E} F(\xi_n) - F(\mu) = \mathbf{E} \partial^1 F(\mu; \xi_n - \mu)$$

$$+ \mathbf{E} \int_0^1 (1-\tau) \partial^2 F(\alpha_n; \tilde{\xi}_n - \mu, \tilde{\xi}_n - \mu) \, d\tau + o(1/n) \tag{3.7}$$

with $\alpha_n = \mu + \tau(\xi_n - \mu)$.

PROOF. In view of the Taylor expansion with the integral remainder term it is sufficient to prove that

$$\Lambda_n = \mathbf{E} \, \partial^2 F(\alpha_n; \xi_n - \mu, \xi_n - \mu)$$
$$= \mathbf{E} \, \partial^2 F(\alpha_n; \tilde{\xi}_n - \mu, \tilde{\xi}_n - \mu) + o(1/n) = \tilde{\Lambda}_n + o(1/n).$$

Let $f_2(\nu; v_1, v_2) = \partial^2 F(\nu; \delta_{v_1}, \delta_{v_2})$. As $\partial^2 F$ is linear in the second and third arguments,

$$\Lambda_n = \mathbf{E} \frac{1}{n^2} \sum_{i \neq j} f_2(\alpha_n; \zeta_i, \zeta_j) + \mathbf{E} \frac{1}{n^2} \sum_i f_2(\alpha_n; \zeta_i, \zeta_i)$$

$$- \frac{2}{n} \mathbf{E} \sum_i f_2(\alpha_n; \zeta_i, \mu) + \mathbf{E} f_2(\alpha_n; \mu, \mu)$$

$$= I_1 + I_2 + I_3 + I_4$$

(we omit the upper index in $\zeta_i^{(n)}$ or $\tilde{\zeta}_i^{(n)}$ for brevity).

Let us rewrite I_1 in another form (I_2 and I_3 may be rewritten in the same manner). Denote $\Delta^{(i)} = \delta_{\tilde{\zeta}_i} - \delta_{\zeta_i}$, $\Delta^{(ij)} = \Delta^{(i)} + \Delta^{(j)}$, $\xi_n^{(ij)} = \xi_n + \Delta^{(ij)}/n$ and

$$\Phi(\nu; \nu_1, x, y) = \frac{df_2(\nu; x, y)}{d\nu}(\nu_1).$$

Using the Taylor expansion of the function $f_2(\alpha_n; \zeta_i, \zeta_j)$ in the first argument at the point $\alpha_n^{(ij)} = \mu + \tau(\xi_n^{(ij)} - \mu)$, we obtain

$$f_2(\alpha_n; \zeta_i, \zeta_j)$$
$$= f_2(\alpha_n^{(ij)}; \zeta_i, \zeta_j) - \frac{\tau}{n} \Phi(\alpha_n^{(ij)}; \Delta^{(ij)}, \zeta_i, \zeta_j) + \frac{\tau}{n} o\left(\|\Delta^{(ij)}\|\right)$$

with $\|\Delta^{(ij)}\| \leq 4$. Now, applying the properties of ζ_i and $\widetilde{\zeta}_i$, we can change our notation and write ζ_i instead of $\widetilde{\zeta}_i$, ξ_n instead of $\xi_n^{(ij)}$, etc. Therefore

$$I_1 = \frac{1}{n^2}\sum_{i\neq j}\mathbf{E}f_2(\alpha_n;\widetilde{\zeta}_i,\widetilde{\zeta}_j)$$
$$+\frac{\tau}{n^3}\sum_{i\neq j}\mathbf{E}\,\Phi(\alpha_n;\Delta^{(ij)},\widetilde{\zeta}_i,\widetilde{\zeta}_j) + o(n^{-1}) = I_1^{(1)} + I_1^{(2)} + o(n^{-1}).$$

Analogously $I_l = I_l^{(1)} + I_l^{(2)} + o(n^{-1})$ for $l = 2, 3$, and $I_1^{(1)} + I_2^{(1)} + I_3^{(1)} + I_4 = \widetilde{\Lambda}_n$. Thus $\Lambda_n = \widetilde{\Lambda}_n + \tau J + o(n^{-1})$, where

$$J = \frac{1}{n^3}\sum_{i\neq j}\mathbf{E}\,\Phi(\alpha_n;\Delta^{(ij)},\widetilde{\zeta}_i,\widetilde{\zeta}_j) + \frac{1}{n^3}\sum_i \mathbf{E}\,\Phi(\alpha_n;\Delta^{(i)},\widetilde{\zeta}_i,\widetilde{\zeta}_i)$$
$$-\frac{2}{n^2}\sum_i \mathbf{E}\int \Phi(\alpha_n;\Delta^{(i)},\widetilde{\zeta}_i,v)\mu(dv) = J_1 + J_2 + J_3.$$

Now we must prove that $J = o(n^{-1})$. Note that $J_2 = o(n^{-1})$. Using the invariant features of Φ under permutations of its last two arguments, we have

$$J_1 = \frac{1}{n^3}\sum_{i,j}\mathbf{E}\,\Phi(\alpha_n;\Delta^{(ij)},\widetilde{\zeta}_i,\widetilde{\zeta}_j) + o(n^{-1})$$
$$= \frac{2}{n^2}\sum_i \mathbf{E}\int \Phi(\alpha_n;\Delta^{(i)},\widetilde{\zeta}_i,v)\widetilde{\xi}_n(dv) + o(n^{-1}). \quad (3.8)$$

Let $(\zeta_1^*,\ldots,\zeta_n^*)$ be another independent copy of $(\zeta_1,\ldots,\zeta_n)$. Denoting

$$\xi_n^* = \frac{1}{n}\sum_{i=1}^n \delta_{\zeta_i^*},$$

we see that $\mathcal{E}(\xi_n^*) = \mu_n$. Let us prove that

$$J_1 = \frac{2}{n^2}\sum_i \mathbf{E}\int \Phi(\alpha_n;\Delta^{(i)},\widetilde{\zeta}_i,v)\mu_n(dv) + o(n^{-1}). \quad (3.9)$$

By (3.8) one can obtain that up to order $o(n^{-1})$

$$J_1 = \frac{2}{n^3}\sum_{i,j}\mathbf{E}\,\Phi(\alpha_n;\Delta^{(i)},\widetilde{\zeta}_i,\widetilde{\zeta}_j)$$
$$= \frac{2}{n^3}\sum_{i\neq j}\mathbf{E}\,\Phi(\alpha_n;\Delta^{(i)},\widetilde{\zeta}_i,\widetilde{\zeta}_j) + \frac{2}{n^3}\sum_i \mathbf{E}\,\Phi(\alpha_n;\Delta^{(i)},\widetilde{\zeta}_i,\widetilde{\zeta}_i)$$
$$= \frac{2}{n^2}\sum_i \mathbf{E}\int \Phi(\alpha_n;\Delta^{(i)},\widetilde{\zeta}_i,v)\xi_n^*(dv)$$

$$+\frac{2}{n^3}\sum_{i\neq j}\mathbf{E}\left(\Phi(\alpha_n;\Delta^{(i)},\widetilde{\zeta}_i,\widetilde{\zeta}_j)-\Phi(\alpha_n;\Delta^{(i)},\widetilde{\zeta}_i,\zeta_j^*)\right)$$

$$+\frac{2}{n^3}\sum_{i}\mathbf{E}\left(\Phi(\alpha_n;\Delta^{(i)},\widetilde{\zeta}_i,\widetilde{\zeta}_i)-\Phi(\alpha_n;\Delta^{(i)},\widetilde{\zeta}_i,\zeta_i^*)\right)$$

$$=\frac{2}{n^2}\sum_{i}\mathbf{E}\int\Phi(\alpha_n;\Delta^{(i)},\widetilde{\zeta}_i,v)\mu_n(dv)+\frac{2}{n^3}\sum_{i\neq j}\Gamma_{ij}^{(n)},$$

where $\Gamma_{ij}^{(n)}=\mathbf{E}\left(\Phi(\alpha_n;\Delta^{(i)},\widetilde{\zeta}_i,\widetilde{\zeta}_j)-\Phi(\alpha_n;\Delta^{(i)},\widetilde{\zeta}_i,\zeta_j^*)\right).$

All we need now is to prove that $\Gamma_{ij}^{(n)}=o(1)$ uniformly in i and j. Denote

$$\phi(\widetilde{\zeta}_i,v)=\mathbf{E}\left(\Phi(\alpha_n;\Delta^{(i)},\widetilde{\zeta}_i,v)\mid\widetilde{\zeta}_i\right).$$

Then by (3.6)

$$\Gamma_{ij}^{(n)}=\mathbf{E}\left(\phi(\widetilde{\zeta}_i,\widetilde{\zeta}_j)-\phi(\widetilde{\zeta}_i,\zeta_j^*)\right)\xrightarrow[n\to\infty]{}0$$

and (3.9) holds.

As $\|\mu_n-\mu\|\to 0$ and Φ is bounded on any compact set, then for any i

$$\mathbf{E}\int\Phi(\alpha_n;\Delta^{(i)},\widetilde{\zeta}_i,v)(\mu_n(dv)-\mu(dv))=o(1)$$

and $J_1=o(n^{-1})$. By the same way one may see that $J_3=o(n^{-1})$. ∎

Let now $\mu\in H$, $\mathcal{L}(\zeta_i^{(n)})=\mu$ and suppose that for any n the mutual distribution of random variables $\zeta_1^{(n)},\ldots,\zeta_n^{(n)}$ is invariant under their permutation. Denote $\mathcal{L}(\zeta_i^{(n)},\zeta_j^{(n)})=\mathcal{P}_2^{(n)}$, $i\neq j$, $\xi_n=\sum_{i=1}^{n}\delta_{\zeta_i^{(n)}}/n$ and $\mathcal{W}_n=\mathcal{P}_2^{(n)}-\mu\otimes\mu$.

Theorem 3.3.1 1. If $F\in\mathbf{C}^3(E)$ and $\|\mathcal{W}_n\|=o(1/n)$, then

$$m_0(F)=\mathbf{E}F(\xi_n)-F(\mu)=\frac{1}{2n}A(F)+o(1/n),\qquad(3.10)$$

where

$$A(F)=\int_D f_2(\mu;v,v)\mu(dv)-\int_{D^2}f_2(\mu;v_1,v_2)\mu(dv_1)\mu(dv_2).\qquad(3.11)$$

2. Let $F\in\mathbf{C}^5(E)$. Suppose that

$$\mathcal{P}_2^{(n)}=\frac{n+1}{n}\mu\otimes\mu-\frac{1}{n}\nu_n+r_n,\qquad(3.12)$$

where $\nu_n\xrightarrow{w}\mathcal{L}(\zeta,\zeta)$, $\mathcal{L}(\zeta)=\mu$ and $\|r_n\|=o(1/n)$. Then $m_0(F)=o(1/n)$.

PROOF. 1. Denote $g_\tau(v_1, v_2) = f_2(\alpha_n; v_1, v_2)$. If we prove that

$$m_0(F) = \frac{1}{2n}A(F) + \int_0^1 (1-\tau) \int_{D^2} \mathbf{E}g_\tau \, d\mathcal{W}_n \, d\tau + o(1/n) \qquad (3.13)$$

for $V_n = \|\mathcal{W}_n\| \to 0$, then the first part of the theorem would be demonstrated.

As $\mathcal{E}(\xi_n) = \mu$, the conditions of Lemma 3.3.1 are fulfilled and (3.7) holds. Moreover $\mathbf{E}\partial^1 F(\mu; \xi_n - \mu) = 0$ and therefore

$$m_0(F) = \mathbf{E} \int_0^1 (1-\tau) \int_{D^2} g_\tau \, d\left((\widetilde{\xi}_n - \mu) \otimes (\widetilde{\xi}_n - \mu)\right) d\tau + o(1/n)$$

$$= \mathbf{E} \int_0^1 (1-\tau) \Big(\frac{1}{n^2} \sum_{i,j} g_\tau(\widetilde{\zeta}_i, \widetilde{\zeta}_j) - \frac{2}{n} \sum_i \int_D g_\tau(\widetilde{\zeta}_i, v) \mu(dv)$$

$$+ \int_{D^2} g_\tau(v_1, v_2) \mu(dv_1) \mu(dv_2) \Big) d\tau + o(1/n)$$

$$= \mathbf{E} \int_0^1 (1-\tau) \Big(\frac{n-1}{n} g_\tau(\widetilde{\zeta}_1, \widetilde{\zeta}_2) + \frac{1}{n} g_\tau(\widetilde{\zeta}_1, \widetilde{\zeta}_1)$$

$$- 2 \int_D g_\tau(\widetilde{\zeta}_1, v) \mu(dv) + \int_{D^2} g_\tau(v_1, v_2) \mu(dv_1) \mu(dv_2) \Big) d\tau + o(1/n).$$

Hence up to order $o(n^{-1})$

$$m_0(F) = \mathbf{E} \int_0^1 (1-\tau) \Big(\int_{D^2} g_\tau(v_1, v_2) \mathcal{W}_n(dv_1 dv_2) +$$

$$+ \frac{1}{n} \Big(\int_D g_\tau(v,v) \mu(dv) - \int_{D^2} g_\tau(v_1, v_2) \mathcal{P}_2^{(n)}(dv_1 dv_2) \Big) \Big) d\tau. \qquad (3.14)$$

As $V_n \to 0$, the expression $\int \mathbf{E}g_\tau d\mathcal{P}_2^{(n)}$ on the right-hand side of (3.14) may be replaced by $\int \mathbf{E}g_\tau d\mu^{\otimes 2}$.

It follows from the same condition $V_n \to 0$ that $\mathcal{L}(\xi_n) \xrightarrow{w} \delta_\mu$ (see Sznitman (1991)) and therefore $\mathbf{E}g_\tau(v_1, v_2) \to f_2(\mu; v_1, v_2)$ for all τ, v_1 and v_2 as $n \to \infty$. Thus we can replace $\mathbf{E}g_\tau$ by f_2 in the second summand on the right-hand side of (3.14), and (3.13) is proved.

2. First, $V_n = O(1/n)$ by (3.12), and (3.13) holds. Let

$$\widetilde{F}(\nu) = \widetilde{F}(\nu; \tau, v_1, v_2) = f_2(\mu + \tau(\nu - \mu); v_1, v_2), \quad \widetilde{F} \in \mathbf{C}^3(E).$$

Accurate analysis of the remainder terms in the proofs of Lemma 3.3.1 and formula (3.13) shows us that

$$|m_0(\widetilde{F})| = |\mathbf{E}g_\tau(v_1, v_2) - f_2(\mu; v_1, v_2)| \leq C/n,$$

where C does not depend on τ, v_1 and v_2, and is determined by supremums of $\widetilde{F}$ and its first three derivatives on the corresponding compact sets. Thus we can replace $\mathbf{E}g_\tau$ by f_2 on the right-hand side of (3.13).

Taking into account (3.12) we obtain that

$$m_0(F) = \frac{1}{2n} A(F) + \frac{1}{2n} \int_{D^2} f_2\, d(\mu \otimes \mu - \nu_n) + o(1/n)$$

$$= \frac{1}{2n} \left(\int_D f_2(\mu; v, v) \mu(dv) - \int_{D^2} f_2\, d\nu_n \right) + o(1/n)$$

and the proof is complete. ∎

The following proposition gives the form for the mean square deviation $s_0(F)$, $s_0^2(F) = \mathbf{E}\big(F(\xi_n) - F(\mu)\big)^2$, under the conditions of Theorem 3.3.1.

Corollary 3.3.1 1. *Under conditions of the first proposition of Theorem 3.3.1,*

$$s_0^2(F) = \frac{1}{n} \left(\int f_1^2(\mu; v) \mu(dv) - \left(\int f_1(\mu; v) \mu(dv) \right)^2 \right) + o(1/n),$$

where $f_1(\mu; v) = \partial^1 F(\mu; \delta_v)$.
2. *Under conditions of the second proposition of Theorem 3.3.1, $s_0^2(F) = o(1/n)$.*

PROOF. As $s_0^2(F) = m_0(F^2) - 2F \cdot m_0(F)$, then by (3.10)

$$s_0^2(F) = \frac{1}{2n} \big(A(F^2) - 2F \cdot A(F) \big) + o(1/n)$$

Denote $\partial^2 F(v_1, v_2) = \partial^2 F(\mu; \delta_{v_1}, \delta_{v_2})$ and $\partial^1 F(v) = \partial^2 F(\mu; \delta_v)$. Then

$$\partial^2 F^2(v_1, v_2) = 2 \partial^1 F(v_1) \partial^1 F(v_2) + 2F \cdot \partial^2 F(v_1, v_2)$$

and therefore

$$A(F^2) = 2 \int_D \big(\partial^1 F(v)\big)^2 \mu(dv) + 2F \cdot \int_D \partial^2 F(v, v) \mu(dv)$$

$$- 2 \Big(\int_D \partial^1 F(v) \mu(dv) \Big)^2 + 2F \cdot \int_{D^2} \partial^2 F(v_1, v_2) \mu(dv_1) \mu(dv_2)$$

$$= 2 \left(\int_D \big(\partial^1 F(v)\big)^2 \mu(dv) - \Big(\int_D \partial^1 F(v) \mu(dv) \Big)^2 \right) - 2F \cdot A(F).$$

The second proposition is evident. ∎

3.4 Stratification

3.4.1 General stratification scheme

The standard sufficient conditions for (3.10) and (3.11) are the conditions for $\zeta_i^{(n)}$ to be i.i.d. Let us formulate the sufficient conditions for (3.12) that give a better order of convergence than in the case of independence.

Let $\mu \in H$ and suppose that for any $n \geq 1$ there exists a decomposition $\mu = \sum_{i=1}^{n} \mu_i^{(n)}/n$ such that $\mu_i^{(n)} \in H$ and $\max_{i \in I_n} \text{diam}(D_{i,n}) \xrightarrow[n \to \infty]{} 0$, where $D_{i,n} = \text{support}(\mu_i^{(n)})$, $I_n \subset \mathcal{N}_n = \{1, 2, \ldots, n\}$ and $\text{card}(\mathcal{N}_n \setminus I_n)/n \to 0$ as $n \to \infty$. Let $\widehat{\zeta}_i^{(n)}$ be independent and $\mathcal{L}(\widehat{\zeta}_i^{(n)}) = \mu_i^{(n)}$.

Corollary 3.4.1 *If $F \in \mathbf{C}^5(E)$, then $m_0(F) = o(1/n)$ for $\xi_n = \sum_{i=1}^{n} \delta_{\widehat{\zeta}_i^{(n)}}/n$.*

PROOF. Let π_n be a random permutation of $1, \ldots, n$ such that π_n does not depend on $(\widehat{\zeta}_1^{(n)}, \ldots, \widehat{\zeta}_n^{(n)})$ and $\mathcal{L}(\pi_n)$ is a uniform distribution on the whole permutation set. Denote $\zeta_i^{(n)} = \widehat{\zeta}_{\pi_n(i)}^{(n)}$. Surely, $\xi_n = \sum_{i=1}^{n} \delta_{\zeta_i^{(n)}}/n$. Moreover,

$$\mathcal{L}(\zeta_k^{(n)}) = \frac{1}{n} \sum_{i=1}^{n} \mathcal{L}(\widehat{\zeta}_i^{(n)}) = \frac{1}{n} \sum_{i=1}^{n} \mu_i^{(n)} = \mu$$

and for $k \neq l$

$$\mathcal{L}\left(\zeta_k^{(n)}, \zeta_l^{(n)}\right) = \sum_{i \neq j} \frac{1}{n(n-1)} \mathcal{L}\left(\widehat{\zeta}_i^{(n)}\right) \otimes \mathcal{L}\left(\widehat{\zeta}_j^{(n)}\right)$$

$$= \frac{1}{n} \sum_{i=1}^{n} \mu_i^{(n)} \otimes \left(\frac{1}{n-1} \sum_{j: j \neq i}^{n} \mu_j^{(n)}\right)$$

$$= \frac{n}{n-1} \mu^{\otimes 2} - \frac{1}{n(n-1)} \sum_{i=1}^{n} \left(\mu_i^{(n)}\right)^{\otimes 2} = \frac{n+1}{n} \mu^{\otimes 2} - \frac{1}{n} \nu_n + o(1/n),$$

where

$$\nu_n = \frac{1}{n} \sum_{i=1}^{n} \left(\mu_i^{(n)}\right)^{\otimes 2}.$$

All we need now is to prove that for any $\phi \in \mathbf{C}(D^2)$

$$\int_{D^2} \phi \, d\nu_n \xrightarrow[n \to \infty]{} \int_{D} \phi(v, v) \mu(dv).$$

Surely, as card$(\mathcal{N}_n \setminus I_n)/n = o(1)$,

$$\left| \int_{D^2} \phi \, d\nu_n - \int_D \phi(v,v) \mu(dv) \right|$$

$$\leq \frac{1}{n} \sum_{i \in I_n} \int_{D^2_{i,n}} \left| \phi(v_1, v_2) - \phi(v_1, v_1) \right| \mu_i^{(n)}(dv_1) \mu_i^{(n)}(dv_2) + o(1)$$

$$\leq \frac{1}{n} \sum_{i \in I_n} \sup_{v_1, v_2 \in D_{i,n}} \left| \phi(v_1, v_2) - \phi(v_1, v_1) \right| + o(1) = o(1).$$

The proof is complete. ∎

3.4.2 Examples

The simplest example of a smooth functional is the linear functional

$$F(\mu) = \int_D f \, d\mu.$$

If $f \in \mathbf{C}(D)$, then, as it may be easily checked, $F \in \mathbf{C}^k(E)$ for all k, $\partial^1 F(\mu, v) = f(v)$ and $\partial^k F \equiv 0$ for $k \geq 2$.

Surely, the first proposition of Theorem 3.3.1 is not interesting in this case, but the second proposition and Corollary 3.4.1 are of interest because there is hope to get more precise results for concrete D here.

Example 3.4.1 (The Lebesgue measure.) The simplest example for Corollary 3.4.1 is the example of $D = [0,1]$, $D_{i,n} = [(i-1)/n, i/n]$, where μ is the Lebesgue measure on $[0,1]$ and $\mu_i^{(n)}$ stands for the uniform distribution on $D_{i,n}$.

Consider the linear functional $F(\mu) = \int_{[0,1]} f \, d\mu$. Under the assumptions that f is measurable and bounded, we have $\mathbf{E}F(\xi_n) = F(\mu)$, and

$$\mathbf{D}F(\xi_n) = \frac{1}{n^2} \sum_{i=1}^n n \int_{(i-1)/n}^{i/n} (f(x) - J_{i,n})^2 dx,$$

where

$$J_{i,n} = n \int_{(i-1)/n}^{i/n} f(x) dx.$$

It may be easily seen that if f is continuous on $[0,1]$ then $\mathbf{D}F(\xi_n) = o(n^{-1})$ (this is just the case of Corollary 3.4.1).

If f has the first continuous derivative, then the variance can be calculated more explicitly. Let $x_i = x_{i,n} = i/n$ and suppose that $x \in D_{i,n}$. Then

$$f(x) = f(x_{i-1}) + (x - x_{i-1})f'(x_{i-1}) + o(n^{-1}),$$

where the remainder term is uniform in $x \in D_{i,n}$ and $1 \leq i \leq n$. Therefore

$$J_{i,n} = f(x_{i-1}) + \frac{1}{2n}f'(x_{i-1}) + o(n^{-1})$$

and

$$n \int_{(i-1)/n}^{i/n} (f(x) - J_{i,n})^2 dx = \frac{1}{12n^2}\left(f'(x_{i-1})\right)^2 + o(n^{-2}).$$

Thus

$$\mathbf{D}F(\xi_n) = \frac{1}{12n^4}\sum_{i=1}^{n}\left(f'(x_{i-1})\right)^2 + o(n^{-3}) = \frac{\sigma^2(F)}{n^3} + o(n^{-3}),$$

where

$$\sigma^2(F) = \frac{1}{12}\int_0^1 \left(f'(x)\right)^2 dx.$$

Note that such results can not be obtained in the general situation of Corollary 3.4.1, as we did not suppose any linear structure for the space D there.

To demonstrate the Central Limit Theorem for this example, it is sufficient to check the Lyapunov conditions. It may be easily seen that

$$n\int_{(i-1)/n}^{i/n}(f(x) - J_{i,n})^4 dx \leq \frac{C}{n^4}$$

with some positive C. Therefore if we take independent random variables $\widehat{\zeta}_i^{(n)}$ with $\mathcal{L}(\widehat{\zeta}_i^{(n)}) = \mu_i^{(n)}$, and denote

$$\alpha_{i,n} = \frac{n^{1/2}\left(f(\widehat{\zeta}_i^{(n)}) - J_{i,n}\right)}{\sigma(F)},$$

then $\sum_{i=1}^{n}\mathbf{E}(\alpha_{i,n})^4 = O(n^{-1}) \underset{n\to\infty}{\longrightarrow} 0$. Thus $\mathcal{L}(\eta_n) \overset{w}{\to} \mathbf{N}(0, \sigma^2(F))$ for

$$\eta_n = n^{3/2}(F(\xi_n) - F(\mu)).$$

Example 3.4.2 (General one-dimensional case.) Let $\mu \in H$ be an arbitrary continuous distribution on $(\mathbf{R}, \mathcal{B})$ with the distribution function $M(x)$. Suppose that $M(x) \neq M(y)$ for $x \neq y$. Let f be a bounded continuous function on $\mathbf{R}$, having finite limits at both $-\infty$ and $+\infty$. We are interested in the integral

$$F(\mu) = \int_{\mathbf{R}} f d\mu.$$

Evidently, the problem may be considered at $\overline{\mathbf{R}} = \mathbf{R} \cup \{-\infty\} \cup \{+\infty\}$, which is a compact in the metric $\rho(x,y) = |M(x) - M(y)|$. The natural extension of f on $\overline{\mathbf{R}}$ is continuous on this compact.

Choose $x_{i,n} \in \mathbf{R}$ such that $\rho(x_{i+1,n}, x_{i,n}) = 1/n$, $x_{0,n} = -\infty$ and $x_{n,n} = +\infty$. Denote $D_i = D_{i,n} = [x_{i-1,n}, x_{i,n}]$ for $1 \leq i \leq n$. Consider independent random variables $\widehat{\zeta}_i^{(n)}$ with distributions $\mu_i^{(n)} = n\mu\big|_{D_i}$. Then, evidently, $\mathbf{E}F(\xi_n) = F(\mu)$ and due to Corollary 3.4.1, $\mathbf{D}F(\xi_n) = o(1/n)$.

Note that this stratification scheme is equivalent to the Lebesgue measure example for estimation of the functional

$$F(\mu) = \int_0^1 f \circ M^{-1}(x) dx.$$

Thus all results of the previous paragraph (including the Central Limit Theorem) hold if $f \circ M^{-1} \in \mathbf{C}^1([0,1])$.

Example 3.4.3 (Discrete distributions.) Let $D = \{1, 2, \ldots, N\}$, $\rho(i,j) = \delta_{ij}$ (δ_{ij} stands for the Kronecker delta) and $\mu(\{i\}) = p_i > 0$, $\sum_i p_i = 1$. Surely, every function $f : D \mapsto \mathbf{R}$ is continuous in metric ρ. Consider

$$F(\mu) = \int_D f d\mu = \sum_i f_i p_i.$$

Let $k_i = [np_i]$ and $m_i = k_1 + \ldots + k_i$. Evidently, $m_N \leq n$. Denote $D_{i,n} = \{j\}$ for $m_{j-1} + 1 \leq i \leq m_j$ and $D_{i,n} = D$ for $m_N + 1 \leq i \leq n$. As for measures $\mu_i^{(n)}$, we take them concentrated at $D_{i,n}$ for $i \leq m_N$ and defined by

$$\mu_i^{(n)}(\{j\}) = \frac{np_j - k_j}{n - m_N}$$

otherwise. Consider independent random variables $\widehat{\zeta}_i^{(n)}$ with $\mathcal{L}(\widehat{\zeta}_i^{(n)}) = \mu_i^{(n)}$ and denote $\xi_n = \sum_i \delta_{\widehat{\zeta}_i^{(n)}}/n$. Surely, $\mathbf{E}F(\xi_n) = F(\mu)$. As $n - m_N = \sum_{i=1}^N \{np_i\} \leq N$, $\mathbf{D}F(\xi_n) = o(1/n)$ due to Corollary 3.4.1.

The simplest example shows that generally there does not hold any traditional variant of the CLT here. Let us take $N = 3$ and $p_i = 1/3$. Then $F(\xi_n) \equiv F(\mu)$ for $n = 3l$,

$$F(\xi_n) = \frac{n-1}{n} F(\mu) + \frac{1}{n} f(\widehat{\zeta}_1^{(n)})$$

for $n = 3l + 1$ and

$$F(\xi_n) = \frac{n-2}{n} F(\mu) + \frac{1}{n}\left(f(\widehat{\zeta}_1^{(n)}) + f(\widehat{\zeta}_2^{(n)})\right)$$

for $n = 3l + 2$, where $\mathbf{P}(\widehat{\zeta}_j^{(n)} = i) = 1/3$, $i = 1, 2, 3$, $j = 1, 2$.

Example 3.4.4 (Multi-dimensional Lebesgue measure.) Suppose $\mu \in H$ is the uniform distribution on $I^d = [0, 1]^d$. Consider the case $d = 2$ as the example and show the way to choose measures $\mu_i^{(n)}$, satisfying the conditions of Corollary 3.4.1. Denote $m = [\sqrt{n}]$, where $[x]$ stands for the entire part of x, and $n_1 = m^2$. Let distributions $\nu_{jk}^{(n)}$, $1 \leq j, k \leq m$, be uniform on

$$J_{jk} = [(j-1)n^{-1/2}, jn^{-1/2}] \times [(k-1)n^{-1/2}, kn^{-1/2}].$$

For $1 \leq i \leq n_1$ we choose $\mu_i^{(n)} = \nu_{jk}^{(n)}$ with some $(j, k) = (j(i), k(i))$ such that $\mu_i^{(n)} \neq \mu_l^{(n)}$ for $i \neq l$. If $n_1 < i \leq n$ we take $\mu_i^{(n)}$ as the uniform distribution on $I^2 \setminus \bigcup_{j_1, j_2 = 1}^{m} J_{j_1 j_2}$. Denote $D_{i,n} = \text{support}(\mu_i^{(n)})$. Surely, $\max_{1 \leq i \leq n_1} \text{diam } D_{i,n} \to 0$ and $n\mu = \sum_{i=1}^{n} \mu_i^{(n)}$. As $n - n_1 = O(n^{1/2})$, the result of Corollary 3.4.1 is valid.

Example 3.4.5 (Stratification without stratification.) Denote $D = (0, 1]$ and $d(x, y) = |x - y|$. Consider the metric space (D, ρ), where $\rho(x, y) = \min\{d(x, y), 1 - d(x, y)\}$. Surely, ρ is the circle metric, and a function, continuous in this metric, is a usual continuous function on $(0, 1]$ with $f(0+) = f(1)$. The space (D, ρ) is a compact space.

For fixed $n > 2$ consider a random vector $\eta^{(n)} = (\zeta_1^{(n)}, \ldots, \zeta_n^{(n)})$ which is uniformly distributed on $B_n \subset D^n$ such that $\rho(x_i, x_j) > 1/2n$ for all pairs $i \neq j$, $(x_1, \ldots, x_n) \in B_n$. Evidently, the distribution $\mathcal{L}(\eta^{(n)})$ does not depend on permutations of random variables.

If we put $x \oplus y = x + y \pmod{1}$, and $\zeta_{i,c}^{(n)} = \zeta_i^{(n)} \oplus c$, $c > 0$, then the mutual distribution of $\zeta_{i,c}^{(n)}$ coincides with $\mathcal{L}(\eta^{(n)})$ and therefore $\zeta_i^{(n)}$ is uniformly distributed on D for any i.

Let $\mathcal{P}_2^{(n)} = \mathcal{L}\left(\zeta_1^{(n)}, \zeta_2^{(n)}\right)$. Denote by U_n the support of $\mathcal{P}_2^{(n)}$. Surely,

$$U_n = \{(x, y) \subset D^2 : \rho(x, y) > 1/2n\}.$$

Moreover $\mathcal{P}_2^{(n)}$ is the uniform distribution on U_n due to the equality $\mathcal{P}_2^{(n)} = \mathcal{L}\left(\zeta_{1,c}^{(n)}, \zeta_{2,c}^{(n)}\right)$.

Let μ be the Lebesgue measure on D. If we demonstrate the identity (3.12), we would have by the second proposition of Theorem 3.3.1, that the variance of the unbiased estimate $F(\xi_n) = \sum_{i=1}^n f(\zeta_i^{(n)})/n$ for $F(\mu) = \int_D f d\mu$ has the order $o(n^{-1})$ if $f \in \mathbf{C}(D, \rho)$.

The two-dimensional Lebesgue measure of U_n is equal to $(n-1)/n$ and

$$\mathcal{P}_2^{(n)} - \frac{n+1}{n} \mu \otimes \mu = \mathcal{P}_2^{(n)} - \frac{n}{n-1} \mu \otimes \mu + o(n^{-1}) = \frac{1}{n-1} \nu_n + o(n^{-1}),$$

where ν_n is the uniform distribution on U_n^C. Surely, $\nu_n \xrightarrow{w} \mathcal{L}(\zeta, \zeta)$, $\mathcal{L}(\zeta) = \mu$.

Let us find a more explicit expression for the variance. Let $f \in \mathbf{C}\big([0,1], d\big)$. As we may suppose that $\mathbf{E}f(\zeta_i^{(n)}) = 0$, then $\mathbf{E}F(\xi_n) = 0$ and

$$\mathbf{D}F(\xi_n) = \frac{1}{n} \int_0^1 f^2(x) dx + \frac{n-1}{n} \mathbf{E}f(\zeta_1^{(n)}) f(\zeta_2^{(n)})$$

with

$$\frac{n-1}{n} \mathbf{E}f(\zeta_1^{(n)}) f(\zeta_2^{(n)}) = \iint_{U_n} f(x)f(y) dx dy = - \iint_{U_n^C} f(x)f(y) dx dy. \quad (3.15)$$

Accurate analysis of the last integral in (3.15) shows that for $f \in \mathbf{C}^2\big([0,1], d\big)$

$$\mathbf{D}F(\xi_n) = \frac{1}{8n^2} \big(f(0) - f(1)\big)^2$$
$$+ \frac{1}{24n^3} \left(\big(f(0) - f(1)\big)\big(f'(0) + f'(1)\big) + \int_0^1 \big(f'(x)\big)^2 dx \right) + o(n^{-3}).$$

Thus the procedure described above is generally worse than that of Example 1 if $f(0) \neq f(1)$ (then $f \notin \mathbf{C}(D, \rho)$), but gives half as much variance if $f(0) = f(1)$.

References

1. Bahvalov, N. S. (1964). On optimal estimation of the convergence rate for quadrature processes and Monte Carlo-type integration methods in functional classes, In *Computational Methods for Differential and Integral Equations and Quadrature Formulas*, pp. 5–63, Moscow: Nauka (in Russian).

2. Cartan, H. (1967). *Calcul Différentiel. Formes Différentielles*, Paris: Hermann.

3. Sznitman, A. S. (1991). Topics in propagation of chaos, *Lecture Notes in Mathematics*, **1464**, Berlin: Springer-Verlag.

4
The Multilevel Method of Dependent Tests

Stefan Heinrich

Universität Kaiserslautern, Kaiserslautern, Germany

Abstract: Approximation properties of the underlying estimator are used to improve the efficiency of the method of dependent tests. A multilevel approximation procedure is developed such that in each level the number of samples is balanced with the level-dependent variance, resulting in a considerable reduction of the overall computational cost. The new technique is applied to the Monte Carlo estimation of integrals depending on a parameter.

Keywords and phrases: Monte Carlo, multi-level method, dependent tests, integration, complexity

4.1 Introduction

The method of dependent tests is a basic way of using Monte Carlo estimates for the approximation of whole functions (as opposed to the approximation of a single function value or a weighted integral as in the classical Monte Carlo approach). The method was developed and studied by Frolov and Chentsov (1962), Sobol (1962, 1973), Ermakov and Mikhailov (1982), Mikhailov (1991), Voytishek (1996, 1997), Prigarin (1995), and others. The aim of the present paper is to propose a multilevel version of this method, based on the ideas developed in Heinrich (1998a). We exploit the approximability of the underlying estimator to decompose it into levels. The number of samples used for each level can be tuned to the variance of the contribution from this level, so that an overall reduction of computational cost is reached. The new method is presented in a general framework, and later on studied in detail for the computation of integrals depending on a parameter. While Heinrich (1998a) and Heinrich and Sindambiwe (1999) consider the class C^r of r-times continuously differentiable functions on the unit cube and provide, besides the algorithm, also a complexity analysis including lower bounds, in this paper we mostly concentrate on the study of the algorithm and give a more general convergence analysis. We consider arbitrary domains instead of the cube and relaxed smoothness assumptions requiring the function to be in a Sobolev class W_p^r with $1 < p < \infty$, and this only with respect to the parameter variable. We study the expected norm error in L_p for $1 < p < \infty$ and we are able to give quantitative results on the convergence. This is new even for the standard, one-level method of dependent tests. For that one, such an approach has so far only been carried out for Hilbert spaces, see, e.g., Mikhailov (1991), Voytishek (1996), Prigarin (1995). For L_p spaces ($p > 2$) only asymptotic results were obtained on the basis of weak convergence, with no information on the speed of convergence (to the Gaussian limit), see, e.g., Frolov and Chentsov (1962), Ermakov and Mikhailov (1982), Prigarin (1995), Voytishek (1997). Finally, the case $1 < p < 2$ is often left out because the usual tools do not work — the involved functions have infinite variance. We are able to study this case too and determine the convergence rates for both one- and multilevel methods. In the end, a few remarks on lower bounds and a comparison between one- and multilevel methods are made. The present paper is an extended version of a note which appeared in the abstract volume of this conference (see Heinrich, 1998b). A multilevel approach with the possibility of independent sampling is developed in Heinrich (1998c).

4.2 The Standard Method of Dependent Tests

Let X be a Banach space. A random variable with values in X is a Borel measurable mapping $\eta : \Omega \to X$ on some probability space (Ω, Σ, μ) such that the values of η are almost surely contained in a separable subspace of X. For $1 \leq p < \infty$ we denote by $L_p(X) = L_p(\Omega, \Sigma, \mu, X)$ the space of all X-valued random variables η on (Ω, Σ, μ) satisfying

$$\mathbb{E}\|\eta\|^p = \int_\Omega \|\eta(\omega)\|^p \, d\mu(\omega) < \infty$$

(see Ledoux and Talagrand, 1991, for details).

Now let $\eta \in L_p(X)$ for some p with $1 \leq p < \infty$. We seek to approximate the expectation

$$u = \mathbb{E}\eta = \int_\Omega \eta(\omega) \, d\mu(\omega) \in X$$

(the integral is meant in the sense of Bochner, see e.g. Ledoux and Talagrand, 1991, p. 42). Usually, X is an infinite dimensional function space, which makes it, in general, impossible to compute u itself. Instead, an estimate for Pu is constructed, where P is some continuous linear finite rank operator (an interpolation or approximation operator, for example), acting from X to another

Banach space Y (as a rule, either X itself or a larger function space, compare section 4.4). We shall assume that X is continuously embedded into Y, that is, there is a continuous injection $J : X \to Y$. In the sequel we shall identify X with $J(X) \subseteq Y$ as sets. The norms will be distinguished by $\|\|_X$ and $\|\|_Y$. Let

$$Px = \sum_{i=1}^{n} \langle x, x_i^* \rangle y_i \qquad (x \in X) \qquad (4.1)$$

be a representation of P, where $x_i^* \in X^*$ (the dual of X) and $y_i \in Y$. The standard method of dependent tests consists of the estimate

$$Pu \approx \theta = \frac{1}{N} \sum_{j=1}^{N} P\eta_j, \qquad (4.2)$$

where $(\eta_j)_{j=1}^{N}$ are independent realizations of η. (We assume that all random variables considered in this paper are defined on the same basic probability space (Ω, Σ, μ).) Combined with (4.1), this gives

$$\sum_{i=1}^{n} \langle u, x_i^* \rangle y_i \approx \theta = \sum_{i=1}^{n} \left(\frac{1}{N} \sum_{j=1}^{N} \langle \eta_j, x_i^* \rangle \right) y_i \qquad (4.3)$$

(which makes it clear that we use the same N samples for the estimation of the whole family of functionals $\langle u, x_i^* \rangle$). As an illustration, we consider integrals depending on a parameter, which will be studied in detail later on. For the moment we do this on an informal level — precise assumptions follow in section 4.4.

Let $G_1 \subset \mathbb{R}^{d_1}, G_2 \subset \mathbb{R}^{d_2}$ and f be a function on $G_1 \times G_2$. We want to approximate

$$u(s) = \int_{G_2} f(s, t) \, dt$$

as a function of the parameter $s \in G_1$. Let X and Y with $X \subseteq Y$ be some spaces of functions on G_1 and let P be an operator acting on $g \in X$ as

$$(Pg)(s) = \sum_{i=1}^{n} g(s_i) \varphi_i(s)$$

(one can think, e.g., of piecewise linear interpolation). Let ξ be a uniformly distributed on G_2 random variable. We set

$$\eta(\omega) = |G_2| f(\,\cdot\,, \xi(\omega)).$$

Then

$$\mathbb{E}\eta = \int_{G_2} f(\,\cdot\,, t) \, dt = u.$$

Now the method of dependent tests approximates

$$Pu \approx \sum_{i=1}^{n} \left(\frac{|G_2|}{N} \sum_{j=1}^{N} f(s_i, \xi_j) \right) \varphi_i$$

with $(\xi_j)_{j=1}^{N}$ being independent realizations of ξ.

4.3 The Multilevel Approach

Assume that we are given a sequence of continuous linear finite rank operators $(P_\ell)_{\ell=1}^{m}$ from X to Y with $P_m = P$ instead of P alone (usually, the approximation operators P belong to such scales in a natural way). Let

$$P_\ell x = \sum_{i=1}^{n_\ell} \langle x, x_{\ell i}^* \rangle y_{\ell i} \qquad (x \in X)$$

($\ell = 1, \ldots, m$) be the respective representations. Choose positive integers $(N_\ell)_{\ell=1}^{m}$ and estimate

$$P_m u = \sum_{\ell=1}^{m} (P_\ell - P_{\ell-1}) u \tag{4.4}$$

(with $P_0 = 0$) by

$$\begin{aligned}
\zeta &= \sum_{\ell=1}^{m} \frac{1}{N_\ell} \sum_{j=1}^{N_\ell} (P_\ell - P_{\ell-1}) \eta_{\ell j} \tag{4.5} \\
&= \sum_{\ell=1}^{m} \left[\sum_{i=1}^{n_\ell} \left(\frac{1}{N_\ell} \sum_{j=1}^{N_\ell} \langle \eta_{\ell j}, x_{\ell i}^* \rangle \right) y_{\ell i} - \sum_{i=1}^{n_{\ell-1}} \left(\frac{1}{N_\ell} \sum_{j=1}^{N_\ell} \langle \eta_{\ell j}, x_{\ell-1, i}^* \rangle \right) y_{\ell-1, i} \right]
\end{aligned}$$

where $(\eta_{\ell j} : j = 1, \ldots, N_\ell, \ell = 1, \ldots, m)$ are independent realizations of η. We set $n_0 = 0$, so for $\ell = 1$ the second term of the last line of (4.5) is to be understood as zero. Observe that the standard (one-level) method corresponds to the case $m = 1$ and $N_1 = N$. For parametric integration the concrete form of (4.5) is given later on — see relation (4.15). Now we shall analyze the error. For $1 \leq p < \infty$ we define the p-th expected norm error of the estimate ζ as

$$e_p(\zeta) = (\mathbb{E} \|u - \zeta\|_Y^p)^{1/p}.$$

By the triangle inequality, $e_p(\zeta)$ can be bounded by a deterministic and a stochastic component:

$$\begin{aligned}
e_p(\zeta) &= (\mathbb{E} \|u - P_m u + P_m u - \zeta\|_Y^p)^{1/p} \\
&\leq \|u - P_m u\|_Y + (\mathbb{E} \|P_m u - \zeta\|_Y^p)^{1/p}. \tag{4.6}
\end{aligned}$$

Next we shall give an upper bound for the stochastic component. For this purpose we let $1 \le p \le 2$ and recall that a Banach space Z is said to be of type p if there is a constant $c > 0$ such that for all $n \in \mathbb{N}$ and $(z_i)_{i=1}^n \subset Z$,

$$(\mathbb{E}\|\sum_{i=1}^n \varepsilon_i z_i\|^p)^{1/p} \le c \left(\sum_{i=1}^n \|z_i\|^p\right)^{1/p}, \tag{4.7}$$

where $(\varepsilon_i)_{i=1}^n$ is a sequence of independent Bernoulli variables with $\mu\{\varepsilon_i = 1\} = \mu\{\varepsilon_i = -1\} = \frac{1}{2}$. We refer to ch. 9.2 of Ledoux and Talagrand (1991) for this definition and background. The smallest possible constant in (4.7) is called the type p constant of Z, denoted by $T_p(Z)$. Let us mention that every Banach space is of type 1 (triangle inequality), and that type p implies type q for $1 \le q < p$. Each finite dimensional space is of type 2, and for $1 \le p < \infty$ the spaces $L_p(\nu)$ (with ν an arbitrary measure) are of type $\min(p,2)$. Clearly, all subspaces U of a type p space Z are of type p themselves, with $T_p(U) \le T_p(Z)$. Now we can present a bound of the stochastic part of the error.

Proposition 4.3.1 *Let $1 < p \le 2$ and assume $\eta \in L_p(X)$. Then*

$$(\mathbb{E}\|P_m u - \zeta\|_Y^p)^{1/p} \le 2T_p(Y_m) \left(\sum_{\ell=1}^m N_\ell^{1-p} \mathbb{E}\|(P_\ell - P_{\ell-1})(u - \eta)\|_Y^p\right)^{1/p},$$

where $Y_m = \text{span}\,(\bigcup_{\ell=1}^m P_\ell(X)) \subset Y$.

PROOF. By (4.4) and the first part of (4.5) we have

$$P_m u - \zeta = \sum_{\ell=1}^m \frac{1}{N_\ell} \sum_{j=1}^{N_\ell} (P_\ell - P_{\ell-1})(u - \eta_{\ell j}). \tag{4.8}$$

Now put

$$\rho_{\ell j} = N_\ell^{-1}(P_\ell - P_{\ell-1})(u - \eta_{\ell j})$$

($\ell = 1, \ldots, m$, $j = 1, \ldots, N_\ell$). These are independent Y_m-valued mean zero random variables with finite p-th moment

$$\mathbb{E}\|\rho_{\ell j}\|_Y^p = N_\ell^{-p} \mathbb{E}\|(P_\ell - P_{\ell-1})(u - \eta)\|_Y^p. \tag{4.9}$$

Proposition 9.11 of Ledoux and Talagrand (1991) states that

$$\mathbb{E}\|\sum_{\ell=1}^m \sum_{j=1}^{N_\ell} \rho_{\ell j}\|_Y^p \le (2T_p(Y_m))^p \sum_{\ell=1}^m \sum_{j=1}^{N_\ell} \mathbb{E}\|\rho_{\ell j}\|_Y^p.$$

Combining this with (4.8) and (4.9) yields the result. ∎

Corollary 4.3.1 *Let $1 < p \le 2$ and assume that $\eta \in L_p(X)$. Then*

$$(\mathbb{E}\|P_m u - \zeta\|_Y^p)^{1/p} \le 2T_p(Y_m)(\mathbb{E}\|u - \eta\|_X^p \sum_{\ell=1}^{m} N_\ell^{1-p}\|P_\ell - P_{\ell-1} : X \to Y\|^p)^{1/p}.$$

PROOF. This follows directly from

$$\mathbb{E}\|(P_\ell - P_{\ell-1})(u - \eta)\|_Y^p \le \|P_\ell - P_{\ell-1} : X \to Y\|^p \mathbb{E}\|u - \eta\|_X^p.$$

∎

4.4 Integrals Depending on a Parameter

Let d_1 and d_2 be positive integers and $G_1 \subset \mathbb{R}^{d_1}$ and $G_2 \subset \mathbb{R}^{d_2}$ be bounded open sets with Lipschitz boundary. Let $1 \le q < \infty$, let r be a positive integer with $r/d_1 > 1/q$, and let $W_q^{r,0}(G_1 \times G_2)$ be the space of all $f \in L_q(G_1 \times G_2)$ such that for each multiindex $\alpha = (\alpha_1, \ldots, \alpha_{d_1})$ with $|\alpha| = \alpha_1 + \ldots + \alpha_{d_1} \le r$ the generalized derivative $D_1^\alpha f$ with respect to the G_1 coordinates exists and belongs to $L_q(G_1 \times G_2)$. Hence, somewhat loosely speaking, we consider functions $f(s,t)$ with smoothness $W_q^r(G_1)$ in the first variable $s \in G_1$ and summability $L_q(G_2)$ in the second variable $t \in G_2$. The norm on $W_q^{r,0}(G_1 \times G_2)$ is defined as

$$\|f\|_{W_q^{r,0}} = \left(\sum_{|\alpha| \le r} \|D_1^\alpha f\|_{L_q(G_1 \times G_2)}^q \right)^{1/q}.$$

(For all notation concerning Sobolev spaces we refer to Adams, 1975). We study the estimation of

$$u(s) = \int_{G_2} f(s,t)\, dt \qquad (4.10)$$

in $L_q(G_1)$, that is, integration over G_2 with parameter domain G_1 and the error measured in the norm of $L_q(G_1)$. To put this into the framework of sections 4.2 and 4.3 we set $X = W_q^r(G_1)$, $Y = L_q(G_1)$ and $p = \min(2, q)$. We let $\xi = \xi(\omega)$ be a uniformly distributed on G_2 random variable on (Ω, Σ, μ) and we define $\eta = \eta(\omega)$ by

$$\eta : \omega \to |G_2| f(\,\cdot\,, \xi(\omega)).$$

Lemma 4.4.1 *The function η is a random variable with values in $X = W_q^r(G_1)$, belongs to $L_q(X)$, $\mathbb{E}\eta = u$ and*

$$(\mathbb{E}\|\eta\|_X^q)^{1/q} \le \|f\|_{W_q^{r,0}}. \qquad (4.11)$$

PROOF. We first verify that the values of η almost surely belong to $W_q^r(G_1)$ and that η is Borel measurable as a mapping into $W_q^r(G_1)$ (note that $W_q^r(G_1)$ is a separable Banach space). Let us denote by f_t the function given by $f_t(s) = f(s,t)$. Then $f_t \in L_q(G_1)$ for almost all t, by Fubini's theorem. Using elementary facts from distribution theory, it is readily checked that for all α with $|\alpha| \leq r$ the weak derivative $(D_1^\alpha f)(\,\cdot\,,t)$ coincides with $D^\alpha f_t$ for almost all t. This implies $f_t \in W_q^r(G_1)$ for almost all t. Moreover, since $W_q^r(G_1)$ is isometric to the subspace of $\oplus_q \sum_{|\alpha|\leq r} L_q(G_1)$ of those $(f_\alpha)_{|\alpha|\leq r}$ with $f_\alpha = D^\alpha f_0$, we can use 6.2.12 of Pietsch (1987) to prove that $t \to (D^\alpha f_t)_{|\alpha|\leq r}$ is Borel measurable as a mapping into the direct sum above, and hence $t \to f_t$ is Borel measurable as a mapping into $W_q^r(G_1)$. We have

$$(\mathbb{E}\|\eta\|_X^q)^{1/q} = \left(\int_{G_2} \|f_t\|_{W_q^r(G_1)}^q \, dt\right)^{1/q} = \|f\|_{W_q^{r,0}} \leq 1.$$

It follows that $\mathbb{E}\eta$ is well-defined, and because of (4.10), equals u. ∎

Now we have to define suitable approximation tools in $W_q^r(G_1)$. There is a vast literature on this subject and a variety of possibilities. Since a review of these tools is not the subject of this paper, we restrict ourselves to formulating the requirements on the approximating operators needed for our purposes and make a few comments on how to satisfy them. Let $P_\ell : W_q^r(G_1) \to L_q(G_1)$ ($\ell = 1, 2 \ldots$) be a sequence of operators of the form

$$P_\ell g = \sum_{i=1}^{n_\ell} g(s_{\ell i}) \varphi_{\ell i} \qquad (4.12)$$

with $s_{\ell i} \in \overline{G}_1$ (the closure of G_1) and $\varphi_{\ell i} \in L_q(G_1)$. The Sobolev embedding theorem guarantees that the point evaluations are well-defined. We assume that there are constants $c_1, c_2, c_3 > 0$ such that for all ℓ

$$c_1 2^{d_1 \ell} \leq n_\ell \leq c_2 2^{d_1 \ell} \qquad (4.13)$$

and, if $I_{r,q}$ denotes the identical embedding of $W_q^r(G)$ into $L_q(G_1)$,

$$\|I_{r,q} - P_\ell : W_q^r(G_1) \to L_q(G_1)\| \leq c_3 2^{-r\ell}. \qquad (4.14)$$

Such sequences can be constructed for many domains, e.g. by using triangular, rectangular or isoparametric finite elements of suitable order. We refer to Ciarlet (1978) for details. For the unit cube and arbitrary r, piecewise multivariate Lagrange interpolation will do (among many others), as described in Heinrich (1998a) and Heinrich and Sindambiwe (1999). For polyhedral domains and $r = 2$ piecewise linear interpolation on successively finer triangulations is a standard approach, the $s_{\ell i}$ being the vertices of the triangles and the $\varphi_{\ell i}$ being the corresponding hat functions.

The restriction to point evaluations of f in (4.12) was just made for notational simplicity. One could also admit values of derivatives $(D^\alpha f)(s)$ with $|\alpha| < r/d_1 - 1/q$. Now the multilevel method of dependent test fixes an m and approximates u according to (4.5) by

$$\zeta = \sum_{\ell=1}^{m} \left[\sum_{i=1}^{n_\ell} \left(\frac{|G_2|}{N_\ell} \sum_{j=1}^{N_\ell} f(s_{\ell i}, \xi_{\ell j}) \right) \varphi_{\ell i} \right.$$
$$\left. - \sum_{i=1}^{n_{\ell-1}} \left(\frac{|G_2|}{N_\ell} \sum_{j=1}^{N_\ell} f(s_{\ell-1,i}, \xi_{\ell j}) \right) \varphi_{\ell-1,i} \right] \quad (4.15)$$

where the N_ℓ ($\ell = 1, \ldots, m$) are positive integers, $\xi_{\ell j}$ ($\ell = 1, \ldots, m$, $j = 1, \ldots, N_\ell$) are independent realizations of the uniformly distributed on G_2 random variable ξ, and for $\ell = 1$ we set $n_0 = 0$, so in this case the second term is zero. To illustrate this procedure, think of $d_1 = d_2 = 1$, $G_1 = G_2 = (0,1)$, f a smooth function on $(0,1)^2$, $\{s_{\ell i}, i = 0, \ldots, 2^\ell\}$ the uniform mesh on $[0,1]$ of size $2^{-\ell}$ and $\varphi_{\ell i}$ the corresponding hat functions (that is, P_ℓ is piecewise linear interpolation). This example also explains the role of the different spaces X and Y: For a smooth function g (in X), $P_\ell g$ needs not to be smooth anymore, but is just piecewise linear (an element of the larger space Y).

Next we provide a bound for the stochastic part of the error of the multilevel method.

Proposition 4.4.1 *Let $1 < q < \infty$, $p = \min(2, q)$ and let $(P_\ell)_{\ell=1}^\infty$ satisfy (4.13) and (4.14). Then there is a constant $c > 0$ such that for all $f \in W_q^{r,0}(G_1 \times G_2)$ with $\|f\|_{W_q^{r,0}} \leq 1$, for all $m \in \mathbb{N}$ and $N_\ell \in \mathbb{N}$ ($\ell = 1, \ldots, m$) the multilevel estimate ζ defined above satisfies*

$$(\mathbb{E}\|P_m u - \zeta\|_{L_q(G_1)}^p)^{1/p} \leq c \left(\sum_{\ell=1}^{m} N_\ell^{1-p} 2^{-rp\ell} \right)^{1/p}. \quad (4.16)$$

PROOF. From Lemma 3 we get $\eta \in L_q(X) \subseteq L_p(X)$ and

$$(\mathbb{E}\|u - \eta\|_X^p)^{1/p} \leq \|u\|_X + (\mathbb{E}\|\eta\|_X^p)^{1/p}$$
$$= \|\mathbb{E}\eta\|_X + (\mathbb{E}\|\eta\|_X^p)^{1/p}$$
$$\leq 2(\mathbb{E}\|\eta\|_X^q)^{1/q} \leq 2\|f\|_{W_q^{r,0}} \leq 2.$$

Moreover, by (4.14)

$$\|P_\ell - P_{\ell-1} : X \to Y\| \leq c 2^{-r\ell}.$$

Finally, since $L_q(G_1)$ is of type p, all of its subspaces have a type p constant not exceeding that of $L_q(G_1)$. Now Corollary 4.3.1 yields the result. ∎

The Multilevel Method of Dependent Tests 55

Remark For the one level method (4.2) we obtain under the assumptions of Proposition 4 with $P = P_k$ for some $k \geq 1$,

$$(\mathbb{E}\|Pu - \theta\|^p_{L_q(G_1)})^{1/p} \leq cN^{1/p-1}. \tag{4.17}$$

For $q = 2$ ($= p$) this is well-known (see, e.g., Voytishek, 1996).

In the following we shall choose the N_ℓ and balance deterministic and stochastic error in such a way that we obtain minimal error at fixed computational cost. In the theorem below 'cost of computing ξ' means the total number of arithmetic operations, random number calls and function evaluations required for the computation of the coefficients of all $\varphi_{\ell i}$ in (4.15).

Theorem 4.4.1 *Let $1 < q < \infty$, $p = \min(2, q)$ and let $(P_\ell)_{\ell=1}^\infty$ satisfy (4.13) and (4.14). Then there exist constants $c_1, c_2 > 0$ such that for each integer $M > 1$ there is a choice of parameters $m, (N_\ell)_{\ell=1}^m$ such that the cost of computing ζ is bounded by $c_1 M$ and for each $f \in W_q^{r,0}(G_1 \times G_2)$ with $\|f\|_{W_q^{r,0}} \leq 1$ the p-th expected norm error (with respect to the norm of $L_q(G_1)$) satisfies*

$$\begin{aligned} e_p(\zeta) &\leq c_2 M^{-r/d_1} && \text{if } r/d_1 < 1 - 1/p, \\ e_p(\zeta) &\leq c_2 M^{1/p-1} \log M && \text{if } r/d_1 = 1 - 1/p, \\ e_p(\zeta) &\leq c_2 M^{1/p-1} && \text{if } r/d_1 > 1 - 1/p. \end{aligned}$$

PROOF. Throughout the proof and in the sequel the same symbol c, c_1, or c_2 is used for possibly different positive constants, not depending on m, M and f. The cost of computing ζ is obviously bounded by

$$c \sum_{\ell=1}^m 2^{d_1 \ell} N_\ell. \tag{4.18}$$

The line of the subsequent proof is the following. For the moment we fix m to be any positive integer with

$$2^{d_1(m-1)} \leq M. \tag{4.19}$$

First we choose the N_ℓ for this fixed m and estimate the stochastic part of the error. Later on we select the final m so that deterministic and stochastic part of the error are in balance.
So let

$$N_\ell = \left\lceil 2^{-(r + d_1/p)\ell - ((1-1/p)d_1 - r)m} M \right\rceil \tag{4.20}$$

if $r/d_1 < 1 - 1/p$,

$$N_\ell = \left\lceil m^{-1} 2^{-d_1 \ell} M \right\rceil \tag{4.21}$$

if $r/d_1 = 1 - 1/p$, and

$$N_\ell = \left\lceil 2^{-(r+d_1/p)\ell} M \right\rceil \quad (4.22)$$

if $r/d_1 > 1 - 1/p$. Although this choice looks complicated, it has an obvious source — this is (up to constants) what we get when minimizing the bound from Proposition 4,

$$\sum_{\ell=1}^{m} N_\ell^{1-p} 2^{-rp\ell}$$

(the variance, for $p = 2$), subject to the condition

$$\sum_{\ell=1}^{m} 2^{d_1\ell} N_\ell \leq M$$

(the cost). Since this aspect is not relevant for the proof, we omit the standard calculation. It is readily checked that from (4.19) and the choices (4.20), (4.21) or (4.22) it follows that

$$\sum_{\ell=1}^{m} 2^{d_1\ell} N_\ell \leq cM.$$

The deterministic part of the error (see (4.6)) satisfies, by (4.14) and (4.11),

$$\|u - P_m u\|_{L_q(G_1)} \leq c 2^{-rm} \|u\|_{W_q^r(G_1)} \leq c 2^{-rm}. \quad (4.23)$$

Next we compute the bounds on the stochastic part of the error in Proposition 4. First we treat the case $r/d_1 < 1 - 1/p$. We have

$$\left(\sum_{\ell=1}^{m} N_\ell^{1-p} 2^{-rp\ell}\right)^{1/p} \leq cM^{1/p-1} 2^{((1-1/p)d_1-r)m}, \quad (4.24)$$

which is a consequence of (4.20) and the following calculation of exponents

$$(1-p)[-(r+d_1/p)\ell - ((1-1/p)d_1 - r)m] - rp\ell$$
$$= ((p-1)(r+d_1/p) - rp)\ell + (p-1)((1-1/p)d_1 - r)m$$
$$= ((1-1/p)d_1 - r)\ell + (p-1)((1-1/p)d_1 - r)m$$
$$= p((1-1/p)d_1 - r)m + ((1-1/p)d_1 - r)(\ell - m).$$

Now we choose m in such a way that

$$c_1 2^{-rm} \leq M^{1/p-1} 2^{((1-1/p)d_1-r)m} \leq c_2 2^{-rm}, \quad (4.25)$$

which means that, up to constants, we equalize the bounds for deterministic and stochastic part of the error, that is, the right hand sides of (4.23) and (4.24). Clearly, (4.25) is equivalent to

$$c_1 2^{d_1 m} \leq M \leq c_2 2^{d_1 m}$$

(different constants!), and it suffices to take m to be the largest integer satisfying

$$2^{d_1(m-1)} \leq M.$$

(4.23), (4.24), (4.25) together with (4.16) and (4.6) yield

$$e_p(\zeta) \leq c M^{-r/d_1}.$$

For $r/d_1 = 1 - 1/p$ we use (4.21) and argue similarly:

$$(1-p)(-d_1 \ell) - rp\ell = p((1-1/p)d_1 - r)\ell = 0$$

and hence

$$\left(\sum_{\ell=1}^{m} N_\ell^{1-p} 2^{-rp\ell} \right)^{1/p} \leq c M^{1/p-1} m.$$

We choose m in such a way that

$$c_1 2^{-rm} \leq M^{1/p-1} m \leq c_2 2^{-rm}.$$

This is equivalent to

$$c_1 m^{1/(1-1/p)} 2^{d_1 m} \leq M \leq c_2 m^{1/(1-1/p)} 2^{d_1 m},$$

and we let m be the largest integer satisfying

$$m^{1/(1-1/p)} 2^{d_1(m-1)} \leq M.$$

We obtain

$$e_p(\zeta) \leq c M^{1/p-1} \log M.$$

Finally, for $r/d_1 > 1 - 1/p$ we have

$$(1-p)(-(r+d_1/p)\ell) - rp\ell = ((1-1/p)d_1 - r)\ell$$

and hence

$$\left(\sum_{\ell=1}^{m} N_\ell^{1-p} 2^{-rp\ell} \right)^{1/p} \leq c M^{1/p-1}.$$

Here we choose m so that

$$c_1 2^{-rm} \leq M^{1/p-1} \leq c_2 2^{-rm},$$

or equivalently

$$c_1 2^{r(1-1/p)^{-1}m} \leq M \leq c_2 2^{r(1-1/p)^{-1}m},$$

and we let m be the largest integer with

$$2^{r(1-1/p)^{-1}(m-1)} \leq M.$$

This yields

$$e_p(\zeta) \leq cM^{1/p-1}$$

and proves the theorem. ∎

Remarks 1. As already mentioned, by 'computation of ζ' we meant the computation of coefficients of the functions $\varphi_{\ell i}$ in (4.15). Having accomplished this, it is often possible to combine these functions in a computationally favorable way. Usually, the spaces $\text{span}\{\varphi_{\ell i} : i = 1, \ldots, n_\ell\}$ are nested, and one can decompose $\varphi_{\ell i}$ successively into combinations of $\varphi_{\ell+1,i}$ until level m is reached. For standard choices of approximation (as e.g. finite elements, piecewise Lagrange polynomials, piecewise linear functions, mentioned above) such a decomposition can be achieved in $cn_m \leq cM$ operations.

Now assume this is done, as well, and we want to compute $\zeta(s)$ for many $s \in G_1$ (e.g., to produce a graph of the approximating function). For each s, this can be carried out in $\leq c$ operations, provided the functions $\varphi_{m,i}$ can be computed in $\leq c$ operations and the supports of these functions are almost disjoint, which means that

$$\sup_m \max_i |\{j : \text{supp}(\varphi_{m,i}) \cap \text{supp}(\varphi_{m,j}) \neq \emptyset\}| < \infty.$$

Again, many known approximation scales, including the above mentioned examples, possess this property.

2. The uniform distribution in the algorithm (4.15) is not really essential. With the appropriate modifications, one can also use a random variable ξ with density $\pi(t)$ on G_2. Then one has to set $\eta = f(\,\cdot\,, \xi(\omega))\pi(\xi(\omega))^{-1}$ and hence, in (4.15) $|G_2|$ has to be replaced by $\pi(\xi_{\ell j})^{-1}$. The proper class of functions is the set of those f with $f\pi^{1/q-1} \in W_q^{r,0}(G_1 \times G_2)$.

Let us finally consider the one-level method and make comparisons. The sum of deterministic and stochastic error (see (4.17)) amounts to

$$c(2^{-rk} + N^{1/p-1}),$$

while the cost M is of the order $2^{d_1 k} N$. Equalizing both terms above, we see that at cost M we can reach an error

$$cM^{\frac{1/p-1}{1+(1-1/p)d_1/r}}. \tag{4.26}$$

This is certainly larger than

$$M^{\max(1/p-1,-rd_1)},$$

which we get (up to the log term) from Theorem 4.4.1. The saving by the multilevel method can be seen better if we compare the cost of reaching an error $\varepsilon > 0$. For the one-level method the cost is

$$\tilde{c}\left(\frac{1}{\varepsilon}\right)^{d_1/r+(1-1/p)^{-1}},$$

while for the multilevel method (up to log's)

$$c\left(\frac{1}{\varepsilon}\right)^{\max(d_1/r,(1-1/p)^{-1})}.$$

The results of Theorem 4.4.1 are optimal in a very general sense: No randomized algorithm of cost M can do better (except for a constant factor independent of M or, perhaps, a log-term in the case $r/d_1 = 1 - 1/p$). We cannot give the required formal framework for such statements here and refer instead to the literature on information-based complexity theory (see Traub, Wasilkowski, and Woźniakowski, 1988, Novak, 1988, Heinrich, 1994, Heinrich and Sindambiwe, 1999). Nevertheless, a few words on these lower bounds seem appropriate. First of all, we now restrict ourselves to the model case $G_1 = [0,1]^{d_1}$, $G_2 = [0,1]^{d_2}$. For the problem of parametric integration of functions from the class $W_q^{r,0}(G_1 \times G_2)$ lower bounds are, in fact, easily derived from known results (quite in contrast to the situation of $C^r(G_1 \times G_2)$ studied in Heinrich and Sindambiwe, 1999). Indeed, by considering the subclass of $W_q^{r,0}(G_1 \times G_2)$ of functions depending only on the second component, i.e. $f(s,t) \equiv g(t)$, we see that the problem is no easier than stochastic integration of $L_q(G_2)$ functions. For this, the lower bound $M^{1/p-1}$ with $p = \min(2,q)$ is known, see Novak (1988, 2.2.9, Proposition 1, and references). Similarly, the subclass of all functions in $W_q^{r,0}(G_1 \times G_2)$ depending only on the first component $f(s,t) \equiv g(s)$, can be identified with $W_q^r(G_1)$, hence the problem is no easier than approximation of functions of $W_q^r(G_1)$ in $L_q(G_1)$, for which the known lower bound for stochastic methods is M^{-r/d_1}, see Heinrich (1994, Thm. 6.1 and references). Thus

$$M^{\max(1/p-1,-r/d_1)} \tag{4.27}$$

is a lower bound, which shows that Theorem 4.4.1 yields, in fact, the optimal rate and hence the minimal Monte Carlo error in the sense of information-based complexity theory (up to a possible log factor in the case $r/d_1 = 1 - 1/p$).

When comparing the one- and the multilevel method, it seemed that we compared only upper bounds. Such a discussion would be meaningless since it does not exclude the existence of better estimates for any of the methods under comparison. In our case, however, this is not so. Looking again at functions depending only on the first or the second variable, it is easy to check directly that the one level method cannot be better than

$$c\left(2^{-rk} + N^{1/p-1}\right),$$

and hence, the rate (4.26) is sharp. Let us finally mention that the lower bound (4.27) also holds for $q = p = 1$, in which case it turns into a positive constant. This shows that no method can have a nontrivial convergence rate for $q = 1$.

References

1. Adams, R. A. (1975). *Sobolev Spaces*, New York: Academic Press.

2. Ciarlet, P. G. (1978). *The Finite Element Method for Elliptic Problems*, Amsterdam: North-Holland.

3. Ermakov, S. M. and Mikhailov, G. A. (1982). *Statistical Modelling*, Moscow: Nauka (in Russian).

4. Frolov, A. S. and Chentsov, N. N. (1962). On the calculation of certain integrals dependent on a parameter by the Monte Carlo method, *Zh. Vychisl. Mat. Mat. Fiz.*, **2**, 714–717 (in Russian).

5. Heinrich, S. (1994). Random approximation in numerical analysis, In *Functional Analysis*, (Eds., Bierstedt, K. D., Pietsch, A., Ruess, W. M. and Vogt, D.), pp. 123–171, New York, Basel, Hong Kong: Marcel Dekker.

6. Heinrich, S. (1998a). Monte Carlo complexity of global solution of integral equations, *J. Complexity*, **14**, 151–175.

7. Heinrich, S. (1998b). A multilevel version of the method of dependent tests, In *Proceedings of the 3rd St. Petersburg Workshop on Simulation* (collection of extended abstracts) (Eds., Ermakov, S. M., Kashtanov, Y. N. and Melas, V. B.), pp. 31–35, St. Petersburg: St. Petersburg University Press.

8. Heinrich, S. (1998c). Wavelet Monte Carlo methods for the global solution of integral equations, In *Proceedings of the Third International Conference on Monte Carlo and Quasi-Monte Carlo Methods, Claremont, 1998* (Eds., Niederreiter, H. and Spanier, J.) (to appear).

9. Heinrich, S. and Sindambiwe, E. (1999). Monte Carlo complexity of parametric integration, *J. Complexity*, (to appear).

10. Ledoux, M. and Talagrand, M. (1991). *Probability in Banach Spaces*, Berlin, Heidelberg, New York: Springer.

11. Mikhailov, G. A. (1991). *Minimization of Computational Costs of Non-Analogue Monte Carlo Methods*, Singapore: World Scientific.

12. Novak, E. (1988). *Deterministic and Stochastic Error Bounds in Numerical Analysis*, Lecture Notes in Mathematics, **1349**, Berlin, Heidelberg, New York: Springer-Verlag.

13. Pietsch, A. (1987). *Eigenvalues and s-Numbers*, Leipzig: Geest and Portig, and Cambridge: Cambridge University Press.

14. Prigarin, S. M. (1995). Convergence and optimization of functional estimates in statistical modelling in Sobolev's Hilbert spaces, *Russian J. Numer. Anal. Math. Modelling*, **10**, 325–346.

15. Sobol, I. M. (1962). The use of ω^2-distribution for error estimation in the calculation of integrals by the Monte Carlo method, *U.S.S.R. Comput. Math. and Math. Phys.*, **2**, 717–723.

16. Sobol, I. M. (1973). *Computational Monte Carlo Methods*, Moscow: Nauka (in Russian).

17. Traub, J. F., Wasilkowski, G. W. and Woźniakowski, H. (1988). *Information-Based Complexity*, New York: Academic Press.

18. Voytishek, A. V. (1996). Discrete-stochastic procedures for the global estimation of an integral which depends on a paramter, *Comp. Maths Math. Phys.*, **36**, 997–1009.

19. Voytishek, A. V. (1997). Using the Strang-Fix approximation in discrete-stochastic numerical procedures, *Monte Carlo Methods and Applications*, **3**, 89–112.

5

Algebraic Modelling and Performance Evaluation of Acyclic Fork-Join Queueing Networks

Nikolai K. Krivulin

St. Petersburg State University, St. Petersburg, Russia

Abstract: Simple lower and upper bounds on service cycle times in stochastic acyclic fork-join queueing networks are derived using a (max, +)-algebra based representation of network dynamics. The behaviour of the bounds under various assumptions concerning the service times in the networks is discussed, and related numerical examples are presented.

Keywords and phrases: (max,+)-algebra, dynamic state equation, acyclic fork-join queueing networks, stochastic dynamic systems, service cycle time

5.1 Introduction

Fork-join networks, as introduced in Baccelli and Makowski (1989), Baccelli et al. (1989), present a class of queueing systems which allow customers (jobs, tasks) to be split into several parts, and to be merged into one when they circulate through the system. The fork-join formalism proves to be useful in the description of dynamical processes in a variety of actual complex systems, including production processes in manufacturing, transmission of messages in communication networks, and parallel data processing in multi-processor computer systems. As a natural illustration of the fork and join operations, one can consider respectively splitting a message into packets in a communication network, each intended for transmitting via separate ways, and merging packets at a destination node of the network to restore the message. Further examples can be found in Baccelli and Makowski (1989).

The usual way to represent the dynamics of fork-join queueing networks relies on the implementation of recursive state equations of the Lindley type [Baccelli and Makowski (1989)]. Since the recursive equations associated with the fork-join networks can be expressed only in terms of the operations of maximum and addition, there is a possibility to represent the dynamics of the networks in terms of the (max,+)-algebra which is actually an algebraic system just supplied with the same two operations [Cuninghame-Green (1979), Baccelli et al. (1992), Maslov and Kolokoltsov (1994)]. In fact, (max,+)-algebra models offer a more compact and unified way of describing network dynamics, and, moreover, lead to equations closely analogous to those in the conventional linear system theory [Baccelli et al. (1992), Krivulin (1994, 1995, 1996a, 1996b)]. In that case, the (max,+)-algebra approach gives one the chance to exploit results and numerical procedures available in the algebraic system theory and computational linear algebra.

One of the problems of interest in the analysis of stochastic queueing networks is to evaluate the service cycle time of a network. Both the cycle time and its inverse which can be regarded as a throughput present performance measures commonly used to describe efficiency of the network operation.

It is often rather difficult to evaluate the cycle time exactly, even though the network under study is quite simple. To get information about the performance measure in this case, one can apply computer simulation to produce reasonable estimates. Another approach is to derive bounds on the cycle time. Specifically, a technique which allows one to establish bounds based on results of the theory of large deviations as well as the Perron-Frobenius spectral theory has been introduced in Baccelli and Konstantopoulos (1991).

In this paper we propose an approach to get bounds on the service cycle time, which exploits the (max,+)-algebra representation of acyclic fork-join network dynamics derived in Krivulin (1996a, 1996b). This approach is essentially based on pure algebraic manipulations combined with application of bounds on extreme values, obtained in Gumbel (1954), Hartly and David (1954).

The rest of the paper is organized as follows. Section 5.2 presents basic (max,+)-algebra definitions and related results which underlie the development of network models and their analysis in the subsequent sections. In Section 5.3, further algebraic results are included which provide a basis for derivation of bounds on the service cycle time.

A (max,+)-algebra representation of the fork-join network dynamics and related examples are given in Section 5.4. Furthermore, Section 5.5 offers some monotonicity property for the networks, which is exploited in Section 5.6 to get algebraic bounds on the service cycle completion time. Stochastic extension of the network model is introduced in Section 5.7. This section concludes with a result which provides simple bounds on the network cycle time. Finally, Section 5.8 presents examples of calculating bounds and related discussion.

5.2 Preliminary Algebraic Definitions and Results

The (max, +)-algebra presents an idempotent commutative semiring (idempotent semifield) which is defined as $\mathbb{R}_{\max} = \langle \underline{\mathbb{R}}, \oplus, \otimes \rangle$ with $\underline{\mathbb{R}} = \mathbb{R} \cup \{\varepsilon\}$, $\varepsilon = -\infty$, and binary operations $\oplus$ and $\otimes$ defined as

$$x \oplus y = \max(x, y), \quad x \otimes y = x + y, \quad \text{for all} \quad x, y \in \underline{\mathbb{R}}.$$

There are the null and identity elements in the algebra, namely ε and 0, to satisfy the conditions $x \oplus \varepsilon = \varepsilon \oplus x = x$, and $x \otimes 0 = 0 \otimes x = x$, for any $x \in \underline{\mathbb{R}}$. The null element ε and the operation $\otimes$ are related by the usual absorption rule involving $x \otimes \varepsilon = \varepsilon \otimes x = \varepsilon$.

As it is easy to see, the operations $\oplus$ and $\otimes$ retain most of the properties of the ordinary addition and multiplication, including associativity, commutativity, and distributivity of multiplication over addition. However, the operation $\oplus$ is idempotent; that is, for any $x \in \underline{\mathbb{R}}$, one has $x \oplus x = x$.

Non-negative integer power of any $x \in \mathbb{R}$ can be defined as $x^0 = 0$, and $x^q = x \otimes x^{q-1} = x^{q-1} \otimes x$ for $q \geq 1$. Clearly, the (max, +)-algebra power x^q corresponds to qx in ordinary notations. We will use the power notations only in the (max, +)-algebra sense.

The (max, +)-algebra of matrices is readily introduced in the regular way. Specifically, for any $(n \times n)$-matrices $X = (x_{ij})$ and $Y = (y_{ij})$, the entries of $U = X \oplus Y$ and $V = X \otimes Y$ are calculated as

$$u_{ij} = x_{ij} \oplus y_{ij}, \quad \text{and} \quad v_{ij} = \bigoplus_{k=1}^{n} x_{ik} \otimes y_{kj}.$$

As the null and identity elements, the matrices

$$\mathcal{E} = \begin{pmatrix} \varepsilon & \cdots & \varepsilon \\ \vdots & \ddots & \vdots \\ \varepsilon & \cdots & \varepsilon \end{pmatrix}, \quad I = \begin{pmatrix} 0 & & \varepsilon \\ & \ddots & \\ \varepsilon & & 0 \end{pmatrix}$$

are respectively taken in the algebra.

The matrix operations $\oplus$ and $\otimes$ possess monotonicity properties; that is, the matrix inequalities $X \leq U$ and $Y \leq V$ result in

$$X \oplus Y \leq U \oplus V, \quad X \otimes Y \leq U \otimes V$$

for any matrices of appropriate size.

Let $X \neq \mathcal{E}$ be a square matrix. In the same way as in the conventional algebra, one can define $X^0 = I$, and $X^q = X \otimes X^{q-1} = X^{q-1} \otimes X$ for any integer $q \geq 1$. However, idempotency leads, in particular, to the matrix identity

$$(X \oplus Y)^q = X^q \oplus X^{q-1} \otimes Y \oplus \cdots \oplus Y^q.$$

As direct consequences of the above identity, one has

$$(X \oplus Y)^q \geq X^p \otimes Y^{q-p}, \qquad (I \oplus X)^q \geq (I \oplus X)^p \geq X^p,$$

for all $p = 0, 1, \ldots, q$.

For any matrix X, one can define the matrix function

$$\|X\| = \bigoplus_{i,j} x_{ij} = \max_{i,j} x_{ij}.$$

Note that the function $\|\cdot\|$ possesses properties similar to those of the ordinary matrix norm. Specifically, for any matrix X, it holds $\|X\| \geq \varepsilon$, and $\|X\| = \varepsilon$ if and only if $X = \mathcal{E}$. Furthermore, we have $\|c \otimes X\| = c \otimes \|X\|$ for any $c \in \mathbb{R}$, as well as additive and multiplicative properties involving

$$\|X \oplus Y\| = \|X\| \oplus \|Y\|, \qquad \|X \otimes Y\| \leq \|X\| \otimes \|Y\|$$

for any two conforming matrices X and Y. For any $c > 0$, we also have $\|cX\| = c\|X\|$. The matrix function $\|\cdot\|$ will be referred to as $(\max,+)$-algebra norm, or simply as norm.

Consider an $(n \times n)$-matrix X with its entries $x_{ij} \in \mathbb{R}$. It can be treated as an adjacency matrix of an oriented graph with n nodes, provided each entry $x_{ij} \neq \varepsilon$ implies the existence of the arc (i, j) in the graph, while $x_{ij} = \varepsilon$ does the lack of the arc.

It is easy to verify that for any integer $q \geq 1$, the matrix X^q has its entry $x_{ij}^{(q)} \neq \varepsilon$ if and only if there exists a path from node i to node j in the graph, which consists of q arcs. Furthermore, if the graph associated with the matrix X is acyclic, we have $X^q = \mathcal{E}$ for all $q > p$, where p is the length of the longest path in the graph. Otherwise, provided that the graph is not acyclic, one can construct a path of any length, lying along circuits, and then it holds that $X^q \neq \mathcal{E}$ for all $q \geq 0$.

Consider the implicit equation in an unknown vector $\boldsymbol{x} = (x_1, \ldots, x_n)^T$,

$$\boldsymbol{x} = U \otimes \boldsymbol{x} \oplus \boldsymbol{v}, \qquad (5.1)$$

where $U = (u_{ij})$ and $\boldsymbol{v} = (v_1, \ldots, v_n)^T$ are respectively given $(n \times n)$-matrix and n-vector. Suppose that the entries of the matrix U and the vector $\boldsymbol{v}$ are either positive or equal to ε. It is easy to verify [see, e.g. Cuninghame-Green (1979), Cohen et al. (1985)] that equation (5.1) has the unique bounded solution if and only if the graph associated with U is acyclic. Provided that the solution exists, it is given by

$$\boldsymbol{x} = (I \oplus U)^p \otimes \boldsymbol{v}, \qquad (5.2)$$

where p is the length of the longest path in the graph.

5.3 Further Algebraic Results

Consider a square matrix X, and denote the adjacency (ε–0)-matrix of the graph associated with X by G. The matrix G is normally referred to as support of X.

Proposition 5.3.1 *For any matrix X, it holds*

$$X \leq \|X\| \otimes G,$$

where G is the support of X.

Proposition 5.3.2 *Suppose that matrices $X_1, \ldots, X_k$ have a common associated acyclic graph, p is the length of the longest path in the graph, and*

$$X = X_1^{m_1} \otimes \cdots \otimes X_k^{m_k},$$

where $m_1, \ldots, m_k$ are nonnegative integers.
   If it holds that $m_1 + \cdots + m_k > p$, then $X = \mathcal{E}$.

PROOF. It follows from Proposition 5.3.1 that

$$X = X_1^{m_1} \otimes \cdots \otimes X_k^{m_k} \leq \|X_1\|^{m_1} \otimes \cdots \otimes \|X_k\|^{m_k} \otimes G^{m_1+\cdots+m_k},$$

where G is the common support of the matrices $X_1, \ldots, X_k$.

Since the graph is acyclic, it holds that $G^q = \mathcal{E}$ for all $q > p$. Therefore, with $q = m_1 + \cdots + m_k > p$, we arrive at the inequality $X \leq \mathcal{E}$ which leads us to the desired result. ∎

Lemma 5.3.1 *Suppose that matrices $X_1, \ldots, X_k$ have a common associated acyclic graph, and p is the length of the longest path in the graph.*
   If $\|X_i\| \geq 0$ for all $i = 1, \ldots, k$, then it holds

$$\left\| \bigotimes_{i=1}^{k} (I \oplus X_i)^{m_i} \right\| \leq \left(\bigoplus_{i=1}^{k} \|X_i\| \right)^p$$

for any nonnegative integers $m_1, \ldots, m_k$.

PROOF. Consider the matrix

$$\begin{aligned} X &= \bigotimes_{i=1}^{k} (I \oplus X_i)^{m_i} = \left(\bigoplus_{i_1=0}^{m_1} X_1^{i_1} \right) \otimes \cdots \otimes \left(\bigoplus_{i_k=0}^{m_k} X_k^{i_k} \right) \\ &= \bigoplus_{i_1=0}^{m_1} \cdots \bigoplus_{i_k=0}^{m_k} X_1^{i_1} \otimes \cdots \otimes X_k^{i_k} \leq \bigoplus_{0 \leq i_1+\cdots+i_k \leq m} X_1^{i_1} \otimes \cdots \otimes X_k^{i_k}, \end{aligned}$$

where $m = m_1 + \cdots + m_k$. From Proposition 5.3.2 we may replace m with p in the last term to get

$$X \leq \bigoplus_{0 \leq i_1 + \cdots + i_k \leq p} X_1^{i_1} \otimes \cdots \otimes X_k^{i_k}.$$

Proceeding to the norm, with its additive and multiplicative properties, we arrive at the inequality

$$\|X\| \leq \bigoplus_{0 \leq i_1 + \cdots + i_k \leq p} \|X_1\|^{i_1} \otimes \cdots \otimes \|X_k\|^{i_k}.$$

Since for all $i = 1, \ldots, k$, it holds $0 \leq \|X_i\| \leq \|X_1\| \oplus \cdots \oplus \|X_k\|$, we finally have

$$\|X\| \leq \bigoplus_{i=0}^{p} (\|X_1\| \oplus \cdots \oplus \|X_k\|)^p = \left(\bigoplus_{i=0}^{k} \|X_i\|\right)^p.$$

∎

5.4 An Algebraic Model of Queueing Networks

We consider a network with n single-server nodes and customers of a single class. The topology of the network is described by an oriented acyclic graph $\mathcal{G} = (\mathbf{N}, \mathbf{A})$, where the set $\mathbf{N} = \{1, \ldots, n\}$ represents the nodes, and $\mathbf{A} = \{(i,j)\} \subset \mathbf{N} \times \mathbf{N}$ represents the arcs determining the transition routes of customers.

For every node $i \in \mathbf{N}$, we denote the sets of its immediate predecessors and successors respectively as $\mathbf{P}(i) = \{j | (j,i) \in \mathbf{A}\}$ and $\mathbf{S}(i) = \{j | (i,j) \in \mathbf{A}\}$. In specific cases, there may be one of the conditions $\mathbf{P}(i) = \emptyset$ and $\mathbf{S}(i) = \emptyset$ encountered. Each node i with $\mathbf{P}(i) = \emptyset$ is assumed to represent an infinite external arrival stream of customers; provided that $\mathbf{S}(i) = \emptyset$, it is considered as an output node intended to release customers from the network.

Each node $i \in \mathbf{N}$ includes a server and its buffer with infinite capacity, which together present a single-server queue operating under the first-come, first-served (FCFS) discipline. At the initial time, the server at each node i is assumed to be free of customers, whereas in its buffer, there may be r_i, $0 \leq r_i \leq \infty$, customers waiting for service. The value $r_i = \infty$ is set for every node i with $\mathbf{P}(i) = \emptyset$, which represents an external arrival stream of customers.

For the queue at node i, we denote the kth arrival and departure epochs respectively as $u_i(k)$ and $x_i(k)$. Furthermore, the service time of the kth customer at server i is indicated by τ_{ik}. We assume that $\tau_{ik} \geq 0$ are given for all $i = 1, \ldots, n$, and $k = 1, 2, \ldots$, while $u_i(k)$ and $x_i(k)$ are considered as unknown state variables. With the condition that the network starts operating at time zero, it is convenient to set $x_i(0) \equiv 0$, and $x_i(k) \equiv \varepsilon$ for all $k < 0$, $i = 1, \ldots, n$.

Algebraic Modelling and Performance Evaluation of Queueing Networks 69

It is easy to set up an equation which relates the system state variables. In fact, the dynamics of any single-server node i with an infinite buffer, operating on the FCFS basis, is described as

$$x_i(k) = \tau_{ik} \otimes u_i(k) \oplus \tau_{ik} \otimes x_i(k-1). \tag{5.3}$$

With the vector-matrix notations

$$\boldsymbol{u}(k) = \begin{pmatrix} u_1(k) \\ \vdots \\ u_n(k) \end{pmatrix}, \quad \boldsymbol{x}(k) = \begin{pmatrix} x_1(k) \\ \vdots \\ x_n(k) \end{pmatrix}, \quad \mathcal{T}_k = \begin{pmatrix} \tau_{1k} & & \varepsilon \\ & \ddots & \\ \varepsilon & & \tau_{nk} \end{pmatrix},$$

we may rewrite equation (5.3) in a vector form, as

$$\boldsymbol{x}(k) = \mathcal{T}_k \otimes \boldsymbol{u}(k) \oplus \mathcal{T}_k \otimes \boldsymbol{x}(k-1). \tag{5.4}$$

5.4.1 Fork-Join queueing networks

In fork-join networks, in addition to the usual service procedure, special join and fork operations are performed in its nodes, respectively before and after service. The join operation is actually thought to cause each customer which comes into node i, not to enter the buffer at the server but to wait until at least one customer from every node $j \in \mathbf{P}(i)$ arrives. As soon as these customers arrive, they, taken one from each preceding node, are united into one customer which then enters the buffer to become a new member of the queue.

The fork operation at node i is initiated every time the service of a customer is completed; it consists in giving rise to several new customers instead of the original one. As many new customers appear in node i as there are succeeding nodes included in the set $\mathbf{S}(i)$. These customers simultaneously depart the node, each being passed to separate node $j \in \mathbf{S}(i)$. We assume that the execution of fork-join operations when appropriate customers are available, as well as the transition of customers within and between nodes require no time.

As it immediately follows from the above description of the fork-join operations, the kth arrival epoch into the queue at node i is represented as

$$u_i(k) = \begin{cases} \displaystyle\bigoplus_{j \in \mathbf{P}(i)} x_j(k - r_i), & \text{if } \mathbf{P}(i) \neq \emptyset, \\ \varepsilon, & \text{if } \mathbf{P}(i) = \emptyset. \end{cases} \tag{5.5}$$

In order to get this equation in a vector form, we first define the number $M = \max\{r_i | r_i < \infty, i = 1, \ldots, n\}$. Now we may rewrite (5.5) as

$$u_i(k) = \bigoplus_{m=0}^{M} \bigoplus_{j=1}^{n} g_{ji}^{(m)} \otimes x_j(k-m),$$

where the numbers $g_{ij}^{(m)}$ are determined by the condition

$$g_{ij}^{(m)} = \begin{cases} 0, & \text{if } i \in \mathbf{P}(j) \text{ and } m = r_j, \\ \varepsilon, & \text{otherwise.} \end{cases} \qquad (5.6)$$

Let us introduce the matrices $G_m = \left(g_{ij}^{(m)}\right)$ for each $m = 0, 1, \ldots, M$. In fact, G_m presents an adjacency matrix of the partial graph $\mathcal{G}_m = (\mathbf{N}, \mathbf{A}_m)$ with $\mathbf{A}_m = \{(i,j)| (i,j) \in \mathbf{A}; r_j = m\}$. Since the graph of the entire network is acyclic, all its partial graphs $\mathcal{G}_m$, $m = 0, 1, \ldots, M$, possess the same property.

With these matrices, equation (5.5) may be written in the vector form

$$\boldsymbol{u}(k) = \bigoplus_{m=0}^{M} G_m^T \otimes \boldsymbol{x}(k-m), \qquad (5.7)$$

where G_m^T denotes the transpose of the matrix G_m.

By combining equations (5.4) and (5.7), we arrive at the equation

$$\boldsymbol{x}(k) = \mathcal{T}_k \otimes G_0^T \otimes \boldsymbol{x}(k) \oplus \mathcal{T}_k \otimes \boldsymbol{x}(k-1)$$
$$\oplus \mathcal{T}_k \otimes \bigoplus_{m=1}^{M} G_m^T \otimes \boldsymbol{x}(k-m). \qquad (5.8)$$

Clearly, it is actually an implicit equation in $\boldsymbol{x}(k)$, which has the form of (5.1), with $U = \mathcal{T}_k \otimes G_0^T$. Taking into account that the matrix $\mathcal{T}_k$ is diagonal, one can prove the following statement (see also Krivulin (1996a, 1996b)).

Theorem 5.4.1 *Suppose that in the fork-join network model, the graph $\mathcal{G}_0$ associated with the matrix G_0 is acyclic. Then equation (5.8) can be solved to produce the explicit dynamic state equation*

$$\boldsymbol{x}(k) = \bigoplus_{m=1}^{M} A_m(k) \otimes \boldsymbol{x}(k-m), \qquad (5.9)$$

with the state transition matrices

$$\begin{aligned} A_1(k) &= (I \oplus \mathcal{T}_k \otimes G_0^T)^p \otimes \mathcal{T}_k \otimes (I \oplus G_1^T), & (5.10) \\ A_m(k) &= (I \oplus \mathcal{T}_k \otimes G_0^T)^p \otimes \mathcal{T}_k \otimes G_m^T, & m = 2, \ldots, M, & (5.11) \end{aligned}$$

where p is the length of the longest path in $\mathcal{G}_0$.

As one can see, with $M = 1$ and $A_1(k) = A(k)$, we have the dynamic equation

$$\boldsymbol{x}(k) = A(k) \otimes \boldsymbol{x}(k-1). \qquad (5.12)$$

Note that this equation differs in appearance from those used in the linear system theory, which normally have the form of $\boldsymbol{x}(k) = A(k-1) \otimes \boldsymbol{x}(k-1)$.

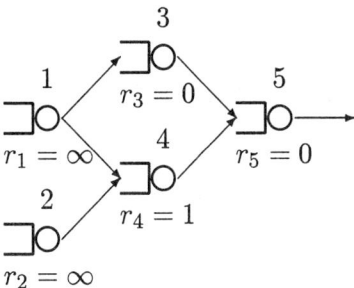

Figure 5.1: An acyclic fork-join network.

5.4.2 Examples of network models

An example of an acyclic fork-join network with $n = 5$ is shown in Fig. 5.1.

Since for the network $M = 1$, we have from (5.6)

$$G_0 = \begin{pmatrix} \varepsilon & \varepsilon & 0 & \varepsilon & \varepsilon \\ \varepsilon & \varepsilon & \varepsilon & \varepsilon & \varepsilon \\ \varepsilon & \varepsilon & \varepsilon & \varepsilon & 0 \\ \varepsilon & \varepsilon & \varepsilon & \varepsilon & 0 \\ \varepsilon & \varepsilon & \varepsilon & \varepsilon & \varepsilon \end{pmatrix}, \quad G_1 = \begin{pmatrix} \varepsilon & \varepsilon & \varepsilon & 0 & \varepsilon \\ \varepsilon & \varepsilon & \varepsilon & 0 & \varepsilon \\ \varepsilon & \varepsilon & \varepsilon & \varepsilon & \varepsilon \\ \varepsilon & \varepsilon & \varepsilon & \varepsilon & \varepsilon \\ \varepsilon & \varepsilon & \varepsilon & \varepsilon & \varepsilon \end{pmatrix}.$$

Taking into account that for the graph $\mathcal{G}_0$, the length of its longest path $p = 2$, we arrive at equation (5.12) with the state transition matrix calculated from (5.10) as

$$A(k) = (I \oplus \mathcal{T}_k \otimes G_0^T)^2 \otimes \mathcal{T}_k \otimes (I \oplus G_1^T)$$

$$= \begin{pmatrix} \tau_{1k} & \varepsilon & \varepsilon & \varepsilon & \varepsilon \\ \varepsilon & \tau_{2k} & \varepsilon & \varepsilon & \varepsilon \\ \tau_{1k} \otimes \tau_{3k} & \varepsilon & \tau_{3k} & \varepsilon & \varepsilon \\ \tau_{4k} & \tau_{4k} & \varepsilon & \tau_{4k} & \varepsilon \\ (\tau_{1k} \otimes \tau_{3k} \oplus \tau_{4k}) \otimes \tau_{5k} & \tau_{4k} \otimes \tau_{5k} & \tau_{3k} \otimes \tau_{5k} & \tau_{4k} \otimes \tau_{5k} & \tau_{5k} \end{pmatrix}.$$

Note that open tandem queueing systems (see Fig. 5.2) can be considered as trivial networks in which no fork and join operations are actually performed.

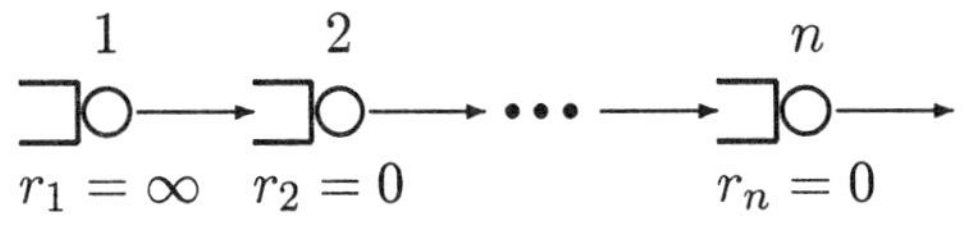

Figure 5.2: Open tandem queues.

For the system in Fig. 5.2, we have $M = 0$, and $p = n - 1$. Its related state transition matrix $A(k)$ has the entries [Krivulin (1994, 1995)]

$$a_{ij}(k) = \begin{cases} \tau_{jk} \otimes \tau_{j+1 k} \otimes \cdots \otimes \tau_{ik}, & \text{if } i \geq j, \\ \varepsilon, & \text{otherwise}. \end{cases}$$

5.5 A Monotonicity Property

In this section, a property of monotonicity is established which shows how the system state vector $x(k)$ may vary with the initial numbers of customers r_i. It is actually proven that the entries of $x(k)$ for all $k = 1, 2, \ldots$, do not decrease when the numbers r_i with $0 < r_i < \infty$, $i = 1, \ldots, n$, are reduced to zero.

As it is easy to see, the change in the initial numbers of customers results only in modifications to partial graphs $\mathcal{G}_m$ and so to their adjacency matrices G_m. Specifically, reducing these numbers to zero leads us to new matrices $\widetilde{G}_0 = G_0 \oplus G_1 \cdots \oplus G_M$, and $\widetilde{G}_m = \mathcal{E}$ for all $m = 1, \ldots, M$.

We start with a lemma which shows that replacing the numbers $r_i = 1$ with $r_i = 0$ does not decrease the entries of the matrix $A_1(k)$ defined by (5.10).

Lemma 5.5.1 *For all $k = 1, 2, \ldots$, it holds*

$$A_1(k) \leq \widetilde{A}(k)$$

with $\widetilde{A}(k) = (I \oplus \mathcal{T}_k \otimes \widetilde{G}_0^T)^q \otimes \mathcal{T}_k$, where $\widetilde{G}_0 = G_0 \oplus G_1$, and q is the length of the longest path in the graph associated with the matrix $\widetilde{G}_0$.

PROOF. Consider the matrix $A_1(k)$ and represent it in the form

$$A_1(k) = ((I \oplus \mathcal{T}_k \otimes G_0^T)^p \otimes \mathcal{T}_k) \oplus ((I \oplus \mathcal{T}_k \otimes G_0^T)^p \otimes \mathcal{T}_k \otimes G_1^T),$$

where p is the length of the longest path in the graph associated with G_0.

As one can see, to prove the lemma, it will suffice to verify both inequalities

$$\widetilde{A}(k) \geq (I \oplus \mathcal{T}_k \otimes G_0^T)^p \otimes \mathcal{T}_k, \tag{5.13}$$
$$\widetilde{A}(k) \geq (I \oplus \mathcal{T}_k \otimes G_0^T)^p \otimes \mathcal{T}_k \otimes G_1^T. \tag{5.14}$$

Let us write the obvious representation

$$(I \oplus \mathcal{T}_k \otimes \widetilde{G}_0^T)^q = \bigoplus_{i=0}^{q} (I \oplus \mathcal{T}_k \otimes G_0^T)^i \otimes (\mathcal{T}_k \otimes G_1^T)^{q-i}.$$

Since $q \geq p$, we get from the representation

$$(I \oplus \mathcal{T}_k \otimes \widetilde{G}_0^T)^q \geq (I \oplus \mathcal{T}_k \otimes \widetilde{G}_0^T)^p = ((I \oplus \mathcal{T}_k \otimes G_0^T) \oplus \mathcal{T}_k \otimes G_1^T)^p \geq (I \oplus \mathcal{T}_k \otimes G_0^T)^p.$$

It remains to multiply both sides of the above inequality by $\mathcal{T}_k$ on the right so as to arrive at (5.13).

To verify (5.14), let us first assume that $q > p$. In this case, we obtain

$$(I \oplus \mathcal{T}_k \otimes \widetilde{G}_0^T)^q \geq (I \oplus \mathcal{T}_k \otimes \widetilde{G}_0^T)^{p+1}$$
$$= ((I \oplus \mathcal{T}_k \otimes G_0^T) \oplus \mathcal{T}_k \otimes G_1^T)^{p+1} \geq (I \oplus \mathcal{T}_k \otimes G_0^T)^p \otimes \mathcal{T}_k \otimes G_1^T.$$

Algebraic Modelling and Performance Evaluation of Queueing Networks   73

Suppose now that $q = p$. Then it is necessary that $G_1 \otimes G_0^p = \mathcal{E}$. If this were not the case, there would be a path in the graph associated with the matrix $\widetilde{G}_0 = G_0 \oplus G_1$, which has its length greater than p, and we would have $q > p$.

Clearly, the condition $G_1 \otimes G_0^p = \mathcal{E}$ results in $(\mathcal{T}_k \otimes G_0^T)^p \otimes \mathcal{T}_k \otimes G_1^T = \mathcal{E}$, and thus we get

$$(I \oplus \mathcal{T}_k \otimes \widetilde{G}_0^T)^q = ((I \oplus \mathcal{T}_k \otimes G_0^T) \oplus \mathcal{T}_k \otimes G_1^T)^p$$
$$\geq (I \oplus \mathcal{T}_k \otimes G_0^T)^{p-1} \otimes \mathcal{T}_k \otimes G_1^T = (I \oplus \mathcal{T}_k \otimes G_0^T)^p \otimes \mathcal{T}_k \otimes G_1^T.$$

Since it holds $(I \oplus \mathcal{T}_k \otimes \widetilde{G}_0^T)^p \otimes \mathcal{T}_k \geq (I \oplus \mathcal{T}_k \otimes \widetilde{G}_0^T)^p$, one can conclude that inequality (5.14) is also valid. ∎

Theorem 5.5.1 *In the acyclic fork-join queueing network model (5.9–5.11), reducing the initial numbers of customers from any finite values to zero does not decrease the entries of the system state vector $\boldsymbol{x}(k)$ for all $k = 1, 2, \ldots$.*

PROOF. Let $\boldsymbol{x}(k)$ be determined by (5.9–5.11). Suppose that the vector $\widetilde{\boldsymbol{x}}(k)$ satisfies the dynamic equation

$$\widetilde{\boldsymbol{x}}(k) = \widetilde{A}(k) \otimes \widetilde{\boldsymbol{x}}(k-1)$$

with

$$\widetilde{A}(k) = \left(I \oplus \mathcal{T}_k \otimes \bigoplus_{m=0}^{M} G_m^T\right)^q \otimes \mathcal{T}_k = (I \oplus \mathcal{T}_k \otimes G^T)^q \otimes \mathcal{T}_k,$$

where q is the length of the longest path in the graph associated with the matrix $G = G_0 \oplus G_1 \oplus \cdots \oplus G_m$.

Now we have to show that for all $k = 1, 2, \ldots$, it holds

$$\boldsymbol{x}(k) \leq \widetilde{\boldsymbol{x}}(k).$$

Since $\boldsymbol{x}(k_1) \leq \boldsymbol{x}(k_2)$ for any $k_1 < k_2$, we have from (5.9)

$$\boldsymbol{x}(k) = \bigoplus_{m=1}^{M} A_m(k) \otimes \boldsymbol{x}(k-m) \leq \left(\bigoplus_{m=1}^{M} A_m(k)\right) \otimes \boldsymbol{x}(k-1).$$

Consider the matrix

$$\widetilde{A}_1(k) = \bigoplus_{m=1}^{M} A_m(k) = (I \oplus \mathcal{T}_k \otimes G_0^T)^p \otimes \mathcal{T}_k \otimes \left(I \oplus \bigoplus_{m=1}^{M} G_m^T\right).$$

By applying Lemma 5.5.1, we have

$$\widetilde{A}_1(k) \leq \left(I \oplus \mathcal{T}_k \otimes G_0^T \oplus \bigoplus_{m=1}^{M} G_m^T\right)^q \otimes \mathcal{T}_k = \widetilde{A}(k).$$

Starting with the condition $\boldsymbol{x}(0) = \widetilde{\boldsymbol{x}}(0)$, we successively verify that the relations
$$\boldsymbol{x}(k) \leq \widetilde{A}_1(k) \otimes \boldsymbol{x}(k-1) \leq \widetilde{A}(k) \otimes \boldsymbol{x}(k-1) \leq \widetilde{A}(k) \otimes \widetilde{\boldsymbol{x}}(k-1) = \widetilde{\boldsymbol{x}}(k)$$
are valid for each $k = 1, 2, \ldots$. ∎

5.6 Bounds on the Service Cycle Completion Time

We consider the evolution of the system as a sequence of service cycles: the 1st cycle starts at the initial time, and it is terminated as soon as all the servers in the network complete their 1st service, the 2nd cycle is terminated as soon as the servers complete their 2nd service, and so on. Clearly, the completion time of the kth cycle can be represented as

$$\max_i x_i(k) = \|\boldsymbol{x}(k)\|$$

with $\boldsymbol{x}(0) = \boldsymbol{0}$.

The next lemma provides simple algebraic lower and upper bounds for the kth cycle completion time.

Lemma 5.6.1 *With the condition that $\boldsymbol{x}(0) = \boldsymbol{0}$, for all $k = 1, 2, \ldots$, it holds*

$$\left\|\sum_{i=1}^{k} \mathcal{T}_i\right\| \leq \|\boldsymbol{x}(k)\| \leq \sum_{i=1}^{k} \|\mathcal{T}_i\| + q\left(\bigoplus_{i=1}^{k} \|\mathcal{T}_i\|\right),$$

where q is the length of the longest path in the network graph.

PROOF. To prove the left inequality first note that
$$\begin{aligned} A_1(k) &= (I \oplus \mathcal{T}_k \otimes G_0^T)^p \otimes \mathcal{T}_k \otimes (I \oplus G_1^T) \\ &= \left(I \oplus \mathcal{T}_k \otimes G_0^T \oplus \cdots \oplus (\mathcal{T}_k \otimes G_0^T)^p\right) \otimes \mathcal{T}_k \otimes (I \oplus G_1^T) \geq \mathcal{T}_k. \end{aligned}$$

With this condition, we have from (5.9)
$$\boldsymbol{x}(k) = \bigoplus_{m=1}^{M} A_m(k) \otimes \boldsymbol{x}(k-m) \geq A_1(k) \otimes \boldsymbol{x}(k-1) \geq \mathcal{T}_k \otimes \boldsymbol{x}(k-1).$$

Now we can write
$$\boldsymbol{x}(k) \geq \mathcal{T}_k \otimes \boldsymbol{x}(k-1) \geq \mathcal{T}_k \otimes \mathcal{T}_{k-1} \otimes \boldsymbol{x}(k-2) \geq \cdots \geq \mathcal{T}_k \otimes \cdots \otimes \mathcal{T}_1 \otimes \boldsymbol{x}(0),$$

where $\boldsymbol{x}(0) = \boldsymbol{0}$. Taking the norm, and considering that $\mathcal{T}_i$, $i = 1, \ldots, k$, present diagonal matrices, we get
$$\|\boldsymbol{x}(k)\| \geq \|\mathcal{T}_k \otimes \cdots \otimes \mathcal{T}_1\| = \|\mathcal{T}_1 + \cdots + \mathcal{T}_k\|.$$

To obtain an upper bound, let us replace the general system (5.9–5.11) with that governed by the equation

$$\widetilde{\boldsymbol{x}}(k) = \widetilde{A}(k) \otimes \widetilde{\boldsymbol{x}}(k-1) \qquad (5.15)$$

with $\widetilde{A}(k) = (I \oplus \mathcal{T}_k \otimes \widetilde{G}^T)^q \otimes \mathcal{T}_k$, where $\widetilde{G} = G_0 \oplus G_1 \oplus \cdots \oplus G_m$, and q is the length of the longest path in the graph associated with $\widetilde{G}$. As it follows from Theorem 5.5.1, one has $\boldsymbol{x}(k) \leq \widetilde{\boldsymbol{x}}(k)$ for all $k = 1, 2, \ldots$.

Let us denote $\widetilde{A}_k = \widetilde{A}(k) \otimes \cdots \otimes \widetilde{A}(1)$. With the condition $\widetilde{\boldsymbol{x}}(0) = \boldsymbol{x}(0) = \mathbf{0}$, we get from (5.15)

$$\|\widetilde{\boldsymbol{x}}(k)\| = \|\widetilde{A}(k) \otimes \cdots \otimes \widetilde{A}(1)\| = \|\widetilde{A}_k\|.$$

With Proposition 5.3.2 we have

$$\widetilde{A}_k = \bigotimes_{i=1}^{k}(I \oplus \mathcal{T}_{k-i+1} \otimes \widetilde{G}^T)^q \otimes \mathcal{T}_{k-i+1} \leq \bigotimes_{i=1}^{k} \|\mathcal{T}_i\| \otimes \bigotimes_{i=1}^{k}(I \oplus \mathcal{T}_{k-i+1} \otimes \widetilde{G}^T)^q.$$

Proceeding to the norm and using Lemma 5.3.1, we arrive at the inequality

$$\|\widetilde{A}_k\| \leq \bigotimes_{i=1}^{k}\|\mathcal{T}_i\| \otimes \left(\bigoplus_{i=1}^{k}\|\mathcal{T}_i\|\right)^q = \sum_{i=1}^{k}\|\mathcal{T}_i\| + q\left(\bigoplus_{i=1}^{k}\|\mathcal{T}_i\|\right).$$

which provides us with the desired result. ■

5.7 Stochastic Extension of the Network Model

Suppose that for each node $i = 1, \ldots, n$, the service times $\tau_{i1}, \tau_{i2}, \ldots$, form a sequence of independent and identically distributed (i.i.d.) non-negative random variables with $\mathbb{E}[\tau_{ik}] < \infty$ and $\mathbb{D}[\tau_{ik}] < \infty$ for all $k = 1, 2, \ldots$.

As a performance measure of the stochastic network model, we consider the service cycle time which can be defined as [Baccelli et al. (1992)]

$$\gamma = \lim_{k \to \infty} \frac{1}{k}\|\boldsymbol{x}(k)\| \qquad (5.16)$$

provided that the above limit exists. Another performance measure of interest is the throughput defined as $\pi = 1/\gamma$.

Since it is frequently rather difficult to evaluate the cycle time exactly, even though the network under study is quite simple, one can try to derive bounds on γ. In this section, we show how these bounds may be obtained based on (max,+)-algebra representation of the network dynamics.

We start with some preliminary results which include properties of the expectation operator, formulated in terms of (max,+)-algebra operations.

5.7.1 Some properties of expectation

Let $\xi_1, \ldots, \xi_k$ be random variables taking their values in $\mathbb{R}$, and such that their expected values $\mathbb{E}[\xi_i]$, $i = 1, \ldots, k$, exist.

First note that ordinary properties of expectation leads us to the obvious relations

$$\mathbb{E}\left[\bigoplus_{i=1}^{k} \xi_i\right] \leq \bigotimes_{i=1}^{k} \mathbb{E}[\xi_i], \quad \text{and} \quad \mathbb{E}\left[\bigotimes_{i=1}^{k} \xi_i\right] = \bigotimes_{i=1}^{k} \mathbb{E}[\xi_i].$$

Furthermore, the next statement is valid.

Lemma 5.7.1 *It holds*

$$\mathbb{E}\left[\bigoplus_{i=1}^{k} \xi_i\right] \geq \bigoplus_{i=1}^{k} \mathbb{E}[\xi_i].$$

PROOF. The statement of the lemma for $k = 2$ follows immediately from the identity

$$x \oplus y = \frac{1}{2}(x + y + |x - y|), \quad \text{for all } x, y \in \mathbb{R}$$

and ordinary properties of expectation. It remains to extend the statement to the case of arbitrary k by induction. ∎

The next result [Gumbel (1954), Hartly and David (1954)] provides an upper bound for the expected value of the maximum of i.i.d. random variables.

Lemma 5.7.2 *Let $\xi_1, \ldots, \xi_k$ be i.i.d. random variables with $\mathbb{E}[\xi_1] < \infty$ and $\mathbb{D}[\xi_1] < \infty$. Then it holds*

$$\mathbb{E}\left[\bigoplus_{i=1}^{k} \xi_i\right] \leq \mathbb{E}[\xi_1] + \frac{k-1}{\sqrt{2k-1}}\sqrt{\mathbb{D}[\xi_1]}.$$

Consider a random matrix X with its entries x_{ij} taking values in $\mathbb{R}$. We denote by $\mathbb{E}[X]$ the matrix obtained from X by replacing each entry x_{ij} by its expected value $\mathbb{E}[x_{ij}]$.

Lemma 5.7.3 *It holds*

$$\mathbb{E}\|X\| \geq \|\mathbb{E}[X]\|.$$

PROOF. It follows from Lemma 5.7.1 that

$$\mathbb{E}\|X\| = \mathbb{E}\left[\bigoplus_{i,j} x_{ij}\right] \geq \bigoplus_{i,j} \mathbb{E}[x_{ij}] = \|\mathbb{E}[X]\|.$$

∎

5.7.2 Existence of the cycle time

In the analysis of the cycle time, one first has to convince himself that the limit at (5.16) exists. As a standard tool to verify the existence of the above limit, the next theorem proposed in Kingman (1973) is normally applied. One can find examples of the implementation of the theorem in the (max, +)-algebra framework in Baccelli and Konstantopoulos (1991), Baccelli et al. (1992).

Theorem 5.7.1 *Let $\{\xi_{lk}|\ l, k = 0, 1, \ldots; l < k\}$ be a family of random variables which satisfy the following properties:*

Subadditivity: $\xi_{lk} \leq \xi_{lm} + \xi_{mk}$ for all $l < m < k$;

Stationarity: both families $\{\xi_{l+1 k+1}|\ l < k\}$ and $\{\xi_{lk}|\ l < k\}$ have the same joint distributions;

Boundedness: for all $k = 1, 2, \ldots$, there exists $\mathbb{E}[\xi_{0k}] \geq -ck$ for some finite number c.

Then there exists a constant γ, such that it holds

1. $\lim_{k \to \infty} \xi_{0k}/k = \gamma$ *with probability* 1,

2. $\lim_{k \to \infty} \mathbb{E}[\xi_{0k}]/k = \gamma$.

For simplicity, we examine the existence of the cycle time for a network with the maximum of the initial numbers of customers in nodes $M \leq 1$. As it follows from representation (5.9–5.11), the dynamics of the system may be described by the equation

$$\boldsymbol{x}(k) = A(k) \otimes \boldsymbol{x}(k-1)$$

with the matrix $A(k) = A_1(k)$ determined by (5.10). Clearly, in the case of $M > 1$, a similar representation can be easily obtained by going to an extended model with a new state vector which combines several consecutive state vectors of the original system.

To prove the existence of the cycle time, first note that τ_{ik} with $k = 1, 2, \ldots$, are i.i.d. random variables for each $i = 1, \ldots, n$, and consequently, $\mathcal{T}_k$ are i.i.d. random matrices, whereas $\|\mathcal{T}_k\|$ present i.i.d. random variables with $\mathbb{E}\|\mathcal{T}_k\| < \infty$ and $\mathbb{D}\|\mathcal{T}_k\| < \infty$ for all k.

Furthermore, since the matrix $A(k)$ depends only on $\mathcal{T}_k$, the matrices $A(1), A(2), \ldots$, also present i.i.d. random matrices. It is easy to verify that $0 \leq \mathbb{E}\|A(k)\| < \infty$ for all $k = 1, 2, \ldots$.

In order to apply Theorem 5.7.1 to stochastic system (5.9) with transition matrix (5.10), one can define the family of random variables $\{\xi_{lk}|\ l < k\}$ with

$$\xi_{lk} = \|A(k) \otimes \cdots \otimes A(l+1)\|.$$

Since $A(i)$, $i = 1, 2, \ldots$, present i.i.d. random matrices, the family $\{\xi_{lk}|\ l < k\}$ satisfies the stationarity condition of Theorem 5.7.1. Furthermore, the multiplicative property of the norm endows the family with subadditivity. The boundedness condition can be readily verified based on the condition that $0 \leq \mathbb{E}[\tau_{ik}] < \infty$ for all $i = 1, \ldots, n$, and $k = 1, 2, \ldots$.

5.7.3 Calculating bounds on the cycle time

Now we are in a position to present our main result which offers bounds on the cycle time.

Theorem 5.7.2 *In the stochastic dynamical system (5.9) the cycle time γ satisfies the double inequality*

$$\|\mathbb{E}[\mathcal{T}_1]\| \leq \gamma \leq \mathbb{E}\|\mathcal{T}_1\|. \tag{5.17}$$

PROOF. Since Theorem 5.7.1 holds true, we may write

$$\gamma = \lim_{k \to \infty} \frac{1}{k} \mathbb{E}\|\boldsymbol{x}(k)\|.$$

Let us first prove the left inequality in (5.17). From Lemmas 5.6.1 and 5.7.3, we have

$$\frac{1}{k}\mathbb{E}\|\boldsymbol{x}(k)\| \geq \frac{1}{k}\mathbb{E}\left\|\sum_{i=1}^{k} \mathcal{T}_i\right\| \geq \left\|\frac{1}{k}\sum_{i=1}^{k} \mathbb{E}[\mathcal{T}_i]\right\| = \|\mathbb{E}[\mathcal{T}_1]\|,$$

independently of k.

With the upper bound offered by Lemma 5.6.1, we get

$$\frac{1}{k}\mathbb{E}\|\boldsymbol{x}(k)\| \leq \mathbb{E}\|\mathcal{T}_1\| + \frac{q}{k}\mathbb{E}\left[\bigoplus_{i=1}^{k} \|\mathcal{T}_i\|\right].$$

From Lemma 5.7.2, the second term on the right-hand side may be replaced by that of the form

$$\frac{q}{k}\left(\mathbb{E}\|\mathcal{T}_1\| + \frac{k-1}{\sqrt{2k-1}}\sqrt{\mathbb{D}\|\mathcal{T}_1\|}\right),$$

which tends to 0 as $k \to \infty$. ∎

5.8 Discussion and Examples

Now we discuss the behaviour of the bounds (5.17) under various assumptions concerning the service times in the network. First note that the derivation of the bounds does not require the kth service times τ_{ik} to be independent for all $i = 1, \ldots, n$. As it is easy to see, if $\tau_{ik} = \tau_k$ for all i, we have $\|\mathbb{E}[\mathcal{T}_1]\| = \mathbb{E}\|\mathcal{T}_1\|$, and so the lower and upper bound coincide.

To show how the bounds vary with strengthening the dependency, we consider the network with $n = 5$ nodes, depicted in Fig. 5.1. Let $\tau_{i1} = \sum_{j=1}^{5} a_{ij}\xi_{j1}$, where ξ_{j1}, $j = 1, \ldots, 5$, are i.i.d. random variables with the exponential distribution of mean 1, and

$$a_{ij} = \begin{cases} a, & \text{if } i = j, \\ \frac{1}{4}(1-a), & \text{if } i \neq j, \end{cases}$$

where a is a number such that $1 \leq a \leq 1/5$.

It is evident that for $a = 1$, one has $\tau_{i1} = \xi_{i1}$, and then τ_{i1}, $i = 1, \ldots, 5$, present independent random variables. As a decreases, the service times τ_{i1} become dependent, and with $a = 1/5$, we will have $\tau_{i1} = (\xi_{11} + \cdots + \xi_{51})/5$ for all $i = 1, \ldots, 5$.

Table 5.1 presents estimates of the cycle time $\hat{\gamma}$ obtained via simulation after performing 100000 service cycles, together with the corresponding lower and upper bounds calculated from (5.17).

Table 5.1: Numerical results for a network with dependent service times.

a	$\|\mathbb{E}[\mathcal{T}_1]\|$	$\hat{\gamma}$	$\mathbb{E}\|\mathcal{T}_1\|$
1	1.0	1.005718	2.283333
1/2	1.0	1.002080	1.481250
1/3	1.0	1.000871	1.213889
1/4	1.0	1.000279	1.080208
1/5	1.0	1.000000	1.000000

Let us now consider the network in Fig. 5.1 under the assumption that the service times τ_{i1} are independent exponentially distributed random variables. We suppose that $\mathbb{E}[\tau_{i1}] = 1$ for all i except for one, say $i = 4$, with $\mathbb{E}[\tau_{41}]$ essentially greater than 1. One can see that the difference between the upper and lower bounds will decrease as the value of $\mathbb{E}[\tau_{41}]$ increases. Table 5.2 shows how the bounds vary with different values of $\mathbb{E}[\tau_{41}]$.

Let us discuss the effect of decreasing the variance $\mathbb{D}[\tau_{i1}]$ on the bounds on γ. Note that if τ_{i1} were degenerate random variables with zero variance, the lower and upper bounds in (5.17) would coincide. One can therefore expect that with decreasing the variance of τ_{i1}, the accuracy of the bounds increases.

As an illustration, consider a tandem queueing system (see Fig. 5.2) with $n = 5$ nodes. Suppose that $\tau_{i1} = \xi_{i1}/r$, where ξ_{i1}, $i = 1, \ldots, 5$, are i.i.d. random variables which have the Erlang distribution with the probability density function

$$f_r(t) = \begin{cases} t^{r-1}e^{-t}/(r-1)!, & \text{if } t > 0, \\ 0, & \text{if } t \leq 0. \end{cases}$$

Table 5.2: Results for a network with a dominating service time.

$\mathbb{E}[\tau_{41}]$	$\|\mathbb{E}[\mathcal{T}_1]\|$	$\widehat{\gamma}$	$\mathbb{E}\|\mathcal{T}_1\|$
1.0	1.0	1.005718	2.283333
2.0	2.0	2.004857	2.896032
3.0	3.0	3.004242	3.685531
4.0	4.0	4.003627	4.554525
5.0	5.0	5.003013	5.465368
6.0	6.0	6.002398	6.400835
7.0	7.0	7.001783	7.351985
8.0	8.0	8.001168	8.313731
9.0	9.0	9.000553	9.282968
10.0	10.0	10.000008	10.257692

Clearly, $\mathbb{E}[\tau_{i1}] = 1$ and $\mathbb{D}[\tau_{i1}] = 1/r$. Related numerical results including estimates $\widehat{\gamma}$ evaluated by simulating 100000 cycles are shown in Table 5.3.

Table 5.3: Results for tandem queues at changing variance.

r	$\|\mathbb{E}[\mathcal{T}_1]\|$	$\widehat{\gamma}$	$\mathbb{E}\|\mathcal{T}_1\|$
1	1.0	1.042476	2.928968
2	1.0	1.026260	2.311479
3	1.0	1.019503	2.045538
4	1.0	1.015637	1.890824
5	1.0	1.013110	1.787242
6	1.0	1.010864	1.711943
7	1.0	1.009920	1.654154
8	1.0	1.008409	1.608064
9	1.0	1.007726	1.570232
10	1.0	1.006657	1.538479

Acknowledgment. Research partially supported by Russian Federation Ministry of Education Programme "The Universities of Russia" under Grant #4233.

References

1. Baccelli, F., Cohen, G., Olsder, G. J. and Quadrat, J.-P. (1992). *Synchronization and Linearity*, Chichester: John Wiley & Sons.

2. Baccelli, F. and Konstantopoulos, P. (1991). Estimates of cycle times in stochastic Petri nets, *Applied Stochastic Analysis*, pp. 1–20, New York: Springer-Verlag, Lecture Notes in Control and Information Sciences, **177**.

3. Baccelli, F. and Makowski, A. M. (1989). Queueing models for systems with synchronization constraints, *Proceedings of the IEEE*, **77**, 138–160.

4. Baccelli, F., Massey, W. A. and Towsley, D. (1989). Acyclic fork-join queueing networks, *Journal of the ACM*, **36**, 615–642.

5. Cohen, G., Dubois, D., Quadrat, J.-P. and Viot, M. (1985). A linear-system-theoretic view of discrete-event processes and its use for performance evaluation in manufacturing, *IEEE Transactions on Automatic Control*, **30**, 210–220.

6. Cuninghame-Green, R. A. (1979). *Minimax Algebra*, Berlin: Springer-Verlag, Lecture Notes in Economics and Mathematical Systems, **166**.

7. Gumbel, E. J. (1954). The maxima of the mean largest value and of the range, *Annals of Mathematical Statistics*, **25**, 76–84.

8. Hartly, H. O. and David, H. A. (1954). Universal bounds for mean range and extreme observations, *Annals of Mathematical Statistics*, **25**, 85–99.

9. Kingman, J. F. C. (1973). Subadditive ergodic theory, *Annals of Probability*, **1**, 883–909.

10. Krivulin, N. K. (1994). Using max-algebra linear models in the representation of queueing systems, *Proceedings of the 5th SIAM Conference on Applied Linear Algebra, Snowbird, UT, June 15-18* (Ed., J. G. Lewis), pp. 155–160.

11. Krivulin, N. K. (1995). A max-algebra approach to modeling and simulation of tandem queueing systems, *Mathematical and Computer Modelling*, **22**, 25–37.

12. Krivulin, N. K. (1996). The max-plus algebra approach in modelling of queueing networks, *Proceedings of the 1996 SCS Summer Computer Simulation Conference, Portland, OR, July 21-25*, pp. 485–490, San Diego: SCS.

PART II
Experimental Designs

6

Analytical Theory of E-optimal Designs for Polynomial Regression

V. B. Melas

St. Petersburg State University, St. Petersburg, Russia

Introduction

E-optimal designs for polynomial regression on an arbitrary segment are the subject of the present paper. The E-optimality criterion was introduced by Ehrenfeld (1955). It has been demonstrated earlier [Kovrigin (1980)] that the E-optimal design for polynomial regression on the segment $[-1, 1]$ is concentrated at the extremum points of a Tchebysheff polynomial of the first kind. Pukelsheim and Studden (1993) proved the same for the truncated E-criterion. Moreover, it was shown by Heiligers (1991), that the statement remains valid for a segment $[a, b]$, where a and b are of the same sign, as well as for a symmetrical segment of sufficiently small length. All these results are based on the simplicity of the minimal eigenvalue of the information matrix of an E-optimal design.

However, Heiligers (1991) established that for symmetrical segments the multiplicity of the eigenvalue can be one or two.

The main purpose of this paper is the investigation of symmetrical segments of sufficiently large segments. Through computations it proves that our technique allows to construct E-optimal designs for all symmetrical segments if degree of the polynomial is no more than four.

We begin with preliminary results for an arbitrary segment. It proves that an E-optimal design is unique. For the case when corresponding minimal eigenvalue is simple we show that the optimal design assumes a Tchebysheff form and can be found explicitly.

The boundary equation for a set of segments, where the optimal design assumes such a form, is derived.

The case of multiplicity 2 is considered for symmetrical segments. For such a case points and weights of E-optimal designs are studied as functions of the segment length. It is demonstrated that these functions are analytical ones and that they can be expanded into Taylor series with respect to inverse degrees of the length in a vicinity of zero. Coefficients of the series can be evaluated by recurrence formulae, which are also deduced, as well as the tables of coefficients. Some results can be found in Russian publications, brought together in the book Melas (1997). But since this book is not available to the general public, all the proofs included are detailed. Some basic results without proofs were published in Melas (1996, 1998a,b).

6.1 Statement of the Problem

Consider the standard model of linear (with respect to the parameters) regression

$$y_j = \Theta^T f(x_j) + \varepsilon_j, \qquad j = 1, 2, \ldots, n,$$

where $y_1, y_2, \ldots, y_n$ are experimental results, $\Theta^T = (\Theta_1, \ldots, \Theta_m)$ is the vector of parameters to be evaluated, $f(x) = (f_1(x), \ldots, f_m(x))^T$ is the vector of known functions which are linearly independent and continuous on a compact linear topological space X, ε_j are random errors such that $E\varepsilon_j = 0$, $E\varepsilon_i\varepsilon_j = \sigma^2 \delta_{ij}$ $(i, j = 1, \ldots, n)$, $\sigma^2 > 0$, and δ_{ij} is the Kronecker symbol.

Let us call by experimental design a discrete probability measure on $\mathcal{X}$ given by the table: $\xi = \{x_1, \ldots, x_n; \mu_1, \ldots, \mu_n\}$, $x_i \in \mathcal{X}$, $x_i \neq x_j$ for $i \neq j$, $\mu_i > 0$, $\sum \mu_i = 1$, where n is an arbitrary natural number. Let

$$M(\xi) = \int_{\mathcal{X}} f(x) f^T(x) \, \xi(dx) = \sum_{i=1}^{n} f(x_i) f^T(x_i) \mu_i$$

be the information matrix. The following E-optimality criterion is of great importance when evaluating the full set of parameters: a design is called E-optimal one, if the minimal eigenvalue of its information matrix attains its maximum over the class of experimental designs, defined above, on it.

Since the set of information matrices is compact [see, for example, Karlin and Studden (1966, ch. 10)], then an E-optimal design for the standard model of linear regression exists. The problem of finding E-optimal design for the model $f_i(x) = x^{i-1}$, $i = 1, 2, \ldots, m$, $\mathcal{X} = [r_1, r_2]$, $r_1 < r_2$ is considered in Sections 6.3–6.10. Duality theorem introduced in Melas (1982), see also Pukelsheim (1980), adduced in the following section, is one of the main tools of analysis.

6.2 Duality Theorem

Let ξ_α be an E-optimal design. Let $\mathcal{P}_\alpha$ denote a linear subset of R^m, spanned by the eigenvectors corresponding to the minimal eigenvalue of $M(\xi_\alpha)$. Denote

Analytical Theory of E-optimal Designs

this minimal eigenvalue by $\lambda_{min}(M(\xi_\alpha))$. Let $\mathcal{P} = \cap \mathcal{P}_\alpha$, where the intersection is realized over all E-optimal designs. Let $\mathbf{N}$ be the class of nonnegative definite $m \times m$ matrices A, such that $tr A = 1$. The results of [Melas (1982)] can be stated in the following way.

Theorem 6.2.1 (duality theorem) *For the model described in Section 6.1 the E-optimality of a design ξ^* is equivalent to the existence of a matrix $A^* \in \mathbf{N}$ such that*

$$\max_{x \in \mathcal{X}} f^T(x) A^* f(x) \leq \lambda_{min}(M(\xi^*)).$$

Moreover,

$$\min_{A \in \mathbf{N}} \max_{x \in \mathcal{X}} f^T(x) A f(x) = \max_{\xi} \lambda_{min}(M(\xi)),$$

where the maximum in the right-hand side is taken over all experimental designs, and equality

$$f^T(x_i^*) A^* f(x_i^*) = \lambda_{min}(M(\xi^*)), \quad i = 1, 2, \ldots, n$$

is valid at the points of the E-optimal design x_i^* ($\xi^* = \{x_1^*, \ldots, x_n^*; \mu_1^*, \ldots, \mu_n^*\}$).

Theorem 6.2.2 *Any matrix A^* from Theorem 6.2.1 is of the form*

$$A^* = \sum_{i=1}^{s} \alpha_i p_{(i)} p_{(i)}^T,$$

where $s = \dim \mathcal{P}$, $\alpha_i \geq 0$, $\sum \alpha_i = 1$, and $\{p_{(i)}\}$ is some orthonormed basis in $\mathcal{P}$.

We will apply these results for $f_i(x) = x^{i-1}$, $\mathcal{X} = [r_1, r_2]$ where $r_1 < r_2$ are arbitrary numbers. Here let us call a model from Section 6.1 the model of polynomial regression (on an arbitrary segment).

6.3 The Number of Design Points

Let $\xi^* = \{x_1^*, \ldots, x_n^*; \mu_1^*, \ldots, \mu_n^*\}$ be an E-optimal design, $\lambda^* = \lambda_{min}(M(\xi^*))$.
If $m = 2$, the following statement is valid.

Lemma 6.3.1 *For $m = 2$:*

1. *If $r_1 r_2 > -1$, then an E-optimal design is unique and is of the form $\{r_1, r_2; \mu_1, \mu_2\}$, where*

$$\mu_1 = \frac{2 + r_2^2 + r_1 r_2}{4 + (r_1 + r_2)^2}, \quad \mu_2 = 1 - \mu_1.$$

Moreover $\lambda^* = r^2/(r^2 + \omega^2)$, where $r = (r_2 - r_1)/2$, $\omega = (r_2 + r_1)/2$;

2. If $r_1 r_2 \leq -1$, then any design of the form $\xi_{a,b} = \{a, b; b/(b-a), -a/(b-a)\}$, where $0 > a \geq r_1, 0 < b \leq r_2, |\ ab\ | \geq 1$ is E-optimal, and also $\lambda^* = 1$. Any E-optimal design can be represented as a convex combination of designs $\xi_{a,b}$.

PROOF. The lemma can be proved by a direct calculation [see also Melas (1997)].

Without loss of generality, let us further assume that

$$r_1 \leq x_1^* < \ldots < x_n^* \leq r_2.$$

Let $\sigma = \sigma_\xi = \{x_1, \ldots, x_n\}$ be the support of design ξ, $\mu = (\mu_1, \ldots, \mu_n)^T$ be the vector of weight coefficients, $\xi = (\sigma, \mu)$. ∎

Definition 6.3.1 The polynomial $g(x) = f^T(x) A^* f(x)$, where the matrix A^* is defined by Theorem 6.2.1, will be called *extremal polynomial*.

Lemma 6.3.2 *If $m > 2$ then all E-optimal designs have the same support of m points with the segment endpoints included. Any extremal polynomial is of the form*

$$g(x) = \lambda^* + \gamma(x - r_1)(x - r_2) \prod_{i=2}^{m-1} (x - x_i^*)^2,$$

where r_1, r_2, x_i^* are the points of E-optimal design, $\gamma > 0$ is a constant, and $\lambda^* < 1$.

PROOF. Fix a matrix A^*. Let ξ^* be an E-optimal design and $x^* \in \sigma_{\xi^*}$. Then according to Theorem 1

$$f^T(x) A^* f(x) \leq \lambda^*, \quad x \in \mathcal{X}; \qquad f^T(x^*) A^* f(x^*) = \lambda^*.$$

Let $g(x) = f^T(x) A^* f(x) \not\equiv \lambda^*$. Note that $g(x)$ is a polynomial of degree $\leq 2m-2$. If $x^* \neq r_1, r_2$, then x^* is a zero of the polynomial $\tilde{g}(x) = \lambda^* - g(x)$ and its multiplicity is no less than 2, otherwise the multiplicity of x^* is 1. Since an E-optimal design includes no fewer than m points (otherwise rang $M(\xi^*) = 0$ and $\lambda_{min}(M(\xi^*)) = 0$), then the polynomial $\tilde{g}(x)$ has degree $2m - 2$ and the same number of zeros with regard to multiplicity. This implies the required representation for $g(x)$. Since the E-optimal design is arbitrary, all E-optimal designs have the same support and $n = m$, $x_1^* = r_1$, $x_m^* = r_2$.

Let us demonstrate that the polynomial $g(x) \equiv \lambda^*$ cannot be an extremal one. Suppose that $g(x) \equiv \lambda^*$ is an extremal polynomial. Then

$$f^T(x) A^* f(x) \equiv \lambda^*. \tag{6.1}$$

Analytical Theory of E-optimal Designs

According to Theorem 6.2.2, the matrix A^* is of the form $\sum_{i=1}^{s} p_{(i)} p_{(i)}^T \alpha_i$, where $\{p_{(i)}\}$ is some orthonormal basis in $\mathcal{P}$, $\alpha_i \geq 0$, $\sum \alpha_i = 1$. Therefore equality (6.1) can be represented in the form

$$\sum_{i=1}^{s} \left(p_{(i)}^T f(x) \right)^2 \alpha_i \equiv \lambda^*.$$

This equality is valid if and only if $e_1 = (1, 0, \ldots, 0)^T \in \mathcal{P}$, and $\lambda^* = 1$. Let us demonstrate that if $m > 2$ then $\lambda^* < 1$ on any segment $\mathcal{X} = [r_1, r_2]$. Indeed,

$$\lambda^* = \lambda_{min}(M(\xi^*)) \leq e_1^T M(\xi^*) e_1 = 1$$

and since $M(\xi^*) e_1 \neq e_1$ for $m > 2$, then $\lambda^* < 1$. ∎

The following result is needed to prove the uniqueness of the E-optimal design.

Lemma 6.3.3 *Let m and s be arbitrary integers. If vectors $b_{(i)} = (b_1^{(i)}, \ldots, b_s^{(i)})$, $i = 1, 2, \ldots, m$ are not all null vectors, there exists a vector $\beta = (\beta_1, \ldots, \beta_s)$, such that $\sum_{j=1}^{s} \beta_j b_j^{(i)} \neq 0$, $i = 1, 2, \ldots, m$.*

PROOF. If $b = (b_1, \ldots, b_s) \neq 0$, the equation $\sum_{j=1}^{s} \beta_j b_j = 0$ is a hyperplane equation. It is obvious, that there exists vector β, which lies outside of m hyperplanes. ∎

Theorem 6.3.1 *For the model of polynomial regression on an arbitrary segment with $m > 2$ an E-optimal design is unique. By Lemma 6.3.2 it is concentrated at m points, including r_1 and r_2.*

PROOF. By Lemma 6.3.2, all E-optimal designs have the same support $\sigma^* = \{x_1^*, \ldots, x_m^*\}$. Let us demonstrate, that the weight vector for E-optimal design is defined uniquely.

According to Theorems 6.2.1 and 6.2.2,

$$\sum_{i=1}^{s} \left(p_{(i)}^T f(x_l^*) \right)^2 \alpha_l - \lambda^*, \qquad l = 1, 2, \ldots, m,$$

where $\{p_{(i)}\}$ is some orthonormal basis in $\mathcal{P}$, $\alpha_l \geq 0$. Without loss of generality, assume that $\alpha_1, \ldots, \alpha_{s'} > 0$, $s' \leq s$. Vectors $b_{(j)} = \left(p_{(1)}^T f(x_j^*), \ldots, p_{(s')}^T f(x_j^*) \right)$ are not all null vectors for $j = 1, 2, \ldots, m$. By Lemma 6.3.3, there exists a vector $\beta = (\beta_1, \ldots, \beta_{s'})$, such that $\sum_{j=1}^{s'} \beta_j b_j^{(l)} \neq 0$ $(l = 1, \ldots, m)$.

Let $p = \sum_{j=1}^{s'} \beta_j p_{(j)}$. Then $p \in \mathcal{P}$ and $M(\xi^*) p = \lambda^* p$, therefore $L\mu^* = \lambda^* p$, where $L = (f_i(x_l^*) d_l)_{i,l=1}^{m}$, $d_l = \sum_{i=1}^{s'} f_i(x_l^*) p_i = \sum_{j=1}^{s'} \beta_j b_j^{(l)} \neq 0$ $(l = 1, 2, \ldots, m)$, $\mu^* = (\mu_1^*, \ldots, \mu_m^*)^T$. Since $\det L = d_1 \ldots d_m \det \left(x_i^{*j-1} \right)_{i,j=1}^{m} \neq 0$, then $\mu^* = \lambda^* L^{-1} p$ is uniquely defined. ∎

6.4 Tchebysheff Designs

Let $m > 2$, $\chi = [r_1, r_2] = [\omega - r, \omega + r]$, $r > 0$.

According to Theorem 6.3.1, the unique E-optimal design is of the form

$$\xi^* = \{x_1^*, \ldots, x_m^*; \mu_1^*, \ldots, \mu_m^*\}, \quad r_1 = x_1^* < x_2^* < \ldots < x_m^* = r_2.$$

If $dim\mathcal{P} = 1$, this design can be found in an explicit form. Let $L = L(r, \omega)$ be a matrix of order $m \times m$, such that

$$f(rx + \omega) = Lf(x)$$

for any real x. Equating the coefficients of x^i, $i = 0, \ldots, m-1$ in the left- and right-hand sides, we derive

$$\begin{aligned} L &= (L_{ij})_{i,j=1}, \quad L_{ij} = 0, \quad j > i, \\ L_{ij} &= C_{i-1}^{j-1} r^{j-1} \omega^{i-j}, \quad j \leq i, \end{aligned}$$

where $C_\nu^k = k!(\nu - k)!/\nu!$ $k \leq \nu$.

Let $W = W(r, \omega) = (L^T(r, \omega) L(r, \omega))^{-1}$ and let $-1 = t_1 < \ldots < t_m = 1$ be extremum points of $T_{m-1}(t)$, which is a Tchebysheff polynomial of the first kind. Let $\hat{q}$ be the coefficient vector of this polynomial: $T_{m-1}(t) = \hat{q}^T f(t) = \sum_{i=0}^{m-1} \hat{q}_i t^i$.

Denote

$$\xi_{r,\omega} = \{rt_1 + \omega, \ldots, rt_m + \omega; \hat{\mu}_1, \ldots, \hat{\mu}_m\}, \quad \hat{\mu} = (\hat{\mu}_1, \ldots, \hat{\mu}_m)^T, \quad (6.2)$$

where $\hat{\mu}$ is equal to

$$\hat{\mu} = \sqrt{\lambda(r,\omega)} \bar{F}^{-1} W \hat{q},$$

$\bar{F} = ((-1)^{j+1} t_j^{i-1})_{i,j=1}^m$, $\lambda(r, \omega) = [\hat{q}^T W \hat{q}]^{-1}$.

It is easy to verify that $0 < \hat{\mu}_i < 1$, $\sum_{i=1}^m \hat{\mu}_i = 1$.

Definition 6.4.1 The design $\xi_{r,\omega}$ will be called the Tchebysheff design.

Theorem 6.4.1 *If $m > 2$ and $dim\mathcal{P} = 1$ then for polynomial regression on an arbitrary segment $[r - \omega, r + \omega]$ the design $\xi_{r,\omega}$ is the unique E-optimal design, and $\lambda^* = \lambda(r, \omega)$.*

PROOF. Since $dim\mathcal{P} = 1$, then by Theorems 6.2.1 and 6.2.2 the matrix A^* is unique and is of the form $A^* = qq^T$, where $||q|| = 1$ and q is the unique normed vector in $\mathcal{P}$. By Theorem 6.2.1,

$$g(x_i^*) = (q^T f(x_i^*))^2 = \max_{x \in \chi} g(x) = \lambda^*, \quad i = 1, \ldots, m.$$

Analytical Theory of E-optimal Designs

Note that the degree of the polynomial $q^T f(x)$ does not exceed $m-1$. Therefore either $g(x) \equiv const$, or $g(x) = \sqrt{\lambda^*} T_{m-1}\left(\frac{x-\omega}{r}\right)$, since the Tchebysheff polynomial of the first kind of degree $m-1$ is the unique polynomial of this degree, whose modulus attains its maximum on the segment $[-1, 1]$ at m points [see, for example, Fedorov (1972, ch. 2)]. Assume that $g(x) \equiv \lambda^*$. Then $q_0^2 = \lambda^*$, $q_i = 0$, $i = 1, 2, \ldots, m-1$. Since $\|q\| = 1$, then $\lambda^* = 1$. But by Lemma 6.3.2 $\lambda^* < 1$ for $m > 2$. Therefore, $g(x) = \sqrt{\lambda^*} T_{m-1}\left(\frac{x-\omega}{r}\right)$ and $x_i^* = rt_i + \omega$, $i = 1, 2, \ldots, m$. Since

$$M(\xi^*)q = \lambda^* q,$$

then

$$\left(\sum_{i=1}^{m} f(rt_i + \omega) f^T(rt_i + \omega) \mu_i^*\right) q = \sqrt{\lambda^*} L \bar{F} \mu^* = \lambda^* (L^{-1})^T \hat{q},$$

whence $\lambda^* = \lambda(r, \omega)$, $\mu^* = \hat{\mu} = W\hat{q}$. ∎

It has been demonstrated in Kovrigin (1980), that $\dim \mathcal{P} = 1$ for $\mathcal{X} \subset [-1, 1]$ and the E-optimal design is a Tchebysheff one (in our terms). According to Pukelsheim and Studden (1993), the Tchebysheff design is generalized E-optimal if $\mathcal{X} = [-1, 1]$ and some set of parameters is to be estimated. Heiligers (1991) demonstrated that $\dim \mathcal{P} = 1$ for any $\mathcal{X} \subset [0, \infty)$ or $\mathcal{X} \subset (-\infty, 0]$. As for $\mathcal{X} = [-r, r]$ (symmetrical segment), it was shown that $\dim \mathcal{P} \leq 2$, where $\dim \mathcal{P} = 1$ for $r < r^*$ and r^* is some critical value.

In the next section we will find an equation for arbitrary segments for which $\dim \mathcal{P} = 1$.

6.5 Boundary Equation

The layout below demonstrates that analysis of space $\mathcal{P}$ dimensionality can be reduced to analyzing the matrix $M(\xi_{r,\omega})$.

Let

$$\Omega - \{(r, \omega);\ r > 0,\ \omega \geq 0,\ \dim \mathcal{P} \geq 2 \text{ for } \chi = [r - \omega, r + \omega]\}.$$

Let A be an arbitrary matrix of order $m \times m$. Let A_- be the matrix of order $(m-1) \times (m-1)$, which is obtained from A by deleting its last row and last column.

Set

$$\lambda_2(r, \omega) = \lambda_{\min}(M(\xi_{r,\omega})_-),$$
$$\psi(r, \omega) = \lambda_2(r, \omega)/\lambda(r, \omega).$$

Theorem 6.5.1 *If $m > 2$ the set Ω is of the form*

$$\Omega = \{(r, \omega); r > 0, \omega \geq 0, \psi(r, \omega) \leq 1\}.$$

PROOF. It is sufficient to show that $dim\mathcal{P} = 1$ if and only if $\psi(r,\omega) > 1$.

Let $dim\mathcal{P} = 1$. Then by Theorem 6.4.1 we have $\xi^* = \xi_{r,\omega}$, $\lambda^* = \lambda(r,\omega)$. Moreover, $dim\mathcal{P} = 1$ implies that the multiplicity of λ^* is equal to 1. Therefore

$$\lambda(r,\omega) = \lambda_{\min}(M(\xi_{r,\omega})) < \lambda_{\min}(M(\xi_{r,\omega})_-) = \lambda_2(r,\omega)$$

hence $\psi(r,\omega) > 1$.

Let $\psi(r,\omega) > 1$. Set $\lambda = \lambda_{\min}(M(\xi_{r,\omega}))$. It is easy to verify that

$$M(\xi_{r,\omega})q(r,\omega) = \lambda(r,\omega)q(r,\omega),$$

where $q(r,\omega) = L^{-1}(r,\omega)\hat{q}$, whence $\lambda(r,\omega) \geq \lambda$.

Let us demonstrate that any eigenvector, corresponding to the minimal eigenvalue of the matrix $M(\xi_{r,\omega})$, has nonzero leading component. Indeed, let q be such a vector, $q_{m-1} = 0$. Set $q_- = (q_0, \ldots, q_{m-2})^T$. Then

$$M(\xi_{r,\omega})_- q_- = \lambda q_-,$$

therefore

$$\lambda \geq \lambda_{\min}(M(\xi_{r,\omega})_-) = \lambda_2(r,\omega) > \lambda(r,\omega) \geq \lambda_{\min}(M(\xi_{r,\omega})) = \lambda,$$

which is the contradiction.

Note that the vector $\hat{q}$ has nonzero leading component, therefore the vector $q(r,\omega)$ possesses the same property. Since $q_{m-1} \neq 0$, there exists $\alpha \neq 0$ such that $(\alpha q - q(r,\omega))_m = 0$.

Let us demonstrate that $\lambda(r,\omega) = \lambda$. Suppose that $\lambda(r,\omega) \neq \lambda$; then $\lambda(r,\omega) > \lambda$. Set $\tilde{q} = q(r,\omega)$, $M = M(\xi_{r,\omega})$. Note that $q^T\tilde{q} = 0$, since q and $\tilde{q}$ are eigenvectors of the matrix M, corresponding to different eigenvalues. Therefore

$$\lambda_2(r,\omega) \leq \frac{(\alpha q - \tilde{q})^T M (\alpha q - \tilde{q})}{(\alpha q - \tilde{q})^T (\alpha q - \tilde{q})}$$
$$= \frac{\alpha\lambda^2 + \lambda(r,\omega)\|\tilde{q}\|^2}{\alpha^2 + \|\tilde{q}\|^2}$$
$$< \lambda(r,\omega),$$

which contradicts $\psi(r,\omega) > 1$.

So, $\lambda(r,\omega) = \lambda = \lambda_{\min}(M(\xi_{r,\omega}))$. Since

$$\max_{x \in \chi}(\tilde{q}^T f(x))^2 = \lambda(r,\omega) = \lambda_{\min}(M(\xi_{r,\omega})),$$

then by Theorem 6.2.1 the design $\xi_{r,\omega}$ is E-optimal.

Let us demonstrate that $dim\mathcal{P} = 1$. If $dim\mathcal{P} = 2$, there exists a vector $q \in \mathcal{P}$ such that $q_{m-1} = 0$. But by definition this vector corresponds to $\lambda_{\min}(M(\xi^*)) = \lambda_{\min}(M(\xi_{r,\omega}))$. According to the above, this is impossible.

Therefore, $\dim \mathcal{P} = 1$. ∎

In fact, we have demonstrated, that an E-optimal design is a Tchebysheff one if and only if $(r, \omega) \bar{\in} Int\Omega$, while proving Theorem 6.5.1.

According to Theorem 6.5.1, the boundary equation of the set Ω can be written in the form

$$\lambda_2(r, \omega) = \lambda(r, \omega). \tag{6.3}$$

Expressions for $\lambda_2(r, \omega)$ can be obtained by means of Maple suite for any fixed m. Note that the equation assumes the following form for $m = 3$:

$$2\omega^8 + (-5r^2 + 16)\omega^6 + (4r^4 - 21r^2 + 36)\omega^4 + (5r^4 - r^6 - 12r^2 + 16)\omega^2 - r^2 + 2 = 0.$$

It can be verified numerically that this equation has no positive solution for $|r| < \sqrt{2}$, and has a unique positive solution $\omega = \omega(r)$ under any fixed $r \geq \sqrt{2}$. The explicit form of this equation for symmetrical segments ($\omega = 0$) and $m = 4, 5, 6$ can be found in Section 6.10.

In all these cases the equation has a unique positive solution.

The layout below is for more detailed analysis of E-optimal designs on symmetrical segments, such that $(r, \omega) \in Int\, \Omega$, i. e. $\dim \mathcal{P} = 2$. In this case the E-optimal design is not a Tchebysheff one but appears to be related to some extremal property of expansions of positive polynomials. This property is the subject of the following section.

6.6 An Extremal Property of Positive Polynomial Representations

Let $h(x)$ be a polynomial of degree $2k + 1$, positive on $[0, \infty)$. It is known from Karlin and Studden (1966, ch. 4), that such a polynomial can be uniquely represented in the form

$$h(x) = \alpha \prod_{i=1}^{k} (x - u_i)^2 + \beta x \prod_{i=1}^{k} (x - v_i)^2, \tag{6.4}$$

where

$$0 < u_1 < v_1 < \ldots < u_k < v_k, \quad \alpha > 0, \quad \beta > 0. \tag{6.5}$$

Definition 6.6.1 The representation (6.4) will be called *the Karlin–Shapley representation*.

Consider the class of representations of the form

$$h(x) = \varphi_1^2(x) + x\varphi_2^2(x), \tag{6.6}$$

where $\varphi_1(x) = p^T f(x) = \sum_{i=0}^{k} p_i x^i$, $\varphi_2(x) = q^T f(x) = \sum_{i=0}^{k} q_i x^i$, $p = (p_0, \ldots, p_k)^T$, $q = (q_0, \ldots, q_k)^T$ are arbitrary polynomials. The Karlin–Shapley representation is a representation of form (6.6), therefore the set of these representations is not empty. On the other hand, considering polynomials $\varphi_1^2(x)$ and $\varphi_2^2(x)$, which have noninterlacing zeros, implies that polynomial $h(x) = \varphi_1^2(x) + x\varphi_2^2(x)$ has at least two representations of form (6.6).

Definition 6.6.2 The representation of the form (6.6) will be called *the maximal* one, if $\|p\|^2 + \|q\|^2$ attains its maximum over all representations of such form for the representation.

Theorem 6.6.1 *The maximal representation for polynomials of $2k+1$th degree, positive on $[0, \infty)$, exists, is unique and coincides with the Karlin–Shapley representation.*

PROOF. The proof of the theorem is based on the lemma on polynomials with fixed absolute value. Let $\varphi(x) = \sum_{i=0}^{k} p_i x^i$ be a polynomial of degree k, such that $0 \le x_0 < x_1 < \ldots < x_k$,

$$\varphi(x_i) = (-1)^i a_i,$$

where $a_i > 0$, $i = 1, 2, \ldots, k$. Let $\tilde{\varphi}(x) = \sum_{i=0}^{k} \tilde{p}_i x^i$ be a polynomial of degree not more than k, such that $|\tilde{\varphi}(x_i)| \le a_i$, $i = 0, 1, \ldots, k$. ∎

Lemma 6.6.1 $|p_i| \ge |\tilde{p}_i|$, $p_i(-1)^{i+k} > 0$, $i = 0, 1, \ldots, k$ *for the polynomials being considered.*

PROOF OF THE LEMMA. Represent the polynomial $\tilde{\varphi}(x)$ in the form

$$\tilde{\varphi}(x) = \sum_{i=0}^{k} \tilde{p}_i x^i = \det \begin{pmatrix} 0 & 1 & x & \ldots & x^k \\ \tilde{a}_0 & 1 & x_0 & \ldots & x_0^k \\ \vdots & \vdots & \vdots & \ddots & \vdots \\ \tilde{a}_k & 1 & x_k & \ldots & x_k^k \end{pmatrix} \Big/ \prod_{j<i} (x_i - x_j),$$

where $\tilde{a}_i = \tilde{\varphi}(x_i)$. Note that

$$\tilde{p}_i = \det \left(\tilde{a}_j \ 1 \ \ldots \ x_j^{i-1} x_j^{i+1} \ \ldots \ x_j^k \right)_{j=0}^{k} \Big/ \delta$$

$$= \sum_{s=0}^{k} \tilde{a}_s (-1)^s \delta_{i,s} / \delta,$$

where

$$\delta_{i,s} = \det \left(\tilde{a}_j \ 1 \ \ldots \ x_j^{i-1} x_j^{i+1} \ \ldots \ x_j^k \right)_{j \ne s, \, j \in 0:k}$$

$$= \prod_{l<j, \, l,j \ne s} (x_j - x_l) \sum x_{j_1} \ldots x_{j_i} > 0,$$

$$\delta = \prod_{j<i} (x_i - x_j) > 0.$$

Since $\varphi(x_i) = (-1)^i a_i$, then $|p_i| = |\sum_{s=0}^{k} a_s(-1)^s \delta_{i,s}/\delta| \geq |\tilde{p}_i|$, $i = 0, 1, \ldots, k$.

Let $h(x) = \tilde{\varphi}_1^2(x) + x\tilde{\varphi}_2^2(x) = \varphi_1^2(x) + x\varphi_2^2(x)$, where φ_1, φ_2 form the Karlin-Shapley representation $\tilde{\varphi}_1(x) = \sum \tilde{p}_i x^i$, $\tilde{\varphi}_2(x) = \sum \tilde{q}_i x^i$, $\varphi_1(x) = \sum p_i x^i$, $\varphi_2(x) = \sum q_i x^i$. Then $|\tilde{\varphi}_1(v_i)| \leq a_i = |\varphi_1(v_i)|$, $i = 0, 1, \ldots, k$, $v_0 = 0$.

According to Lemma 6.6.1, $|p_i| \geq |\tilde{p}_i|$, $i = 0, 1, \ldots, k$.

The leading coefficients of the polynomials $\varphi_2(x)$ and $\tilde{\varphi}_2(x)$ coincide with one another; moreover

$$|\tilde{\varphi}_2(u_i)| \leq |\varphi_2(u_i)|, \quad i = 1, \ldots, k,$$

since $\varphi_1(u_i) = 0$. Consider the polynomials

$$\varphi(x) = x^k \varphi_2(1/x), \quad \tilde{\varphi}(x) = x^k \tilde{\varphi}_2(1/x).$$

They satisfy the hypothesis of Lemma 6.6.1 for $x_0 = u_0 = 0$, $x_i = u_i$, $i = 1, \ldots, k$. Therefore, the coefficients of x^i in $\varphi_2(x)$ are not less (in absolute value) than those in $\tilde{\varphi}_2(x)$. Therefore, $\sum p_i^2 + \sum q_i^2 \geq \sum \tilde{p}_i^2 + \sum \tilde{q}_i^2$. ∎

The similar result for the polynomials of even degree also can be derived.

6.7 Differential Equation

Let us call the model from Section 6.3 with $-r_1 = r_2 = r$ the model of polynomial regression on a symmetrical segment. The case $m = 2$ hase been already considered in Section 6.3. Let $m > 2$.

Lemma 6.7.1 *For $m > 2$, an E-optimal design for polynomial regression on a symmetrical segment is concentrated at points, equidistant from zero, particularly, for $m = 2k$: $-x_i^* = x_{2k+1-i}^*$, $i = 1, \ldots, k$, for $m = 2k+1$: $-x_i^* = x_{2k+2-i}^*$, $i = 1, \ldots, k$, $x_{k+1}^* = 0$. Moreover, the weights of such design have symmetrical values: $\mu_i^* = \mu_{2k+1-i}^*$, $i = 1, 2, \ldots, k$ for $m = 2k$ and $\mu_i^* = \mu_{2k+2-i}^*$, $i = 1, 2, \ldots, k$ for $m = 2k+1$.*

PROOF. Let $\xi^* = \{x_1^*, \ldots, x_m^*; \mu_1^*, \ldots, \mu_m^*\}$ be an E-optimal design. Let $m = 2k$ (the proof for $2k+1$ is similar). Consider the design

$$\tilde{\xi} = \{\tilde{x}_i, \ldots, \tilde{x}_m; \tilde{\mu}_1, \ldots, \tilde{\mu}_m\}: \tilde{x}_i = -x_{2k-i}^*, \tilde{\mu}_i = \mu_{2k-i}^*.$$

The matrix $M(\xi)$ for the design $\xi = (\xi^* + \tilde{\xi})/2$ with even rows and columns moved to the first positions assumes the form

$$\begin{pmatrix} M_1 & 0 \\ 0 & M_2 \end{pmatrix}.$$

It is known that the minimal eigenvalue of any nonnegative definite matrix M can be represented in the form

$$\min_{\|p\|=1} p^T M p.$$

Therefore

$$\lambda_{\min}\begin{pmatrix} M_1 & 0 \\ 0 & M_2 \end{pmatrix} \geq \lambda_{\min}\begin{pmatrix} M_1 & C \\ C^T & M_2 \end{pmatrix}$$

for $M_1, M_2 \geq 0$ and any matrix C; whence

$$\lambda_{\min}(M(\xi)) \geq \lambda_{\min}(M(\xi^*))$$

for design $\xi = (\xi^* + \tilde{\xi})/2$; i.e. ξ is E-optimal design. If the points of the design ξ^* are not equidistant from zero, then $\xi \neq \xi^*$. Since by Theorem 6.3.1 the E-optimal design is unique for $m > 2$, the proof is complete. ■

As has been proved by Heiligers (1991), $\dim \mathcal{P} \leq 2$ for polynomial regression on a symmetrical segment. The case $\dim \mathcal{P} = 1$ had been investigated already (for arbitrary segments with $\dim \mathcal{P} = 1$ an E-optimal design is a Tchebysheff one).

Let us consider the case $\dim \mathcal{P} = 2$. We confine our consideration by the case $m = 2k$ since the case $m = 2k + 1$ can be studied in a similar way.

First of all let us elaborate the results of section 6.3 for our case.

Definition 6.7.1 *The orthonormal basis $\{p_{(i)}\}_{i=1}^s$ of the space $\mathcal{P}$ from theorem 6.3.2 be called an extremal basis.*

Let $M = M(\xi^*)$. Since $\dim \mathcal{P} = 2$ then minimal eigenvalues of matrices M_1 and M_2 from Lemma 6.7.1 coincide with each other and equal λ^*. Denote by $p^* = (p_0^*, \ldots, p_{k-1}^*)^T$ and $q^* = (q_0^*, \ldots, q_{k-1}^*)^T$ normed lowest eigenvectors of matrices M_1 and M_2, respectively. It is obvious that the vectors $\bar{p} = (p^*, 0, p_1^*, 0, \ldots, p_{k-1}^*, 0)^T$ and $\bar{q} = (0, q_0^*, \ldots, 0, q_{k-1}^*)^T$ are orthogonal and generate a basis of $\mathcal{P}$. Let us prove that $\{\bar{p}, \bar{g}\}$ is an extremal basis.

Lemma 6.7.2 *For $m = 2k > 2$, $\mathcal{X} = [-r, r]$ if $\dim \mathcal{P} = 2$ then any matrix A^* from Theorem 6.3.1, 6.3.2 is of the form*

$$A^* = \alpha \bar{p}\bar{p}^T + (1-\alpha)\bar{q}\bar{q}^T,$$

where $0 \leq \alpha < 1$ is uniquely determined.

PROOF. Consider the extremal polynomial $q(x) = f^T(x) A^* f(x)$. By Lemma 6.3.2 this polynomial possesses the representation

$$g(x) = \lambda^* + \gamma(x^2 - r^2) \prod_{i=2}^{m-1}(x - x_i^*)^2 =$$

$$= \lambda^* + \gamma(x^2 - r^2) \prod_{i=1}^{k-1}(x^2 - x_i^{*2})^2, \gamma > 0.$$

From this representation we conclude that $g(x)$ is a polynomial from x^2 of degree $2k - 1$.

Note that arbitrary orthonormal basis of $\mathcal{P}$ is of the form $\{p_{(1)}, p_{(2)}\}$, $p_{(1)} = \delta \bar{p} + \sqrt{1 - \delta^2} \bar{q}$, $p_{(2)} = \sqrt{1 - \delta^2} \bar{p} - \delta \bar{q}$ for some δ, $a \leq \delta \leq 1$. Let $\{p_{(1)}, p_{(2)}\}$ be an extremal basis. Then any matrix A^* has the form

$$A^* = \alpha p_{(1)} p_{(1)}^T, \ 0 \leq \alpha \leq 1.$$

Since $q(x) = f^T(x) A^* f(x)$ is a polynomial of x^2 then

$$\alpha \sum_{l=0}^{j} p_{(1)l} p_{(1)j-l} + (1 - \alpha) \sum_{l=0}^{J} p_{(2)l} p_{(2)j-l} = 0, \ j = 1, 3, \ldots, 2k - 1.$$

Therefore

$$\delta \sqrt{1 - \delta^2}(1 - 2\alpha) \sum_{l=0}^{j} p_l^* q_{j-l}^* = 0, \ j = 0, 1, \ldots, k - 1.$$

From this if $\alpha \neq \frac{1}{2}$ then $\delta - 1$ or 0. Note that with $\delta - 0, 1$ we have $\{p_{(1)}, p_{(2)}\} = \{\bar{p}, \bar{q}\}$.

If $\alpha = \frac{1}{2}$ then

$$A^* = \frac{1}{2} p_{(1)} p_{(1)}^T + \frac{1}{2} p_{(2)} p_{(2)}^T = \frac{1}{2} \bar{p} \bar{p}^T + \frac{1}{2} \bar{q} \bar{q}^T,$$

that is the matrix A^* possesses the required representation.

Note that $\alpha \neq 1$ since $\bar{p}^T f(x)$ is a polynomial of $2k - 2$ degree from x^2 and $q(x)$ is a polynomial of $2k - 1$ degree from x^2. Note that $A^* \bar{p} = \alpha \bar{p}$ and therefore α is uniquely determined. ∎

It will be proved in the next section that the polynomials $\bar{p}^T f(x)$ and $\bar{q}^T(x)$ has no common roots and $\alpha > 0$ (so that the extremal polynomial is positive for every x) for sufficiently large r. Also for sufficiently large r it was shown in Heiligers (1991) that an E-optimal design is not Tchebysheff one. Therefore $\dim \mathcal{P} = 2$ for sufficiently large r.

Let r^* be such that with $r > r^*$ the extremal polynomial is positive for every x (in particular it means that $\dim \mathcal{P} = 2$).

The representation of Lemma 6.7.2. can be rewrite in the following form

$$g(x) = \tilde{g}(y) = \lambda^* + \gamma(y - r^2) \prod_{i=1}^{k-1}(y - y_i^*)^2 =$$

$$= \varphi_1^2(y) + y\varphi_2^2(y),$$

where $y_i^* = x_i^{*2}$, $\varphi_1(y) = \alpha(p^{*T}\tilde{f}(y))^2$, $\varphi_2(y) = (1-\alpha)(q^{*T}\tilde{f}(y))^2$, $\tilde{f}(y) = (1, y, \ldots, y^{k-1})^T$.

This representation will be called the dual representation since it is connected with the duality theorem.

Lemma 6.7.3 *For $r > r^*$ the dual representation of the extremal polynomial coincides with its Karlin–Shapley representation.*

sc Proof. Consider all possible representations of $\tilde{g}(y) = g(x)$, $y = x^2$ of the following form

$$\tilde{g}(y) = \varphi_1^2(y) + y\varphi_2^2(y), \tag{6.7}$$

where $\varphi_q(y) = p^T\tilde{f}(y)$, $\varphi_2(y) = 1^T\tilde{f}(y)$ and p, q are arbitrary vectors. At least one of such representations exists and is the dual representation. Let us integrate both sides of equality (6.7) by the measure $\tilde{\xi}^*(dy) = \xi^*(dx)$. Then we receive

$$\lambda^* = p^T M_1(\xi^*)p + q^T M_2(\xi^*)q,$$

where matrices M_1 and M_2 defined in the proof of Lemma 6.7.1. From this

$$||p||^2 + ||q||^2 \leq 1$$

and the equality takes place for $p = \alpha p^*$, $q = (1-\alpha)q^*$. Since the Karlin–Shapley representation is the unique one which maximizes $||p||^2 + ||q||^2$ it coincides with the dual representation. ∎

Introduce vectors $p = (p_0, \ldots, p_{k-1})^T$, $q = (q_0, \ldots, q_{k-1})^T$, $\pi = (p^T, q^T)^T$, $\pi_\alpha = (\alpha p^T, (1-\alpha)q^T)^T$, where $0 \leq \alpha \leq 1$, and the matrix

$$P_\pi = \begin{pmatrix} p_0 & p_1 & \cdots & p_{k-1} & 0 & 0 & \cdots & 0 \\ 0 & p_0 & p_1 & \cdots & p_{k-1} & 0 & \cdots & 0 \\ \vdots & \vdots & \vdots & \ddots & \vdots & \vdots & \ddots & \vdots \\ 0 & \cdots & 0 & p_0 & p_1 & \cdots & p_{k-1} & 0 \\ 0 & q_0 & q_1 & \cdots & q_{k-1} & 0 & \cdots & 0 \\ 0 & 0 & q_0 & q_1 & \cdots & q_{k-1} & \cdots & 0 \\ \vdots & \vdots & \vdots & \vdots & \ddots & \vdots & \ddots & \vdots \\ 0 & 0 & \cdots & 0 & q_0 & q_1 & \cdots & q_{k-1} \end{pmatrix}.$$

(The order of matrix P_π is $2k \times 2k$.)

Analytical Theory of E-optimal Designs

An immediate calculation gives the following representation

$$g(x) = \tilde{g}(y) = \pi_\alpha^{*T} P_{\pi_{\alpha^*}^*} f(y),$$

where $\pi_\alpha^* = (\alpha p^{*T}, (1-\alpha^*)q^{*T})$ and $\alpha \in (0,1)$ is such that $A^* = \alpha^* \bar{p}\bar{p}^T + (1-\alpha^*)\bar{q}\bar{q}^T$.

Note that solutions of direct and dual problems can be expressed in terms of

$$\begin{aligned}
y_1^* &= (x_{k+1}^*)^2, \ldots, y_{k-1}^* = x_{2k-1}^{*2}, \\
\nu_2^* &= 2\mu_{k+1}^*, \ldots, \nu_k^* = 2\mu_{2k}^*,
\end{aligned}$$

$$p_0^*, \ldots, p_{k-1}^*, q_0^*, \ldots, q_{k-2}^*.$$

Introduce vectors $\nu^* = (\nu_2^*, \ldots, \nu_k^*)^T, y^* = (y_1^*, \ldots, y_{k-1}^*)^T$. Set $z = 1/r^2$. Introduce the new variables $y_1 = y_1^*, \tilde{y}_i = y_i^*/z, \tilde{\nu}_i = \nu_i^*/z, i = 2, \ldots, k (y_k^* = 1)$, $\pi_{\alpha^*}^* \to \tilde{\pi} = (\tilde{p}^T, \tilde{q}^T)^T$, where

$$\tilde{\pi} = \left(Z_1^{-1} \pi_{\alpha^*}^* / ((1-\alpha^*)q_{k-1}^*)\right) \sqrt{z} z^{k-1}),$$

$$Z_1 = diag\{1, z, \ldots, z^{k-1}, \sqrt{z}, \ldots, \sqrt{z} z^{k-1}\}.$$

Definition 6.7.2 The vector

$$\tilde{\theta} = \tilde{\theta}(z) = (\tilde{p}_0, \ldots, \tilde{p}_{k-1}, \tilde{q}_0, \ldots, \tilde{q}_{k-2}, \tilde{\nu}_2, \ldots, \tilde{\nu}_k, y_1, \tilde{y}_2, \ldots, \tilde{y}_{k-1})^T$$

will be called the parameter vector of solutions of two dual problems on a symmetrical segment.

Let $\theta = (p_0, \ldots, p_{k-1}, q_0, \ldots, q_{k-2}, \nu_2, \ldots, \nu_k, y_1, \ldots, y_{k-1})^T$ be an arbitrary vector and let the relations $q_{k-1} = 1, y_k = 1$ be satisfied. Set $Z = Z_1^2$,

$$\lambda^* = \lambda^*(z) = \lambda_{min}(M(\xi^*)), \lambda(z, \theta) = \frac{\pi^T \Gamma_\pi c}{\pi^T Z \pi},$$

where $c = \sum_{i=2}^k (f(\tilde{y}_i) - f(y_1 z))\tilde{\nu}_i z + f(y_1 z)$.

Lemma 6.7.4 *If $r > r^*, m > 2$, the relations*

$$\lambda(z, \tilde{\theta}) = \lambda^*(z),$$

$$\lambda'_{\theta_i}(z, \tilde{\theta}) = 0, i = 1, 2, \ldots, 4k-3 \tag{6.8}$$

are satisfied.

PROOF. Note that the equality
$$\lambda'_\pi(z,\tilde\theta) = 0 \tag{6.9}$$
is equivalent to the equality $P_\pi c = \lambda \pi$, where $\lambda = \lambda(z,\tilde\theta)$, and the last one is equivalent to the statement $M_1 p = \lambda p$, $M_2 q = \lambda q$, i.e. multiplicity of eigenvalue $\lambda = \lambda(z,\tilde\theta)$ is 2. Thus, (6.9) is a necessary condition for E-optimality of the design ξ^*.

Next, the equalities
$$\lambda'_y(z,\tilde\theta) = 0, \lambda'_\nu(z,\tilde\theta) = 0 \tag{6.10}$$
are equivalent to the equalities
$$\pi^T P_\pi f'(y_i) = 0, \quad \pi^T P_\pi(y_1) = \pi^T P_\pi f(y_{i+1}), \tag{6.11}$$
where $\pi = \pi^*_{\alpha^*}$, $y_i = y_i^*$, $i = 1, 2, \ldots, k-1$. Since $\pi^{*T}_{\alpha^*} P_{\pi^*_{\alpha^*}} f(y)$ is an extremal polynomial, conditions (6.10) are the necessary conditions for the design ξ^* to be E-optimal.

Equality $\lambda(z,\tilde\theta) = \lambda^*$ follows from (6.11). ∎

Consider the vector $\tilde\theta = \tilde\theta(z)$.

Lemma 6.7.5 *For any $z \in (0, 1/r^{*2})$ vector $\tilde\theta(z)$ is determined uniquely.*

PROOF. According to Theorem 6.3.1, the vectors $y = y(z)$ and $\nu = \nu(z)$ are determined uniquely. And according to Lemma 6.7.5 the vector $\pi^*_{\alpha^*}$ is uniquely determined. Therefore, the vector $\tilde\pi$ is also uniquely determined. ∎

According to Lemma 6.7.5, the vector $\tilde\theta = \tilde\theta(z)$ is implicitly defined by the relations
$$\lambda'_{\theta_i} = 0, \quad i = 1, 2, \ldots, 4k - 3.$$

Introduce the matrix
$$J(z,\theta) = \left(\lambda''_{\theta_i \theta_j}(z,\theta)\right)_{i,j=1}^{4k-3}$$
and the vector
$$J_z(z,\theta) = -\left(\lambda''_{\theta_i z}(z,\theta)\right)_{i=1}^{4k-3}.$$

By the Implicit Function Theorem, for any point $z_0 > 0$, such that $\det J(z_0, \tilde\theta(z_0)) \neq 0$, there exists a vicinity, in which a continuously differentiable vector function $\theta(z)$ is defined, such that $\theta(z_0) = \tilde\theta(z_0)$ and the differential equation
$$J(z,\theta(z))\theta'(z) = J_z(z,\theta(z)) \tag{6.12}$$
is satisfied. Denote $J = J(z) = J(z,\tilde\theta(z))$, $J_z = J_z(z,\tilde\theta(z))$.

Analytical Theory of E-optimal Designs

Applying the explicit representation of $\lambda(z, \theta)$ we derive

$$J = J(z) = \frac{2}{\pi^T Z \pi} \begin{pmatrix} \tilde{M} - \lambda Z & P_\pi Y \\ (P_\pi Y)^T & \mathcal{D} \end{pmatrix}_-, \qquad (6.13)$$

$$J_z = J_z(z) = \frac{2}{\pi^T Z \pi}(b^T(z), 0, \ldots, 0)^T,$$

$$b(z) = \left(\lambda Z' \pi - \lambda Z \pi \frac{\pi^T Z' \pi}{\pi^T Z \pi} - P_\pi \tilde{c} - P_\pi f'(y_1 z)(1 - \sum \tilde{\nu}_i z) z\right)_-,$$

where the symbol "$-$" right from a matrix means, that its $2k$th row and $2k$th column are rejected, $\tilde{c} = \sum_{i=2}^{k}(f(\tilde{y}_i) - f(\tilde{y}_1))\tilde{\nu}_i$, $\tilde{y}_1 = y_1 z$,

$$\tilde{M} = \begin{pmatrix} M_1(\tilde{\xi}) & 0 \\ 0 & M_2(\tilde{\xi}) \end{pmatrix},$$

$$\tilde{\xi} = \{x_1^* \sqrt{z}, \ldots, x_m^* \sqrt{z}, z^{-1}\mu_1^*, \ldots, z^{-1}\mu_m^*\}, \lambda = \lambda(z, \tilde{\theta}),$$

$$Y = (Y_\nu \vdots Y_y), Y_\nu = ((f(\tilde{y}_i) - f(\tilde{y}_1))z)_{i=2}^{k}.$$

$$Y_y = (f'(\tilde{y}_1)z\tilde{\nu}_1 \vdots (f'(\tilde{y}_i)\tilde{\nu}_i z)_{i=2}^{k-1}),$$

$$\mathcal{D} = \begin{pmatrix} 0 & 0 \\ 0 & E \end{pmatrix},$$

$$E = \frac{1}{2}\mathrm{diag}\{\pi^T P_\pi f''(\tilde{y}_1)z^2 \tilde{\nu}_1, \pi^T P_\pi f''(y_2)\tilde{\nu}_2 z, \ldots, \pi^T P_\pi f''(y_{k-1})\tilde{\nu}_k z\},$$

$$\tilde{\nu}_1 = 1 - \sum_{i=2}^{k} \tilde{\nu}_i z, \quad \pi = \tilde{\pi}.$$

In this representation vanishing terms, e.g. $\pi^T P_\pi f'(\tilde{y}_i)$ are omitted. Set $z^* = 1/r^{*2}$.

Lemma 6.7.6 $\det J(z) \neq 0$ for $z \in (0, z^*)$.

PROOF. Let A be the matrix $\frac{\pi^T Z \pi}{2} J$ with omitted kth column and kth row, a be apart column with omitted kth element, $a^* = (\frac{\pi^T Z \pi}{2} J)_{kk}$. Note that the matrix A is of the following form:

$$A = \begin{pmatrix} G & H \\ H^T & \mathcal{D} \end{pmatrix}, \qquad (6.14)$$

where G is the matrix $\tilde{M} - \lambda Z$ with omitted kth and $2k$th columns and kth and $2k$th rows, H is the matrix $P_{\tilde{\pi}} Y$ with omitted kth and $2k$th rows, matrix $\mathcal{D}$ is the same as in formula (6.13) to constant precision.

Since the multiplicity of the minimal eigenvalue of the matrix $Z_1 \tilde{M} Z_1$ is not more than two, and there are no zeros in the vector π^*, the matrix G is positive definite.

Since the Vandermonde determinant is not zero, the matrix Y is of full rank. Since the polynomials $\hat{p}^T f(x)$ and $\hat{q}^T f(x)$ have no common factors, and P_π is the resultant matrix of these polynomials, then $\det P_\pi \neq 0$ [see Van der Waerden (1967)]. Therefore the matrix H is of full rank.

By Frobenius formula [see Fedorov (1972, ch.1)]

$$\det A = \det G \det(\mathcal{D} - H^T G^{-1} H).$$

Since the matrix H is of full rank, $H^T G^{-1} H > 0$.

All the elements of the matrix $\mathcal{D}$ are nonpositive and $-\mathcal{D} \geq 0$, whence it follows that

$$\det A > 0,$$

since the matrices $\mathcal{D}$ and $H^T G^{-1} H$ are of order $(2k-1) \times (2k-1)$.

Let us demonstrate that $\det J(z) \neq 0$ for $z \in (0, z^*)$. Multiplying 2nd, 3rd, ..., and kth rows of the matrix $J(z)$ by $p_1, \ldots, p_{k-1}$, respectively, adding them to the first one, multiplied by p_0, and doing the same thing with the columns gives us the matrix

$$\begin{pmatrix} 0 & b_{(1)}^T & b_{(2)}^T \\ b_{(1)} & G & H \\ b_{(2)} & H^T & \mathcal{D} \end{pmatrix},$$

where $b_{(1)}^T = (0, \ldots, 0)$,

$$b_{(2)}^T = \Big((p^T f(y_2))^2 - (p^T f(y_1))^2, \ldots, (p^T f(y_k))^2 - (p^T f(y_1))^2, \\ \big((p^T f(y_2))^2\big)', \ldots, \big((p^T f(y_k))^2\big)' \Big).$$

Note that the vector $b_{(2)} \neq 0$, since otherwise $p^T f(y)$ is a Tchebysheff polynomial, what is impossible. Therefore $\det J(z) = b^T A^{-1} b \neq 0$. ∎

Lemma 6.7.7 *The vector function $\tilde{\theta}(z)$ is a real analytical vector function for $z \in (0, z^*)$.*

PROOF. Since $\det J(z) \neq 0$ for $z \in (0, z^*)$, then by the Implicit Function Theorem, for any $z_0 \in (0, z^*)$ there exists some $\epsilon > 0$, such that system of equations (6.8) has the unique solution in $z \in V = (z_0 - \epsilon, z_0 + \epsilon)$ (let $\hat{\theta}(z)$ be it), such that $\hat{\theta}(z_0) = \tilde{\theta}(z_0)$.

Moreover, $\hat{\theta}(z)$ is a real analytical vector function on V (by the same theorem). Since the vector function $\tilde{\theta}(z)$ satisfies the system of equations (6.8) and is determined uniquely, then $\tilde{\theta}(z) = \hat{\theta}(z)$ for $z \in V$. Thus, the vector function $\tilde{\theta}(z)$ is analytical in some vicinity of any point $z_0 \in (0, z^*)$. ∎

Analytical Theory of E-optimal Designs 103

According to the results derived and the Implicit Function Theorem, the following theorem is valid.

Theorem 6.7.1 *For polynomial regression on a symmetrical segment with $m = 2k > 2, r > r^*$, the parameter vector of solutions of two dual problems $\hat{\theta}(z)$, where $z = 1/r^2$, is an analytical real vector function, satisfying the differential equation (6.12).*

6.8 Limiting Design

Let us prove that limit of $\hat{\theta}(z)$ at $z \to +0$ exists and find it. At first, consider the vector
$$\hat{\theta}(z) = \hat{\theta} = (\tilde{p}_0, \ldots, \tilde{p}_{k-1}, \tilde{q}_0, \ldots, \tilde{q}_{k-2}, \tilde{y}_1, \ldots, \tilde{y}_{k-1}, \tilde{\nu}_2 z, \ldots, \tilde{\nu}_k z)^T.$$
By Lemma 6.7.3, zeros of the polynomials $\tilde{p}^T \tilde{f}(y)$ and $\tilde{q}^T \tilde{f}(y)$ belong to the segment $[0, 1]$. Since $\tilde{q}_{k-1} \equiv 1$, all the variables $\tilde{q}_i = \tilde{q}_i(z)$ are bounded for $z \in (0, z^*)$. Moreover, by definition $\tilde{y}_i \in [0, 1]$, $i = 1, \ldots, k-1$. Equating the coefficients of y^{2k-1} in the two representations (Lemmas 6.3.2 and 6.7.5) of an extremal polynomial, we derive
$$\tilde{p}_{k-2} = -2\tilde{q}_{k-1} - 2\sum_{i=1}^{k-1} \tilde{y}_i - z,$$
whence it follows that $\tilde{p}_{k-2}$, and therefore all the variables $\tilde{p}_i$, $i = 0, \ldots, m-1$ are finite at $z \in (0, z^*)$. By definition, $0 \leq \tilde{\nu}_i z = \nu_i^* \leq 1$, $i = 2, \ldots, k$. Thus, all the components of the vector $\hat{\theta}(z)$ are bounded at $z \in (0, z^*)$. Therefore, there exists a sequence of such vectors, converging while $z \to +0$. And $\lim_{z \to +0} \hat{\theta}(z)$ exists, if all convergent sequences $\hat{\theta}(z_j)$, $z \to +0$ tend to the same limit at $j \to \infty$.

Let us introduce the notations
$$\psi_{(0)} = \lim_{z \to +0} \psi(z), \quad \psi_{(n)} = \lim_{z \to +0} \frac{\psi^{(n)}(z)}{n!}, \quad n = 1, 2, \ldots$$
for any function (scalar, vector or matrix) $\psi(z)$, if these limits exist.

Let $U_n(x)$ be the Tchebysheff polynomial of the second kind of nth degree, $\tilde{y} = (\tilde{y}_1, \ldots, \tilde{y}_k)^T$, $\tilde{y}_k \equiv 1$, $\tilde{p} = (\tilde{p}_0, \ldots, \tilde{p}_{k-1})^T$, $\tilde{q} = (\tilde{q}_0, \ldots, \tilde{q}_{k-1})^T$.

Lemma 6.8.1 *For $m = 2k > 2$ the vectors $\tilde{y}(z), \tilde{p}(z), \tilde{q}(z)$ converge while $z \to +0$ and*
$$\tilde{y}_{(0)i} = t_{k+i-1}^2, \quad i = 1, 2, \ldots, k,$$
where $0 = t_k < t_{k+1} < \ldots < t_{2k-1} = 1$ are the extremum points of the polynomial $T_{2k-2}(t)$ on segment $[0, 1]$, $\{\tilde{p}_{(0)i}\}_{i=0}^{k-1}$ are nonzero coefficients of the polynomial $T_{2k-2}(t)$, $\{\tilde{q}_{(0)i}\}_{i=0}^{k-1}$ are nonzero coefficients of the polynomial $(t^2 - 1)U_{2k-3}(t)$.

PROOF. Consider the following problem: to find

$$\max_{\sigma} \min_{v,u,\alpha} \lambda(z;\sigma,v,u,\alpha),$$

where $\lambda(z;\sigma,v,u,\alpha)$ is equal to value λ in the relation

$$\lambda + \gamma(y - \frac{1}{z})\prod_{i=1}^{k-1}(y-y_i)^2 = \alpha \prod_{i=1}^{k-1}(y-v_i)^2 + \gamma y \prod_{i=1}^{k-1}(y-u_i)^2 = $$
$$= (p_0 + p_1 y + \ldots + p_{k-1} y^{k-1})^2 + y(q_0 + q_1 y + \ldots + q_{k-1} y^{k-1})^2, \quad (6.15)$$

such that $z = 1/r^2$, $y = x^2$, $y_i = x_i^2$, $i = 1, \ldots, k-1$, $\gamma > 0$, p_i and q_i are real number, $p_0^2 + \ldots + p_{k-1}^2 + q_0^2 + \ldots + q_{k-1}^2 = 1$, $x_i \leq r$. Due to previous results the solution of this problem corresponds to finding $\tilde{y}(z)$, $\tilde{p}(z)$, $\tilde{q}(z)$.

Devide all sides of (6.15) by $\gamma = q_{k-1}^2$ and make the substitution $\hat{p}_i = p_i/q_{k-1}$, $i = 1, \ldots, k-1$, $\hat{q}_i = q_i/q_{k-1}$, $i = 1, \ldots, k-2$. Then we receive

$$\frac{\lambda}{\gamma} + (y - 1/z)\prod_{i=1}^{k-1}(y-y_i)^2 = \beta \prod_{i=1}^{k-1}(y-v_i)^2 + y \prod_{i=1}^{k-1}(y-u_i)^2 = $$
$$= (\hat{p}_0 + \hat{p}_1 y + \ldots + \hat{p}_{k-1} y^{k-1})^2 + y(\hat{q}_0 + \quad (6.16)$$
$$+ \hat{q}_1 y + \ldots + \hat{q}_{k-2} y^{k-2} + y^{k-1})^2,$$

where $\gamma = \dfrac{1}{\pi^T \pi}$, $\pi^T = (\hat{p}_0, \hat{p}_1, \ldots, \hat{p}_{k-1}, \hat{q}_0, \hat{q}_1, \ldots, \hat{q}_{k-2}, 1)$.

Make now the substitution: $\tilde{y} = yz$, $\tilde{y}_i = y_i z$, $\tilde{v}_i = v_i z$, $\tilde{u}_i = u_i z$, $i = 1, \ldots, k-1$, $\tilde{\beta} = \beta z$. Multiplying sides (6.16) by z^{2k-1}, receive

$$\frac{\lambda}{\gamma} z^{2k-1} + (\tilde{y} - 1)\prod_{i=1}^{k-1}(\tilde{y}-\tilde{y}_i)^2 = \tilde{\beta}\prod_{i=1}^{k-1}(\tilde{y}-\tilde{v}_i)^2 + \tilde{y}\prod_{i=1}^{k-1}(\tilde{y}-\tilde{u}_i)^2 = $$
$$= (\hat{p}_0 z^{k-1}\sqrt{z} + \ldots + \hat{p}_{k-1}\sqrt{z}\tilde{y}^{k-1})^2 + \tilde{y}(\hat{q}_0 z^{k-1} + \ldots + \tilde{y}^{k-1})^2. \quad (6.17)$$

Denote $\tilde{p}_0 = \hat{p}_0 z^{k-1}\sqrt{z}$, $\tilde{p}_1 = \hat{p}_1 z^{k-2}\sqrt{z}$, ..., $\tilde{p}_{k-1} = \hat{p}_{k-1}\sqrt{z}$; $\tilde{q}_0 = \hat{q}_0 z^{k-1}$, $\tilde{q}_1 = \hat{q}_1 z^{k-2}$, ..., $\tilde{q}_{k-2} = \hat{q}_{k-2} z$. Then we can rewrite (6.17) in the form:

$$\frac{\lambda}{\gamma} z^{2k-1} + (\tilde{y} - 1)\prod_{i=1}^{k-1}(\tilde{y}-\tilde{y}_i)^2 = \tilde{\beta}\prod_{i=1}^{k-1}(\tilde{y}-\tilde{v}_i)^2 + \tilde{y}\prod_{i=1}^{k-1}(\tilde{y}-\tilde{u}_i)^2 = $$
$$= (\tilde{p}_0 + \tilde{p}_1 \tilde{y} + \ldots + \tilde{p}_{k-1}\tilde{y}^{k-1})^2 + \tilde{y}(\tilde{q}_0 + \tilde{q}_1 \tilde{y} + \ldots + \tilde{y}^{k-1})^2. \quad (6.18)$$

Equating the coefficients of $\tilde{y}^0$ and $\tilde{y}^{2k-2}$ in (6.17) and (6.18) receive:

$$\frac{\lambda}{\gamma} z^{2k-1} - \prod_{i=1}^{k-1}\tilde{y}_i^2 = \tilde{\beta}\prod_{i=1}^{k-1}\tilde{v}_i^2 = \hat{p}_0^2 z^{2k-1} = \tilde{p}_0^2, \quad \tilde{\beta} = 2\sum_{i=1}^{k-1}(\tilde{u}_i - \tilde{y}_i) - 1.$$

Analytical Theory of E-optimal Designs

From this we have the expression for λ:

$$\lambda = \frac{\tilde{\beta}\prod_{i=1}^{k-1}\tilde{v}_i^2 + \prod_{i=1}^{k-1}\tilde{y}_i^2}{z^{2k-1}(\hat{p}_0^2 + \hat{p}_1^2 + \ldots + \hat{p}_{k-1}^2 + \hat{q}_0^2 + \ldots + \hat{q}_{k-2}^2 + 1)} =$$

$$= \frac{\tilde{p}_0^2 + \prod_{i=1}^{k-1}\tilde{y}_i^2}{\tilde{p}_0^2 + \tilde{p}_1^2 z^2 + \ldots + \tilde{p}_{k-1}^2 z^{2k-2} + \tilde{q}_0^2 z + \tilde{q}_1^2 z^3 + \ldots + z^{2k-1}}.$$

Since $\lambda^*(z) < 1$ due to Lemma 6.3.2 then $\lim_{z\to 0}\lambda(z) \le 1$, and we receive $\tilde{y}_{(0)} = 0$ and with $z \to 0$

$$\lambda = 1 - \frac{\tilde{q}_{0_{(0)}}^2}{\tilde{p}_{0_{(0)}}^2}z + \ldots + o(z).$$

Therefore,

$$\lambda'(0) = -\frac{\tilde{q}_{0_{(0)}}^2}{\tilde{p}_{0_{(0)}}^2} = -\frac{\prod_{i=1}^{k-1}\tilde{u}_{i_{(0)}}^2}{\tilde{\beta}_{(0)}\prod_{i=1}^{k-1}\tilde{v}_{i_{(0)}}^2} \quad \text{where } \tilde{\beta}_{(0)} = 2\sum_{i=1}^{k-1}(\tilde{u}_{i_{(0)}} - \tilde{y}_{i_{(0)}}) - 1.$$

Equating coefficients at first degree of $\tilde{y}$ in (6.17), receive

$$\prod_{i=1}^{k-1}\tilde{y}_i^2 + \sum_{j=1}^{k-1}2\tilde{y}_j\prod_{i\ne j}\tilde{y}_i^2 = -\tilde{\beta}\sum_{j=1}^{k-1}2\tilde{v}_j\prod_{i\ne j}\tilde{v}_i^2 + \prod_{i=1}^{k-1}\tilde{u}_i^2.$$

From this $\prod_{i=1}^{k-1}\tilde{u}_{i_{(0)}}^2 = 2\tilde{\beta}_{(0)}\sum_{j=1}^{k-1}\tilde{v}_{j_{(0)}}\prod_{i\ne j}\tilde{v}_{i_{(0)}}^2$. And we have

$$-\lambda'(0) = \frac{\prod_{i=1}^{k-1}\tilde{u}_{i_{(0)}}^2}{\tilde{\beta}_{(0)}\prod_{i=1}^{k-1}\tilde{v}_{i_{(0)}}^2} = \frac{2\tilde{\beta}_{(0)}\sum_{j=1}^{k-1}\tilde{v}_{j_{(0)}}\prod_{i\ne j}\tilde{v}_{i_{(0)}}^2}{\tilde{\beta}_{(0)}\prod_{i=1}^{k-1}\tilde{v}_{i_{(0)}}^2} = 2\sum_{j=1}^{k-1}\frac{1}{\tilde{v}_{j_{(0)}}}.$$

The problem of finding $\max_\sigma \min_{v,u,\alpha}\lambda(z;\sigma,v,u,\alpha)$ is reduced to

$$\min_\sigma \max_{v,u}(-\lambda'(0)) = \min_\sigma \max_{v,u} 2\sum_{i=1}^{k-1}\tilde{v}_{i_{(0)}}^{-1},$$

under the condition that

$$\tilde{\beta}_{(0)}\prod_{i=1}^{k-1}\tilde{v}_{i_{(0)}}^2 + y^2(y-1)\prod_{i=2}^{k-1}(y-\tilde{y}_{i_{(0)}})^2 = \tilde{\beta}_{(0)}\prod_{i=1}^{k-1}(y-\tilde{v}_{i_{(0)}})^2 + y\prod_{i=1}^{k-1}(y-\tilde{u}_{i_{(0)}})^2.$$

Introduce the polynomial $\varphi(y) = \dfrac{\prod_{i=1}^{k-1}(y - \tilde{v}_{i_{(0)}})}{\prod_{i=1}^{k-1} \tilde{v}_{i_{(0)}}}$. Substitute in the last equation values $y = \tilde{y}_{1_{(0)}}, \ldots, \tilde{y}_k$, while $\tilde{y}_{1_{(0)}} = 0$, $\tilde{y}_k = 1$, receive

$$\tilde{\beta}_{(0)} \prod_{i=1}^{k-1} \tilde{v}_{i_{(0)}}^2 = \tilde{\beta}_{(0)} \prod_{i=1}^{k-1} (\tilde{y}_j - \tilde{v}_{i_{(0)}})^2 + \tilde{y}_j \prod_{i=1}^{k-1}(\tilde{y}_j - \tilde{u}_{i_{(0)}})^2.$$

Then $\varphi^2(\tilde{y}_j) = \dfrac{\prod_{i=1}^{k-1}(\tilde{y}_j - \tilde{v}_{i_{(0)}})^2}{\prod_{i=1}^{k-1} \tilde{v}_{i_{(0)}}^2} = 1 - \psi(\tilde{y}_j)$, where $\psi(y) = \dfrac{y \prod_{i=1}^{k-1}(y - \tilde{u}_{i_{(0)}})^2}{\tilde{\beta}_{(0)} \prod_{i=1}^{k-1} \tilde{v}_{i_{(0)}}^2} \geq 0$

with $y \geq 0$.

Due to Lemma 6.6.1 the absolute value of each coefficient of polynomial φ attains its maximum along v_i under fixed y_i, u_i ($i = 1, \ldots, k-1$), if

$$\varphi(y_i) = (-1)^i \sqrt{1 - \psi(y_i)}, \quad i = 1, \ldots, k. \tag{6.19}$$

Since $\varphi'(y) = \dfrac{\sum_{j=1}^{k-1} \prod_{i \neq j}(y - \tilde{v}_{i_{(0)}})}{\prod_{i=1}^{k-1} \tilde{v}_{i_{(0)}}}$, $\varphi'(0) = (-1)^k \sum_{j=1}^{k-1} \tilde{v}_{j_{(0)}}^{-1}$, then $|\lambda'(0)| = |2\varphi'(0)|$;

Since $\lambda'(0) = (2\varphi'(0))$, and $\varphi'(0)$ is equal to the coefficient of y in the polynomial $\varphi(y)$, $\lambda'(0)$ attains its maximum along v_i, if (6.19) is satisfied. It is evident that $\lambda'(0)$, where v_i is selected to satisfy (6.19), attains its maximum over $\{u_i\}$ if $\psi(y_i) = 0$, i.e. on $u_i = y_{i+1}, i = 1, 2, \ldots, k-1$. Thus, the equality

$$\varphi(y_i) = (-1)^i, \quad i = 1, 2, \ldots, k$$

is satisfied.

Consider the polynomial $h(x) = \varphi(x^2)$. According to the latter relations

$$h(x_i) = (-1)^i, \quad i = 1, 2, \ldots, 2k-1,$$

where $x_{i+k-1}^2 = y_i, i = 1, 2, \ldots, k$, $x_i = -x_{2k-i}, i = 1, 2, \ldots, k-1$.

At the same time, it follows from the condition, that

$$|h(x)| \leq 1.$$

Really, since $\tilde{u}_{i_{(0)}} = \tilde{y}_{i+1_{(0)}}$, $i = 1, 2, \ldots, k-1$, then $\tilde{\beta}_{(0)} = 2 \sum_{i=1}^{k-1}(\tilde{u}_{i_{(0)}} - \tilde{y}_{i_{(0)}}) - 1 = 2 \left(\sum_{i=2}^{k} \tilde{y}_{i_{(0)}} - \sum_{i=1}^{k-1} \tilde{y}_{i_{(0)}} \right) - 1 = 2(\tilde{y}_k - \tilde{y}_{1_{(0)}}) - 1 = 1$,

$$y \prod_{i=1}^{k-1}(y - \tilde{u}_{i_{(0)}})^2 - y^2(y-1) \prod_{i=2}^{k-1}(y - \tilde{y}_{i_{(0)}})^2 = y(1-y) \prod_{i=2}^{k-1}(y - \tilde{y}_{i_{(0)}})^2. \tag{6.20}$$

Analytical Theory of E-optimal Designs   107

Note that $\varphi^2(y) = 1 - \dfrac{y(1-y)\prod\limits_{i=2}^{k-1}(y-\tilde{y}_{i_{(0)}})^2}{\prod\limits_{i=1}^{k-1}\tilde{v}_{i_{(0)}}^2} \leq 1$ for $y \in [0,1]$.

Thus, the polynomial $h(x)$ has degree $2k-2$ and attains its maximal absolute value on the segment $[-1,1]$ at $2k-1$ points. Therefore $h(x)$ coinsides with the Tchebysheff polynomial $T_{2k-2}(x)$.

$$T_{2k-2}(x) = h(x) = \varphi(x^2) = \varphi(y).$$

Thus

$$\tilde{p}_{(0)}^T(1, x^2, \ldots, x^{2k-2}) = const\, T_{2(k-1)}(x) \text{ where } const = \tilde{p}_{0_{(0)}}.$$

Hence, zeros of polynomial $\varphi(y)$ are $\tilde{v}_{i_{(0)}} = s_i^2$, $i = 1, 2, \ldots, k-1$, where s_i are positive zeros of Tchebysheff polynomial T_{2k-2}.

Consider nonnegative points in which $T_{2k-2}(x)$ attains its maximum on $[0,1]$: $0 = t_k < t_{k-1} < \ldots < t_{2k-1} = 1$. We have

$$\tilde{y}_{1_{(0)}} = t_k^2 = 0, \quad \tilde{y}_{i_{(0)}} = t_{k+i-1}^2, \quad i = 1, 2, \ldots, k-1, \quad \tilde{y}_k = t_{2k-1}^2 = 1.$$

Deviding both sides of the condition by $\tilde{\beta}_{(0)} \prod\limits_{i=1}^{k-1}\tilde{v}_{i_{(0)}}^2 = \tilde{p}_{0_{(0)}}^2$, receive

$$1 + \dfrac{y^2(y-1)\prod\limits_{i=2}^{k-1}(y-\tilde{y}_{i_{(0)}})^2}{\tilde{p}_{0_{(0)}}^2} = \dfrac{\prod\limits_{i=1}^{k-1}(y-\tilde{v}_{i_{(0)}})^2}{\tilde{p}_{0_{(0)}}^2} + \dfrac{y\prod\limits_{i=1}^{k-1}(y-\tilde{u}_{i_{(0)}})^2}{\tilde{p}_{0_{(0)}}^2}. \qquad (6.21)$$

Using (6.20) rewrite relation (6.21) in the form

$$1 = \dfrac{\prod\limits_{i=1}^{k-1}(y-\tilde{v}_{i_{(0)}})^2}{\tilde{p}_{0_{(0)}}^2} + y(1-y)\dfrac{\prod\limits_{i=2}^{k-1}(y-\tilde{y}_{i_{(0)}})^2}{\tilde{p}_{0_{(0)}}^2},$$

That is $1 = h_1^2(x) + (1-x^2)h_2^2(x)$, where $h_1(x) = h(x)$, $h_2(x) = \psi_1(y)$ with $y = x^2$, $\psi_1(y) = \sqrt{y}\dfrac{\prod\limits_{i=2}^{k-1}(y-\tilde{y}_{i_{(0)}})}{\tilde{p}_{0_{(0)}}}$.

As it is known [e.g., see Karlin, Studden (1966)], the identity is valid on the set of polynomials if and only if $h_1(x)$ and $h_2(x)$ are Tchebysheff polynomials of the first and second kinds, respectively. Thus, $h_2(x) = U_{2k-3}(x)$.

Since

$$(y-1)\psi_1(y) = \dfrac{\sqrt{y}}{\tilde{p}_{0_{(0)}}}\prod_{i=2}^{k}(y-\tilde{y}_{i_{(0)}}) = \dfrac{\sqrt{y}}{\tilde{p}_{0_{(0)}}}\prod_{i=1}^{k-1}(y-\tilde{u}_{i_{(0)}}) = \dfrac{\sqrt{y}}{\tilde{p}_{0_{(0)}}}\sum_{i=0}^{k-1}\tilde{q}_{i_{(0)}}y^i,$$

then we have
$$\tilde{q}_{0_{(0)}}x + \tilde{q}_{1_{(0)}}x^3 + \ldots + x^{2k-1} = \tilde{p}_{0_{(0)}}(x^2 - 1)U_{2k-3}(x).$$

∎

Now let us prove existence of limits of $(\nu(z))_{(0)} = (\tilde{\nu}(z)z)_{(0)}$, $(\tilde{\nu}(z))_{(0)}$ and $y_{1(0)}$, where $\tilde{\nu}(z) = (\tilde{\nu}_1(z), \ldots, \tilde{\nu}_k(z))^T$ and find them.

Consider the matrix
$$\tilde{M} = \tilde{M}(z) = \begin{pmatrix} M_1(\xi_z) & 0 \\ 0 & M_1(\xi_z) \end{pmatrix}.$$

It can be checked, that
$$\tilde{M}(z)\tilde{\pi}(z) = \lambda^*(z)Z\tilde{\pi}(z). \tag{6.22}$$

Since $\sum_{i=1}^{k} \tilde{\nu}(z)z = 1$, the variables $\tilde{\nu}(z)z$, $i = 1, 2, \ldots, k$ are bounded. Let $(\tilde{\nu}(z)z)_{(0)}$ be the limit of some convergent sequence $\tilde{\nu}(z_j)z_j$, $z_j \to +0$. Passing to the limit in equality (6.22), we derive
$$\tilde{M}_{(0)}\tilde{\pi}_{(0)} = \lambda_{(0)}\tilde{p}_{(0)}e_1 = \tilde{p}_{0(0)}e_1, \quad e_1 = (1, 0, \ldots, 0)^T.$$

Set $F = (\tilde{y}_{i(0)}^j(-1)^i)_{i=1,j=0}^{k,k-1}$.

After some computations we can establish that $\tilde{M}_{(1)}$, $\tilde{\nu}_{(0)}$ and $\tilde{\pi}_{(1)}$ are uniquely determined and
$$\tilde{\nu}_{(0)} = F^{-1}\left(\frac{\tilde{p}_{1(0)}}{\tilde{p}_{0(0)}}e_1 - e_2\right), \tag{6.23}$$

where $e_2 = (0, 1, 0, \ldots, 0)^T$. So, we can state the following theorem.

Theorem 6.8.1 *If $m = 2k > 2$, vector $\tilde{\theta}(z)$ at $z \to +0$ tends to the limit vector $\tilde{\theta}_{(0)}$, which components are defined by Lemma 6.8.1 and formulas*
$$\lim_{z \to +0} y_1(z) = 1, \quad \lim_{z \to +0} \tilde{\nu}_i(z) = \tilde{\nu}_{(0)}, \quad i = 2, \ldots, k,$$

where $\tilde{\nu}_{i(0)}$ are the components of vector $\tilde{\nu}_{(0)}$, defined by formula (6.23).

From this Theorem we receive that the polynomials $\bar{p}^T f(x)$ and $\bar{q}^T f(x)$ have no common roots for sufficiently large r since the corresponding limiting polynomials have no common roots.

Note that this result does not depend on the differentiability of $\tilde{\theta}(z)$ and thus we have not a logical circle.

6.9 Taylor Expansion

Let us expand the vector function $\tilde{\theta}(z)$ into a Taylor series.

It is easy to verify that all derivatives

$$\frac{\partial}{\partial \theta_l}(J(z,\theta))_{ij}, \frac{\partial}{\partial \theta_l}(J_z(z,\theta))_i, l,i,j = 1,2,\ldots,4k-3$$

at point $\theta = \tilde{\theta}(z)$ tend to zero while $z \to +0$. Therefore $J_{(0)} = 0$, $J_{z(0)} = 0$.

Set

$$J_1 = (J)'_z = -(J_z)'_\theta, J_{z^2} = (J_z)'_z, J_\theta = \sum_{i=1}^{4k-3} J'_{\theta_i}\theta'_i.$$

Differentiating equation (6.12) with respect to z, derive

$$J\theta'' + J_\theta \theta' + 2J_1\theta' = J_{z^2}. \tag{6.24}$$

By Theorem 6.5.1, the function $\tilde{\theta} = \tilde{\theta}(z)$ is an analytical one at $z \in (0, z^*)$, as well as the function $J(z) = J(z, \tilde{\theta}(z))$, $J_{z^2}(z) = J_{z^2}(z, \tilde{\theta}(z))$.

Similar to Lemma 6.7.6, one can verify that $\det J_{1(0)} \neq 0$. Therefore, by the Implicit Function Theorem, complex version, all the above functions are analytical in some vicinity of zero.

Expanding them into Taylor series and passing to the limit at $z \to +0$, we derive from equation (6.24) that

$$\theta_{(n+1)} = \frac{1}{(n+1)(n+2)} J_{1(0)}^{-1} \Big(J_{z^2(n)} - (J_{\theta(n)} + 2J_{1(n)})\theta_{(1)}$$

$$- \sum_{s=1}^{n-1} \frac{n-s+1}{s+1}(J_{\theta(s)}(n+1) + (n+s+2)J_{1(s)})\theta_{(n-s+1)}\Big),$$

$$n = 0, 1, \ldots,$$

$$\tilde{\theta}(z) = \sum_{n=0}^{\infty} \theta_{(n)} z^n. \tag{6.25}$$

Thus, the following theorem is valid.

Theorem 6.9.1 *The vector function $\tilde{\theta}(z)$ can be analytically continued in some vicinity of $z = 0$, its Taylor series converge in this vicinity, and the coefficients of the series are defined by formulas (6.25).*

With the help of computer calculations we verified that for $\omega = 0$ and $3 \leq m \leq 10$ equation (6.3) has a unique positive solution, denote it by $\hat{r} = \hat{r}(m)$. Since for $r = 1$ $dim\mathcal{P} = 1$ then $dim\mathcal{P} = 1$ for $r < \hat{r}$ and for all segments $[-r,r]$ with $r \leq \hat{r}$ an E-optimal design is Tchebysheff. For $r > \hat{r}$ an E-optimal design can be calculated from the vector $\tilde{\theta} = \tilde{\theta}(z)$, $z = 1/r^2$. If the minimum of radiuses of convergence of the Taylor series of functions $\tilde{\theta}_i(z)$, $i = 1, 2, \ldots, 4k - 3$ is no less than $\hat{z} = 1/\hat{r}^2$ then the vector $\tilde{\theta}(z)$ with a fixed $z < \hat{z}$ can be calculated by formula (6.25) with arbitrary given precision. And if the minimal radius is less than $\hat{z}$, then it can be constructed a sequence of reexpansions of vector function $\tilde{\theta}(z)$ into Taylor series along with the general procedure of analytical expansion [see, for example, Titchmarsh (1939)].

Thus, at least for $m \leq 10$ the problem can be solved with any given precision in the way, described above. In the next section we will present the solution for $m = 3, 4, 5, 6$.

6.10 Particular Cases

Let us consider polynomial regression on an arbitrary symmetrical segments with $m = 3, 4, 5, 6$.

For $m = 3$ an E-optimal design can be found in an explicit form given by the following result.

Theorem 6.10.1 *For $m = 3$, $\chi = [-r, r]$ an E-optimal design is unique and is of the form*
$$\xi^* = \{-r, 0, r; \mu, 1 - 2\mu, \mu\},$$
where $\mu = 1/(4 + r^4)$ for $r \leq \sqrt{2}$ and $\mu = (r^2 - 1)/(2r^4)$ for $r \geq \sqrt{2}$. In the first case $\lambda^* = r^4/(4 + r^4)$, in the second case $\lambda^* = (r^2 - 1)/r^2$.

PROOF. The proof of the theorem can be performed through an elementary study of the characteristic polynomial of the matrix $M(\xi)$.

Note that equation (6.3) with $\omega = 0$ in this case can be written in the form
$$r^2 - 2 = 0.$$

Besides it can be easily checked that the minimal eigenvalue of matrix $M(\xi_{r,0})$ is $2r^2/(r^4 + 4)$ for $r \geq \sqrt{2}$ and tends to zero with $r \to \infty$. This means that the Tchebysheff design is very bad for large r. At the same time $\lambda^* \to 1$ with $r \to \infty$.

For $m = 4, 5, 6$ we have the following results obtained with the help of *Maple* software package.

6.10.1 Boundary equation

Let $z = 1/r^2$. For $m = 4$ and $m = 5$ the boundary equation is, respectively,
$$64z^5 - 32z^4 + 60z^3 - 30z^2 + 11z - 3 = 0,$$

and
$$64z^5 - 32z^4 + 96z^3 - 44z^2 + 33z - 12 = 0.$$

The roots of the first equation are:

$$z_1 = 0.381425, z_{2,3} = -0.035509 \pm 0.859247i, z_{4,5} = 0.094796 \pm 0.396465i,$$

The roots of the second equation are:

$$z_1 = 0.396787, z_{2,3} = -0.007778 \pm 0.988626i, z_{4,5} = 0.059385 \pm 0.692766i.$$

In the case $m = 6$ the boundary equation is

$$1048576z^{13} - 524288z^{12} + 4587520z^{11} - 2195456z^{10} + 6631424z^9 - 2959360z^8$$
$$+3474944z^7 - 1408000z^6 + 418320z^5 - 156064z^4 + 17880z^3 - 3900z^2 + 225z$$
$$-25 = 0.$$

The roots of this equation are

$$\begin{aligned}
z_1 &= 0.376998, \\
z_{2,3} &= 0.040212 \pm 0.10205i, \\
z_{4,5} &= 0.040026 \pm 0.951596i, \\
z_{6,7} &= -0.021924 \pm 0.320835i, \\
z_{8,9} &= -0.005953 \pm 1.400837i, \\
z_{10,11} &= 0.009297 \pm 1.149805i, \\
z_{12,13} &= -0.000157 \pm 0.147012i.
\end{aligned}$$

6.10.2 Matrices $J_{1_{(0)}}$ and vectors $J_{z^2_{(0)}}$

For $m = 4$:

$$J_{1_{(0)}} = \begin{pmatrix} 4 & 2 & 0 & 1 & 1 \\ 2 & 1 & 0 & 1/2 & -1/2 \\ 0 & 0 & 1 & 0 & -1 \\ 1 & 1/2 & 0 & 0 & 0 \\ 1 & 1/2 & -1 & 0 & 0 \end{pmatrix}, \quad J_{z^2_{(0)}} = \begin{pmatrix} -6 \\ -1 \\ 4 \\ 0 \\ 2 \end{pmatrix}.$$

For $m = 5$:

$$J_{1_{(0)}} = \begin{pmatrix} 9 & 3 & 0 & 0 & -1/4 & -1/4 & -6 \\ 3 & 3/2 & 0 & 0 & 0 & 0 & -3/2 \\ 0 & 0 & 2 & 3/2 & -1/8 & 1/4 & -2 \\ 0 & 0 & 3/2 & 9/8 & -1/32 & 1/4 & -3/2 \\ -1/4 & 0 & -1/8 & -1/32 & 0 & 0 & 0 \\ -1/4 & 0 & 1/4 & 1/4 & 0 & 0 & 0 \\ -6 & -3/2 & -2 & -3/2 & 0 & 0 & -3/2 \end{pmatrix},$$

$$J_{z_{(0)}^2} = \begin{pmatrix} 28 \\ -5/2 \\ 27/2 \\ 0 \\ 0 \\ 0 \\ 0 \end{pmatrix}.$$

For $m = 6$:

$$J_{1_{(0)}} = \begin{pmatrix} 16 & 4 & 2 & 0 & 0 & 0 & -1/4 & -1 & 0 \\ 4 & 2 & 3/2 & 0 & 0 & 1/8 & -1/16 & 1/8 & -1/2 \\ 2 & 3/2 & 5/4 & 0 & 0 & 1/8 & -1/32 & 0 & -1/2 \\ 0 & 0 & 0 & 3 & 2 & 0 & 0 & 1/2 & -1 \\ 0 & 0 & 0 & 2 & 3/2 & 0 & 0 & 0 & -1/2 \\ 0 & 1/8 & 1/8 & 0 & 0 & 0 & 0 & 0 & 0 \\ -1/4 & -1/16 & -1/32 & 0 & 0 & 0 & 0 & 0 & 0 \\ -1 & 1/8 & 0 & 1/2 & 0 & 0 & 0 & 0 & 0 \\ 0 & -1/2 & -1/2 & -1 & -1/2 & 0 & 0 & 0 & -1/2 \end{pmatrix},$$

$$J_{z_{(0)}^2} = \begin{pmatrix} 32 \\ 5/4 \\ -1/4 \\ -8 \\ -1 \\ 0 \\ 0 \\ 1/2 \\ 0 \end{pmatrix}.$$

6.10.3 Tables of coefficients

Coefficients of the Taylor series are given in the Tables 6.1 – 6.3.

Table 6.1: Table of coefficients ($m = 4$).

n	0	1	2	3	4	5	6	7	8	9	10
$\tilde{p}_0$	$-\frac{1}{2}$	1	$-\frac{1}{2}$	$\frac{5}{2}$	$-\frac{19}{4}$	8	9	$-\frac{239}{4}$	$\frac{3533}{16}$	$-\frac{2645}{8}$	$-\frac{2277}{16}$
$\tilde{p}_1$	1	-2	1	-3	$\frac{1}{2}$	2	-38	$\frac{167}{2}$	$-\frac{1297}{8}$	$-\frac{931}{4}$	$\frac{14225}{8}$
$\tilde{q}_0$	-1	1	-2	5	-8	4	31	-147	362	-348	-1464
$\tilde{\nu}_2$	1	-2	1	0	8	-38	78	2	-579	2064	-3218
y_1	1	-1	0	1	-2	6	-13	11	58	-350	1022
λ	1	-4	8	-8	-4	24	-8	-132	404	-364	-1328

Table 6.2: Table of coefficients ($m = 5$).

n	0	1	2	3	4	5	6
$\tilde{p}_0$	0.25	−0.4167	0.27778	0.2592593	−1.802469	4.69135803	−8.482853
$\tilde{p}_1$	−1.25	0.8333	−0.5555	−0.962963	6.5679012	−18.271605	34.54595
$\tilde{q}_0$	−0.75	1.25	0.20833	−0.430556	5.4594907	−10.52662	27.585455
$\tilde{q}_1$	1	−1.6667	−0.2778	0.5740741	−6.291667	8.438787	−22.56936
$\tilde{\nu}_1$	8	−76.444	563.358	−3691.545	22832.071	−136732.03	803114.49
$\tilde{\nu}_2$	1	−4.556	23.309	−132.455	753.707	−4276.56	24258.808
$\tilde{y}_1$	0.25	0.8333	−0.5556	−0.074074	0.6419753	−0.4938272	0.175583
λ	1	−9	56	−320	1808	−10232.89	57991.111

Table 6.3: Table of coefficients ($m = 6$).

n	0	1	2	3	4	5	6	7
$\tilde{p}_0$	0.125	−0.5	−1	−14.375	69.5625	260.625	−9329.3758	79074.5625
$\tilde{p}_1$	−1	4	−19	65	168.5	−5777	59989	−205578.5
$\tilde{p}_2$	1	−4	19	−55	−313.5	6501	−56727	116764.5
$\tilde{q}_0$	0.5	−1.5	12	−73.5	137.5	3083.5	−45059	271037.5
$\tilde{q}_1$	−1.5	2	−16	98	−184.5	−4077	59699.5	−360356
$\tilde{\nu}_1$	1	−8	69	−658	6876	−69426	645626	−5756116
$\tilde{\nu}_2$	4	−44	472	−5052	53624	−536240	4950020	−43677308
y_1	1	−5	6	87	−946	8058	−38051	−242621
$\tilde{y}_2$	0.5	1	−6	27	10.5	−1687	19417.5	−86027
λ	1	−16	176	−1696	15792	−149792	1457984	−14280464

6.10.4 Studying of convergence radius

The minimal radius of convergence was studied with the help of computer calculations.

It proved that for $m = 4, 5$ this radius is close to the magnitude $\hat{z}(m)$, which is the unique positive solution of the boundary equation: $\hat{z}(4) \approx 0.381$, $\hat{z}(5) \approx 0.396$.

Thus for $m = 4, 5$ we have the full analytical solution of the problem.

For $m = 6$ the minimal radius of convergency is proved to be approximately equal to 0.1, whereas $\hat{z}(6) = 0.376$. Therefore to obtain the solution of the problem under the condition $0.1 < r < 0.376$ at least one reexpansion of the vector function $\tilde{\theta}(z)$ is needed at a vicinity of the point $z = 0.1$. Our calculations showed that no more than two reexpansions will be needed.

A theoretical study of the minimal radius as well as a generalization of the results obtained for the case of arbitrary nonsymmetrical segments the author intend to present in a next paper.

Note that for $m = 3$ and arbitrary segments E-optimal designs were constructed in Melas and Krylova (1998).

Acknowledgment. The present paper is partly supported by grant of RFBR 98-01-00342 and Russian University grant N 4233.

References

1. Ehrenfeld, E. (1955). On the efficiency of experimental design, *Annals of Math. Stat.*, **26**, 247–255.

2. Fedorov, V. V. (1972). *Theory of Optimal Experiments*, New York: Academic Press.

3. Heiligers, B. (1991). E-optimal polynomial regression designs Habilitationssrift, RWTH, Aahen, 89 p.

4. Karlin, S. and Studden, W. (1966). *Tchebysheff Systems: With Application in Analysis and Statistics*, New York: John Wiley & Sons.

5. Kovrigin, A. B. (1980). Construction of E-optimal designs, Vestnik Leningrad University, **19**, 120, Abstract.

6. Melas, V. B. (1982). A duality theorem and E-optimality, *Industrial Laboratory*, **48**, 295–296 (translated from Russian).

7. Melas, V. B. (1996). A study of E–optimal designs for polynomial regression In *Proceedings in computational statistics - 12th symposium held in Barcelona, Spain, 1996/COMPSTAT* (Ed., Albert Prat), pp. 101–110, Heidelberg: Physica-Verlag.

8. Melas, V. B. (1997). E-optimal experimental designs, St. Petersburg University Publishers (in Russian).

9. Melas, V. B. (1998a). Analytical theory of E-optimal designs for polynomial regression on a segment, In *MODA-5 – Advances in Model-Oriented Data Analysis and Experimental Design* (Ed., A.C. Atkinson, L. Pronzato and H.P. Wynn), pp. 51–58, Physica-Verlag.

10. Melas, V. B. (1998b). On the functional approach to E-optimal designs for polynomial regression on arbitrary segments In *Proceedings of the 3rd St. Petersburg workshop on simulation* (Eds., S.M. Ermakov, Y.N. Kashtanov and V.B. Melas), St. Petersburg, Russia: St.Petersburg University Press.

11. Melas, V. B. and Krylova, L. A. (1998). E-optimal designs for quadratic regression on arbitrary segments, *Vestnik Sankt-Peterburgskogo Universiteta*, ser. 1, No. 15, 44–49 (Russian).

12. Pukelsheim, F. (1980). On linear regression designs which maximize information, *Journal of Statistical Planning and Inference*, **4**, 339–364.

13. Pukelsheim, F. and Studden, W. (1993). E-optimal designs for polynomial regression, *Annals of Statistics*, **21**, No. 1, 402–415.

14. Titchmarsh, E. C. (1939). *The Theory of Functions*, Oxford University Press.

15. Van der Waerden, B .L. (1967). *Algebra*, Berlin: Springer-Verlag.

7

Bias Constrained Minimax Robust Designs for Misspecified Regression Models

Douglas P. Wiens

University of Alberta, Edmonton, AB, Canada

Abstract: We exhibit regression designs and weights which are robust against incorrectly specified regression responses and error heteroscedasticity. The approach is to minimize the maximum integrated mean squared error of the fitted values, subject to an unbiasedness constraint. The maxima are taken over broad classes of departures from the 'ideal' model. The methods yield particularly simple treatments of otherwise intractable design problems. This point is illustrated by applying these methods in a number of examples including polynomial and wavelet regression and extrapolation. The results apply to generalized M-estimation as well as to least squares estimation. Two open problems - one concerning designing for polynomial regression and the other concerning lack of fit testing - are given.

Keywords and phrases: Extrapolation, generalized M-estimation, lack of fit, Legendre polynomials, optimal design, polynomial regression, wavelets, weighted least squares

7.1 Introduction

In this article we synthesize some recent findings concerning the interplay between the choice of design points for regression based data analyses, and the weights used in weighted least squares or generalized M-estimation. The regression model which we envisage is described in Section 7.2, and is one for which the ordinary least squares estimates are biased due to uncertainty in the specification of the response function by the experimenter. Furthermore, we allow for the possibility of error heteroscedasticity of a very general form.

The experimenter seeks robustness in the form of protection from the bias and efficiency loss engendered by model misspecification. The recommendations made here are that these robustness issues be addressed both at the design stage and at the estimation stage. We exhibit designs and regression weights which minimize the maximum loss, with the maximum evaluated over broad classes of departures from the fitted response function and from homoscedasticity, subject to a side condition of unbiasedness. The designs also leave degrees of freedom to allow for the exploration of regression responses other than the one initially fitted.

An appealing consequence of our methods is that some otherwise very intractable design problems become amenable to simple remedies. This latter point is illustrated by applying these methods in a number of examples. The case of multiple linear regression, over ellipsoidal or rectangular design spaces, is addressed in Section 7.3. Section 7.4 covers polynomial regression, a particularly difficult problem in robust design theory. Wavelet regression is treated in Section 7.5; in Section 7.6 we consider designs for the extrapolation of the regression estimates beyond the design space. In Section 7.7 we discuss post-design inference, in particular lack of fit testing. Section 7.8 treats generalized M-estimation. Two intriguing open problems are given - one in Section 7.4 and the other in Section 7.7.

7.2 General Theory

We suppose that the experimenter is to take n uncorrelated observations on a random variable Y whose mean is thought to vary in an approximately linear manner with p-dimensional regressors $\mathbf{z}(\mathbf{x})$: $E[Y|\mathbf{x}] \approx \mathbf{z}^T(\mathbf{x})\theta$. The sites $\mathbf{x}_i$ are chosen from $\mathcal{S}$, a Euclidean design space with finite volume defined by $\int_\mathcal{S} d\mathbf{x} = \Omega^{-1}$. We define the "true" value of θ by requiring the linear approximation to be most accurate in the L^2-sense:

$$\theta := \arg\min_{\mathbf{t}} \int_\mathcal{S} (E[Y|\mathbf{x}] - \mathbf{z}^T(\mathbf{x})\mathbf{t})^2 d\mathbf{x}. \qquad (7.1)$$

We then define $f(\mathbf{x}) = E[Y|\mathbf{x}] - \mathbf{z}^T(\mathbf{x})\theta$ and $\epsilon(\mathbf{x}) = Y(\mathbf{x}) - E[Y|\mathbf{x}]$; these definitions together with (7.1) imply that

$$\text{(i) } Y(\mathbf{x}) = \mathbf{z}^T(\mathbf{x})\theta + f(\mathbf{x}) + \epsilon(\mathbf{x}), \text{ (ii) } \int_\mathcal{S} \mathbf{z}(\mathbf{x}) f(\mathbf{x}) d\mathbf{x} = \mathbf{0}. \qquad (7.2)$$

We allow for the possibility that the variance of $\epsilon(\mathbf{x})$ is proportional to a function $g(\mathbf{x})$, with g varying over some class $\mathcal{G}$: $\text{VAR}[\varepsilon(\mathbf{x})] = \sigma^2 g(\mathbf{x})$. We take the bound

$$\int_\mathcal{S} g^2(\mathbf{x}) d\mathbf{x} \leq \Omega^{-1}, \qquad (7.3)$$

implying that $\sigma^2 = sup_{g \in \mathcal{G}} \left(\int_{\mathcal{S}} \text{VAR}^2[\epsilon(\mathbf{x})]\Omega d\mathbf{x} \right)^{1/2}$. Note that (7.3) allows for homoscedastic errors $\mathcal{G} = \{\mathbf{1}\}$, where $\mathbf{1}(\mathbf{x}) \equiv 1$.

We require a further condition in order that errors due to bias not swamp those due to variance. In keeping with the L^2-nature of (7.1), we assume that

$$\int_{\mathcal{S}} f^2(\mathbf{x})d\mathbf{x} \leq \eta^2, \tag{7.4}$$

for some positive constant η^2. Neither σ^2 nor η^2 need be known to the experimenter in order for our results to be applied.

The balancing of bias and variance can be achieved in other ways than (7.4). Pesotchinsky (1982) and Li and Notz (1982) among others take

$$|f(\mathbf{x})| \leq \phi(\mathbf{x}) \tag{7.5}$$

for a known function $\phi(\mathbf{x})$ satisfying various assumptions. An apparently unavoidable consequence of this rather thin neighbourhood structure is that the resulting 'robust' designs have their mass concentrated at a small number of extreme points of the design space, and thus afford little opportunity to test the model for lack of fit or to fit alternate models. The neighbourhoods implied by (7.4) can on the other hand be viewed as somewhat broad, since for them any design with finite maximum loss must be absolutely continuous, hence must be approximated at the implementation stage. For details and discussion see Wiens (1992), where the conclusion is reached that *"Our attitude is that an approximation to a design which is robust against more realistic alternatives is preferable to an exact solution in a neighbourhood which is unrealistically sparse."* Indeed, simulation studies carried out to compare the implementations of the continuous designs with some common competitors have consistently shown these designs to be very successful at reducing the mean squared error, against realistic departures which are sufficiently large to destroy the performance of the classical procedures yet small enough to be generally undetectable by the usual tests. See Wiens (1998, 1999) and Fang and Wiens (1999).

We propose to estimate θ by least squares, possibly weighted with nonnegative weights $w(\mathbf{x})$. Let ξ be the design measure, i.e. the distribution placing mass n^{-1} at each of $\{\mathbf{x}_1, \ldots, \mathbf{x}_n\}$, and denote by $K(\mathbf{x})$ the corresponding distribution function. Define matrices $\mathbf{A}, \mathbf{B}, \mathbf{D}$ and a vector $\mathbf{b}$ by

$$\mathbf{A} = \int_{\mathcal{S}} \mathbf{z}(\mathbf{x})\mathbf{z}^T(\mathbf{x})d\mathbf{x}, \qquad \mathbf{B} = \int_{\mathcal{S}} w(\mathbf{x})\mathbf{z}(\mathbf{x})\mathbf{z}^T(\mathbf{x})\xi(d\mathbf{x}),$$

$$\mathbf{D} = \int_{\mathcal{S}} w^2(\mathbf{x})g(\mathbf{x})\mathbf{z}(\mathbf{x})\mathbf{z}^T(\mathbf{x})\xi(d\mathbf{x}), \quad \mathbf{b} = \int_{\mathcal{S}} w(\mathbf{x})\mathbf{z}(\mathbf{x})f(\mathbf{x})\xi(d\mathbf{x}).$$

In a more familiar regression notation these are

$$\mathbf{B} = n^{-1}\mathbf{Z}^T\mathbf{W}\mathbf{Z} \quad \text{and} \quad \mathbf{D} = n^{-1}\mathbf{Z}^T\mathbf{W}\mathbf{G}\mathbf{W}\mathbf{Z},$$

where $\mathbf{Z}$ is the $n \times p$ model matrix with rows $\mathbf{z}^T(\mathbf{x}_i)$ and $\mathbf{W}, \mathbf{G}$ are the $n \times n$ diagonal matrices with diagonal elements $w(\mathbf{x}_i)$ and $g(\mathbf{x}_i)$ respectively. The motivation for writing these quantities as integrals with respect to ξ will become apparent below, where we broaden the class of allowable design measures to include continuous designs. Note also that although it is mathematically convenient to treat ξ as a probability measure, we do so only in the formal sense of a non-negative measure with a total mass of unity - there is no implication that the $\mathbf{x}_i$ are measured with error.

Assume that $\mathbf{A}$ and $\mathbf{B}$ are nonsingular. The mean vector and covariance matrix of the weighted least squares estimate $\hat{\theta} = \mathbf{B}^{-1} \cdot n^{-1} \mathbf{Z}^T \mathbf{W} \mathbf{Y}$ (where $\mathbf{Y}$ is the data vector) are

$$E[\hat{\theta}] - \theta = \mathbf{B}^{-1} \mathbf{b}, \quad COV[\hat{\theta}] = \frac{\sigma^2}{n} \mathbf{B}^{-1} \mathbf{D} \mathbf{B}^{-1}. \tag{7.6}$$

We estimate $E[Y|\mathbf{x}]$ by $\hat{Y}(\mathbf{x}) = \mathbf{z}^T(\mathbf{x})\hat{\theta}$, and consider the resulting integrated mean squared error: $IMSE = \int_\mathcal{S} E[(\hat{Y}(\mathbf{x}) - E[Y|\mathbf{x}])^2] d\mathbf{x}$. This splits into terms due solely to estimation bias, estimation variance, and model misspecification:

$$IMSE(f, g, w, \xi) = ISB(f, w, \xi) + IV(g, w, \xi) + \int_\mathcal{S} f^2(\mathbf{x}),$$

where, with $\mathbf{H} = \mathbf{B}^{-1} \mathbf{A} \mathbf{B}^{-1}$, the Integrated Squared Bias (ISB) and Integrated Variance (IV) are

$$ISB(f, w, \xi) = \int_\mathcal{S} \left(E[\hat{Y}(\mathbf{x})] - \mathbf{z}^T(\mathbf{x})\theta \right)^2 d\mathbf{x} = \mathbf{b}^T \mathbf{H} \mathbf{b},$$

$$IV(g, w, \xi) = \int_\mathcal{S} VAR[\hat{Y}(\mathbf{x})] d\mathbf{x} = \frac{\sigma^2}{n} \cdot trace(\mathbf{H}\mathbf{D}).$$

We adopt the viewpoint of *approximate* design theory, and allow as a design measure ξ any distribution on $\mathcal{S}$. It can be shown - a formal proof can be based on that of Lemma 1 of Wiens (1992) - that if either of $\sup_f ISB(f, w, \xi)$ or $\sup_g IV(g, w, \xi)$ is to be finite, then ξ must necessarily be absolutely continuous. The resulting designs may be approximated and implemented by placing the design points at appropriately chosen quantiles of $K(\mathbf{x})$.

We say that the pair (ξ, w) is *unbiased* if it satisfies $\sup_f ISB(f, w, \xi) = 0$. Equivalently, $E[\hat{\theta}] = \theta$ for all f. The pair is *minimum variance unbiased* (MVU) if it minimizes $\sup_{f,g} IMSE(f, g, w, \xi)$ subject to being unbiased. The following theorem, which can be established as in Wiens (1998) by standard variational methods, gives a necessary and sufficient condition for unbiasedness, and minimax weights. Before stating it we require some definitions. Let $k(\mathbf{x}) = K'(\mathbf{x})$ be the design density, and define $m(\mathbf{x}) = k(\mathbf{x})w(\mathbf{x})$. Assume, without loss of generality, that the average weight $\int_\mathcal{S} w(\mathbf{x})\xi(d\mathbf{x})$ is 1. Then m is a density on $\mathcal{S}$ and each of $\mathbf{b}, \mathbf{B}$ depends on (k, w) only through m. Rather than optimize over (k, w) we may optimize over (m, w) subject to the constraint $\int_S m(\mathbf{x})/w(\mathbf{x})d\mathbf{x} = 1$. Define also $l_m(\mathbf{x}) = \mathbf{z}^T(\mathbf{x})\mathbf{H}\mathbf{z}(\mathbf{x})$.

Theorem 7.2.1 *a) The pair (ξ, w) is unbiased iff $m(\mathbf{x}) \equiv \Omega$.*
b) For fixed $m(\mathbf{x})$, maximum Integrated Variance is

$$\sup_g IV(g, w, \xi) = \frac{\sigma^2}{n} \Omega^{-1/2} \left(\int_S (w(\mathbf{x})l_m(\mathbf{x})m(\mathbf{x}))^2 d\mathbf{x} \right)^{1/2},$$

attained at the least favourable variance function $g_{m,w}(\mathbf{x}) \propto w(\mathbf{x})l_m(\mathbf{x})m(\mathbf{x})$. Maximum IV is minimized by weights $w_m(\mathbf{x}) \propto (l_m^2(\mathbf{x})m(\mathbf{x}))^{-1/3} I(m(\mathbf{x}) > 0)$.
c) MVU designs and weights (ξ_*, w_*) for heteroscedastic errors are given by

$$k_*(\mathbf{x}) = \frac{(\mathbf{z}(\mathbf{x})^T \mathbf{A}^{-1} \mathbf{z}(\mathbf{x}))^{\frac{2}{3}}}{\int_S (\mathbf{z}(\mathbf{x})^T \mathbf{A}^{-1} \mathbf{z}(\mathbf{x}))^{\frac{2}{3}} d\mathbf{x}}, \quad (7.7)$$

$$w_*(\mathbf{x}) = \Omega/k_*(\mathbf{x}). \quad (7.8)$$

The least favourable variances satisfy $g_*(\mathbf{x}) = w_*(\mathbf{x})^{-1/2}$. If the errors are homoscedastic ($\mathcal{G} = \{1\}$) the exponents $2/3$ in (7.7) are replaced by $1/2$.

Part c) of Theorem 7.2.1 is an immediate consequence of parts a) and b), since $m(\mathbf{x}) \equiv \Omega$ implies that $l_m(\mathbf{x}) \propto \mathbf{z}(\mathbf{x})^T \mathbf{A}^{-1} \mathbf{z}(\mathbf{x})$. Note that under heteroscedasticity the minimax weights $w_*(\mathbf{x})$ are equal to $g_*(\mathbf{x})^{-2}$; recall that if $g(\mathbf{x})$ is *known* then the *efficient* weights are proportional to $g(\mathbf{x})^{-1}$.

In our consideration of the special cases in the following sections, we will take mathematically convenient, canonical forms of the design spaces. This is justified in each case by the following lemma.

Lemma 7.2.2 *Suppose that the variables $\mathbf{x} \in S$ are subjected to an affine transformation $\mathbf{x} \to \mathbf{M}\mathbf{x} + \mathbf{c} =: \tilde{\mathbf{x}}$ with $\mathbf{M}$ nonsingular, that $\tilde{S} := \{\tilde{\mathbf{x}} | \mathbf{x} \in S\}$ and that the regressors are equivariant in that $\mathbf{z}(\tilde{\mathbf{x}}) = \mathbf{P}\mathbf{z}(\mathbf{x})$ for some nonsingular matrix $\mathbf{P}$. If (ξ_*, w_*) are MVU for the design problem with regressors $\mathbf{z}(\mathbf{x})$, $\mathbf{x} \in S$ then the induced design $\tilde{\xi}_*$ with distribution function $\tilde{K}_*(\tilde{\mathbf{x}}) = K_*(\mathbf{M}^{-1}(\tilde{\mathbf{x}} - \mathbf{c}))$, and weights $\tilde{w}_*(\tilde{\mathbf{x}}) = w_*(\mathbf{M}^{-1}(\tilde{\mathbf{x}} - \mathbf{c}))$, are MVU for the design problem with regressors $\mathbf{z}(\tilde{\mathbf{x}})$, $\tilde{\mathbf{x}} \in \tilde{S}$.*

Lemma 7.2.2 is established by checking that (7.7) and (7.8) hold when applied to $(\tilde{K}_*, \tilde{w}_*)$.

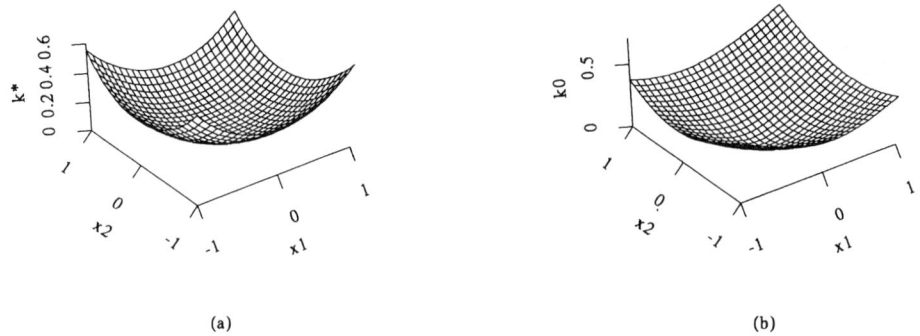

Figure 7.1: MVU design densities over the square $\mathcal{S} = [-1, 1] \times [-1, 1]$, robust against heteroscedasticity as well as response uncertainty. (a) Design for estimating a second order response. (b) Design for extrapolation of a first order fit to $\mathcal{T} = \{[-1, 2] \times [-1, 2]\} \setminus \mathcal{S}$.

7.3 Fitting a Second Order Response in Several Regressors

The following two examples illustrate the ease with which the MVU approach can be applied to otherwise quite intractable problems. Unconstrained, i.e. without the requirement of unbiasedness, minimax designs were obtained by Wiens (1990) for a greatly simplified version of the model in Section 7.3.2: $\mathbf{z}(\mathbf{x}) = (1, x_1, x_2, x_1 x_2)$, $\mathcal{S} = [-1, 1] \times [-1, 1]$, $\mathcal{G} = \{1\}$; even then the solution was very complex.

7.3.1 $\mathcal{S}$ an ellipsoid

Let $\mathbf{x} = (x_1, ..., x_q)^T$. Suppose that the experimenter anticipates fitting a full second order model, so that $\mathbf{z}(\mathbf{x})$ contains the elements $1, x_i, x_i^2, x_i x_j$ ($1 \leq i < j \leq q$), and that $\mathcal{S}$ is a q-dimensional ellipsoid. By virtue of Lemma 7.2.2 we may assume that $\mathcal{S} = \{\mathbf{x} | \, ||\mathbf{x}|| \leq 1\}$. We then find that the MVU design has density

$$k_*(\mathbf{x}) \propto \left(1 + \frac{||\mathbf{x}||^2}{\mu_1} + \left(\sum_{i=1}^{q} x_i^4\right)\left(\frac{1}{\mu_2} - \frac{1}{2\mu_3}\right) + ||\mathbf{x}||^4 \left(\frac{1}{2\mu_3} - \frac{\mu_1}{\mu_2(\mu_2 + q\mu_1)}\right)\right)^{\frac{2}{3}}, \quad (7.9)$$

where

$$\mu_1 = \Omega \int_S x_1^2 d\mathbf{x} = \frac{1}{q+2},$$
$$\mu_2 = \Omega \int_S x_1^4 d\mathbf{x} = \frac{3}{(q+2)(q+4)},$$
$$\mu_3 = \Omega \int_S x_1^2 x_2^2 d\mathbf{x} = \frac{1}{(q+2)(q+4)}.$$

7.3.2 S a q-dimensional rectangle

If $\mathbf{x}$ is as in Section 7.3.1 but S is a q-dimensional rectangle, which we may take as $\{\mathbf{x}|\ |x_i| \leq 1,\ i = 1,...,q\}$ by appealing to Lemma 7.2.2, then $k_*(\mathbf{x})$ is as in (7.9) with $\mu_1 = 1/3, \mu_2 = 1/5, \mu_3 = 1/9$. See Figure 7.3(a) for a plot of $k_*(\mathbf{x})$ with $q = 2$.

7.4 Fitting a Polynomial Response

The problem of designing for a polynomial fit, when it is assumed that the fitted polynomial form is exactly correct, has a long and rich history - see Pukelsheim (1993). Considerations of robustness against incorrectly specified response functions have entered the literature relatively recently. Stigler (1971) proposed a robustness criterion of designing for a polynomial fit of particular degree, whilst maintaining a lower bound on the efficiency if the true response is a higher order polynomial of fixed maximum degree. He gave particular solutions in the case of a linear fit with a quadratic alternative. Studden (1982) characterized this problem in terms of canonical moments and gave solutions for a linear fit against an alternative of arbitrary maximum degree. More recent approaches have focussed on minimizing a scalar-valued function of the covariance matrix of the estimates of the coefficients of the fitted polynomial, subject to a bound on the bias under alternative models. One can also seek to minimize bias while bounding variance. See Montepiedra and Fedorov (1997) for applications of this approach with polynomial alternatives and Liu and Wiens (1997) for arbitrary alternatives in neighbourhoods defined by (7.5).

Unconstrained minimax solutions for models given by (7.2) and (7.4), with $\mathbf{z}(x) = (1, x, x^2, \ldots x^q)^T$, have been notoriously elusive. It is a straightforward matter to maximize the loss over f - see Theorem 1 of Wiens (1992). However, even in the case $q = 2$ the problem of finding a design to minimize the resulting maximum loss remains unsolved. For $q = 1$ the solution was obtained by Huber (1975). Rychlik (1987) addressed concerns over the required continuity, hence lack of implementability, of Huber's (1975) design by proposing that one

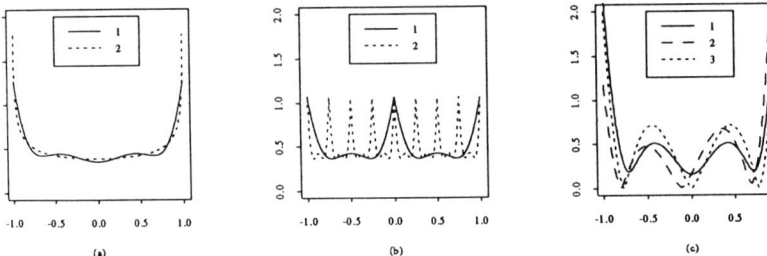

Figure 7.2: MVU design densities on $[-1, 1]$, robust against heteroscedasticity as well as response uncertainty. (a) Polynomial regression. 1: $k_*(x, q = 3)$; 2: $k_*(x, \infty)$. (b) Wavelet regression. 1: $k_{3,0}(x)$; 2: $k_{3,2}(x)$. (c) Extrapolation of a cubic fit. 1: $k_0(x, t_0 = 1.5)$, symmetric extrapolation; 2: $k_0(x, t_1 = 1.5)$, asymmetric extrapolation; 3: $k_0(x, \infty)$.

require f to be a polynomial of fixed maximum degree, but to otherwise satisfy (7.2) and (7.4). For IMSE loss he showed that, for large values of η, any design is minimax as long as its moments agree with those of Huber's design, up to a certain order. Thus he was able to construct discrete minimax designs. Heo (1998) extended these results to determinant loss.

In contrast, the MVU designs and weights are easily evaluated, for polynomial fits of any degree. By Lemma 7.2.2 we may take $\mathcal{S} = [-1, 1]$. For q^{th} degree polynomial regression we write the optimal design density k_* as $k_*(x; q)$. This density turns out to be most conveniently expressed in terms of orthogonal polynomials, and this leads to an interesting connection to the classical D-optimal design ξ_D, i.e. the discrete measure minimizing the determinant of the covariance matrix of the ordinary least squares estimate $\hat{\theta}$. In the following result we denote by $P_q(x)$ the qth degree Legendre polynomial on $\mathcal{S}$, normalized by $\int_{-1}^{1} P_q^2(x) dx = (q + .5)^{-1}$.

Theorem 7.4.1 *Define a density on $[-1, 1]$ by $h_q(x) = (q+1)^{-1}\mathbf{z}^T(x)\mathbf{A}^{-1}\mathbf{z}(x)$. Then the design density $k_*(x; q)$, MVU for polynomial regression with heteroscedastic errors, satisfies*

$$k_*(x;q) \propto h_q(x)^{\frac{2}{3}} = .5(P_q(x)P'_{q+1}(x) - P'_q(x)P_{q+1}(x))^{\frac{2}{3}}.$$

See Wiens (1998) for a proof. The exponents $2/3$ are replaced by $1/2$ for homoscedastic errors.

It can be shown that the local maxima of $h_q(x)$, hence those of $k_*(x; q)$, are the zeros of $(1 - x^2)P'_q(x)$. But these are precisely the points of support of ξ_D. In this sense, $k_*(\cdot; q)$ is a smoothed version of ξ_D, which has the limiting density $(1-x^2)^{-1/2}/\pi = \lim_{q\to\infty} h_q(x)$. The limiting MVU density is

$$k_*(x;\infty) = \frac{(1-x^2)^{-\frac{1}{3}}}{2^{\frac{1}{3}}\beta(\frac{2}{3},\frac{2}{3})}.$$

See Figure 7.4(a) for plots of $k_*(x;3)$ and $k_*(x;\infty)$.

Placing designs points at the modes of $k_*(x;q)$ would recover the classically optimal but non-robust design ξ_D. We recommend placing design points at the n quantiles $x_i = K_*^{-1}\left(\frac{i-1}{n-1};q\right)$, $i = 1, ..., n$. Of course replication at a smaller number of locations is also an option, and is the subject of current research.

7.5 Wavelet Regression

The flexibility of wavelets in function representation methods has in recent years stimulated interest in wavelet approximations of regression response functions for the analysis of experimental data - see e.g. Antoniadis, Gregoire and McKeague (1994) and Benedetto and Frazier (1994). A primary attraction of wavelets in regression is their ability to approximate response functions which lack the smoothness properties of, e.g., polynomial responses. For instance if one assumes only that the response function is square integrable, with no further smoothness assumptions, then a *Haar* wavelet approximation is suitable. When the response is smoother, but not necessarily continuous, then a *multi-wavelet* (Alpert 1992) approximation may be more appropriate. The class of multiwavelets contains the Haar wavelets.

Related design questions have arisen concurrently. For models in which the response function can be represented exactly as a linear combination of Haar wavelets see Herzberg and Traves (1994) with extensions by Xie (1998). Both unconstrained and constrained minimax design problems for wavelet approximations of the response function have been studied by Oyet (1997, 1998). Here we summarize the constrained solutions; details may be found in Oyet (1997) and Oyet and Wiens (1999).

Take $\mathcal{S} = [0,1)$ and assume that the regression response $E[Y|x]$ is in the space $L^2(\mathcal{S})$ of square integrable functions, so that it may be approximated arbitrarily closely by linear combinations of multiwavelets. Denote by $\mathbf{z}_{N,m}(x)$ the $N \cdot 2^{m+1} \times 1$ vector consisting of the wavelets $\{\phi_l(x),\ _N\nu_l^{-j,k}(x) \mid j = 0,...,m,\ k = 0,...,2^j-1,\ l = 0,...,N-1\}$ in some order. The elements of $\mathbf{z}_{N,m}$ form an orthogonal basis for $L^2(\mathcal{S})$ as $m \to \infty$. When $N = 1$ they coincide with the Haar wavelets; in general they are described as follows. Recall from Section 7.4 that P_l denotes a normalized Legendre polynomial. In this notation $\phi_l(x) = \sqrt{2l+1}P_l(2x-1)I_{[0,1)}(x)$; also $_N\nu_l^{-j,k}(x) = 2^{j/2}\,_N\nu_l(\{2^j x\})I([2^j x] = k)$, where

$\{x\} = x - [x]$ denotes the *fractional* part of x. The *primary wavelets* $_N\nu_l$ can in turn be developed recursively; examples are $_1\nu_0(x) = I_{[0,1/2)}(x) - I_{[1/2,1)}(x)$ and

$$\begin{aligned} _2\nu_0(x) &= \sqrt{3}(4|x-1/2|-1) \cdot I_{[0,1)}(x), \\ _2\nu_1(x) &= 2(1-3|x-1/2|) \cdot (I_{[0,1/2)}(x) - I_{[1/2,1)}(x)). \end{aligned}$$

By virtue of the orthogonality of $\mathbf{z}_{N,m}(x)$, Theorem 7.2.1c) can be reduced to a particularly simple form in this case.

Theorem 7.5.1 *(Oyet and Wiens 1999) For the multiwavelet approximation the MVU design density $k_{N,m}(x)$ for homoscedastic errors is*

$$\kappa_N \cdot \left[\phi_{N-1}(\{2^{m+1}x\})\phi'_N(\{2^{m+1}x\}) - \phi'_{N-1}(\{2^{m+1}x\})\phi_N(\{2^{m+1}x\}) \right]^{\frac{1}{2}},$$

where the normalizing constant is

$$\kappa_N = \left(\int_0^1 [\phi_{N-1}(x)\phi'_N(x) - \phi'_{N-1}(x)\phi_N(x)]^{\frac{1}{2}} dx \right)^{-1}.$$

The exponents $1/2$ are replaced by $2/3$ for heteroscedastic errors. The limiting density (for homoscedasticity) is

$$k_{\infty,m}(x) = \frac{\{2^{m+1}x\}^{-1/4}(1 - \{2^{m+1}x\})^{-1/4}}{\beta(\frac{3}{4}, \frac{3}{4})}.$$

A comparison with Theorem 7.4.1 reveals that $k_{N,m}(x)$ is a scaled and dilated copy of $k_*(x; N-1)$, extended periodically over $\mathcal{S}$ with period $length(\mathcal{S})/2^{m+1}$. Some particular cases are $k_{2,m}(x) = 2.51 \cdot [(\{2^{m+1}x\} - 1/2)^2 + 1/12]^{1/2}$ and $k_{3,m}(x) = 8.00 \cdot [((\{2^{m+1}x\} - 1/2)^2 - 1/20)^2 + 1/100]^{1/2}$. See Figure 7.4(b) for plots of $k_{3,0}(x)$ and $k_{3,2}(x)$ for heteroscedasticity, scaled to $x \in \mathcal{S} = [-1,1)$ for purposes of comparison with the densities in (a) and (c).

7.6 Extrapolation Designs

Spruill (1984) and Dette and Wong (1996) constructed extrapolation designs for polynomial regression, robust against various misspecifications of the degree of the polynomial. Draper and Herzberg (1973) extended the methods of Box and Draper (1959) to extrapolation under response uncertainty. In their approach one estimates a first order model but designs with the possibility of a second order model in mind; the goal is extrapolation to one fixed point outside of the spherical design space. Huber (1975) obtained designs for extrapolation of a

Bias Constrained Minimax Robust Designs

response, assumed to have a bounded derivative of a certain order but to be otherwise arbitrary, to one point outside of the design interval. These results were corrected and extended by Huang and Studden (1988).

If the goal is extrapolation to a *range* of values outside of $\mathcal{S}$, robust against broad classes of alternate models, then the conclusions of Theorem 7.2.1 apply with only minor changes and yield easily implemented procedures. Suppose that one is to extrapolate the estimates of the regression response to a region $\mathcal{T}$ disjoint from $\mathcal{S}$, with (7.2)(i) holding on $\mathcal{T} \cup \mathcal{S}$. In the expressions of Section 7.2 replace the range $\mathcal{S}$ of the integrals defining IMSE, ISB and IV by $\mathcal{T}$, obtaining in this way integrated mean squared *prediction* error, etc. Redefine $\mathbf{H}$ to be $\mathbf{B}^{-1}\mathbf{A}_{\mathcal{T}}\mathbf{B}^{-1}$, where $\mathbf{A}_{\mathcal{T}} = \int_{\mathcal{T}} \mathbf{z}(\mathbf{x})\mathbf{z}^T(\mathbf{x})d\mathbf{x}$. Then Theorem 7.2.1a),b) apply to the minimization of the resulting IMPSE, subject to $\sup IPB = 0$. The minimax design density under heteroscedasticity is (Fang and Wiens 1999)

$$k_0(\mathbf{x}) = \frac{(\mathbf{z}(\mathbf{x})^T \mathbf{A}^{-1} \mathbf{A}_{\mathcal{T}} \mathbf{A}^{-1} \mathbf{z}(\mathbf{x}))^{\frac{2}{3}}}{\int_{\mathcal{S}} (\mathbf{z}(\mathbf{x})^T \mathbf{A}^{-1} \mathbf{A}_{\mathcal{T}} \mathbf{A}^{-1} \mathbf{z}(\mathbf{x}))^{\frac{2}{3}} d\mathbf{x}} \tag{7.10}$$

(the exponents are 1/2 for homoscedastic errors) and correspondingly optimal weights are $w_0(\mathbf{x}) = \Omega/k_0(\mathbf{x})$.

7.6.1 Extrapolation of a polynomial fit

For q^{th} degree polynomial regression we find that

$$k_0(x) \propto \left(\sum_{i,j} \alpha_{ij} P_i(x) P_j(x) \right)^{2/3}, \quad x \in \mathcal{S} = [-1, 1],$$

where $\alpha_{ij} = (i + .5)(j + .5) \int_{\mathcal{T}} P_i(x) P_j(x) dx$ for $0 \leq i, j \leq q$. For example, for quadratic regression and a symmetric extrapolation region, i.e. $\mathcal{T} = [-t_0, t_0] \setminus \mathcal{S}$, we find that $k_0(x) = k_0(x; t_0)$ is given by

$$k_0(x; t_0) \propto [5t_0^3(t_0+1)(3x^2-1)^2 - t_0(t_0+1)(5x^4-22x^2+5)+4(1-2x^2+5x^4)]^{\frac{2}{3}},$$

(an even function of x) while for one-sided extrapolation ($\mathcal{T} = [1, t_1]$) we find that

$$k_0(x; t_1) \propto [5t_1^4(3x^2-1)^2 + 5t_1^3(3x-1)(x+1)(3x^2-1) - t_1^2(5x^4-30x^3-22x^2$$
$$+10x+5) - t_1(x+1)(5x^3-15x^2-7x+5) + 2(10x^4+5x^3-4x^2+x+2)]^{\frac{2}{3}}.$$

For both types of extrapolation region and for arbitrary but fixed q, $k_0(x; \infty) \propto \left(P_q^2(x)\right)^{2/3}$. See Figure 7.4(c) for plots of $k_0(x; 1.5)$ in both the symmetric and asymmetric cases, and of $k_0(x; \infty)$.

7.6.2 Extrapolation of a first order response in several variables

S an ellipsoid

Suppose that S is as in Section 7.3.1 and has been transformed to a sphere of unit radius, that T is, after this transformation, the annulus $\{\mathbf{x} | 1 < ||\mathbf{x}|| \leq t_2\}$ and that $\mathbf{z}(\mathbf{x}) = (1, x_1, ..., x_q)^T$. Evaluating (7.10) gives the optimal unbiased extrapolation design density

$$k_0(\mathbf{x}, t_2) \propto \left\{ 1 + (q+2) \frac{t_2^{q+2} - 1}{t_2^q - 1} ||\mathbf{x}||^2 \right\}^{2/3}.$$

S a q-dimensional rectangle

Suppose that again $\mathbf{z}(\mathbf{x}) = (1, x_1, ..., x_q)^T$ but that S is the q-dimensional cube $[-1, 1]^q$, and that the extrapolation region is the possibly asymmetric perimeter $T = [-t_3, t_4]^q \backslash S$, where $t_3, t_4 \geq 1$. One of t_3, t_4 may be unity, for one-sided extrapolation. We find that

$$k_0(\mathbf{x}, \mathbf{t}) \propto \left\{ \left(1 + 3\mu_1 \sum_{i=1}^{q} x_i \right)^2 + 9 \left(\mu_2 - \frac{1}{3\mu_3^q} \right) ||\mathbf{x}||^2 - \frac{1}{\mu_3^q} \right\}^{2/3}, \quad (7.11)$$

where $\mu_1 = (t_4 - t_3)/2$, $\mu_2 = (t_4 + t_3)^2/12$ and $\mu_3 = (t_4 + t_3)/2$. For symmetric extrapolation $t_3 = t_4$ and $\mu_1 = 0$.

See Figure 7.3(b) for a plot of $k_0(\mathbf{x})$ of (7.11) with $t_3 = 1$, $t_4 = 2$.

7.7 Lack of Fit Testing

If weighted least squares is used for reasons other than to address heteroscedasticity, then degrees of freedom are lost in estimating the error variance. The following result is Theorem 3.2 of Wiens (1999).

Lemma 7.7.1 *Suppose that the data obey the linear model $\mathbf{Y} = \mathbf{Z}\theta + \varepsilon$, where $\mathbf{Z}_{n \times p}$ has rank p and the elements of ε are uncorrelated random errors with mean 0 and variance σ^2. The weighted least squares estimate $\hat{\theta} = \left(\mathbf{Z}^T \mathbf{W} \mathbf{Z} \right)^{-1} \mathbf{Z}^T \mathbf{W} \mathbf{Y}$ has mean θ and covariance matrix $\sigma^2 \mathbf{C}$, where*

$$\mathbf{C} = \left(\mathbf{Z}^T \mathbf{W} \mathbf{Z} \right)^{-1} \left(\mathbf{Z}^T \mathbf{W}^2 \mathbf{Z} \right) \left(\mathbf{Z}^T \mathbf{W} \mathbf{Z} \right)^{-1}$$

and $\mathbf{W}$ is the diagonal matrix of regression weights. Let $\mathbf{P_V}$ be the projector onto the column space of $\mathbf{V} := (\mathbf{Z} \vdots \mathbf{WZ})$ and denote the rank of $\mathbf{P_V}$ by r.

Then $S^2 = ||(\mathbf{I} - \mathbf{P_V})\mathbf{Y}||^2/(n-r)$ is an unbiased estimate of σ^2. The vector $(\mathbf{I} - \mathbf{P_V})\mathbf{Y}$ is uncorrelated with $\hat{\theta}$. If the errors are normally distributed then $(n-r)S^2 \sim \sigma^2 \chi^2_{n-r}$, independently of $\hat{\theta} \sim \mathbf{N}(\theta, \sigma^2\mathbf{C})$.

The projector $\mathbf{P_V}$ will typically have rank $r = 2p$ when the weights are non-constant, so that p degrees of freedom are lost in the estimate of σ^2, relative to ordinary least squares. Note that S^2 is easily obtained as the mean square of the residuals in a regression of $\mathbf{Y}$ on the columns of $\mathbf{V}$. Inferences on linear functions of θ are then carried out in the usual way, with the required change in the degrees of freedom.

As an example, consider the standard test for lack of fit, based upon groups of replicated observations and assuming the errors to be homoscedastic. The experimenter takes n_i observations Y_{ij} at each of locations $\mathbf{x}_i$, $i = 1, ..., c$. He computes the Pure Error estimate of σ^2: $S^2_{PE} = \sum (n_i - 1) S^2_i/(n-c)$, where S^2_i is the sample variance of $\{Y_{i,1}, ..., Y_{i,n_i}\}$, and computes also the regression estimate S^2 of Lemma 7.7.1. Then the test of the hypothesis that $f \equiv 0$ in (7.2) consists of rejecting for large values of

$$T = \frac{(n-r)S^2 - (n-c)S^2_{PE}}{(c-r)S^2_{PE}}.$$

Under the null hypothesis both estimates of σ^2 are unbiased. Under (7.2) S^2_{PE} is unbiased but S^2 is positively biased, with bias $E[S^2 - \sigma^2] = ||(\mathbf{I} - \mathbf{P_V})\mathbf{f}||^2/(n-r)$. Here $\mathbf{f} = (f(\mathbf{x}_1), ..., f(\mathbf{x}_1), ..., f(\mathbf{x}_c), ..., f(\mathbf{x}_c))^T$ with $f(\mathbf{x}_i)$ appearing n_i times. If the errors are normal then T has a non-central F^{c-r}_{n-c} distribution (central under the null hypothesis). In this case the power of the test is an increasing function of the non-centrality parameter, which is in any event an intuitive measure of the quality of the procedure. This non-centrality parameter is $\lambda^2 = n\mathcal{B}(f, w, \xi)/\sigma^2$, where $\mathcal{B}(f, w, \xi) = n^{-1}||(\mathbf{I} - \mathbf{P_V})\mathbf{f}||^2$. A somewhat more informative expression is determined as follows. Let ξ be the design measure placing mass n_i/n at $\mathbf{x}_i$ and define r-vectors $\mathbf{q}_w(\mathbf{x})$ and $\mathbf{d}_{f,w,\xi}$, and an $r \times r$ matrix $\mathbf{Q}$ by

$$\mathbf{q}_w(\mathbf{x}) = \left(\mathbf{z}^T(\mathbf{x}), \mathbf{z}^T(\mathbf{x})w(\mathbf{x})\right)^T,$$

$$\mathbf{d}_{f,w,\xi} = \int_S \mathbf{q}_w(\mathbf{x}) f(\mathbf{x}) \xi(d\mathbf{x}),$$

$$\mathbf{Q}_{w,\xi} = \int_S \mathbf{q}_w(\mathbf{x}) \mathbf{q}_w^T(\mathbf{x}) \xi(d\mathbf{x}).$$

Assume that $\mathbf{Q}_{w,\xi}$ has full rank. Then

$$\mathcal{B}(f, w, \xi) = \int_S f^2(\mathbf{x}) \xi(d\mathbf{x}) - \mathbf{d}^T_{f,w,\xi} \mathbf{Q}^{-1}_{w,\xi} \mathbf{d}_{f,w,\xi}$$

$$= \int_S \left(f(\mathbf{x}) - \mathbf{q}^T_w(\mathbf{x}) \mathbf{Q}^{-1}_{w,\xi} \mathbf{d}_{f,w,\xi}\right)^2 \xi(d\mathbf{x}).$$

From this last expression we see that $\mathcal{B}(f, w, \xi)$ is the squared $L^2(\xi)$-distance from f to the closest linear combination of the elements of $\mathbf{q}_w(\mathbf{x})$.

Wiens (1991) showed that for ordinary least squares procedures, i.e. $w \equiv 1$ and $\mathbf{q}_w(\mathbf{x}) = \mathbf{z}(\mathbf{x})$, the continuous uniform design with density $k(\mathbf{x}) \equiv \Omega$ has the property of maximizing the minimum value of λ^2, hence of the power of the test under the normality assumption, as f ranges over the class

$$\mathcal{F}^+ = \left\{ f(\mathbf{x}) \mid \int_S \mathbf{z}(\mathbf{x}) f(\mathbf{x}) d\mathbf{x} = \mathbf{0}, \int_S f^2(\mathbf{x}) d\mathbf{x} \geq \eta^2 \right\}.$$

The inequality in the definition of $\mathcal{F}^+$ serves to separate the null and alternate hypotheses. A related and open problem in the current context is to choose maximin weights w_ξ for a fixed design ξ, i.e. $w_\xi = \arg\min_w \min_{f \in \mathcal{F}^+} \mathcal{B}(f, w, \xi)$.

7.8 Generalized M-Estimation

For a Mallows-type generalized (or 'Bounded Influence') M-estimate defined by

$$\hat{\theta}_{GM} = arg min_\theta \int_S \rho\left(\frac{Y(\mathbf{x}) - \mathbf{z}^T(\mathbf{x})\theta}{\sigma}\right) w(\mathbf{x}) \xi(d\mathbf{x})$$

the asymptotic bias of $\sqrt{n}\, \hat{\theta}_{GM}$ is $\mathbf{B}^{-1}\mathbf{b}$ and, with $\psi = \rho'$, the asymptotic covariance matrix is $\nu \mathbf{B}^{-1} \mathbf{D} \mathbf{B}^{-1}$ where $\nu = \sigma^2 E[\psi^2(\epsilon/\sigma)]/E[\psi'(\epsilon/\sigma)]^2$. See Hampel *et al.* (1986) and Wiens (1996) for background material and details of the asymptotics, respectively. A comparison with (7.6) shows that the optimality properties of the MVU designs and weights, derived under the assumption that the estimation is to be carried out by weighted least squares, in fact hold for the case of generalized M-estimation as well.

Acknowledgments. This research was carried out with the support of the Natural Sciences and Engineering Research Council of Canada. We are also grateful for the careful review by the referee.

References

1. Alpert, B. K. (1992). Wavelets and Other Bases for Fast Numerical Linear Algebra, In *Wavelets: A Tutorial in Theory and Applications* (Ed., C. K. Chui), pp. 181–216, Boston: Academic Press.

2. Antoniadis, A., Gregoire, G. and McKeague, I. W. (1994). Wavelet methods for curve estimation, *Journal of the American Statistical Association*, **89**, 1340–1352.

3. Benedetto, J. J. and Frazier, M. W. (eds.) (1994). *Wavelets: Mathematics and Applications*, Boca Raton, Fla.: CRC Press.

4. Box, G. E. P. and Draper, N. R. (1959). A basis for the selection of a response surface design, *Journal of the American Statistical Association*, **54**, 622–654.

5. Dette, H., and Wong, W. K. (1996). Robust optimal extrapolation designs, *Biometrika*, **83**, 667–680.

6. Draper, N. R. and Herzberg, A. (1973). Some designs for extrapolation outside a sphere, *Journal of the Royal Statistical Society, Series B*, **35**, 268–276.

7. Fang Z. and Wiens, D. P. (1999). Robust extrapolation designs and weights for biased regression models with heteroscedastic errors, *The Canadian Journal of Statistics* (to appear).

8. Hampel, F. R., Ronchetti, E., Rousseeuw, R. J. and Stahel, W. (1986). *Robust Statistics: The Approach Based on Influence Functions*, New York: John Wiley & Sons.

9. Heo, G. (1998). Optimal Designs for Approximately Polynomial Regression Models, *Ph.D. Dissertation*, University of Alberta, Department of Mathematical Sciences.

10. Herzberg, A. M. and Traves, W. N. (1994). An optimal experimental design for the Haar regression model, *Canadian Journal of Statistics*, **22**, 357–364.

11. Huang, M. N. L. and Studden, W. J. (1988). Model robust extrapolation designs, *Journal of Statistical Planning and Inference*, **18**, 1–24.

12. Huber, P. J. (1975). Robustness and Designs, In *A Survey of Statistical Design and Linear Models* (Ed., J. N. Srivastava), pp. 287–303, Amsterdam: North Holland.

13. Li, K. C. and Notz, W. (1982). Robust designs for nearly linear regression, *Journal of Statistical Planning and Inference*, **6**, 135–151.

14. Liu, S. X. and Wiens, D. P. (1997). Robust designs for approximately polynomial regression, *Journal of Statistical Planning and Inference*, **64**, 369–381.

15. Montepiedra, G. and Fedorov, V. V. (1997). Minimum bias designs with constraints, *Journal of Statistical Planning and Inference*, **63**, 97–111.

16. Oyet, A. J. (1997). Robust Designs for Wavelet Approximations, *Ph.D. Dissertation*, University of Alberta, Department of Mathematical Sciences.

17. Oyet, A. J. (1998). Wavelets and designs for heteroscedastic error regression models, *Technical Report*, Memorial University of Newfoundland, Department of Mathematics and Statistics.

18. Oyet, A. J. and Wiens, D. P. (1999). Robust designs for wavelet approximations of regression models, revisions requested by the *Journal of Nonparametric Statistics*.

19. Pesotchinsky, L. (1982). Optimal robust designs: Linear regression in R^k, *Annals of Statistics*, **10**, 511–525.

20. Pukelsheim, F. (1993). *Optimal Design of Experiments*, New York: John Wiley & Sons.

21. Rychlik, T. (1987). Robust Experimental Designs: A Comment on Huber's Result, *Zastosowania Matematyki Applicationes Mathematicae*, **19**, 93–107.

22. Spruill, M. C. (1984). Optimal designs for minimax extrapolation, *Journal Multivariate Analysis*, **15**, 52–62.

23. Stigler, S. (1971). Optimal experimental design for polynomial regression, *Journal of the American Statistical Association*, **66**, 311–318.

24. Studden, W. J. (1982). Some robust-type D-optimal designs in polynomial regression, *Journal of the American Statistical Association*, **77**, 916–921.

25. Wiens, D. P. (1990). Robust Minimax Designs for Multiple Linear Regression, *Linear Algebra and Its Applications, Second Special Issue on Linear Algebra and Statistics*, **127**, 327–340.

26. Wiens, D. P. (1991). Designs for Approximately Linear Regression: Two Optimality Properties of Uniform Designs, *Statistics and Probability Letters*, **12**, 217–221.

27. Wiens, D. P. (1992). Minimax Designs for Approximately Linear Regression, *Journal of Statistical Planning and Inference*, **31**, 353–371.

28. Wiens, D. P. (1996). Asymptotics of Generalized M-Estimation of Regression and Scale With Fixed Carriers, in an Approximately Linear Model, *Statistics and Probability Letters*, **30**, 271–285.

29. Wiens, D. P. (1998). Minimax Robust Designs and Weights for Approximately Specified Regression Models with Heteroscedastic Errors," *Journal of the American Statistical Association*, **93**, 1440–1450.

30. Wiens, D. P. (1999). Robust Weights and Designs for Biased Regression Models: Least Squares and Generalized M-estimation, *Journal of Statistical Planning and Inference* (to appear).

31. Xie, M.-Y. (1998). Some Optimalities of Uniform Designs and Projection Uniform Designs Under Multi-Factor Models, *Ph.D. Dissertation*, Hong Kong Baptist University, Department of Mathematics.

8

A Comparative Study of MV- and SMV-Optimal Designs for Binary Response Models

J. López-Fidalgo and W. K. Wong

University of Salamanca, Salamanca, Spain
University of California, Los Angeles, CA

Abstract: In this paper, we investigate the robustness properties of minimax designs for binary models. We focus on *SMV* and *MV*-optimal designs for the logistic and double exponential models and report on the efficiency changes in both types of designs when the nominal values of the parameters are misspecified.

Our results show that while *SMV*-optimal designs may appear as a more rational criterion, *MV*-optimal designs can be less sensitive to misspecification in the nominal value of the parameters.

Keywords and phrases: Double-exponential model, efficiency, logistic model, locally optimal design, robustness, standardized criteria

8.1 Introduction

Recently, there appears to be a renewed interest in constructing minimax or maximin types of optimal designs. Examples can be found in Wiens and Zhou (1996), Schwabe (1997), Dette and Sahm (1998), Mueller and Pazman (1998), King and Wong (1998a,b), Wiens (1998) and Brown and Wong (1999). The approach in these references treat designs as probability measures on the design space and so these designs are called approximate designs. When the sample size is fixed, usually by cost or physical constraints, determining the optimal measure yields the optimal approximate design. Background material and the advantages of using approximate designs are given in design monographs, such as Fedorov (1972) and Pazman (1986).

Two early examples of minimax type of designs are E and MV-optimal designs. These designs, respectively, aim to minimize the maximum of the variances of any unitary linear combination of the parameters and minimize the largest of the variances of the parameter estimates. Early proponents of these criteria are Elfving (1952) and Ehrenfeld (1955). A generalization of these criteria for linear models is given in Dette and Studden (1994), where they considered optimality criterion of the form

$$\Phi_{|\cdot|}(I(\xi)) = \max\{c^T I^{-1}(\xi) c : c \in \Re^m, |c| = 1\},$$

where $|\cdot|$ is any norm on the Euclidean space and $I(\xi)$ is the normalized Fisher information matrix for the design ξ. In particular, for the ℓ_1-norm we have the MV-optimality criterion:

$$\Phi_{MV}(I(\xi)) = \Phi_{|\cdot|_1}(I(\xi)) = \max_{i \in \{1,\ldots,m\}} e_i^T I^{-1}(\xi) e_i = \max_{i \in \{1,\ldots,m\}} \operatorname{var}_\xi(\hat{\alpha}_i).$$

Here α_i is the ith parameter in the model.

López-Fidalgo (1992) determined the gradient of the MV-criterion function on a partition of the information matrix space and provided MV-optimal designs for the simple linear model for all compact intervals (Torsney and López-Fidalgo, 1995). The work was later expanded to include simple linear models with heteroscedastic errors (López-Fidalgo, Torsney and Ardanuy, 1998).

Dette (1997) proposed to standardize the MV-criterion because of the problem of different magnitude in the design variables. He called this the SMV-criterion, where S stands for standardized. The new criterion seeks to minimize the maximum inefficiency of the design. For nonlinear models, a SMV-optimal design is one which minimizes the following function over the set of all designs:

$$\max_{i \in \{1,\ldots,m\}} \frac{e_i^T I^-(\theta,\xi) e_i}{\min_\eta e_i^T I^-(\theta,\eta) e_i},$$

where $I^-(\theta,\xi)$ is a generalized inverse of the information matrix for the design ξ. This optimality criterion, just like the MV-optimality criterion, is convex over the set of all designs defined on the design space. Consequently, standard approximate design theory can be used to study the optimal designs. In particular, equivalence theorems can be used to find candidate designs and verify if a design is optimal. Details on the use of equivalence theorems are given in Fedorov (1972) and Pazman (1986).

Optimal experimental designs for models with a binary response have been studied using the frequentist and Bayesian approaches. For example, Ford, Torsney and Wu (1988) used a frequentist approach and found locally D- and c-optimal designs for a class of binary models and, Chaloner and Larntz (1989) found Bayesian D-optimal designs and Bayesian c-optimal designs for the logistic regression model. Recently, minimax optimal designs have been constructed for binary models as well. Some examples are Sitter (1992), King and Wong

(1998a) and Dette and Sahm (1998). The first two sets of authors assumed that there is an uncertainty rectangular region around the true values of the parameters and sought a design to minimize the largest possible volume of the confidence ellipsoid for the parameters.

In this paper, we study the robustness properties of MV and SMV-optimal designs with respect to misspecification in the nominal values of the parameter values. This is an important practical issue and has been discussed in many papers; some recent ones are Gaudard et al. (1993), Rasch (1995) and, King and Wong (1998a), for example. The parametrization we used for our models in this paper is one of the two considered in Dette and Sahm (1998) and one we believe is the more popular of the two. Specifically, we suppose the probability of observing a response at x is given by

$$p(x, \theta) = F(b(x - a)), \quad \theta^T = (a, b), \quad x \in \Re, \qquad (8.1)$$

where F is a known cdf with symmetric density function f and differentiable for $x \neq 0$.

Under this parametrization, the normalized information matrix for the parameter $\theta^T = (a, b)$ using design ξ is

$$I(\theta, \xi) = \int h(b(x-a)) \begin{pmatrix} b^2 & -b(x-a) \\ -b(x-a) & (x-a)^2 \end{pmatrix} d\xi(x),$$

where the function $h(x)$ is a symmetric function defined by

$$h(x) = \frac{f^2(x)}{F(x)(1 - F(x))}.$$

López-Fidalgo, Torsney and Ardanuy (1998) showed that a function $h(x)$ is of this kind if, and only if $h(x) \geq 0$ and $\int_{-\infty}^{\infty} \sqrt{h(x)} dx = \pi$, whereupon it follows that $F(x) = \frac{1}{2}\left(1 - \cos\left(\int_{-\infty}^{x} \sqrt{h(z)} dz\right)\right)$. This result, which does not require $h(x)$ to be symmetric, comes from integrating the differential equation

$$\sqrt{h(x)} = \frac{F'(x)}{\sqrt{F(x)(1 - F(x))}}.$$

In practice, it is customary to assume that the function $h(x)$ satisfies (i) $\max_{t \in \Re} h(t) = h(0)$ and (ii) $k^2 = \max_{t \in \Re} t^2 h(t) = c^2 h(c)$ for some critical value $c > 0$. Geometrically, the value of c corresponds to a critical point of the Elfving set where the lines to form the convex hull cut the original curve. Figure 8.1 shows the Elfving sets for the logistic and double exponential models for two values of the parameter b; $b = 1$ and $b = 3$. Additional critical values for other models can be obtained from Table 4 in Ford, Torsney and Wu (1992).

The Elfving set just alluded to is described in detail in Elfving (1952) and Elfving (1959). Further illustrations on the use of this set to find an optimal

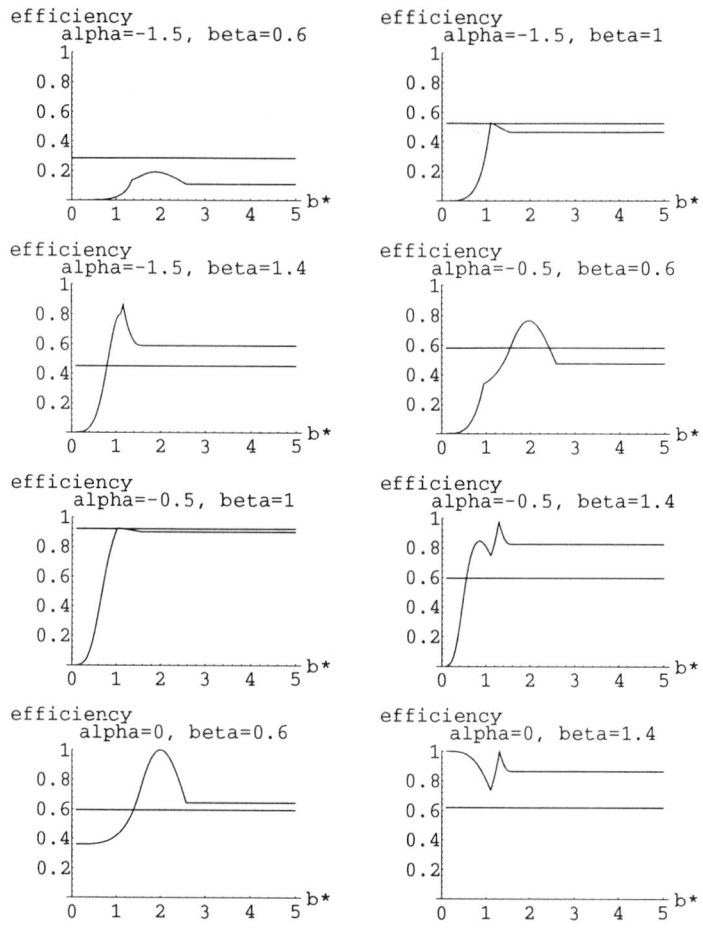

Figure 8.1:

design for estimating a linear combination of the parameters and the resulting variance of the estimate are given in Chernoff (1972, pp. 12–16). Geometrically, the variance of the estimate is simply the ratio of certain distances determined from the Elfving set. When specialized to estimating each parameter in the model, the Elfving's method provide the optimal designs for estimating a and b in (8.1), and the variances of the estimates. These variances can be shown to be respectively given by

$$\min_\eta e_1^T I^-(\theta,\eta) e_1 = \frac{1}{b^2 h(0)} \quad \text{and} \quad \min_\eta e_2^T I^-(\theta,\eta) e_2 = \frac{b^2}{k^2}.$$

Dette and Sahm (1998) gave a general procedure for finding MV and SMV-optimal designs for binary response models. Examples of MV and SMV-optimal designs were worked out for the logistic and double exponential models using one of the two parameterization for each model. Additionally, they found that SMV-optimal designs tend to distribute their mass more evenly than MV-optimal designs.

The algorithm of Dette and Sahm (1998) for determining MV-optimal design proceeds first with a two-point design. If it is determined that a 2-point design cannot be MV-optimal, the algorithm moves on to find the optimal design within the class of all 3-point designs. They showed that MV-optimal designs require at most four support points and so their procedure must find the optimal design within the class of all designs with four or less support points. In each cycle, a candidate design is found and verified for its optimality using the equivalence theorem. Dette and Sahm (1998) provides details. The SMV-optimal design is then computed from the MV-optimal design $\xi_{0,s}(\frac{b}{s}(x-a))$, where $\xi_{0,s}$ is the MV-optimal design assuming the nominal values for a and b are 0 and $s = k^{1/2}h(0)^{-1/4}$ respectively.

In the next few sections, we focus on the logistic and double exponential models. In Section 8.2, we review the local MV and SMV-optimal designs for both models and evaluate the efficiencies of the optimal designs for estimating each parameter. In section 8.3 we compare the robustness properties of both types of optimal designs to misspecification in the nominal values.

8.2 MV- and SMV-Optimal Designs

8.2.1 Logistic model

For this model, we have $F(x) = \frac{1}{1+e^{-x}}$ and a direct calculation shows $h(x) = \frac{e^x}{(1+e^x)^2}$. The values of c and k are 2.39936 and 0.66274 respectively. The MV-optimal design for this parameterization is not given in Dette and Sahm (1998) but can be directly calculated to be

$$\xi_{MV} = \begin{cases} \begin{pmatrix} a-b & a+b \\ \frac{1}{2} & \frac{1}{2} \end{pmatrix} & \text{if } 0 \leq b^2 \leq c \\ \begin{pmatrix} a-\frac{c}{b} & a+\frac{c}{b} \\ \frac{1}{2} & \frac{1}{2} \end{pmatrix} & \text{if } b^2 \geq c. \end{cases}$$

The first case is an optimal equal variance design and the second one is the optimal design for estimating the parameter b. The optimal design for estimating a is concentrated at the point a. Both these assertions follow easily from Elfving's method (1952).

The SMV-optimal design is

$$\xi_{SMV} = \begin{pmatrix} a - \frac{s^2}{b} & a + \frac{s^2}{b} \\ \frac{1}{2} & \frac{1}{2} \end{pmatrix},$$

where $s^2 = kh(0)^{-1/2} = 2.28911$. Taking into account that

$$\min_\eta e_1^T I^-(\theta,\eta) e_1 = \frac{1}{b^2 h(0)} = \frac{4}{b^2} \quad \text{and} \quad \min_\eta e_2^T I^-(\theta,\eta) e_2 = \frac{b^2}{k^2},$$

the values for both the criteria under these designs are

$$\Phi_{MV}(\xi_{MV}) = \begin{cases} \frac{1}{b^2 h(b^2)} & \text{if } 0 < b^2 < c \\ \frac{b^2}{k^2} & \text{if } b^2 \geq c \end{cases}$$

$$\Phi_{MV}(\xi_{SMV}) = \begin{cases} \frac{1}{b^2 h(s^2)} & \text{if } 0 < b^2 < s^2 \\ \frac{b^2}{s^4 h(s^2)} & \text{if } b^2 \geq s^2 \end{cases}$$

$$\Phi_{SMV}(\xi_{MV}) = \begin{cases} \frac{c^2 h(c)}{b^4 h(b^2)} & \text{if } 0 < b^2 \leq s^2 \\ \frac{1}{4h(b^2)} & \text{if } s^2 \leq b^2 \leq c \\ \frac{1}{4h(c)} & \text{if } b^2 \geq c \end{cases}$$

$$\Phi_{SMV}(\xi_{SMV}) = \frac{1}{4h(s^2)}$$

The efficiencies of these optimal designs for estimating each parameter can be easily calculated to be

$$\text{eff}_a(\xi_{MV}) = \frac{\min_\eta e_1^T I^-(\theta,\eta) e_1}{e_1^T I^-(\theta,\xi_{MV}) e_1} = \begin{cases} 4h(b^2) & \text{if } 0 < b^2 < c \\ 4h(c) & \text{if } b^2 \geq c \end{cases}$$

$$\text{eff}_b(\xi_{MV}) = \frac{\min_\eta e_2^T I^-(\theta,\eta) e_2}{e_2^T I^-(\theta,\xi_{MV}) e_2} = \begin{cases} \frac{b^4 h(b^2)}{k^2} & \text{if } 0 < b^2 < c \\ 1 & \text{if } b^2 \geq c \end{cases}$$

$$\text{eff}_a(\xi_{SMV}) = \frac{\min_\eta e_1^T I^-(\theta,\eta) e_1}{e_1^T I^-(\theta,\xi_{SMV}) e_1}$$
$$= 4h(s^2) = \text{eff}_b(\xi_{SMV}) = \frac{\min_\eta e_2^T I^-(\theta,\eta) e_2}{e_2^T I^-(\theta,\xi_{SMV}) e_2}.$$

Because there is symmetry with respect to b, it is enough to consider the case $b > 0$. Figure 8.2 shows that these efficiencies are equal at the point $b = s$ and around this point the efficiencies for MV- and the SMV-optimal designs are quite similar. When $b = 0$, the efficiencies of the MV-optimal design are on opposite ends; the design is 100% efficient for estimating a but has 0% efficient

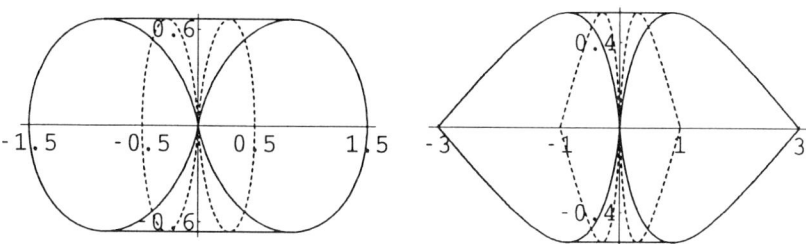

Figure 8.2:

for estimating b. On the other hand, the SMV-optimal design is about 65% efficient for estimating both parameters for all values of b. For values larger than $b = \sqrt{c}$, the efficiencies of MV-optimal design for estimating either of the parameters stablize and do not change as b varies. It has 100% efficiencey for estimating b and about 30% for estimating a.

One could also compare the relative merits of both optimal designs. The one with the higher efficiency is more robust under a change of criterion. We have

$$\text{eff}_{SMV}(\xi_{MV}) = \frac{\Phi_{SMV}(\xi_{SMV})}{\Phi_{SMV}(\xi_{MV})} = \begin{cases} \frac{b^4 h(b^2)}{4k^2 h(s^2)} & \text{if } 0 < b^2 \leq s^2 \\ \frac{h(b^2)}{h(s^2)} & \text{if } s^2 \leq b^2 < c \\ \frac{h(c)}{h(s^2)} & \text{if } b^2 \geq c \end{cases}$$

$$\text{eff}_{MV}(\xi_{SMV}) = \frac{\Phi_{MV}(\xi_{MV})}{\Phi_{MV}(\xi_{SMV})} = \begin{cases} \frac{h(s^2)}{h(b^2)} & \text{if } 0 < b^2 \leq s^2 \\ \frac{s^4 h(s^2)}{b^4 h(b^2)} & \text{if } s^2 \leq b^2 < c \\ \frac{s^4 h(s^2)}{c^2 h(c)} & \text{if } b^2 \geq c \end{cases}$$

Figure 8.3 compares the performance of the MV-optimal design under the SMV criterion and vice versa. When $b = s$ the MV-efficiency for the SMV-optimal design and the SMV-efficiency of the MV-optimal are 100%. Between $b = s$ and $b=\sqrt{c}=1.33379$, the SMV-efficiency of the MV-optimal design is higher than the MV-efficiency of the SMV-optimal but outside this interval, the MV-efficiency of the SMV-optimal design is consistently higher than the SMV-efficiency of the MV-optimal design. Beyond the point $b = 1.33379$, the SMV-efficiency of the MV-optimal design is 45% and the MV-efficiency of the SMV-optimal design is about 65%.

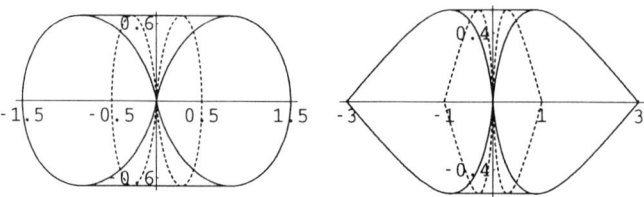

Figure 1. Elving sets for the logistic and double exponential model for b=1 (interior) and b=3 (exterior).

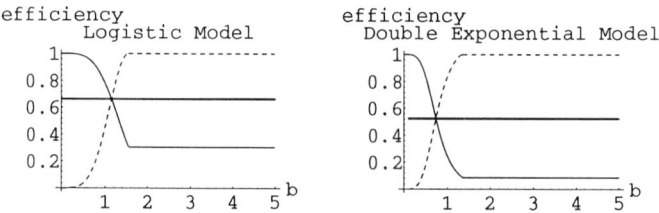

Figure 2. Efficiencies of MV-optimal designs for estimating b (dashed line) and for estimating a (continuous line), and efficiencies of SMV-optimal designs for estimating either of the parameters (thick line) in the two models.

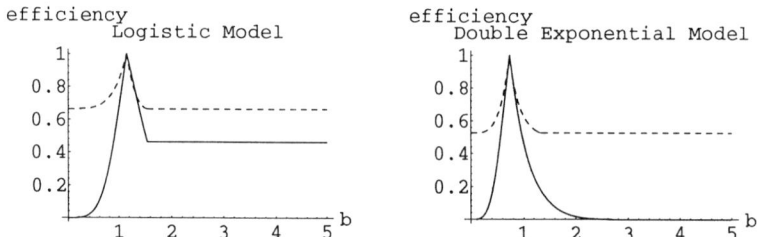

Figure 3: MV-efficiencies of SMV-optimal designs (dashed line) and MV-efficiencies of SMV-optimal designs (continuous line) in the logistic and double exponential model.

Figure 8.3:

8.2.2 Double exponential model

Here we have $F(x) = \frac{1+\text{sign}(x)}{2} - \frac{e^{-|x|}\text{sign}(x)}{2}$ and a straightforward calculation shows that $h(x) = (2e^{|x|} - 1)^{-1}$. The values of c and k are 1.84141 and 0.54040 respectively, and MV-optimal design is given by

$$\xi_{MV} = \begin{cases} \begin{pmatrix} a - v_0/b & a & a + v_0/b \\ (1-w)/2 & w & (1-w)/2 \end{pmatrix} & \text{if } b^2 \leq v_0 \\[2ex] \begin{pmatrix} a - b & a + b \\ 1/2 & 1/2 \end{pmatrix} & \text{if } v_0 < b^2 \leq c \\[2ex] \begin{pmatrix} a - c/b & a + c/b \\ 1/2 & 1/2 \end{pmatrix} & \text{if } b^2 \geq c. \end{cases}$$

[Dette and Sahm (1998)]. Here $w = \frac{(v_0^2 - b^4)h(v_0)}{h(v_0)(v_0^2 - b^4) + b^4}$ and $v_0 = 1.59362$. From this design, the SMV-optimal design is found to be

$$\xi_{SMV} = \begin{pmatrix} a - v_0/b & a & a + v_0/b \\ (1-\tilde{w})/2 & \tilde{w} & (1-\tilde{w})/2 \end{pmatrix},$$

where $\tilde{w} = 0.4653$. An easy calculation gives the values of both criteria evaluated at these designs:

$$\Phi_{MV}(\xi_{MV}) = \begin{cases} \frac{b^2}{(1-w)h(v_0)v_0^2} & 0 < b^2 < v_0 \\ \frac{1}{h^2 h(b^2)} & v_0 \leq b^2 \geq c \\ \frac{b^2}{k^2} & b^2 > c \end{cases}$$

$$\Phi_{MV}(\xi_{SMV}) = \begin{cases} \frac{1}{b^2((1-\tilde{w})h(v_0)+\tilde{w})} & 0 < b^2 < k \\ \frac{b^2}{(1-\tilde{w})h(v_0)v_0^2} & b^2 \geq k \end{cases}$$

$$\Phi_{SMV}(\xi_{MV}) = \begin{cases} \frac{k^2}{(1-w)h(v_0)v_0^2} & 0 < b^2 \leq k \\ \frac{1}{(1-w)h(v_0)+w} & k \leq b^2 \leq v_0 \\ \frac{1}{h(b^2)} & v_0 \leq b^2 \leq c \\ 1 & b^2 \geq c \end{cases}$$

$$\Phi_{SMV}(\xi_{SMV}) = 1.90199$$

We now evaluate the efficiencies of the MV and SMV-optimal designs for estimating each parameter. As before, these efficiencies depend only on the parameter b. Again, because of the symmetry with respect to b, it is enough to consider the case $b > 0$. Figure 8.2 shows these efficiencies exhibit a similar pattern observed earlier for the logistic model. The efficiencies coincide at $b = \sqrt{k}$

and near this point, the efficiencies for MV- and SMV-optimal designs are quite similar. For values greater than $b = \sqrt{c}$, the efficiencies stabilize and the MV-optimal design is optimum for estimating b, regardless of its value. However, this design is only about 12% efficient for estimating a. In constrast, the efficiencies of the SMV-optimal design for estimating either of the parameters are about 50% for all values of b.

Figure 8.3 shows the efficiencies of each optimal design under the other criterion. The MV-efficiency of the SMV-optimal design is always over 50%. At $b = \sqrt{k}$ the MV-efficiency of the SMV-optimal design and the SMV-efficiency of the MV-optimal are 100%. Near this point, both efficiencies are similar, but beyond that the SMV-efficiency of the MV-optimal design decreases rapidly to 0.

8.3 Robustness Properties of MV- and SMV-Optimal Designs

In this section, we study if MV and SMV- optimal designs are robust when the nominal values of the parameters are misspecified. To do this, we first define the efficiency of the MV- and SMV-optimal design under misspecification of the nominal values. Suppose the nominal values of the parameters are (a, b) and the true values of the parameters are $(a^\star, b^\star)$. Following Fedorov and Hackl (1997, p. 100), we define

$$\text{eff}_{MV}((a,b),(a^\star,b^\star)) = \frac{\Phi_{MV}(\xi_{MV}^{(a^\star,b^\star)}, (a^\star,b^\star))}{\Phi_{MV}(\xi_{MV}^{(a,b)}, (a^\star,b^\star))}$$

and

$$\text{eff}_{SMV}((a,b),(a^\star,b^\star)) = \frac{\Phi_{SMV}(\xi_{SMV}^{(a^\star,b^\star)}, (a^\star,b^\star))}{\Phi_{SMV}(\xi_{SMV}^{(a,b)}, (a^\star,b^\star))}.$$

These efficiencies may be used to measure how the choice of the initial values affect the goodness of the designs. To do this, we use two measures for calibrating departures from the true values: $\alpha = b(a - a^\star)$ and $\beta = b/b^\star$, see also Gaudard et al (1993) where they used these measures to assess the robustness of locally D-optimal designs for the power logistic quantal response model. Thus α is the error in the choice of the nominal value for the parameter a multiplied by the magnitude of the assumed slope. The ratio of the nominal value of b to the true slope is β. With these two measures, the above efficiency formulas now contain three variables: α, β and $b^\star$ instead of four: a, b, $a^\star$ and $b^\star$. Consequently, if we fix α and β as in Gaudard et. al. (1993), the above efficiencies become one dimensional functions of $b^\star$ alone. The sensitivities of the optimal designs to initial nominal values can now be studied conveniently using graphs.

Figures 8.4 and 8.5 display the efficiencies of MV- and SMV- optimal designs for selected values of α and β, for the logistic and the double exponential models respectively. An interesting feature in these figures is that the SMV-efficiency is not affected by the value of $b^\star$. This is because of the following reason; we have $b^\star(a - a^\star \pm \delta/b) = (\alpha \pm \delta)/\beta$ for any δ and this implies the normalized information matrix is now a sum of matrices of the form,

$$\gamma h(\frac{\alpha+\delta}{\beta}) \begin{pmatrix} b^{\star 2} & -\frac{\alpha+\delta}{\beta} \\ -\frac{\alpha+\delta}{\beta} & \left(\frac{\alpha+\delta}{\beta b^\star}\right)^2 \end{pmatrix},$$

where (δ, γ) varies over the pairs $(s^2, 1/2), (z, (1 - \tilde{w}_0)/2), (0, \tilde{w}_0), (\tilde{z}_1, \tilde{w}_1/2)$ and $(\tilde{z}_2, (1 - \tilde{w}_1)/2)$. Since each of these pairs do not depend on $b^\star$, the inverse of the normalized information matrix is

$$\begin{pmatrix} C/b^{\star 2} & D \\ D & Eb^{\star 2} \end{pmatrix}$$

where C, D and E depend only on α and β. The variances of the optimal designs for estimating the parameters using the nominal values are $\frac{1}{b^{\star 2} h(0)}$ and $\frac{b^2}{k^2}$. Therefore, once α and β are fixed, the SMV-efficiency defined above do not depend on $b^\star$.

The above observation is actually a general result which applies to any models, which can be parameterized as (8.1). Furthermore, a similar argument can be made to show that

$$\Phi_{SMV}(\xi_{SMV}) = \max\{\frac{1}{\text{eff}_a(\xi_{SMV})}, \frac{1}{\text{eff}_b(\xi_{SMV})}\} = \max\{h(0)/A, k^2/B\},$$

where A and B are independent of a and b. The equality of both efficiencies can be easily seen for each of the three possible designs using the definition $s^2 = kh(0)^{-1/2}$. This explains why the efficiencies of the SMV-optimal design for estimating the two parameters a and b do not depend on the values of a and b (Figure 8.2).

Tables 8.1 and 8.2 add further information on the behavioral changes in the efficiency for the two models when the nominal values are misspecified. The magnitude of the departure of the nominal values from the true values studied here are the same as those used in Gaudard et al. (1993). The first column under the heading $b^\star = 0.5$ contains the efficiencies of the MV-optimal design when $b^\star$ is near 0, say at 0.5. The other numbers on the same row as $b^\star$ represent values of $b^\star$ for which there is a change in the behavior of the MV-efficiency. For instance, consider the logistic model when $\beta = 0.6$ and $\alpha = 0$ corresponding to the case when the nominal value of a is correctly specified but the nominal

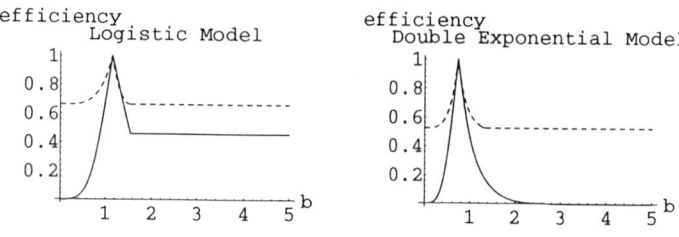

Figure 8.4:

value of β is incorrectly assumed to be 0.6 times its true value. The other numbers on the same row are $\sqrt{c}$ and $\sqrt{c}/\beta$. Comparing these numbers with the last graph on the first column of Figure 8.4, we see that the change points occur precisely at these locations. The other cases are similarly interpreted.

8.4 Conclusions

In summary, SMV-optimality seems to be convenient and useful from a theoretical point of view especially when both parameters in the model are equally important. Our numerical study gives support that if nominal values of the parameters are quite accurate, SMV-optimal designs are more desirable than MV-optimal designs for the logistic and the double exponential model. This is because the SMV-optimal designs have higher efficiencies under both the SMV and MV-optimality criteria for almost all values of b (Figure 8.3).

If there is uncertainty in the nominal values, MV-optimal designs appear more robust on average than SMV-optimal designs. For a given misspecification of the nominal values (in terms of α and β), the MV-optimal designs seem to suffer smaller loss in efficiency than the SMV-optimal designs (see Figure 8.4 and Figure 8.5). This means that MV-optimal designs are sometimes more robust to misspecification of the nominal values than SMV-optimal designs. In our examples, this observation is especially true for the double exponential model.

Acknowledgments. The research of López-Fidalgo is supported by a grant from Secretaría de Estado de Universidades, Investigación y Desarrollo. The paper was written while the first author was visiting Wong and he wishes to thank the Department of Biostatistics at UCLA for its hospitality. The research of Wong is partially supported by a NIH research grant R29 AR44177-01A1.

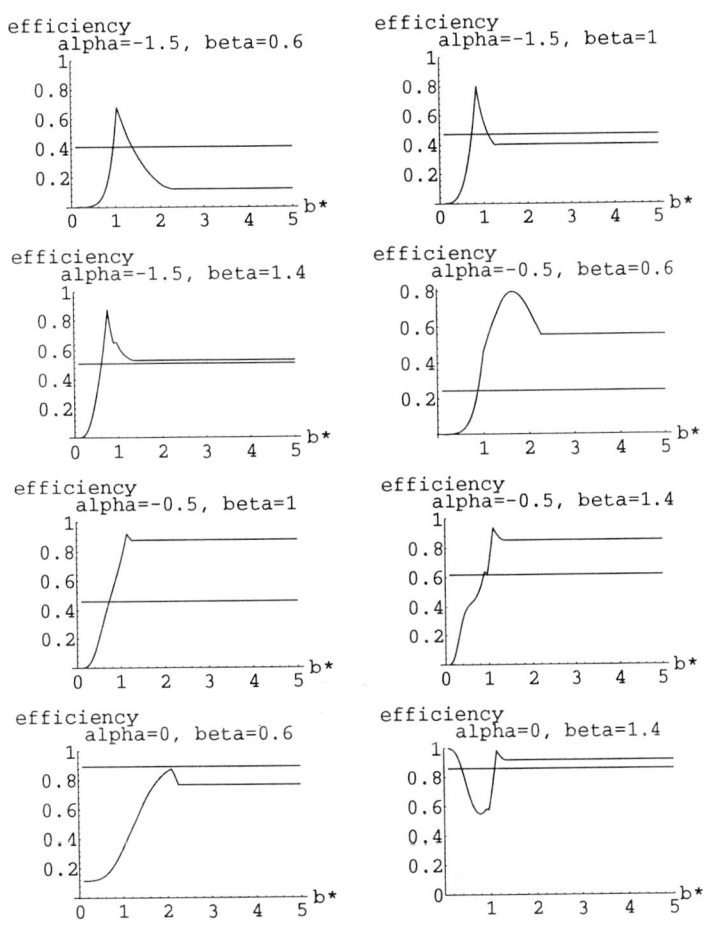

Figure 8.5:

Table 8.1: Sensitivities of MV- and SMV-optimal designs to the nominal values of the parameters in the logistic model. The value of $\sqrt{c}$ is 1.33379.

β	$|\alpha|$						
1.4	1.5	$b^\star$	0.5		$\sqrt{c}/\beta$	1.17	$\sqrt{c}$
		$|a-a^\star|$	2.14		0.97	0.92	0.69
		eff$_{MV}$	7.9		79.9	98.4	8 6.2
		eff$_{SMV}$	45.0		45.0	45.0	4 5.0
	0.5	$b^\star$	0.5	0.85	$\sqrt{c}/\beta$	1.29	$\sqrt{c}$
		$|a-a^\star|$	1.00	0.59	0.32	0.28	0.23
		eff$_{MV}$	49.8	84.8	74.6	91.5	90.2
		eff$_{SMV}$	59.7	59.7	59.7	59.7	59.7
	0	$b^\star$	0.5		$\sqrt{c}/\beta$	1.31	$\sqrt{c}$
		$|a-a^\star|$	0		0	0	0
		eff$_{MV}$	98.5		73.7	100.00	86.5
		eff$_{SMV}$	62.0		62.0	62.0	62.0
0.6	1.5	$b^\star$	0.5	1.36	$\sqrt{c}$	1.89	$\sqrt{c}/\beta$
		$|a-a^\star|$	5.00	1.83	1.61	1.32	0.97
		eff$_{MV}$	0.1	13.6	16.1	19.0	10.8
		eff$_{SMV}$	28.6	28.6	28.6	28.6	28.6
	0.5	$b^\star$	0.5	0.95	$\sqrt{c}$	1.97	$\sqrt{c}/\beta$
		$|a-a^\star|$	1.67	0.88	0.54	0.42	0.32
		eff$_{MV}$	2.7	34.0	58.3	76.3	47.5
		eff$_{SMV}$	58.2	58.2	58.2	58.2	58.2
	0	$b^\star$	0.5		$\sqrt{c}$		$\sqrt{c}/\beta$
		$|a-a^\star|$	0		0		0
		eff$_{MV}$	36.4		73.1		64.4
		eff$_{SMV}$	59.7		59.7		59.7
1	1.5	$b^\star$	0.5	1.10	$\sqrt{c}$		
		$|a-a^\star|$	3.00	1.36	0.97		
		e ff$_{MV}$	1.6	53.3	46.6		
		eff$_{SMV}$	52.5	52.5	52.5		
	0.5	$b^\star$	0.5	1.03	$\sqrt{c}$		
		$|a-a^\star|$	1	0.48	0.32		
		eff$_{MV}$	19.7	92.7	92.1		
		eff$_{SMV}$	92.1	92.1	92.1		

Table 8.2: Sensitivities of MV- and SMV-optimal designs to the nominal values of the parameters in the double exponential model. The values of $\sqrt{c}$ and $\sqrt{v_0}$ are 1.35699 and 1.26239 respectively.

β	$|\alpha|$								
1.4	1.5	$b^\star$	0.5	0.77	$\sqrt{v_0}/\beta$	$\sqrt{c}/\beta$		$\sqrt{v_0}$	$\sqrt{c}$
		$|a-a^\star|$	2.14	1.40	1.19	1.11		0.85	0.79
		eff$_{MV}$	28.9	87.7	64.9	65.5		54.0	53.1
		eff$_{SMV}$	50.9	50.9	50.9	50.9		50.9	50.9
	0.5	$b^\star$	0.5		$\sqrt{v_0}/\beta$	$\sqrt{c}/\beta$	1.10	$\sqrt{v_0}$	$\sqrt{c}$
		$|a-a^\star|$	0.71		0.40	0.37	0.32	0.28	0.26
		eff$_{MV}$	39.0		63.7	62.4	93.8	86.9	85.5
		eff$_{SMV}$	61.7		61.7	61.7	61.7	61.7	61.7
	0	$b^\star$	0.5	0.78	$\sqrt{v_0}/\beta$	$\sqrt{c}/\beta$	1.15	$\sqrt{v_0}$	$\sqrt{c}$
		$|a-a^\star|$	0	0	0	0	0	0	0
		eff$_{MV}$	73.0	55.0	58.0	57.8	98.3	93.4	91.8
		eff$_{SMV}$	86.1	86.1	86.1	86.1	86.1	86.1	86.1
0.6	1.5	$b^\star$	0.5	1.04	$\sqrt{v_0}$	$\sqrt{c}$		$\sqrt{v_0}/\beta$	$\sqrt{c}/\beta$
		$|a-a^\star|$	5.00	2.41	1.98	1.84		1.19	1.11
		eff$_{MV}$	1.4	68.3	49.2	42.7		13.9	12.3
		eff$_{SMV}$	41.3	41.3	41.3	41.3		41.3	41.3
	0.5	$b^\star$	0.5	1.01	$\sqrt{v_0}$	$\sqrt{c}$	1.64	$\sqrt{v_0}/\beta$	$\sqrt{c}/\beta$
		$|a-a^\star|$	1.67	0.83	0.66	0.61	0.51	0.40	0.37
		eff$_{MV}$	1.7	46.6	65.2	79.9	70.9	63.6	55.7
		eff$_{SMV}$	24.3	24.3	24.3	24.3	24.3	24.3	24.3
	0	$b^\star$	0.5		$\sqrt{v_0}$	$\sqrt{c}$		$\sqrt{v_0}/\beta$	$\sqrt{c}/\beta$
		$|a-a^\star|$	0		0	0		0	0
		eff$_{MV}$	13.5		50.8	57.1		87.9	76.7
		eff$_{SMV}$	89.4		89.4	89.4		89.4	89.4
1	1.5	$b^\star$	0.5	0.84	$\sqrt{v_0}$	$\sqrt{c}$			
		$|a-a^\star|$	3.00	1.78	1.19	1.11			
		eff$_{MV}$	10.5	80.2	39.9	40.5			
		eff$_{SMV}$	47.5	47.5	47.5	47.5			
	0.5	$b^\star$	0.5	1.15	$\sqrt{v_0}$	$\sqrt{c}$			
		$|a-a^\star|$	1	0.44	0.40	0.37			
		eff$_{MV}$	20.1	92.1	87.6	87.8			
		eff$_{SMV}$	45.9	45.9	45.9	45.9			

References

1. Brown, L. D. and Wong, W. K. (1999). An algorithmic construction of optimal minimax designs for heteroscedastic linear models, *Journal of Statistical Planning and Inference* (in press).

2. Chaloner, K. and Larntz, K. (1989). Optimal Bayesian design applied to logistic regression experiments, *Journal of Statistical Planning and Inference*, **21**, 191–208.

3. Chernoff, H. (1972). *Sequential Analysis and Optimal Design*, CBMS-NSF, Regional Conference Series in Applied Mathematics, SIAM.

4. Dette, H. (1997). Designing Experiments with Respect to 'Standardized' Optimality Criteria, *Journal of the Royal Statistical Society*, **B 59(1)**, 97–110.

5. Dette, H. and Sahm, M. (1998). Minimax Optimal Designs in Nonlinear Regression Models, *Statistica Sinica*, **8(4)**, 4249–4264.

6. Dette, H. and Studden, W. J. (1994). Optimal designs with respect to Elfving's partial minimax criterion in polynomial regression, *Annals of the Institute of Statistical Mathematics*, **46**, 389–403.

7. Ehrenfeld, S. (1955). On the efficiency of experimental designs, *Annals of Mathematics and Statistics*, **26**, 247–255.

8. Elfving, G. (1952). Optimum allocation in linear regression theory, *Annals of Mathematics and Statistics*, **23**, 255–262.

9. Elfving, G. (1959). Design of Linear Experiments, In *Probability and Statistics. The Harald Cramér Volume* (Ed., U. Grenander, ed.), pp. 58-74, Stockholm: Almagrist & Wiksell.

10. Fedorov, V. V. (1972). *Theory of Optimal Experiments*, Translated and edited by W. J Studden and E. M Klimko, New York; Academic Press.

11. Fedorov, V.V. and Hackl, P. (1997). *Model-Oriented Design of Experiments*, NewYork: Springer-Verlag.

12. Ford, I., Torsney, B. and Wu, C. F. J. (1992). The use of a canonical form in the construction of locally optimal designs for non-linear problems, *Journal of the Royal Statistical Society*, **B, 54(2)**, 569–583.

13. Gaudard, M. A., Karson, M. J., Linde,r E. and Tse, S. K. (1993). Efficient Designs for Estimation in the Power Logistic Quantal Response Model. *Statistica Sinica*, **3**, 233–243.

14. King, J. and Wong, W. K. (1998a). Robust designs for the power logistic model, *Joural of Computational Statistics and Data Analysis*, under revision.

15. King, J. and Wong ,W. K. (1998b). Optimal minimax designs for prediction in heteroscedastic models, *Journal of the Statistical Planning and Inference*, **69**, 371–383.

16. López-Fidalgo, J. (1992). Minimizing the Largest of the Parameters Variances. $V(\beta)$-optimality, *Proceedings of MODA 3*, pp. 71–79, New York: Springer-Verlag.

17. López-Fidalgo, J., Torsney, B. and Ardanuy, R. (1998). MV-optimization in weighted linear regression, *Proceedings of MODA 5*, pp. 39–50 , New York: Springer-Verlag.

18. Mueller, C. and Pazman, A. (1998). Applications of necessary and sufficient conditions for maximin efficient designs, *Metrika*, **48**, 1–19.

19. Pazman, A. (1986). *Foundations of Optimum Experimental Design*, D. Reidel.

20. Rasch, D. (1995). The robustness against parameter variation of exact locally optimum experimental designs in growth models - A case study, *Computational Statistics and Data Analysis*, **20**, 441–453.

21. Sitter, R. R. (1992). Robust design for binary data, *Biometrics*, **48**, 1145–1155.

22. Schwabe, R. (1997). Maximin efficient designs: Another view at D-optimality, *Statistics and Probability Letters*, **35**, 109–114.

23. Torsney, B. and López-Fidalgo, J. (1995). MV-optimization in Simple Linear Regression, *Proceedings of MODA 4*, pp. 57–69, New York: Springer-Verlag.

24. Wiens, D. P. and Zhou, J. (1996). Minimax regression designs for approximately linear models with autocorrelated errors, *Journal of Statistical Planning and Inference*, **55**, 95–106.

25. Wiens, D. P. (1998). Minimax robust designs with weights for approximately specified regression models with heteroscedastic errors, *Journal of the American Statistical Association*, **93**, 1440–1450.

9

On the Criteria for Experimental Design in Nonlinear Error-In-Variables Models

Silvelyn Zwanzig

University of Hamburg, Hamburg, Germany

Abstract: Unlike nonlinear regression models the design points of functional relation models are subject to measurement error. However the experimenter has some influence on the design. Different strategies for selecting the set up of the experiment will be presented. The goal is to obtain consistent estimators with minimal covariance matrix of their asymptotic distribution.

Keywords and phrases: Experimental design, error-in-variables, functional relations, copolymerization, orthogonal regression, total least squares, Hajek bound

9.1 Introduction

In functional relation models the design points are unknown and considered as nuisance parameters. Otherwise the asymptotic properties of statistical estimation procedures depend mainly on the asymptotic design. However, in some fields of application, especially in chemistry, the experimenter does not know the single design points, but still has some influence on the design.

An error-in-variables model was needed for copolymerization at the University of Hamburg. Copolymerization is the reaction of two monomers M_1, M_2 to a polymer consisting of both. The concentration of the monomer j at time point t is denoted by $M_j(t)$. Experiments were carried out for determination of the copolymerization parameters r_1, r_2:

$$r_1 = \frac{k_{11}}{k_{12}}, \quad r_2 = \frac{k_{22}}{k_{21}},$$

where k_{jj} is the reaction rate for polymerization of the monomer M_j and k_{jk} is the reaction rate when the polymerization changes from the monomer M_j to M_k. Using a first order Markovian model, the copolymerization reaction is described by the following differential equation [see König (1990)]

$$\frac{dM_1(t)}{dM_2(t)} = \frac{M_1(t)}{M_2(t)} \frac{r_1\left(\frac{M_1(t)}{M_2(t)}\right) + 1}{\left(\frac{M_1(t)}{M_2(t)}\right) + r_2}, \tag{9.1}$$

where $dM_j(t)$ denotes the first derivative $\frac{d}{dt}M_j(t)$, for $j = 1, 2$. The molar fraction in the polymer at a constant reaction time and for a fixed mixture of concentrations in the solvent solution is given by

$$\eta = \frac{dM_1(t)}{dM_1(t) + dM_2(t)}.$$

The molar fraction in the feed at a constant reaction time and for a fixed mixture of concentrations in the solvent solution is given by

$$\xi = \frac{M_1(t)}{M_1(t) + M_2(t)}.$$

Application of (9.1) yields the following relation between η and ξ:

$$\eta = \frac{r_1\xi^2 + \xi(1-\xi)}{r_1\xi^2 + 2\xi(1-\xi) + r_2(1-\xi)^2}.$$

The experiment is repeated for n different mixtures i of the concentration of the monomers in the solvent solution and constant reaction time t_c. Let us denote by η_i the molar fraction in the polymer and by ξ_i the molar fraction in the feed for the mixture i after the reaction time t_c. Both are measured with errors ε_{1i} and ε_{2i} by

$$y_i = \eta_i + \varepsilon_{1i}$$
$$x_i = \xi_i + \varepsilon_{2i}.$$

We describe the copolymerization by a nonlinear functional relation model

$$y_i = g(\xi_i, r_1, r_2) + \varepsilon_{1i}, \tag{9.2}$$
$$x_i = \xi_i + \varepsilon_{2i}, \quad i = 1, ..., n,$$

with

$$g(\xi, r_1, r_2) = \frac{r_1\xi^2 + \xi(1-\xi)}{r_1\xi^2 + 2\xi(1-\xi) + r_2(1-\xi)^2}.$$

The influence of the copolymerization parameter r_1 is demonstrated in Figure 9.1 below, where the regression function $g(\xi, r_1, r_2)$ is plotted as function of ξ for $r_2 = 0.2$ and three different values of $r_1 : r_1 = 1, 2$ and 3.

Experimental Design in Nonlinear Error-In-Variables Models 155

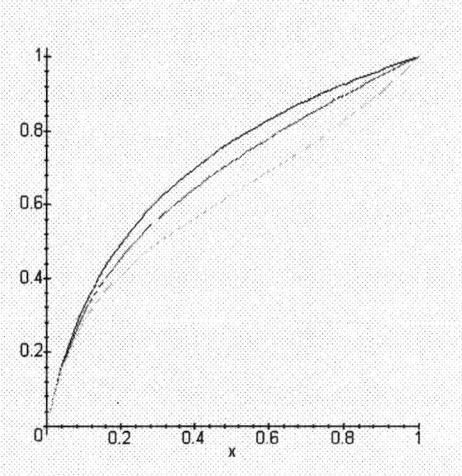

Figure 9.1:

In Figure 9.2 the regression function is plotted for $r_1 = 3$ and three different values of r_2 : $r_2 = 0.001, 0.01$ and 0.2.

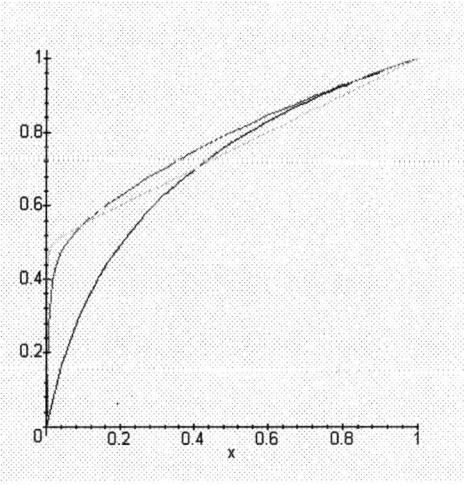

Figure 9.2:

We see that different regions of the design variable ξ have different effects on the estimation of r_1 and r_2. Small values of ξ provide more information on r_2, whereas large values give more information on r_1.

The choice of the starting concentrations of the monomers in the feed solution has an effect on the design $\xi = (\xi_1, ..., \xi_n)$.

The first paper on experimental design in error-in-variable models (EVM) also comes from chemical applications: Keeler and Reilly (1992). It is the only one I have been able to find in the wide range of published literature on error-in-variables models. The design points were usually assumed as absolutely unknown and no one imagined that it was possible to modify the position of the design points. Keeler and Reilly (1992) considered another approach, the Bayesian model, and proposed a new EVM optimality criterion for its design.

In this paper we will discuss the impact of the consistency conditions given in Zwanzig (1997), for the least squares estimator (l.s.e.) on the experimental design. If the l.s.e. is inconsistent, see Kukush and Zwanzig (1996). Alternative estimators are proposed in Kukush and Zwanzig (1997), that can be applied to some chemical models. The asymptotic efficiency in the Hajek sense of both types of estimator is compared in Zwanzig (1997). The l.s.e. is also the asymptotic best estimator in cases where it is consistent. Basing on Zwanzig (1997) the l.s.e. should be used in all cases where the design conditions imply its consistency and the alternative estimator should be used in the other cases.

Three different approaches for the choice of the design can be considered. The goal of the first approach is to optimize the consistency rate. That of the second is to minimize the Hajek bound, and the goal of the last is to provide a good alternative estimator with a small degree of covariance of the asymptotic distribution.

9.2 Error-In-Variables Model

Consider the implicit model

$$G(\eta_i, \xi_i, \beta) = 0, \quad y_i = \eta_i + \varepsilon_{1i}, \quad x_i = \xi_i + \varepsilon_{2i} \quad (9.3)$$

where $i = 1, .., n$, and $G : \mathrm{R}^2 \times \mathrm{R}^p \longrightarrow \mathrm{R}$ is a known smooth function of η_i and ξ_i (unknown in $[0,1]$) and $\beta \in \mathcal{B} \subset \mathrm{R}^p$, where β is the vector of the parameters of interest. $\mathcal{B}$ is compact . The errors ε_{ji} are independent r. v. with expected value zero and variance $\sigma_{ji}^2 > 0$, with $\max_{ji} \sigma_{ji}^2 \leq const \min_{ji} \sigma_{ji}^2$, $E \exp\left(\sigma_{ji}^{-1} |\varepsilon_{ji}|\right) \leq const$. The explicit functional relation model is a special case for (9.3) in which $G(\eta_i, \xi_i, \beta) = \eta_i - g(\xi_i, \beta)$. In the case

$$g(\xi_i, \beta) = \frac{g_1(\xi_i, \beta)}{g_2(\xi_i, \beta)}, \quad \inf_{\xi_1} \inf_{\beta} g_2(\xi_i, \beta) \geq d > 0, \quad (9.4)$$

we can reformulate (9.3) into

$$G_0(\eta_i, \xi_i, \beta) = (g_2(\xi_i, \beta) \eta_i - g_1(\xi_i, \beta)) = 0. \quad (9.5)$$

Experimental Design in Nonlinear Error-In-Variables Models

Here, we are especially interested in models for chemical applications in which

$$g_1(\xi_i, \beta_1) = \sum_{k=0}^{K_1} \beta_{1k}(\xi_i)^k, \quad g_2(\xi_i, \beta_2) = \sum_{k=0}^{K_2} \beta_{2k}(\xi_i)^k. \qquad (9.6)$$

9.3 The Total Least Squares Estimator

The weighted least squares estimator in the explicit measurement error model is defined by

$$\hat{\beta} \in \arg\min_{\beta} \min_{\xi} \sum_{i=1}^{n} w_{1i}(y_i - g(\xi_i, \beta))^2 + w_{2i}(x_i - \xi_i)^2, \qquad (9.7)$$

where w_{ji} are known weights fulfilling $w_{ji} > 0$, $\max_{ji} w_{ji} \leq const \min_{ji} w_{ji}$, $\sum_{ji} w_{ji} = 1$. Under the assumption of normal distribution for the errors ε_{ji} and $w_{ji}^* = v\sigma_{ji}^{-2}$, v such that $\sum_{ji} w_{ji}^* = 1$, the least squares estimator is the maximum likelihood one. If the nuisance parameters ξ_i vary independently of each other then we can modify (9.7) to yield

$$\hat{\beta} \in \arg\min_{\beta} \sum_{i=1}^{n} \min_{\xi_i} \left[w_{1i}(y_i - g(\xi_i, \beta))^2 + w_{2i}(x_i - \xi_i)^2 \right]. \qquad (9.8)$$

The estimator (9.8) is well known in the numerical literature. The numerical procedures for solving (9.8) converge, and they are already implemented in software packages; see Boggs *et al.* (1987). The estimator is usually called a total least squares estimator, orthogonal regression estimator or minimal distance estimator. However, the estimator (9.8) is not consistent for nonlinear functions g; see Kukush, Zwanzig (1996).

In Zwanzig (1997) the consistency of the least squares estimator (9.7) is shown using additional conditions for the design. Here we consider the case

$$\mathcal{X}^{(n)} = \left\{ \xi = (\xi_1, ..., \xi_n)^T : 0 \leq \xi_1 \leq \xi_2 \leq ... \leq \xi_{n-1} \leq \xi_n \leq 1 \right\}. \qquad (9.9)$$

At first glance, this assumption seems to be artificial. It is, however, useful in chemical applications, where the unknown design points ξ_i stand for different levels of concentration. The experimenter can guarantee with high reliability that the concentration level in the next experiment will be higher.

Our recommendation is: use the numerical procedure for solving (9.8) and then check the order of the estimated design points: $0 \leq \hat{\xi}_1 \leq ... \leq \hat{\xi}_n \leq 1$. Then we can apply the following consistency result described in Zwanzig (1997), which can be considered as special case for Theorem 7.16 (page 113) using the following values for $\varepsilon(n)$ and $r_0(n)$ in Theorem 7.16: $\varepsilon(n) = r_0(n) = \varepsilon$.

Theorem 9.3.1 *For the regression function g, let us assume that $\exists n_0 \, \exists L_1, L_2 < \infty$, $\forall n \geq n_0 \, \exists a_n, a_n > n^{-1} \quad \forall \beta, \beta' \in \mathcal{B} \quad \forall \xi, \xi' \in \mathcal{X}^{(n)}$*

$$a_n \|\beta - \beta'\|^2 \leq \sum_{i=1}^n w_{1i} \left(g\left(\xi_i^0, \beta\right) - g\left(\xi_i^0, \beta'\right)\right)^2 \leq L_2 \|\beta - \beta'\|^2$$

$$\text{and} \quad \sum_{i=1}^n w_{1i} \left(g\left(\xi_i, \beta\right) - g\left(\xi_i', \beta\right)\right)^2 \leq L_1 \sum_{i=1}^n w_{2i} \left(\xi_i - \xi_i'\right)^2. \tag{9.10}$$

This implies the existence of a positive constant C_0 such that for all ε and all $n \geq n_0$

$$P_{\xi^0, \beta^0}\left(a_n \|\widehat{\beta} - \beta^0\|^2 > \varepsilon\right) \leq \exp\left(-n\varepsilon^2 C_0\right), \tag{9.11}$$

where P_{ξ^0, β^0} is the probability measure of $(y_i, x_i)_{i=1...n}$ fulfilling the explicit functional relation with $\xi = \xi^0 = (\xi_1^0, ..., \xi_1^0)$ and $\beta = \beta^0$.

Short note on proof of consistency: The first step is to derive the proof as a result of sums of some non i.i.d. random vectors $\varepsilon_{ji} h(\xi_i, \beta)$ using following inequality.

$$P_{\xi^0, \beta^0}\left(a_n \|\widehat{\beta} - \beta^0\|^2 > \varepsilon\right) \leq P_{\xi^0, \beta^0}\left(\sup_{\beta \in \mathcal{B}} \sup_{\xi \in \mathcal{X}^{(n)}} \sum_{i=1}^n w_{ji} \varepsilon_{ji} h(\xi_i, \beta) > \varepsilon\right). \tag{9.12}$$

The main problem is that the supremum is taken over a set with increasing dimension. In order to estimate the right-hand side of (9.12) we applied empirical processes theory-based methods in Theorem 7.16 (Zwanzig (1997)), which require an entropy condition on $\mathcal{X}^{(n)}$. This entropy condition is essential for consistency. It ensures the consistency of the estimator of nuisance parameters. Lemma 7.22 (Zwanzig (1997)) shows that $\mathcal{X}^{(n)}$ in (9.9) fulfills the entropy condition of Theorem 7.16. Zwanzig (1997) also shows that $\mathcal{X}^{(n)} = [0, 1]^n$ violates this entropy condition.

The first proposal for a design criterion is related to that consistency result. The value of a_n in (9.11) depends mainly on the design. The separation property of the regression function is better for large values of a_n. Design μ is considered to be an arbitrary probability measure on $[0, 1]$.

Definition 9.3.1 *The design μ^* is provides* better local separation *than μ iff*

$$\min_{\|\beta - \beta^0\| > \varepsilon} \int \left(g(t, \beta) - g\left(t, \beta^0\right)\right)^2 d\mu^* \geq \min_{\|\beta - \beta^0\| > \varepsilon} \int \left(g(t, \beta) - g\left(t, \beta^0\right)\right)^2 d\mu. \tag{9.13}$$

For smooth regression functions, this criterion is related to the E-*optimality* in nonlinear regression. In nonlinear regression the information matrix is

$$I(\beta) = \sum_{i=1}^n \frac{1}{\sigma_{1i}^2} g^\beta(\xi_i, \beta) g^\beta(\xi_i, \beta)^T, \tag{9.14}$$

Experimental Design in Nonlinear Error-In-Variables Models

which is the inverse covariance matrix of the weighted least squares estimator in the asymptotic normal distribution, where $g^\beta(\xi_i, \beta)$ denotes the p-dimensional vector of the first derivatives of $g(\xi_i, .)$. We have

$$\sum_{i=1}^n \frac{1}{\sigma_{1i}^2} (g(\xi_i, \beta) - g(\xi_i, \beta'))^2$$
$$\approx (\beta - \beta')^T I(\beta)(\beta - \beta') \geq \lambda_{\min}(I(\beta)) \|\beta - \beta'\|^2.$$

9.3.1 Asymptotic normality and the Hajek bound

In order to get an upper bound for the risk of Hajek type we require that for all ξ_i all partial derivatives up to the second order of $g(\xi_i, .)$ exist and are continuous and for all $\beta \in \Theta$ all partial derivatives up to the second order of $g(., \beta)$ exist and are continuous. Let $g^\xi(\xi_i, \beta)$ denote the first derivatives of $g(., \beta)$ at ξ_i. The following inequality is given in Zwanzig (1989).

Theorem 9.3.2 *Under the above model assumptions for the explicit model in (9.3) with $\varepsilon_{ji} \sim N\left(0, \sigma_{ji}^2\right)$ and $\lim_{n \to \infty} \lambda_0(n) = \infty$, with $\lambda_0(n) = \lambda_{\min} \Phi(\xi)$, for any bounded, symmetric loss function l, with $l(0) = 0$, for $\mu(\xi, \xi') = \sum_{i=1}^n (\xi_i - \xi_i')^2$ and for any estimator $\tilde{\beta}$ it holds*

$$\lim_{\delta \to 0} \varliminf_{n \to \infty} \sup_{\{t: \|t-\beta\| \leq \delta\}} \sup_{\{f: \mu(\xi,f) \leq \delta \lambda_0(n)\}} E_{t,f}\left(l\left(\Phi(\xi)^{\frac{1}{2}} (\tilde{\beta} - t)\right)\right) \geq \int l dN, \tag{9.15}$$

with

$$\Phi(\xi) = \sum_{i=1}^n \frac{1}{\sigma_{1i}^2 + \sigma_{2i}^2 (g^\xi(\xi_i, \beta))^2} g^\beta(\xi_i, \beta) g^\beta(\xi_i, \beta)^T. \tag{9.16}$$

Under strong regularity and technical conditions the weighted least squares estimator with the optimal weights w_{ji}^* attains the bound in (9.15); see Zwanzig (1997), page 167. Using this theorem, one can base $D-optimality$ in error-in-variables models on the maximization of the determinate of Φ. Note that this quite different from that of $D-optimality$ in nonlinear regression because we have another information matrix in which the design has another influence. Generally, $\Phi(\xi) \prec I(\beta)$.

Definition 9.3.2 *We call a design ξ^* better in the sense of D than ξ iff*

$$\det \Phi(\xi^*) \geq \det \Phi(\xi). \tag{9.17}$$

The criteria, (9.13), (9.17), are mainly based on the least squares concept and include more or less the idea of consistency of the least squares estimator.

The problem is that if there is no additional knowledge like (9.9) then the l.s.e. is inconsistent, see Kukush, Zwanzig (1996). This is why it is reasonable to look for an alternative estimator and to construct a design criterion for this new estimator.

9.4 The Alternative Estimator

Suppose $\mathcal{X}^{(n)} = [0,1]^n$. The alternative estimator is defined as solution of the minimization problem

$$\widetilde{\beta} = \arg\min_{\beta} \sum_{i=1}^{n} q(y_i, x_i, \beta), \qquad (9.18)$$

where the new estimation function $q(y_i, x_i, \beta)$ does not depend of the nuisance parameters. The idea behind is that the function $q(.,.,.)$ is chosen such that the expected value $q(\eta_i + \varepsilon_{1i}, \xi_i + \varepsilon_{2i}, \beta)$ with respect to ε_{1i} and ε_{2i} is equal to a given contrast $c(\eta_i, \xi_i, \beta)$:

$$\int\int q(\eta_i + \varepsilon_{1i}, \xi_i + \varepsilon_{2i}, \beta) p(\varepsilon_{1i}) p(\varepsilon_{2i}) d\varepsilon_{1i} d\varepsilon_{2i} = c(\eta_i, \xi_i, \beta). \qquad (9.19)$$

Generally, we have no explicit solution $q(.,.,.)$ of the integral equality (9.19). In the case of model (9.4) we take $c(\eta_i, \xi_i, \beta) = (g_2(\xi_i, \beta)\eta_i - g_1(\xi_i, \beta))^2$ and for regression functions of polynomial form (9.6) we have (9.19) for $q(y_i, x_i, \beta) =$

$$\left(y_i^2 - \sigma_i^2\right) \sum_{l_1,l_2=0}^{K_2} \beta_{2l_1}\beta_{2l_2} k_{l_1+l_2} - 2y_i \sum_{l_1=0}^{K_2}\sum_{l_2=0}^{K_1} \beta_{2l_1}\beta_{1l_2} k_{l_1+l_2} + \sum_{l_1,l_2=0}^{K_1} \beta_{1l_1}\beta_{1l_2} k_{l_1+l_2},$$

where $k_m = k_m(x_i)$, such that $E k_m(x_i) = \xi_i^m$. For the construction of k_m we need the knowledge of the m'th moments μ_{mi} of ε_{2i} up to the order $2\max(K_1, K_2)$. The $k_m(x_i)$ are given iteratively through

$$k_0(x_i) = 1, \quad k_1(x_i) = x_i, \quad k_m(x_i) = (x_i)^m - \sum_{l=1}^{m} \binom{m}{l} k_{m-l}(x_i) \mu_{li}.$$
(9.20)

Applying this approach to (9.2) with normally distributed errors we get

$$q(y, x, r_1, r_2) = q_0 + 2r_1 q_1 + 2r_2 q_2 + r_1^2 q_{11} + r_1 r_2 q_{12} + r_2^2 q_{22},$$

with

$$k_2(x) = x^2 - \sigma_2^2, \quad k_3(x) = x^3 - 3\sigma_2^2 x, \quad k_4(x) = x^4 - 6\sigma_2^2 x^2 + 3\sigma_2^4$$

$$q_{11} = k_4(x)(1 - 2y + k_2(y))$$
$$q_{12} = (-y + k_2(y))(k_2(x) - 2k_3(x) + k_4(x))$$
$$q_{22} = k_2(y)(1 - 4x + 6k_2(x) - 4k_3(x) + k_4(x))$$
$$q_1 = (k_4(x) - k_3(x))(-1 + 3y - 2k_2(y))$$
$$q_2 = (-y + 2k_2(y))(k_2(x) - 2k_3(x) + k_4(x))$$

For proving asymptotic normality we need additional regularity conditions, which are fulfilled for the model (9.4), (9.6). Let be $q^\beta(x_i, y_i, \beta)$ the p–dimensional vector of first derivatives of $q(x_i, y_i, .)$ and $q^{\beta\beta}(x_i, y_i, \beta)$ the $p \times p$–dimensional matrix of second derivatives of $q(x_i, y_i, .)$. We quote the result from Zwanzig (1997), page 198.

Theorem 9.4.1 *Assume the model (9.4), (9.6). Then it holds*

$$\sqrt{n} A_n^{\frac{1}{2}} \left(\tilde{\beta} - \beta \right) \longrightarrow N_p(0, I_p), \text{ with } A_n = A_n(\beta, \xi), A_n^{-1} = V_n^{-1} D_n V_n^{-1}, \text{ and}$$

$$D_n = \frac{1}{n} \sum_{i=1}^n Cov_{\xi_i \beta} \left(q^\beta(x_i, y_i, \beta) \right), \quad V_n = \frac{1}{n} \sum_{i=1}^n E_{\xi_i \beta} q^{\beta\beta}(x_i, y_i, \beta). \quad (9.21)$$

The elements of q^β and of $q^{\beta\beta}$ are polynomials in x_i, y_i. Using the higher moments of x_i and y_i, the matrices in (9.21) can be computed explicitly. For (9.2) we have

$$E_{\xi_i \beta} q^{\beta\beta}(x_i, y_i, \beta) = \begin{pmatrix} 2\xi_i^2(1-\eta_i)^2 & \eta_i \xi_i^2(\eta_i - 1)(1 - \xi_i)^2 \\ \eta_i \xi_i^2(\eta_i - 1)(1 - \xi_i)^2 & 2\eta_i^2(1 - \xi_i)^2 \end{pmatrix},$$

and the elements of $Cov_{\xi_i \beta} \left(q^\beta(x_i, y_i, \beta) \right)$ are polynomials of fourth order in η_i and of $8'th$ order in ξ_i and linear functions of r_1 and r_2.

Relating to this theorem we will compare the design by the covariance matrix in (9.21).

Definition 9.4.1 *We call a design ξ^* better for the alternative estimator than the design ξ iff*

$$\det(A_n(\beta, \xi^*)) \geq \det(A_n(\beta, \xi)). \quad (9.22)$$

The application of the new estimator needs some further investigations. The alternative approach has the advantage, that the estimation function is quadratic in the parameters. In that case we obtain an explicit formula for the estimator. But unlike the total least squares estimator the alternative estimating function has no geometrical justification it only bases on the adjusting condition (9.19).

9.5 Conclusions

Let us briefly return to the introductory example of copolymerization. In Figure 9.3 the regression function (9.2) is plotted for $r_1 = 2, r_2 = 0.2$ together with

$$I_{11}(\xi) = (g^{r_1}(\xi, 2, 0.2))^2 \text{ and } \Phi_{11}(\xi) = \frac{1}{1+(g^\xi)^2}(g^{r_1}(\xi, 2, 0.2))^2.$$

Note $I_{11}(\xi)$ is the curve with the larger values, because $I_{11}(\xi) \geq \Phi_{11}(\xi)$. The regression curve is the bold line.

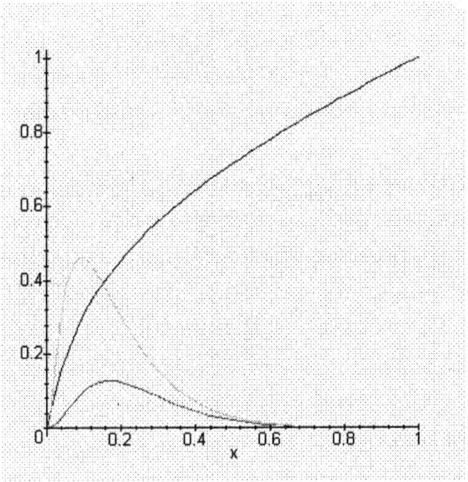

Figure 9.3:

Figure 9.3 illustrates that, unlike the separation -criterion (9.13) the D-criterion (9.17) favours higher design points, because errors in small concentrations have an especially negative influence on the orthogonal fit of the regression curves.

This paper may provide some useful, preliminary ideas for developing a theory for optimal experimental design of a new model. There are a lot of unsolved problems - for instance, the mathematical quantitation of the degree of influence that the experimenter has on the design. But error-in-variables models need different designs than nonlinear regression models. The development of such a theory seems to be promising.

Acknowledgment. The author would like to thank an anonymous referee for his or her excellent suggestions for improvement of the manuscript.

References

1. Boggs, P. T., Byrd, H. R. and Schnabel, R. B. (1987). A stable and efficient algorithm for nonlinear distance regression, *SIAM Journal on Scientific Statistical Computing*, **8**, 1052–1078.

2. Keeler, S. E. and Reilly, P. M. (1992). The design of experiments when there are errors in all the variables, *Canadian Journal of Chemical Engineering*, **70**, 774–778.

3. König, L. J. (1990). Chemical Microstructure of Polymer Chains, New York: John Wiley & Sons.

4. Kukush, A. and Zwanzig, S. (1996). On inconsistency of the least squares estimator in nonlinear functional error-in-variables models, *Technical Report 96-12*, Institute of Mathematical Stochastics, University of Hamburg.

5. Kukush, A. and Zwanzig, S. (1997). On a corrected contrast estimator in the implicit nonlinear functional relation model, *Technical Report 97-11*, Institute of Mathematical Stochastics, University of Hamburg.

6. Zwanzig, S. (1989). On an asymptotic minimax result in nonlinear errors-in-variables models, In *Proceedings of the Fourth Prague Symposium on Asymptotic Statistics*, pp. 549–558.

7. Zwanzig, S. (1997). Estimation in nonlinear functional error-in-variables models, *Habilitationsschrift*, Institute of Mathematical Stochastics, University of Hamburg.

10

On Generating and Classifying All q^{n-m-l} Regularly Blocked Factional Designs

P. J. Laycock and P. J. Rowley

University of Manchester Institute of Science and Technology, Manchester, U.K.

Keywords and phrases: Fractional factorials, block designs, vector spaces, algorithms, confounded interactions, resolution numbers, alias sets

10.1 Introduction

This paper is concerned with methods for systematically generating and classifying all regularly blocked versions of q^{-m} th fractions of q^n factorial designs (for q prime or a power of a prime), in the conventional sense of such designs as defined, for example, by Finncy (1960, p73) or as displayed in the classic set of NBS tables (1957, 1959). Following a standard notation, we refer to these as q^{m-n-l} designs, implying a division of the selected fraction into q^l blocks each of size $q^{n-m}/q^l = q^{n-m-l}$. The need for, and practical examples of such design structures can be found, for example, in Davies (1956, p465 on) or Logothetis and Wynn (1991).

Our relatively straightforward algorithmic procedure supplies a unique numbering scheme for all such designs, and these numbers are easily decoded so as to recover their originating design. This means, for instance, that tables of 'optimal' or 'good' designs, relative to one particular criterion, can be presented as a short list of design numbers. Alternatively, random selection of one design from a set of designs which satisfy a particular criterion is easily programmed. The simple but all purpose nature of our procedure should also help make the use of balanced designs an easier option in appropriate circumstances for applied scientists running large simulations of complex mathematical models, when the alternative might be an inefficient random search procedure.

A regular q^{-m}th fraction of a q^n factorial divides the q^n estimable effects or interactions into q^m alias sets each with q^{n-m} members. It is the sums of the members of these sets which are estimated by the corresponding contrasts of data means in the usual linear model formulation. An effect is said to be *estimable* provided all other members of its alias set (its aliases) are of higher order (and so assumed negligible relative to that particular effect) and provided all effects of a lower order which involve the factors in that effect are also estimable. An effect which is not estimable in this sense is said to be *aliased*. A subset of q^l of these alias sets (including the aliases of the grand mean) will be confounded with block effects when the fraction is divided into q^l blocks for the purpose of eliminating block effects. This may affect the set of estimable effects.

An important number used to describe and discriminate amongst such designs is the *resolution number*, **r**, which is the minimum number of factors involved in any of the aliases of the grand mean. A blocked design of resolution **r** will not confound any (**r**−**2**)th, or lower order, interactions; a fractional design of resolution r contains complete factorials in any **r** − **1** (or fewer) factors and no f-factor interactions will be aliased whenever **r** exceeds $2f$. When such a fraction is blocked, there will be a corresponding local *resolution number*, which will be the minimum number of treatment factors to be found in any of those effects (other than the grand mean) which appear in all those alias sets which are confounded with the blocking factor. This is discussed in section 10.2. Note that it is a design convention to present the resolution number of a design using Roman numerals, but for convenience within this paper we have used Arabic numerals.

The analysis and construction of such fractions utilizing the formal structures of group theory has an intermittent history, but dates back to the origins of the subject with Finney (1945) and Kempthorne (1947). More recently there are the books by Raktoe *et al.* (1981), Hald (1985), Street and Street (1987) and Constantine (1987). We shall emphasize the vector space structure of these fractions and blocks rather than their group theoretic properties.

In this setting, given in detail in sections 10.2 and 10.3, the elements of the full factorial are identified with the n-dimensional vector space V over the *Galois Field* of q elements. The principal fraction, **U**, is then an $(n - m)$-dimensional subspace of **V**, whilst the principal block of this blocked fraction, **W**, is an $(n - m - l)$-dimensional subspace of **U** (and of **V**). The other blocks are cosets of this subspace generated by a set of coset representatives chosen from **U**. The orthogonal complement to **U** (in **V**) forms another subspace which consists of the zero vector plus all the *defining contrasts* of the fraction. This subspace along with its cosets forms the collection of alias sets for the fractional design. The orthogonal complement to **W**, within **V**, forms another subspace of **V**, which can be used to identify the set of alias sets whose union forms the set of confounded interactions for the chosen blocking arrangement. Any one of these subspaces, (constituting the principal block of the blocking arrangement or confounded alias structure) can be constructed from a set of d

generators, where d is the dimension of the subspace. One further generator per coset is then sufficient to generate the cosets (which constitute the other blocks or confounded alias sets).

The number of possible designs or subspaces increases rapidly with n, and the popular methods of generating specific designs are usually 'target specific'. Such as the algorithm of Franklin (1985), the 'design keys' of Patterson(1976), Patterson and Bailey (1978), or the generalized cyclic methods of John and Dean (1975), and John (1987). There are, for example, 200787 4-dimensional subspaces of an 8-space when $q = 2$, and 75913222 when $q = 3$. This latter number represents about the practical limit when searching through all designs when using our algorithm Laycock and Rowley (1991) on a desktop PC. It should be noted that designs are said to be 'equivalent', in an obvious sense, if one can be obtained from the other by rearranging the labelling of the factors. Such designs lie in identical orbits of the symmetric group on n symbols, S_n, acting by permuting the symbols. Where designs are symmetric in all their factors this equivalencing is completely transparent to the experimenter; otherwise the user must choose his labelling so as to match his or her own particular preferences amongst the effects to be estimated or confounded. This asymmetry can also be used in replicates so as to induce partial confounding amongst the confounded interactions.

This reduction by equivalencing can still leave a large number of non-equivalent designs having the same resolution number. The idea of 'minimum aberration', introduced by Fries and Hunter (1980), can be used to rank designs having the same resolution, and so further reduce the set of "optimal" designs (relative to this particular criterion) which an experimenter need consider without omitting possibly important design patterns. The aberration of a fractional design is the number of aliases of the grand mean which achieve that design's resolution number. Alternatively, and equivalently, the aberration number of a blocked design is the count of those confounded interactions which achieve that design's resolution number. Rowley and Laycock (1991) give general results on the properties and construction of minimum aberration q^{n-m} designs and some specific results for the case $q = 2$. Tables listing properties, design numbers and orbit generators for $q = 2, 3, 4, 5, 7, 8, 9, 11$ and 13, were given for various combinations of n and $k = n - m$.

In this paper we will show how to generate systematically and efficiently all possible blockings of fractional designs in the class q^{n-m-l}, for given n, $k = n - m$, $j = n - m - l$, and q; along with their coset representatives, confounded interactions, aliases, resolution numbers, aberration numbers, estimable effects and design generation numbers. When an aberration is calculated over the confounding generated by the blocking of a selected fraction, we shall refer to this as local aberration, with a corresponding local minimum aberration number. Blocked fraction designs can alternatively be graded according to the estimability of specified effects. An effect is said to be unestimable when it is aliased with effects of the same or lower order, or else when it is in a confounded alias set. In particular, a count can be made of the number of estimable first-order (two factor) effects. Such criteria are easily automated with our algorithm for a search over all possible blocked fraction designs, enabling counts of, and selection of, designs as sorted by such criteria.

10.2 The Algorithm

First we establish some notation. We let F denote the Galois field $GF(q)$, where $q = p^a$, $a \in \mathbf{N}$ and p is a prime number, and let $\mathbf{V}$ be an n-dimensional space over F. We shall view $\mathbf{V}$ as the set of column vectors of length n with entries in F (so in effect we have chosen some fixed basis for $\mathbf{V}$). Let $\mathbf{U}$ be a fixed k-dimensional subspace where $1 \leq k \leq n$.

According to Laycock and Rowley (1990, §4)

$$\mathbf{U} = \beta\omega_\mathbf{i}\mathbf{V_k} = \{\beta\omega_\mathbf{i}\mathbf{v} | \mathbf{v} \in \mathbf{V_k}\}$$

where $\mathbf{V_k}$ is the following k-dimensional subspace of V

$$\mathbf{V_k} = \left\{ \begin{pmatrix} * \\ \vdots \\ * \\ 0 \\ \vdots \\ 0 \end{pmatrix} \begin{matrix} \downarrow k \\ \end{matrix} \right\}$$

(By the asterisks $*$ we mean that the indicated positions are to be filled with elements from F in all possible ways). The β and ω_i are certain $n \times n$ matrices over F and for a description of these matrices we refer the reader to Laycock and Rowley (1990, §3). The important point here is that both β and ω_i are determined explicitly by the algorithm in Laycock and Rowley (1990) in the course of obtaining $\mathbf{U}$.

Now we describe a procedure whereby we may list all the j-dimensional subspaces of $\mathbf{U}$, where $1 \leq j \leq k$. Since $\beta\omega_i$ is an invertible matrix, every j-dimensional $\mathbf{W}$ of $\mathbf{U}$ is of the form

$$\mathbf{W} = \beta\omega_\mathbf{i}\mathbf{Y} = \{\beta\omega_\mathbf{i}\mathbf{y} | \mathbf{y} \in \mathbf{Y}\}$$

for some j-dimensional subspace $\mathbf{Y}$ of $\mathbf{V_k}$.

Observe that we may identify $\mathbf{V_k}$ with the set of column vectors of length k with entries in F. A further application of the algorithm in Laycock and Rowley

Regularly Blocked Factional Designs 169

(1990) allows us to produce all the j-dimensional subspaces of $\mathbf{V_k}$ [so we have k and j playing, respectively, the roles of n and k in Laycock and Rowley(1990)]. By re-instating the $j = n - k$ 0's to each of the vectors in these subspaces we obtain all the possibilities for $\mathbf{Y}$ and hence for $\mathbf{W}$.

For $\mathbf{X}$ a subspace of $\mathbf{V}$ we set

$$\mathbf{X}^\perp = \{\mathbf{v} \in \mathbf{V} | (\mathbf{v}, \mathbf{x}) = \mathbf{0} \text{ for all } \mathbf{x} \in \mathbf{X}\}.$$

The inner product (x, y) where $x, y \in \mathbf{V}$ and $x = (x_1, x_2, ..., x_n)^T$, $y = (y_1, y_2, ...y_n)^T$ is defined by

$$(x, y) = x_1 y_1 + ... + x_n y_n.$$

Let $v \in \mathbf{V}$. Then $v = (\lambda_1, ..., \lambda_n)$ where $\lambda_i \in \mathbf{F}$. Following Rowley and Laycock (1991) we define

$$\#(v) = |\{i | \lambda_i \neq 0\}|$$

and for $\mathbf{X}$ a subspace of $\mathbf{V}$ and $d \in \mathbf{N}$

$$\omega_d(X) = |\{x \in X | \#(x) = d\}|;$$

$$\rho(X) = \min\{d \in \mathbf{N} | \omega_d(X) \neq 0\}; \text{ and}$$

$$\sigma(X) = |\{x \in X | \#(x) = \rho(X)\}|.$$

We recall that the resolution of $\mathbf{X}$, $r(\mathbf{X})$, is defined to be $\rho(\mathbf{X}^\perp)$ and the aberration count of $\mathbf{X}$, $s(\mathbf{X})$, is defined to be $\sigma(\mathbf{X}^\perp)$.

Our attention now focuses upon certain extremal subspaces contained in the fixed k-dimensional subspace $\mathbf{U}$. Put $j = k - l$, where $1 \leq l \leq k$. Next we introduce

$$r(\mathbf{U}) = \max\{r(\mathbf{X}) | \mathbf{X} \text{ is a } j - \text{dimensional subspace of } \mathbf{U}\}; \text{ and}$$

$$s_j(\mathbf{U}) = \min\{s(\mathbf{X}) | \mathbf{X} \text{ is a } j - \text{dimensional subspace of } \mathbf{U} \text{ and } r(\mathbf{X}) = r_j(\mathbf{U})\}.$$

We say that a j-dimensional subspace $\mathbf{W}$ of $\mathbf{U}$ is a local minimum aberration design (with respect to $\mathbf{U}$) if $r(\mathbf{W}) = \mathbf{r_j}(\mathbf{U})$ and $s(\mathbf{W}) = \mathbf{s_j}(\mathbf{U})$.

In order to pinpoint the local minimum aberration designs of $\mathbf{U}$ we must examine all the vectors in $\mathbf{X}^\perp$ for each j-dimensional subspace $\mathbf{X}$ of $\mathbf{U}$. Now the algorithm in Laycock & Rowley(1990, 1991) also outputs a set of aliases, $\mathcal{A}(\mathbf{U})$, for $\mathbf{U}$. Since

$$\mathbf{X}^\perp = \cup\left\{\mathbf{U}^\perp + \mathbf{v} | \mathbf{v} \in \mathbf{A}(\mathbf{U}) \text{ and } \mathbf{v} \in \mathbf{X}^\perp\right\},$$

we only need to check whether each $v \in A(\mathbf{U})$ is in $\mathbf{X}^\perp$ in order to compute a list of the elements of $\mathbf{X}^\perp$.

With regard to blocking the fraction **U** using a particular j-dimensional subspace **W** of **U**, it is necessary to determine a set of coset representatives for **W** in **U**. From the above

$$\mathbf{W} = \beta\omega_{\mathbf{i}}\mathbf{Y}$$

for some k-dimensional subspace **Y** of $\mathbf{V_k}$. Applying Laycock and Rowley (1990) (again with k and j playing, respectively, the roles of n and k) and adding on the requisite $n-k$ **O**'s to each vector produces a set of coset representatives for **Y** in $\mathbf{V_k}$. Then left multiplying by $\beta\omega_i$ gives us a set of coset representatives for **W** in **U**.

10.3 Some Specimens

In this section we present a small selection of interesting q^{n-m-l} regularly blocked fractional designs which illustrate the scope of the algorithm given in section 10.2. Our first example improves upon the local aberration of the design given as Plan 8.10.32 in the NBS tables, whilst leaving all 2-factor effects estimable, as in that plan.

10.3.1 $n = 10$, $k = 7$, $l = 2$, $q = 2$

We choose the minimum aberration design number 2104000, from Rowley and Laycock (1990, section 4.3), to be the selected fraction. So we have

$$U = <(1100000000), (1011100000), (0011010000), (1011001000),$$

$$(1001000100), (1010000010), (1010000001)>.$$

Running the above algorithm we discover, in 27 seconds on a Pentium 133 PC running under Windows, that of the 2667 blocked fractional designs, 432 have the (maximum) local resolution of 3. Of these 32 have the minimum local aberration of 2. Below, we give details of one of these local minimum aberration designs

$$U_{1001} = <(1011100000), (1111010000), (1110001100),$$

$$(0110000010), (1101001001)>.$$

(The subscript 1001 is the local design number for the blocked fractions of U when $l = 2$.)

The first two of the 32 entries in the 4 sub-blocks are therefore

1	2	3	4
0000000000	1011001000	1100000000	0111001000
1101001001	0110000001	0001001001	1010000001

Regularly Blocked Factional Designs 171

The NBS Plan 8.10.32 also has local resolution 3, but with a local aberration number of 5. (In their notation, these locally confounded effects for their selected fraction are : GHJ, EHK, ABE, BCF and CDH).

10.3.2 $n = 8$, $k = 6$, $l = 3$, $q = 2$

Here we take the minimum aberration design

$$U = <(11000000),(00110000),(00101000),(10100100),$$
$$(10100010),(10100001)>$$

which has design number 5931, as in Rowley and Laycock (1990, section 4.3). We also recall from this same reference, Lemma 2.5, that these minimum aberration designs for $n = 8$, $k = 6$, $q = 2$ form one orbit under the action of S_8. Also, from Rowley and Laycock (1990, §4.1) we have that

$$H := \text{Stab}_{s_8} U = <(12),(34),(45),(67),(78),(36)(47)(58)>, \text{ with } |H| = 2^4 3^2.$$

The algorithm in section 10.2 above reveals that there are 90 local minimum aberration designs and that $r(U) = 2 = s_3(U)$. Now it is clear that H permutes this set of 90 local minimum aberration designs and, as we shall verify below, H has two orbits on the 90 local minimum aberration designs. One of these local minimum aberration designs is

$$U_{283} = <11110000),(10001100),(00101011)>.$$

(Again, the subscript 283 is the local design number for the blocked fractions of U when $l = 3$.) Put $K = \text{Stab}_H U_{283}$. We wish to determine K, and begin by listing all the vectors of U_{283}:

	1	2	3	4	5	6	7	8
$y_1 =$	0	0	0	0	0	0	0	0
$y_2 =$	0	0	1	0	1	0	1	1
$y_3 =$	1	0	0	0	1	1	0	0
$y_4 =$	1	0	1	0	0	1	1	1
$y_5 =$	1	1	1	1	0	0	0	0
$y_6 =$	1	1	0	1	1	0	1	1
$y_7 =$	0	1	1	1	1	1	0	0
$y_8 =$	0	1	0	1	0	1	1	1

(The first row of numbers labels the eight factors.) Noting that y_3 is the only vector in U_{283} having three 1's and five 0's, we see that K must fix y_7. Observe that H on the labels (1,2,3,4,5,6,7,8) has orbits $\{1,2\}$ and $\{3,4,5,6,7,8\}$ with H acting as S_3 upon both $\{3,4,5\}$ and $\{6,7,8\}$ together with a permutation interchanging $\{3,4,5\}$ and $\{6,7,8\}$. Hence each element of K must fix 1 and

2 (as y_3 has a 1 in the first position and a 0 in the second). Because y_2 and y_5 are the only vectors in U_{283} having four 1's and four 0's and y_2 has 0,0 in the first two positions we must have K fixing both y_2 and y_5. So (looking at y_5) no element of K interchanges $\{3,4,5\}$ and $\{6,7,8\}$. Therefore elements of K must leave $\{3,5\}$ and $\{7,8\}$ invariant (looking at y_2) and $\{3,4\}$ invariant (looking at y_5). Hence $K \leq\, <(78)>$. Plainly $(78) \in K$ and so $K =\, <(78)>$. Thus the number of local minimum aberration designs in the H-orbit of U_{283} is

$$\frac{|H|}{|K|} = \frac{2^4.3^2}{2} = 2^3.3^2 = 72$$

We next examine the local minimum aberration design

$$U =\, < 11110000), (00101110), (11101101) > .$$

(This is not in the H-orbit of U_{283} since U_{403} possesses no vectors with three 1's and five 0's.) Subjecting U_{403} to an analysis similar to that for U_{283} given above it can be shown that

$$\text{Stab}_H U_{403} =\, < (34)(78), (38)(47)(56), (12) > .$$

In particular $|\text{Stab}_H U_{403}| = 8$ and consequently the number of local minimum aberration designs in the H-orbit of U_{403} is

$$\frac{|H|}{|\text{Stab}_H U_{403}|} = \frac{2^4.3^2}{8} = 2.3^2 = 18.$$

Since $72 + 18 = 90$, we conclude that H has two orbits on the local minimum aberration designs.

10.3.3 $n = 7$, $k = 4$, various l, $q = 3$

Let $n = 7$, $k = 4$ and $q = 3$; from Rowley and Laycock (1990, section 4.2), we select the fractional design whose number is 301600. Thus

$$U =\, < (1110000), (1201100), (0101010), (2002001) > .$$

We now block U, taking in turn $l = 1, 2, 3$. In each case we give tables detailing the number of blocked fractional designs with particular local resolution and local aberration numbers. For each of $l = 1, 2$ and 3 we see that there are three local minimum aberration designs, and we give an example of each of these, together with the corresponding blocking. The following interesting points also emerge :- for $l = 1$, respectively $l = 3$, all three local minimum aberration designs have 30, respectively 22, estimable two factor effects, the largest possible. While for $l = 2$, all 3 local minimum aberration designs have 28 estimable two factor effects, whereas the maximum possible is 30. There are respectively 18, 6 and 3 designs, for $l = 1, 2$ and 3, which achieve the maximum possible number of estimable two factor effects.

Regularly Blocked Factional Designs

(i) $l = 1$

local resolution, $r_1(U)$	1	2	3
blocked fraction count	7	28	5
local aberration, $s_1(U)$	2	2	10
$s_1(U)$ occurs this often	7	15	3

$$U_{22} =< (1201100), (2021010), (1222001) >$$

and the first two entries out of the 27 in the three sub-blocks are

block 1	block 2	block 3
0000000	1110000	2220000
2111002	0221002	1001002

(ii) $l = 2$

local resolution, $r_2(U)$	1	2
blocked fraction count	70	60
local aberration, $s_2(U)$	2	6
$s_2(U)$ occurs this often	49	3

$$U_{79} =< (1211010), (1101201) >$$

and the first two entries out of the nine in each of the nine sub-blocks are therefore

block 1	block 2	block 3	block 4	block 5	block 6
0000000	1201100	2102200	1110000	2011100	0212200
1101201	2002001	0200101	2211201	0112001	1010101

block 7	block 8	block 9
2220000	0121100	1022200
2110221	0011021	1212121

(iii) $l = 3$

local resolution, $r_3(U)$	1	2
blocked fraction count	37	3
local aberration, $s_3(U)$	2	42
$s_3(U)$ occurs this often	8	3

$$U_{27} =< (1111111) >$$

and the first seven sub-blocks out of the 27 sub-blocks are therefore:

block 1	block 2	block 3	block 4	block 5	block 6	block 7
0000000	0101010	0202020	1201100	1002110	1100120	2102200
1111111	1212121	1010101	2012211	2110221	2211201	0210011
2222222	2020202	2121212	0120022	0221002	0022012	1021122

10.3.4 Some $q = 4$ examples

Here $F = \{0, 1, \omega, \omega^2\}$; for the blockings given below we set $0 = 0$, $1 = 1$, $\omega = 2$ and $\omega = 3$.

(a) $n = 4$, $k = 2$, $q = 4$, $l = 1$

Let U be the fractional design whose number is 188. So

$$U = < (1110), (1\omega01) >$$

local resolution, $r_1(U)$	1	2
blocked fraction count	4	1
local aberration, $s_1(U)$	3	18
$s_1(U)$ occurs this often	4	1

The one and only local minimum aberration design is

$$U_5 = < (\omega 1 \omega^2 1) >,$$

with blocks as follows

block 1	block 2	block 3	block 4
0000	1110	2220	3330
2131	3021	0311	1201
3212	2302	1032	0122
1323	0233	3103	2013

(b) $n = 5$, $k = 3$, $q = 4$

Let U be the fractional design whose number is 3073. So

$$U = < (11100), (1\omega010), (1\omega^2001) >$$

(i) $l = 1$

local resolution, $r_1(U)$	1	2	3
blocked fraction count	5	15	1
local aberration, $s_1(U)$	3	6	30
$s_1(U)$ occurs this often	5	15	1

The unique local minimum aberration design is

$$U_{20} = < (\omega 1 \omega^2 10), (\omega^2 1 \omega 01) >$$

and the first two entries in the 4 sub=blocks are:

block 1	block 2	block 3	block 4
00000	11100	22200	33300
31201	20301	13001	02101

(ii) $l = 2$

local resolution, $r_2(U)$	1	2
blocked fraction count	15	6
local aberration, $s_2(U)$	3	30
$s_2(U)$ occurs this often	5	6

One of the local aberration designs, together with the blocking, is as follows

$$U = < (\omega 11\omega 1) >$$

and the first eight sub-blocks of the sixteen are:

block 1	block 2	block 3	block 4	block 5	block 6	block 7	block 8
00000	12010	23020	31030	11100	03110	32120	20130
21121	33131	02101	10111	30021	22031	13001	01011
32232	20222	11212	03202	23332	31322	00312	12302
13313	01303	30333	22323	02213	10203	21233	33223

References

1. Constantine, G. M. (1987). *Combinatorial Theory and Statistical Design*, New York: John Wiley & Sons.

2. Davies, O. L. (1956). *The design and analysis of industrial experiments*, 2d ed. rev., London: Oliver and Boyd.

3. Finney, D. J. (1945). The fractional replication of factorial arrangements, *Ann. Eugen., London*, **12**, 291–301.

4. Finney, D. J. (1963). *An introduction to the theory of experimental design*, 1st ed. 3rd imp. Chicago University Press.

5. Franklin, M. F. (1985). Selecting Defining Contrasts and Confounded Effects in p^{n-m} Factorial Experiments, *Technometrics*, **27**, 165–172.

6. Fries, A. and Hunter, W. G. (1980). Minimum aberration 2^{k-p} designs, *Technometrics*, **22**, 601–608.

7. Hald (1985).

8. John, J. A. (1987). *Cyclic Designs*, London: Chapman and Hall.

9. John, J. A. and Dean, A. M. (1975). Single replicate factorial experiments in generalized cyclic designs. I, Symmetrical arrangements, *Journal of the Royal Statistical Society, Series B*, **37**, 72–76.

10. Kempthorne, O. (1947). A simple approach to confounding and fractional replication in factorial experiments, *Biometrika*, **34**, 255–272.

11. Laycock, P. J. and Rowley, P. J. (1992). A method for generating and classifying all regular fractions and blocks for q^{n-m} designs, *Accepted for publication by Journal of the Royal Statistical Society, Series B*.

12. Laycock, P. J. and Rowley, P. J. (1991). A FORTRAN algorithm for generating all regular fractions and blocks for q^{n-m} designs, *Technical Report, No. 204*, Mathemathics Department, UMIST, Manchester UK.

13. Logothetis, N. and Wynn, H. P. (1989) *Quality Through Design*, Oxford: Clarendon Press.

14. National Bureau of Standards (1957). *Fractional Factorial Experiment Designs for Factors at Two Levels*, Applied Mathematics Series, No. 48, Washington, D.C.: NBS.

15. National Bureau of Standards (1959). *Fractional Factorial Experiment Designs for Factors at Three Levels*, Applied Mathematics Series, No. 54. Washington, D.C.: NBS.

16. Patterson, H. D. (1976). Generation of Factorial Designs, *Journal of the Royal Statistical Society, Series B*, **38**, 175–179.

17. Patterson, H. D. and Bailey, R. A. (1978). Design Keys for Factorial Experiments, *Applied Statistics*, **27**, 335–343.

18. Raktoe, B. L., Hedayat, A. and Federer, W. T. (1981). *Factorial Designs*, New York: John Wiley & Sons.

19. Rowley, P. J. and Laycock, P. J. (1990). Minimum aberration designs, *Technical Report No. 201*, Department of Mathematics, UMIST, Manchester M60 1QD, *Submitted for publication*, 1993.

20. Street, A. P. and Street, D. J. (1987). *Combinatorics of Experimental Design*, Oxford: Clarendon Press.

11

Locally Optimal Designs in Non-Linear Regression: A Case Study of the Michaelis-Menten Function

E. P. J. Boer, D. A. M. K. Rasch and E. M. T. Hendrix

Department of Agricultural, Environmental and Systems Technology, Wageningen, The Netherlands

11.1 Introduction

It is well known that optimal (given criterion Φ) designs for a non-linear regression function $f(x, \theta)$, with measurements $x_i \in B \subset \mathbb{R}^1$ and parameters $\theta^T = (\theta_1, \ldots, \theta_p) \in \Omega \subset \mathbb{R}^p$, depend on a priori values for θ and are therefore *locally* optimal. An exact optimal design can be seen as a choice of n points $x_1, \ldots, x_n$ out of a set B. Two sets will be considered in this paper: an interval $B = B_I = [x_l, x_u]$ and a set $B = B_N = \{x_1, \ldots, x_N\}$ which consists of a finite number of candidate points. If the set B_N is used for the calculation of the optimal design, the design will be called replicationfree because replicated measurements at the same measurement point x_i are not allowed. The selection of designs in B_I allows several measurements at the same measurement point x_i^* and will be called unrestricted design in this paper. The notation of such a design is:

$$\begin{pmatrix} x_1^* & x_2^* & \ldots & x_k^* \\ r_1 & r_2 & \ldots & r_k \end{pmatrix} \quad (11.1)$$

with $x_i^* \in B$ ($i = 1, 2, \ldots, k$) as design points and r_i as the number of measurements at point x_i^* ($\sum_{i=1}^{k} r_i = n$).

The aim of this paper is to get more insight into the locally Φ-optimal choice of the points x_i, $i = 1, \ldots, n$ for some θ. Research starts with the restricted set B_N. If N is not too large, it will be possible to calculate the locally optimal replicationfree design by full enumeration (Boer, 1995; Rasch *et al*, 1997). Secondly, (sub)-optimal unrestricted designs chosen from set B_I are calculated by solving a non-linear programming (NLP) problem. This optimisation problem can be simplified by use of the optimal replicationfree design. The Φ-criteria used are introduced in Section 11.2.

For illustrative purposes the following parametrisation of the Michaelis-Menten function

$$f(x,\theta) = \frac{\alpha x}{1+\beta x}, \quad \theta^T = (\alpha, \beta) \qquad (11.2)$$

will be used to show the principles of calculating the locally exact optimal (unrestricted and replicationfree) design for a non-linear regression function.

11.2 Calculation of Optimal Designs

Optimal designs for the Michaelis-Menten function are calculated using three optimality criteria: D-, C_α- and C_β-optimality. These three optimality criteria are all functions of the asymptotic covariance matrix $V(\hat\theta)$. For the Michaelis-Menten function (11.2) this matrix can be written as:

$$V(\hat\theta) = \frac{\sigma^2}{\Delta}\begin{pmatrix} C & B \\ B & A \end{pmatrix}$$

with $A = \sum_{i=1}^{n} \frac{x_i^2}{(1+\beta x_i)^2}$, $B = \sum_{i=1}^{n} \frac{\alpha x_i^3}{(1+\beta x_i)^3}$, $C = \sum_{i=1}^{n} \frac{\alpha^2 x_i^4}{(1+\beta x_i)^4}$

and $\Delta = AC - B^2$.

The three optimality criteria are defined as:

$$D-\text{optimality}: \quad \min\{K_1 = \det(\frac{1}{\sigma^2}V(\hat\theta)) = \frac{1}{\Delta}\} \qquad (11.3)$$

$$C_\alpha-\text{optimality}: \quad \min\{K_2 = \frac{1}{\sigma^2}V(\hat\alpha) = \frac{C}{\Delta}\} \qquad (11.4)$$

$$C_\beta-\text{optimality}: \quad \min\{K_3 = \frac{1}{\sigma^2}V(\hat\beta) = \frac{A}{\Delta}\} \qquad (11.5)$$

11.2.1 Replicationfree designs

Finding exact optimal replicationfree designs is a combinatorial problem. Namely, it concerns the selection of a subset of n points from the set B_N. To find the optimal design one should calculate the value of criterion K_i ($i = 1, 2, 3$) for each possible combination (full enumeration). However, this approach works only if N, the number of points in set B_N, is relatively small (say ≤ 35). Therefore search algorithms have been developed (Rasch et al, 1997) which are able to find a (sub)-optimal design. In this paper we will keep N small, so that we can find the optimal replicationfree design by enumeration.

11.2.2 Unrestricted designs

From the theory of optimal designs in non-linear regression it is known that an optimal unrestricted design has a number of so-called support points which is not less than the number p of parameters in the function and not larger than $p(p+1)/2$. That means that we have a number of support points x_i^*, with r_i measurements (replications) at each support point. For D-optimality it is known (Ermakov and Zigljavskij, 1987; Rasch, 1990) that the optimal unrestricted design for the Michaelis-Menten function is a two point design. It has two support points which are independent of n and of the parametrisation of the function. Given the two point design, the sums A, B and C of the asymptotic covariance matrix can be written as:

$$A = r_1 \frac{x_1^{*2}}{(1+\beta x_1^*)^2} + r_2 \frac{x_2^{*2}}{(1+\beta x_2^*)^2}$$
$$B = r_1 \frac{\alpha x_1^{*3}}{(1+\beta x_1^*)^3} + r_2 \frac{\alpha x_2^{*3}}{(1+\beta x_2^*)^3}$$
$$C = r_1 \frac{\alpha^2 x_1^{*4}}{(1+\beta x_1^*)^4} + r_2 \frac{\alpha^2 x_2^{*4}}{(1+\beta x_2^*)^4}$$

The problem of determining the optimal design can be written as a mixed-integer non-linear programming (MINLP) problem. For D-optimality this is the following problem:

$$\min\{K_1(x_1^*, x_2^*, r_1, r_2) = \frac{1}{AC-B^2}\}$$
under the conditions :
$$r_1 + r_2 = n$$
$$r_1, r_2 \geq 1, \text{ integer}$$
$$x_l \leq x_1^* < x_2^* \leq x_u$$
(11.6)

The values of x_1^*, x_2^*, r_1 and r_2 are chosen in such away, that the criterion value of D-optimality is minimised. Note that x_1^* and x_2^* can be any number within the range $[x_l, x_u]$, but r_1 and r_2 have to be chosen from the set $\{1, 2, \ldots, n-1\}$. These kind of problems are in general hard to solve. An alternative formulation of problem (11.6) is given by the following fully continuous nonlinear programming problem:

$$\min\{K_1(x_1, x_2, \ldots, x_n) = \frac{1}{AC - B^2}\} \quad (11.7)$$
$$\text{under the condition}:$$
$$x_l \leq x_1 \leq x_2 \leq \ldots \leq x_n \leq x_u$$

In Section 11.3.3 we will show that the NLP formulation (11.7) is not more advantageous with respect to MINLP problem (11.6). Problems (11.6) and (11.7) can be solved by existing programs for non-linear optimisation.

For D-optimality the solution of MINLP problem (11.6) can be calculated analytically. By substituting the sums A, B and C in the function K_1 and rewriting the function to a simpler form, we get:

$$K_1(x_1^*, x_2^*, r_1, r_2) = \frac{(1 + \beta x_1^*)^4 \, (1 + \beta x_2^*)^4}{\alpha^2 \, (x_1^* - x_2^*)^2 \, x_1^{*2} \, x_2^{*2} \, r_1 \, r_2} \quad (11.8)$$

It can easily be seen that when minimising (11.8), r_1 and r_2 have to be chosen as equal as possible. The optimal values do not depend on the parameters nor on the choice of the support points. The choice of x_1^* and x_2^* is more complicated. By analysing the derivatives of (11.8) the following D-optimal design can be found for $x \in [0, x_u]$, $x_u > 0$ [Ermakov and Zigljavskij (1987)]

$$\begin{pmatrix} \frac{x_u}{2+x_u\beta} & x_u \\ r_1 & r_2 \end{pmatrix} \quad (11.9)$$

with $r_1 + r_2 = n$. For $n = 2r$, r_1 equals r_2 and $n = 2r+1$ applies $r_1 = r, r_2 = r+1$ or $r_1 = r+1, r_2 = r$.

We were not able to find an analytical solution for the C_α- and C_β-optimal unrestricted designs. These designs depend - contrary to D-optimal designs Rasch (1990) - on the parametrisation of the function. We will restrict ourselves on the class of two-point designs. Because locally optimal designs with three support points can not be excluded, the C_α- and C_β-designs are called (sub)-optimal. A solution was only found numerically by solving the MINLP problem (11.6) using K_2 or K_3.

11.3 Results

An application of Equation (11.2) can be found in the original paper of Michaelis and Menten (1913). They investigated the kinetic of the invertase effect. We used experiment 5 of their Table I as a test case. In this experiment the change

Locally Optimal Designs in Non-Linear Regression

Table 11.1: Measurements of Michaelis and Menten (1913) of the change of rotation y depending on time x.

x_i	0	1	6	17	27	38	62	95	1372	1440
y_i	0	0.025	0.117	0.394	0.537	0.727	0.877	1.023	1.136	1.178

in rotation y depending on time x was measured. Table 11.1 presents 10 measurements for y and x.

With the SAS procedure PROC NLIN the parameters α and β of Equation (11.2) can be estimated. For this case the estimates are: $\alpha = 0.04135$ and $\beta=0.03392$. Figure 11.1 shows the measurements and the fitted function.

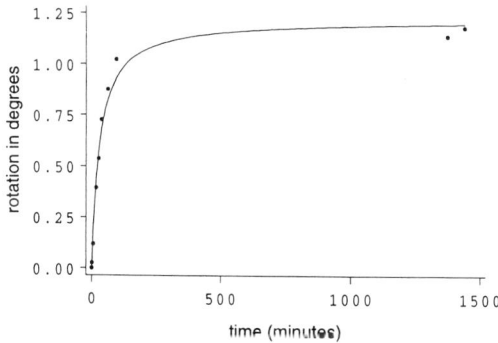

Figure 11.1: Measurements and the fitted Michaelis-Menten function.

Given these parameters, we can calculate the optimal unrestricted and replicationfree designs. First we will calculate the optimal replicationfree designs, followed by the optimal unrestricted designs.

11.3.1 Optimal replicationfree designs

Suppose we take the following (arbitrary) set B_N of (candidate) measurement points:

$B_N = \{$ 15(1)35, 50, 500, 800, 1000, 1200, 1300, 1435(1)1440 $\}$

in which 15(1)35 is a notation for the set $\{15, 16, \ldots, 35\}$ and analogous to 1435(1)1440.

Suppose one can take 10 measurements ($n = 10$ as in the original experiment), then the D-, C_α- and C_β-optimal replicationfree design can be calculated by full enumeration. Table 11.2 presents the optimal replicationfree designs for all three criteria.

Table 11.2: Optimal replicationfree designs for the three different criteria, with the criterion value of the optimal design and the criterion value of the original design of Michaelis and Menten.

		Criterion value	
Criterion	Opt. replicationfree design	Opt. design	Orig. design
K_1	26(1)30 1436(1)1440	$6.45 \cdot 10^{-7}$	$16.44 \cdot 10^{-7}$
K_2	18(1)25 1439 1440	$2.85 \cdot 10^{-3}$	$5.44 \cdot 10^{-3}$
K_3	18(1)24 1438(1)1440	$2.87 \cdot 10^{-3}$	$5.24 \cdot 10^{-3}$

The optimal replicationfree designs are, for all three criteria, clustered round two points. One point somewhere at the beginning of the interval and one point at the end of the interval. The clusters are of different size for the three criteria.

11.3.2 Optimal unrestricted designs

In the unrestricted case, the set B is taken as the interval $B_I = [0, 1440]$. The 10 points x_i can be chosen freely out of this interval. In the first place, a D-optimal unrestricted design will be calculated, followed by C_α- and C_β-optimal designs.

The D-optimal design can be calculated exactly by Equation (11.9). In this test case the optimal design is equal to:

$$\begin{pmatrix} 28.32 & 1440 \\ 5 & 5 \end{pmatrix}$$

with a criterion value of $K_1 = 6.44 \cdot 10^{-7}$. The same result can be found numerically by running the non-linear optimisation program GINO (General INteractive Optimizer), based on a version of the generalised reduced gradient [Liebman et al. (1985)]. Notice that r_1 and r_2 take integral values automatically.

The MINLP problem (11.6) can be simplified by fixing r_1 and r_2, $r_1 = r_2 = 5$. Now it is possible to make a contour plot of the minimisation surface of the variables x_1^* and x_2^* which represent the support points. Figure 11.2 shows the contour plot of the D-optimality criterion value.

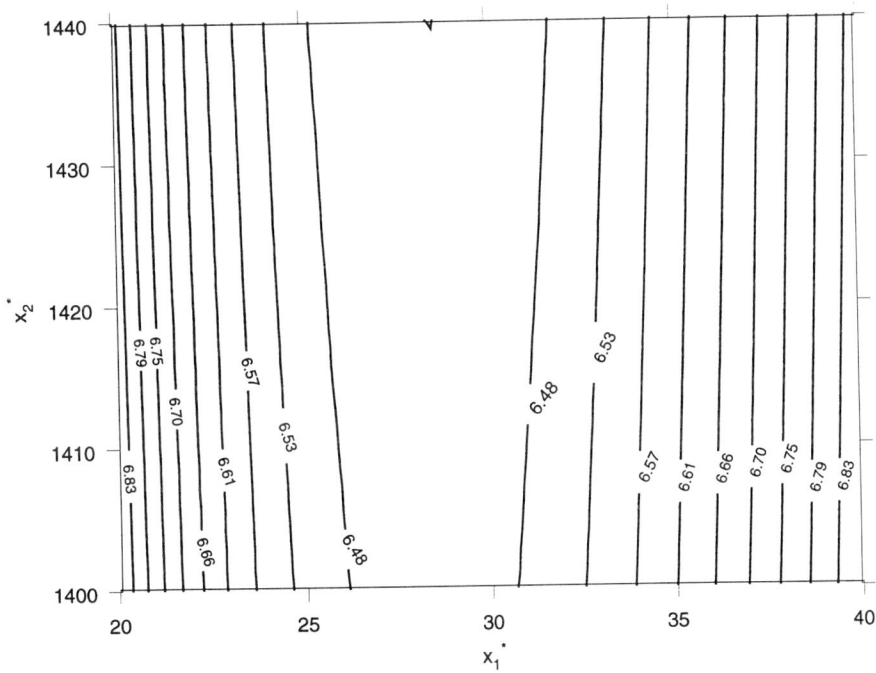

Figure 11.2: Contour plot of the minimisation problem of x_1^* and x_2^* for D-optimality, $r_1 = r_2 = 5$. The plotted criterion values are equal to $K_1 \cdot 10^7$.

Figure 11.2 shows that the criterion value of the D-optimality is more sensitive to the choice of x_1^* than to x_2^*. The optimal value of x_2^* is found at the boundary: $x_2^* = 1440$. This minimum is equal to that of the analytical solution.

Optimal designs for the C_α-criterion can not be calculated analytically. Therefore the same non-linear programming algorithm GINO was used to find designs for the C_α-criterion. Use of GINO results into the following optimal discrete [Fedorov (1972)] design:

$$\begin{pmatrix} 20.14 & 1440 \\ 8.54 & 1.46 \end{pmatrix}$$

with criterion value $K_2 = 2.79 \cdot 10^{-3}$. Notice that for this criterion r_1 and r_2 do not automatically take an integral value. All different combinations of r_1 and

r_2 have to be checked to find the (integral) optimum. Figure 11.3 shows the results of the three most important combinations of the number of replications on the two support points.

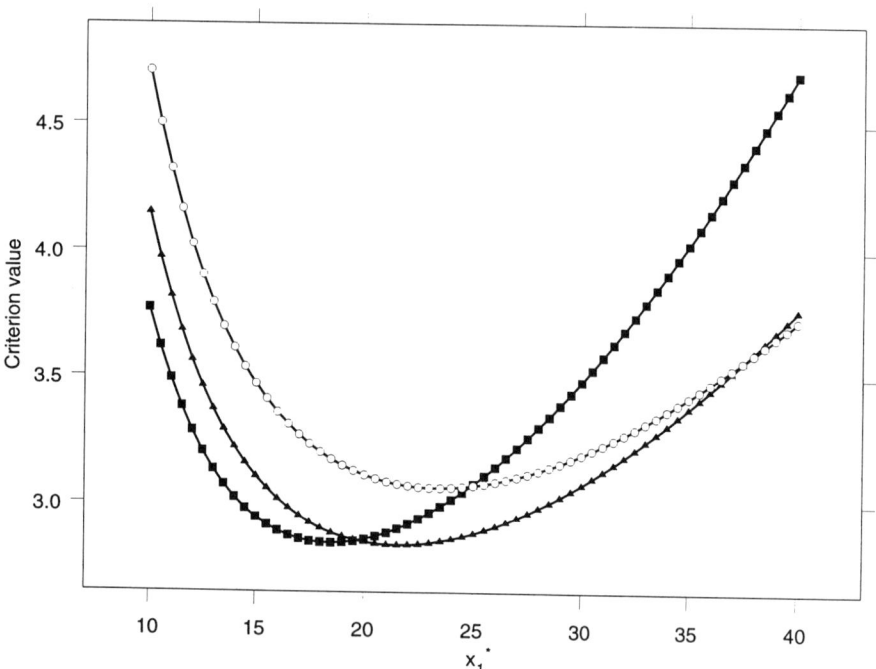

Figure 11.3: Criterion K_2 for different combinations of (r_1, r_2): ○ *corresponds to* $(7,3)$, ▲ *to* $(8,2)$ and ■ *to* $(9,1)$. Criterion value: $K_2 \cdot 10^3$.

The combination $r_1 = 8$ and $r_2 = 2$ gives the following (sub)-optimal design

$$\begin{pmatrix} 21.55 & 1440 \\ 8 & 2 \end{pmatrix}$$

with criterion value $K_2 = 2.83 \cdot 10^{-3}$.

The same analysis leads to the (sub)-optimal solution for the C_β-optimal unrestricted design

$$\begin{pmatrix} 20.29 & 1440 \\ 7 & 3 \end{pmatrix}$$

with a criterion value of $K_3 = 2.85 \cdot 10^{-3}$.

11.3.3 Non-convexity of the continuous NLP formulation

In Section 11.3.2 the optimal unrestricted two-point designs for all three criteria K_i are found as a solution of a mixed-integer NLP problem. One may think that the problem can be solved easier (without checking all different combinations of r_1 and r_2) by a continuous formulation of the NLP, as given by (11.7). However, the continuous NLP is not convex which will be shown for D-optimality.

For D-optimality, Equation (11.8) shows that the choice of x_1^* and x_2^* is independent of r_1 and r_2. For the test case: $x_1^* = 28.32$ and $x_2^* = 1440$. Suppose we take the sub-optimal design with $(r_1, r_2) = (6, 4)$ as starting values of the continuous NLP so that the design is given by $x_1 = x_2 = \ldots = x_6 = 28.32$ and $x_7 = x_8 = x_9 = x_{10} = 1440$. Both GINO and the solver of EXCEL could not find the optimal design ($r_1 = r_2 = 5$). The solution found by both programs is equal to the chosen starting values (sub-optimal design). For the optimal design, $(r_1, r_2) = (5, 5)$, x_6 of the sub-optimal design has to be changed from 28.32 to 1440. Figure 11.4 shows the criterion value depending on the value of x_6 ranging from 28.32 to 1440.

From Figure 11.4 it is clear that the continuous NLP is not convex. It also shows that improving several starting designs by non-linear programming methods leads to different not necessarily (global) optimal designs. The optimal design (global minimum) is hard to be found because of the existing sub-optimal two-points designs. So, for the continuous NLP it is also necessary to check different combinations of the number of measurements on each support point. For the C_α and C_β-criterion the same problems are encountered.

11.4 Conclusions

For a test case of the Michaelis-Menten function, exact optimal replication-free and unrestricted designs have been calculated. From the present results it can be seen that the result of the numerical calculation of the D-optimal unrestricted design is equal to the optimal design found analytically. Given the optimal unrestricted design, the D-optimal replicationfree design corresponds with ones intuition: two clusters of 5 points are found around the two support points. These results are promising for calculating the C_α- and C_β-(sub)-

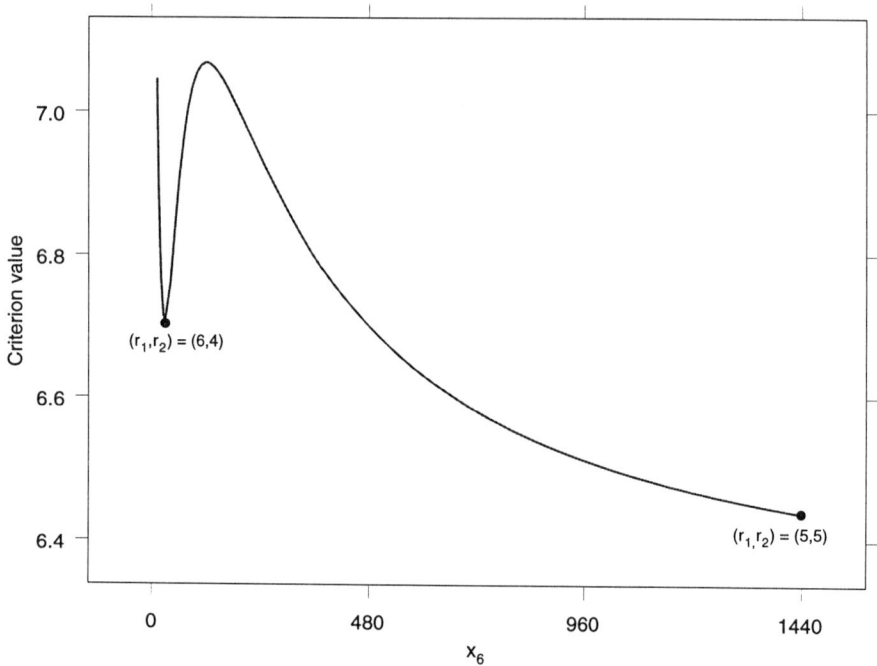

Figure 11.4: Non-convexity of the continuous NLP formulation of the optimal unrestricted design problem. Criterion value: $K_1 \cdot 10^7$.

optimal designs, since there are no analytical solutions for the corresponding optimisation problems.

The C_α- and C_β-(sub)-optimal unrestricted designs which were found numerically, correspond with the optimal replicationfree designs. For this reason we are hopeful that the unrestricted two-point designs found are globally optimal. However, absolute certainty can only be guaranteed by an analytical solution.

The mixed-integer formulation is preferred over the continuous NLP formulation because of the following reasons. In the first place, the MINLP problem has only 4 variables for two-point designs, independent of n. The continuous NLP problem has n variables. Secondly, the mixed-integer formulation makes it more clear that we have to check all different combinations (given n) of the number of measurements at each support point (r_1 and r_2). Using several (random) starting designs for a non-linear programming improvement with respect to the NLP formulation does not guarantee that the best design will be found (Hendrix, 1998). Furthermore, the D-optimal design was found directly by the mixed-integer formulation.

In this paper we focused on exact optimal designs for one test case. These designs are indicative of designs for the Michaelis-Menten function more generally. For the Michaelis-Menten function, as for all intrinsically non-linear functions, different values of the parameters can be investigated to find out how robust exact optimal designs for the Michaelis-Menten function are against misspecification of the parameters. This will be done in future research in analogy to Rasch (1995) who showed, in a case study of the three parametric Bertalanffy function, that a fair robustness of optimal designs can be assumed in a reasonable part of the parameter space.

References

1. Boer, E. P. J. (1995). *Snelle algoritmen voor het vinden van optimale herhalingsvrije proefopzetten*, Afstudeervak in de Wiskundige Statistiek, Vakgroep Wiskunde, Landbouwuniversiteit Wageningen.

2. Ermakov S. M. and Zigljavskij A. A. (1987). *Matematitscheskaja teorija optimalnich experimentov*, Moskva: Nauka.

3. Fedorov, V. V. (1972). *Theory of Optimal Experiments*, New York: Academic Press.

4. Hendrix, E. M. T. (1998). Global Optimization at Work, *Thesis*, Wageningen Agricultural University.

5. Liebman, J. S., Schrage, L., Lasdon, L. S. and Waren, A. D. *GINO, General Interactive Optimizer*, Chicago: Lindo Systems.

6. Michaelis, L. and Menten, M. L. (1913). Die Kinetik der Invertinwirkung. *Biochem. Zeit.*, **49**, 333–369.

7. Pukelsheim, F. (1993). *Optimal Design of Experiments*, New York: John Wiley & Sons.

8. Rasch, D. A. M. K. (1990). Optimum experimental design in nonlinear regression. *Communications in Statistics—Theory and Methods*, **19**, 4789–4806.

9. Rasch, D. A. M. K. (1995). The Robustness against Parameter Variation of Exact Locally Optimum Experimental Designs in Growth Models - A Case Study, *Computational Statistics and Data Analysis*, **20**, 441–453.

10. Rasch, D. A. M. K., Hendrix, E. M. T. and Boer E. P. J. (1997). Replication-free optimal design in regression analysis, *Computational Statistics*, **12**, 19–52.

12
D-Optimal Designs for Quadratic Regression Models

E. E. M. van Berkum, B. Pauwels and P. M. Upperman

Eindhoven University of Technology, Eindhoven, The Netherlands
University of Antwerp, Antwerpen, Belgium
Quality Engineering Consultancy, Oosterhesselen, The Netherlands

Abstract: In this paper, D-optimal designs are given for some incomplete quadratic models in linear regression. The models we consider contain all first-order interactions, all linear terms and some quadratic terms. The experimental region is a hypercube.

Key words and phrases: D-optimal designs, approximate theory, linear regression, incomplete quadratic models

12.1 Introduction

In recent years many optimal designs have been constructed for the case that the expected value of the response variable is related to explanatory variables by a linear model. In the models considered all (quantitative) explanatory variables have the same degree (for example quadratic or cubic). However, in practice it is not always possible that all, say q, explanatory variables have the same degree. It might be the case that some have two levels and other have three levels. This paper considers an incomplete quadratic model in the sense that some, say k, variables have linear and quadratic terms, but the other $(q-k)$ variables, corresponding with $(q-k)$ two-level factors, have only linear terms. All $\binom{q}{2}$ interactions of two factors are however included in the model. So we have the model

$$Y = \beta_0 + \beta_{11}x_1^2 + \ldots + \beta_{kk}x_k^2 + \beta_1 x_1 + \ldots + \beta_q x_q +$$
$$+\beta_{12}x_1 x_2 + \ldots + \beta_{q-1,q}x_{q-1}x_q + \varepsilon, \quad (12.1)$$

where each x_i corresponds to a factor of the design, Y is the response, and ε is the error term of the response variable in the model. The experimental region $\mathcal{X}$ is defined by

$$\mathcal{X} = \{x \in \mathbb{R}^q | x = (x_1, ..., x_q)', -1 \leq x_i \leq 1 \text{ for all } i = 1, ..., q\} \;.$$

The experiment consists of N runs. The number of parameters p equals

$$p = q + \binom{q}{2} + 1 + k = 1 + k + \tfrac{1}{2}q(q+1) \;.$$

The design can be seen as a probability distribution giving weights N^{-1} to N not necessarily distinct elements of the experimental region $\mathcal{X}$. Such a design is called 'exact' because it can be realized in practice. The derivation of optimal designs is simplified by using the so-called approximate theory in which the N-trial design is replaced by a measure ξ over $\mathcal{X}$. In the sequel we present D-optimal designs, i.e. designs with a maximal determinant of the information matrix, for model (12.1).

12.2 D-Optimal Designs

Define S_l as the set of $\binom{k}{l}2^{q-l}$ elements of $\mathcal{X}$, where l of the k quadratic variables have the value zero, and all other variables have the value -1 or 1. S_0 contains for example the 2^q vertices of the q-dimensional cube. If $l > k$, then the set S_l is empty.

Definition 12.2.1 We define a design $\xi(\alpha_0, \alpha_r, \alpha_s)$ consisting of the subsets S_0, S_r and S_s where

   i. the elements of S_0 are given weights α_0,

   ii. the elements of S_r are given weights α_r,

   iii. the elements of S_s are given weights α_s,

$1 \leq r \leq k$ and $s > r$. Note that if $k = 1$, then S_s is empty.
The spectrum of this design consists of N points, where

$$N = 2^q + \binom{k}{r}2^{q-r} + \binom{k}{s}2^{q-s} \;.$$

Now we shall specify r, s and weights α_0, α_r and α_s for which $\xi(\alpha_0, \alpha_r, \alpha_s)$ of Definition 12.2.1 is a D-optimal design.

D-Optimal Designs

Theorem 12.2.1 *First define*

$$\alpha_0^* = \frac{(k^2 - k(r+s) + rs) + (k - 2k^2 + k(r+s))u^* + k(k-1)v^*}{rs2^q},$$

$$\alpha_r^* = \frac{(r-1)!(k-r)![(k-s) + (1 - 2k + s)u^* + (k-1)v^*]}{(k-1)!(r-s)2^{q-r}},$$

$$\alpha_s^* = \frac{(s-1)!(k-s)![(k-r) + (1 - 2k + r)u^* + (k-1)v^*]}{(k-1)!(s-r)2^{q-s}},$$

where

$$xu^* = \frac{2q(k+1) + k + 7 + (k-1)\sqrt{4q^2 + 12q + 17}}{4(q+2)(2qk - k^2 + 3k + 2)/(2q - k + 3)}, \quad (12.2)$$

$$v^* = \frac{(4q^2 + 8q + 9)k + 2q - 5 + (2qk + k + 3)\sqrt{4q^2 + 12q + 17}}{8(q+2)^2(2qk - k^2 + 3k + 2)/(2q - k + 3)}.$$

The design $\xi(\alpha_0^*, \alpha_r^*, \alpha_s^*)$ as defined in Definition 12.2.1 is D-optimal for values r and s that satisfy

$$1 \leq r \leq [k + (1 - 2k)u^* + (k-1)v^*]/(1 - u^*) < s \leq \max\{2, k\}. \quad (12.3)$$

Corollary 12.2.1 *A design consisting of all the vertices, of the centers of edges that are related to the 'quadratic' variables and of the center points of the simplices corresponding to the k 'quadratic' variables, with appropriate weights, is D-optimal. More formally: the design $\xi(\alpha_0^*, \alpha_1^*, \alpha_k^*)$ of Theorem 12.2.1 is D-optimal. For $2 \leq k \leq q$ this design consists of the subsets S_0, S_1 and S_k. For $k = 1$ this design consists of the two subsets S_0 and S_1. The values of the weights of these designs are given in Table 12.1.*

12.3 Optimality of the Designs

We shall prove the D-optimality of the design $\xi(\alpha_0^*, \alpha_r^*, \alpha_s^*)$ of Theorem 12.2.1 using the equivalence theorem of Kiefer and Wolfowitz (1960). For the information matrix of the design $\xi(\alpha_0, \alpha_r, \alpha_s)$ we use the notation $M(\xi)$ and for the standardized variance of the estimated response in a point $x \in \mathcal{X}$

$$d(x, \xi) = (f(x))'M^{-1}(\xi)f(x), \quad (12.4)$$

where

$$(f(x))' = (1, x_1^2, ..., x_k^2, x_1, ..., x_q, x_1x_2, ..., x_{q-1}x_q).$$

First we shall derive that the best design of the type $\xi(\alpha_0, \alpha_r, \alpha_s)$ given in Definition 12.2.1 has the weights α_0^*, α_r^* and α_s^*, as given in (12.2). However,

Table 12.1: Values determining the D-optimal designs of Corollary 12.2.1.

q	k	u^*	v^*	sums of weights for subsets $2^q\alpha_0^*$	$k2^{q-1}\alpha_1^*$	$2^{q-k}\alpha_k^*$
1	1	0.6667	–	0.6667	0.3333	–
2	1	0.7500	–	0.7500	0.2500	–
	2	0.7435	0.5832	0.5832	0.3206	0.0962
3	1	0.8000	–	0.8000	0.2000	–
	2	0.7970	0.6549	0.6549	0.2842	0.0609
	3	0.7930	0.6516	0.5102	0.4242	0.0656
4	1	0.8333	–	0.8333	0.1667	–
	2	0.8317	0.7055	0.7055	0.2524	0.0421
	3	0.8297	0.7038	0.5779	0.3777	0.0444
	4	0.8271	0.7016	0.4505	0.5021	0.0474
5	1	0.8571	–	0.8571	0.1429	–
	2	0.8562	0.7432	0.7432	0.2260	0.0308
	3	0.8550	0.7421	0.6293	0.3386	0.0321
	4	0.8536	0.7409	0.5156	0.4507	0.0337
	5	0.8518	0.7394	0.4021	0.5622	0.0358
6	1	0.8750	–	0.8750	0.1250	–
	2	0.8744	0.7723	0.7723	0.2041	0.0236
	3	0.8736	0.7717	0.6697	0.3060	0.0244
	4	0.8728	0.7709	0.5671	0.4075	0.0253
	5	0.8717	0.7700	0.4647	0.5088	0.0265
	6	0.8705	0.7689	0.3624	0.6097	0.0279
7	1	0.8889	–	0.8889	0.1111	–
	2	0.8885	0.7955	0.7955	0.1859	0.0186
	3	0.8880	0.7951	0.7022	0.2787	0.0191
	4	0.8874	0.7946	0.6089	0.3714	0.0198
	5	0.8867	0.7940	0.5157	0.4639	0.0205
	6	0.8860	0.7933	0.4225	0.5561	0.0213
	7	0.8850	0.7924	0.3295	0.6481	0.0224
8	1	0.9000	–	0.9000	0.1000	–
	2	0.8997	0.8144	0.8144	0.1705	0.0105
	3	0.8994	0.8141	0.7289	0.2557	0.0154
	4	0.8990	0.8138	0.6434	0.3408	0.0158
	5	0.8985	0.8134	0.5579	0.4257	0.0163
	6	0.8980	0.8129	0.4725	0.5106	0.0169
	7	0.8974	0.8124	0.3871	0.5953	0.0176
	8	0.8967	0.8117	0.3019	0.6798	0.0183

D-Optimal Designs

this does not prove that the design is D-optimal. It might be better (meaning a larger $\det(M(\xi))$ to use other points of the experimental region. But then we show that $\max_{x \in \mathcal{X}} d(x, \xi(\alpha_0^*, \alpha_r^*, \alpha_s^*)) \leq p$ from which we conclude that the design is G-optimal and hence D-optimal.

Consider a design $\xi(\alpha_0, \alpha_r, \alpha_s)$ as in Definition 12.2.1. The sum of weights of this design equals

$$2^q \alpha_0 + \binom{k}{r} 2^{q-r} \alpha_r + \binom{k}{s} 2^{q-s} \alpha_s = 1 . \tag{12.5}$$

The information matrix $M(\xi)$ is equal to

$$\begin{bmatrix} A & & & & \\ & uI_k & & & \\ & & I_{q-k} & & \\ & & & vI_{\frac{1}{2}k(k-1)} & \\ & & & & uI_{k(q-k)} \\ & & & & & I_{\frac{1}{2}(q-k)(q-k-1)} \end{bmatrix} , \tag{12.6}$$

where all the other entries are zero, I_m is the $m \times m$ identity matrix, J_m is the $m \times m$ matrix consisting of ones,

$$A = \begin{bmatrix} 1 & u \ldots u \\ \begin{matrix} u \\ \vdots \\ u \end{matrix} & (u-v)I_k + vJ_k \end{bmatrix} ,$$

and u and v are defined by

$$u = 2^q \alpha_0 + \binom{k-1}{r} 2^{q-r} \alpha_r + \binom{k-1}{s} 2^{q-s} \alpha_s , \tag{12.7}$$

$$v = 2^q \alpha_0 + \binom{k-2}{r} 2^{q-r} \alpha_r + \binom{k-2}{s} 2^{q-s} \alpha_s . \tag{12.8}$$

From equalities (12.5), (12.7) and (12.8), we can see that

$$0 < v < u < 1 \text{ and } 0 < 1 - 2u + v < u - v < 1 - u . \tag{12.9}$$

If one maximizes $\det(M(\xi))$ with respect to u and v under the conditions $u > 0$ and $v > 0$, then one obtains $u = u^*$ and $v = v^*$, where u^* and v^* are the values given in (12.2). Solving the equations (12.5), (12.7) and (12.8), we can determine the values of α_0, α_r and α_s. This yields the values α_0^*, α_r^* and α_s^* of (12.2). Now we will show that

$$\max_{x \in \mathcal{X}} d(x, \xi(\alpha_0^*, \alpha_r^*, \alpha_s^*)) \leq p \tag{12.10}$$

from which we conclude D-optimality using the equivalence theorem of Kiefer and Wolfowitz (1960). We compute $d(x, \xi^*) = d(x, \xi(\alpha_0^*, \alpha_r^*, \alpha_s^*))$ from (12.4) and (12.6), and we obtain

$$d(x, \xi^*) = \frac{u^* + (k-1)v^*}{u^* + (k-1)v^* - ku^{*2}} + \qquad (12.11)$$

$$+ \frac{u^* + (k-1)v^* - ku^{*2} - v^* + u^{*2}}{(u^* - v^*)(u^* + (k-1)v^* - ku^{*2})} \sum_{i=1}^{k} x_i^4 - \frac{2u^*}{u^* + (k-1)v^* - ku^{*2}} \sum_{i=1}^{k} x_i^2 +$$

$$- \frac{2(v^* - u^{*2})}{(u^* - v^*)(u^* + (k-1)v^* - ku^{*2})} \sum_{i=1}^{k-1} \sum_{j=i+1}^{k} x_i^2 x_j^2 + \frac{1}{u^*} \sum_{i=1}^{k} x_i^2 + \sum_{j=k+1}^{q} x_j^2 +$$

$$+ \frac{1}{v^*} \sum_{i=1}^{k-1} \sum_{j=i+1}^{k} x_i^2 x_j^2 + \frac{1}{u^*} \sum_{i=1}^{k} \sum_{j=k+1}^{q} x_i^2 x_j^2 + \sum_{i=k+1}^{q-1} \sum_{j=i+1}^{q} x_i^2 x_j^2 .$$

First we note that the x_j^2 (for $k+1 \leq j \leq q$) only occur in positive terms. Therefore we can find an upper bound for $d(x, \xi^*)$ by substituting $x_j^2 = 1$ in (12.11), $k+1 \leq j \leq q$. Using equation $\frac{\partial \det M(\xi)}{\partial v} = 0$, one can show that the coefficient of $\sum \sum_{1 \leq i < j \leq k} x_i^2 x_j^2$ is equal to zero, and using equation $\frac{\partial \det M(\xi)}{\partial u} = 0$ that the coefficient of $\sum_{1 \leq i \leq k} x_i^2$, say c, is equal to the coefficient of $-\sum_{1 \leq i \leq k} x_i^2$. Simplifying constant terms, we conclude

$$d(x, \xi^*) \leq k + 1 + \frac{q(q+1)}{2} + c \sum_{i=1}^{k} (x_i^2 - x_i^4) , \qquad (12.12)$$

with

$$c = \frac{(q-k+1)(u^* + (k-1)v^* - ku^{*2}) - 2u^{*2}}{u^*(u^* + (k-1)v^* - ku^{*2})},$$

or

$$c = \frac{4(q+2)^2(-2 - 3k + k^2 - 2kq)}{(3 - k + 2q)(7 + k + 2q + 2kq + (k-1)\sqrt{4q^2 + 12q + 17})}.$$

Since $1 \leq k \leq q$, one can see that in (12.12) $c < 0$, which proves (12.10). The last thing that has to be verified is that the weights are nonnegative. It is easily verified that $\alpha_0^* > 0$. From 12.2 we conclude that $\alpha_s^* \geq 0$ implies that $(k - r) + (1 - 2k + r)u + (k - 1)v > 0$. So the second inequality of 12.3 is a necessary condition with equality if and only if $k = 1$. From $\alpha_r^* > 0$ it follows that the third inequality of 12.3 is a necessary condition. That these conditions are also sufficient follows from the fact that (12.10) holds. Furthermore it can easily be verified that there is always an s, satisfying the third inequality of 12.3, that also satisfies the fourth inequality, i.e. $s \leq k$ if $k \geq 2$ and $s = 2$ if $k = 1$.

12.4 Conclusion

We derived D-optimal approximate designs for some quadratic models. In general these designs can not be realized in practice. To compare designs with respect to the D-criterion, the D-efficiency has been introduced in literature.

Definition 12.4.1 The D-efficiency of a design ξ is defined as

$$D\text{-efficiency} = (\det(M(\xi))/\det(M(\xi^*)))^{(1/p)}, \qquad (12.13)$$

where ξ^* is a D-optimal design.

It is possible to construct exact designs (i.e. designs that can be realized in practice) with efficiency $1-\eta$ for any small positive value of η [Fedorov (1972, Theorem 3.1.1)]. For such a design, since the product of the weights and the number of observations must be an integer, in general a large number of observations has to be chosen. Such designs are not very useful for practical applications. Sometimes one is lucky. For $q=3$ and $k=1$ one can construct an exact D-optimal design with only 20 observations as follows. At each of the 8 vertices two observations are taken and in each of 4 out of the 8 edge centers one observation (see Definition 12.2.1). Indeed we have $\alpha_0 = \frac{2}{20} = 0.10$ and $\alpha_1 = \frac{1}{20} = 0.05$. In other cases one can use the D-optimal approximate designs to construct exact designs with a good efficiency. This method, among other methods, has been used in Upperman (1993) to construct designs with a small number of runs for factorial experiments.

References

1. Fedorov, V. V. (1972). *Theory of Optimal Experiments*, London, New York: Academic Press Inc..

2. Kiefer, J. and Wolfowitz, (1960). The equivalence of two extremum problems, *Canadian Journal of Mathematics*, **12**, 363–366.

3. Upperman, P. M. (1993). Designs with a small number of runs for factorial experiments, *Dissertation*, Eindhoven: Technische Universiteit Eindhoven.

13

On the Use of Symmetry in Optimal Design of Experiments

Vladimir Soloviov

Moscow State Social University, Russia

Abstract: The paper deals with symmetry considerations in convex optimization problems with applications to optimal experimental design. In contrast to the common use of symmetry no group structure is assumed for the underlying invariance generating transformations.

Keywords and phrases: Symmetry, convex optimization problems, invariance generating transformations, change of variables, optimal design of experiments, symmetric designs, polynomial regression, MV-optimality, L-optimality

13.1 Symmetry in Convex Optimization Problems

Analytic solutions to extremal problems may be obtained explicitly only in rare cases. But it is often possible to simplify extremal problems upon taking into account considerations of symmetry [Kiefer (1970), Pukelsheim (1993) and Solov'ev (1990, 1991)].

We shall first consider the extremal problem

$$\inf_{x \in A} f(x) \qquad (13.1)$$

in which the set A and the functional $f(x)$ are invariant with respect to the change in variables $\tilde{x} = \tilde{x}(x)$, in the sense that

$$\tilde{x}(x) \in A \quad \forall x \in A$$

and

$$f(\tilde{x}) \leq f(x) \quad \forall x \in A.$$

Theorem 13.1.1 *If there exists a unique solution x_* to the problem (13.1), then it is invariant with respect to the change in variables, that is*

$$\tilde{x}(x_*) = x_*.$$

PROOF. Show that the set X_* of all solutions to the problem (13.1) is invariant with respect to the change in variables $\tilde{x}$. If x_* is a solution, i.e. $f(x) \geq f(x_*)$ $\forall x \in A$, then

$$f(x) \geq f(x_*) \geq f(\tilde{x}(x_*)) \quad \forall x \in A,$$

and therefore $\tilde{x}(x_*)$ is also a solution. If this solution is unique, then $\tilde{x}(x_*)$ must coincide with x_*. Anyway, the set X_* is invariant with respect to the change in variables. ∎

Theorem 13.1.2 *If the functional $f(x)$ is convex and lower semi-continuous on a convex compact subset A of a locally convex linear topological space and the mapping $\tilde{x}$ is continuous, then there exists a solution x_* to the problem (13.1) which is invariant with respect to the change in variables $\tilde{x}$.*

PROOF. It follows from the Weierstrass theorem that the set X_* is nonempty and compact; and from convexity of the functional $f(x)$ and the set A it follows that X_* is convex. As we have just proved, the mapping $\tilde{x}$ is invariant on this set, i.e. it maps all points of the set into the same set. Since the mapping is continuous, then it follows from Tichonoff's fixed point theorem that there exists a point $x_* \in X_*$ such that $\tilde{x}(x_*) = x_*$. ∎

In contrast to the common use of symmetry no group structure is assumed for the underlying invariance generating transformations. Here is a simple example where the transformation $\tilde{x}$ does <u>not</u> induce any group (because it is degenerate).

$$f(x_1, x_2) = |x_1| + |x_2|, \quad A = \{-4 \leq x_1 \leq 3, \; -1 \leq x_2 \leq 2\}$$

$$\tilde{x}_1 = x_1, \quad \tilde{x}_2 = 0.$$

13.2 Optimal Design of Experiments

Now we shall consider the following problem of optimal design of experiments

$$\Phi_{min} := \inf_{\xi} \Phi(M(\xi)^{-1}). \tag{13.2}$$

In (13.2) ξ runs over all discrete probability measures (designs of experiment) on a compact set X, such that the corresponding information matrix

$$M(\xi) = \int_X f(x)f^T(x)\xi(dx)$$

is not degenerate. In other words, all designs ξ in (13.2) are supposed to be nonsingular. The regression function $f(x) : X \to \mathbf{R}^m$ is assumed to be continuous on X.

Let the set $\mathcal{M}$ of all the information matrices $M(\xi)$ contain at least one nondegenerate matrix. Then the set

$$\mathcal{M}_0 := \{M \in \mathcal{M} : \quad M > 0\}$$

is not empty. Almost all optimality criteria $\Phi(D), D = M^{-1}$ from the theory of optimal design of experiments meet the following conditions.

1. $\Phi(\alpha D) = \alpha \Phi(D) > 0$ for all $\alpha > 0, D > 0$;

2. The function $\Phi(M^{-1})$ is convex in M, provided $M > 0$.

Lemma 13.2.1 *The function*

$$k(M) := 1/\Phi(M^{-1}), \quad M > 0$$

is positively homogeneous of degree $p = 1$ and concave.

PROOF. The proof is based on the fact that the level sets $k(M) \geq 1$ and $\Phi(M^{-1}) \leq 1$ are the same. Since the function $\Phi(M^{-1})$ is convex in M, provided $M > 0$, the latter is convex (to be more precise, its intersection with the convex cone of all positive definite matrices is convex). Because the function $\Phi(M^{-1})$ is positively homogeneous of degree $p = -1$, then it follows that the function $k(M)$ is positively homogeneous of degree $p = 1$. Hence the set

$$\{ (\alpha, M) : 0 < \alpha \leq k(M), \quad M > 0 \ \}$$

is a convex cone generated by the convex set $k(M) \geq 1$ and therefore the function $k(M)$ is concave. ∎

We shall assume a bit more restrictive hypothesis that the function $k(M)$, $M \geq 0$ is a concave gauge (i.e. upper semi-continuous, positively homogeneous and concave). From homogeneity considerations we get

$$\Phi^{-1}_{min} = \sup_{M \in \mathcal{M}_0} 1/\Phi(M^{-1}) = \sup_{M \in \mathcal{M}_0} k(M).$$

Then it follows from Lemma 13.3.1 in [Bakhshiyan and Solov'ev (1998)] that

$$\Phi^{-1}_{min} = \sup_{M \in \mathcal{M}} k(M). \tag{13.3}$$

Since the set $\mathcal{M}$ of all the information matrices $M(\xi)$ is compact [Pukelsheim (1993)], then the solution of the problem (13.3) will always exist, but in general it may correspond to a singular design ξ_*. Modern studies on optimal design of experiments often start directly from the problem (13.3). This leads to no loss in generality, because the classical problem (13.2) reduces to (13.3), as we have just seen. This important point seems to be missed out from the theory of optimal design of experiments.

Taking into account this consideration, one can deduce from the general equivalence theorem [see Pukelsheim (1993, p. 177)] that

$$\Phi_{min} = \max_{D \geq 0} \{k^0(D): \max_{x \in X} <Df(x), f(x)> \leq 1\} \qquad (13.4)$$

where

$$k^0(D) := \inf_{k(M) \geq 1} <D, M>, \quad <D, M> := SpDM \qquad (13.5)$$

and that a solution D_* of the dual problem (13.4) exists. If the matrix $M_* \in \mathcal{M}$ is a solution of the problem (13.3), then the function $<D_* f(x), f(x)>$ attains its maximal value on the set X at the support points x_s^* of the optimal design ξ_*.

13.3 Optimal Designs for Polynomial Regression

In this section we need some definitions. A matrix L is called even, if the elements l_{ij} of the matrix L are equal to zero, when $i+j$ is odd. A design ξ is called even, if $\xi(x) = \xi(-x))$ for all $x \in X$.

Lemma 13.3.1 *The criteria of A-optimality $\Phi(D) = SpD$, E-optimality $\Phi(D) = \lambda_{max}(D)$, MV-optimality $\Phi(D) = \max_{i=1,...,m} d_{ii}$, and L-optimality $\Phi(D) = <L, D>$ with an even matrix L are invariant with respect to the change in variables*

$$\tilde{m}_{ij} = (-1)^{i+j} m_{ij}. \qquad (13.6)$$

PROOF. In the matrix terms we have

$$\tilde{M} = \Lambda M \Lambda, \quad \Lambda = diag\{\lambda_i\}, \quad \lambda_i = (-1)^i, \quad i = 1,\ldots,m, \quad \Lambda^{-1} = \Lambda.$$

Therefore $\tilde{M}^{-1} = \Lambda M^{-1} \Lambda$, i.e. the inverse matrix is also invariant with respect to this change of variables. Since its diagonal elements do not change, the criteria of A-, MV- and L-optimality with a diagonal matrix L are invariant with respect to this change of variables.

For E-criterion

$$\Phi(M^{-1}) = \lambda_{max}(M^{-1}) = 1/\lambda_{min}(M)$$

this follows from the fact that the matrix $\tilde{M}^{-1} = \Lambda^{-1}M^{-1}\Lambda$ has the same characteristic values as the matrix M^{-1}. And for L-criterion with an even matrix L this follows from the equalities

$$SpL\tilde{M}^{-1} = SpL(\Lambda M^{-1}\Lambda) = Sp(\Lambda L \Lambda)M^{-1} = SpLM^{-1}.$$

∎

The rest of the paper relates to the polynomial regression

$$f(x) = (b_1, b_2 x, b_3 x^2, \cdots, b_m x^{m-1})^T$$

in which $b_1, \cdots, b_m$ are fixed numbers.

Theorem 13.3.1 *For polynomial regression on the symmetric interval $[-a, a]$ and the optimality criteria which were indicated above there exist an even optimal design ξ_* and even solutions M_* and D_* of the dual problems (13.3) and (13.4) (for which $m_{ij}^* = d_{ij}^* = 0$, if $i + j$ is odd).*

PROOF. We shall use the duality theorem (13.4) and Lemma 13.3.1. To this end, we have to check that both the functions $k^0(D)$ and

$$\max_{x \in X} <Df(x), f(x)> = \max_{|x| \leq a} \sum_{i,j=1}^{m} d_{ij} b_i b_j x^{i+j-2} \tag{13.7}$$

are invariant with respect to the following change of variables

$$\tilde{d}_{ij} = (-1)^{i+j} d_{ij}.$$

From Lemma 13.3.1 and the definition of the concave polar gauge $k^0(D)$ [see (13.5)] we obtain

$$k^0(\tilde{D}) = k^0(\Lambda D \Lambda) = \inf_{k(M) \geq 1} <\Lambda D \Lambda, M> = \inf_{k(\Lambda M \Lambda) \geq 1} <D, \Lambda M \Lambda> = k^0(D)$$

Similarly, if we substitute $\tilde{d}_{ij} = (-1)^{i+j-2} d_{ij}$ in (13.7) instead of d_{ij}, then we get

$$\max_{|x| \leq a} \sum_{i,j=1}^{m} d_{ij} b_i b_j (-x)^{i+j-2}$$

which is the same as the right-hand side of (13.7). By Theorem 13.1.2 there exists a solution D_* of the dual problem (13.4) which is invariant with respect to this change in variables, so that $d_{ij}^* = -d_{ij}^*$ if $i+j$ is odd, i.e. $d_{ij}^* = 0$ if $i+j$ is odd.

Then the function $< D_* f(x), f(x) >$ is an even polynomial [see (13.7)] which attains its maximal value on the symmetric interval $[-a, a]$ at the support points x_s^* of an optimal design ξ_*. Hence, the support points of an optimal design ξ_* are symmetric with respect to zero. ∎

Now we fix the support points x_s^* of designs ξ and consider the convex extremal problem
$$\inf_{M \in \mathcal{M}} -k(M)$$
in which optimization is carried out with respect to the weights $\xi(x_s^*)$. Because the support points are symmetric with respect to zero, the set $\mathcal{M}$ and the functional $-k(M)$ are invariant with respect to the next change in variables $\tilde{\xi}(x_s^*) = \xi(-x_s^*)$ (which in fact leads to the change in variables of the form [13.6]). Once again, by Theorem 13.1.2 (applied to the weights) there exists an even optimal design.

Thus, the elements m_{ij}^* of the information matrix M_* are sums of the terms $p b_i b_j x^{i+j-2}$ which occur in pairs: one time with $x = x_s^*$, $p = \xi(x_s^*)$ and next time with $x = -x_s^*$, $p = \xi(x_s^*)$. These sums are equal to zero, if $i+j$ is odd.

Remarks

1. In Theorem 13.1.1 the set A may be arbitrary.

2. In Theorem 13.1.2 the assumptions of lower semi-continuity and convexity of the functional $f(x)$ may be replaced with those of convexity and closedness of all the level sets $\{x : f(x) \leq \alpha\}$.

3. For A- and E-criteria Theorem 13.3.1 is well known [Pukelsheim (1993) and Melas (1997)]. For E-criterion it follows immediately from Theorem 13.1.1, because E-optimal design is unique for $m > 2$ [Melas (1997)].

4. It was the aim of the present paper to provide a reader with a finite-dimensional proof of the existence of symmetric optimal designs. But Theorem 13.1.2 also applies to the set of all Radon probability measures. This leads to infinite-dimensional considerations similar to those from [Kiefer (1970)].

One more application of the modern duality theory to the theory of optimal design of experiments is mentioned in [Solov'ev (1997)].

Acknowledgments. This work is supported by RFBR, grant N97-01-01005. The author is grateful to a referee for helpful comments on this paper.

References

1. Bakhshiyan, B. Ts. and Solov'ev, V. N. (1998). Theory and Solution Algorithms for the Problems of L- and MV- Optimal Design of Experiments, *Automation and Remote Control*, N8, 80-96 (in Russian).

2. Kiefer, J. C. (1970). The Role of Symmetry and Approximation in Exact Design Optimality, *Statistical Decision Theory and Related Topics. Proceedings of a Symposium, Purdue University* (Eds., S. S. Gupta and J. Yackel), pp. 109-118. New York: Academic Press.

3. Melas, V. B. (1997). *E-optimal Design of Experiments*. St-Petersburg State University (in Russian).

4. Pukelsheim, F. (1993). *Optimal Design of Experiments*. New York: John Wiley & Sons.

5. Solov'ev, V. N. (1990). On the Use of Symmetry in Smooth Extremal Problems, *Proceedings of the Steklov Inst. Math.*, **185**, 263-268 (translated from *Trudy Mat. Inst. Steklova*, 1988, **185**, 236-241).

6. Solov'ev, V. N. (1991). Simplification of Dual Extremal Problems Invariant with Respect to Change in Variables, *Math. Notes*, **49**, N5-6, 514-518 (translated from *Mat. Zametki*, 1991, **49**, N5, 104-109).

7. Solov'ev, V. N. (1997). Dual Extremal Problems and Their Applications to the Problems of Minimax Estimation, *Uspechi Mat. Nauk*, **52**, N4, 49-86 (translated in *Russian Mathematical Surveys*).

PART III
STATISTICAL INFERENCE

14

Higher Order Moments of Order Statistics from the Pareto Distribution and Edgeworth Approximate Inference

Aaron Childs, K. S. Sultan and N. Balakrishnan

McMaster University, Hamilton, ON, Canada
Al-Azhar University, Nasr City, Cairo, Egypt
McMaster University, Hamilton, ON, Canada

Abstract: In this paper, we first derive exact explicit expressions for the triple and quadruple moments of order statistics from the Pareto distribution. Also, we establish recurrence relations for single, double, triple and quadruple moments of order statistics from the Pareto distribution. These relations will enable one to find all moments (of order up to four) of order statistics for all sample sizes in a simple recursive manner. We then use these results to determine the mean, variance, and coefficients of skewness and kurtosis of certain linear functions of order statistics. These are then utilized to develop approximate confidence intervals for the Pareto parameters using the Edgeworth approximation. Finally, we extend the recurrence relations to the case of the doubly truncated Pareto distribution.

Keywords and phrases: Order statistics, exact moments, single moments, double moments, triple moments, quadruple moments, Pareto distribution, doubly truncated distribution, recurrence relations, Edgeworth approximation, coefficients of skewness and kurtosis, approximate confidence interval, pivotal quantity

14.1 Introduction

Let $X_1, X_2, \ldots, X_n$ be a random sample of size n from the Pareto population with probability density function (p.d.f.)

$$f(x; \theta, \sigma) = \nu \sigma^\nu (x - \theta)^{-(\nu+1)}, \quad x \geq \sigma + \theta, \ \nu > 0, \ \sigma > 0 \qquad (14.1)$$

and cumulative distribution function (c.d.f.)

$$F(x;\theta,\sigma) = 1 - \sigma^\nu(x-\theta)^{-\nu}, \quad x \geq \sigma + \theta, \; \nu > 0, \; \sigma > 0. \tag{14.2}$$

Let $X_{1:n} \leq X_{2:n} \leq \ldots \leq X_{n:n}$ be the order statistics obtained by arranging the above sample in increasing order of magnitude. Notice from (14.1) and (14.2) that

$$f(x) = \frac{\nu}{x-\theta}\{1 - F(x)\}. \tag{14.3}$$

Balakrishnan and Joshi (1982) used the above differential equation to derive some recurrence relations satisfied by the single and the double moments of order statistics from the Pareto model. These recurrence relations allow one to compute means, variances and covariances of all order statistics from the Pareto distribution for all sample sizes n in a simple recursive manner. Exact explicit expressions for the single and double moments of order statistics have been derived by Malik (1966) and Huang (1975); see also Arnold, Balakrishnan and Nagaraja (1992). For a detailed discussion of various aspects of the Pareto distribution, one may refer to Arnold (1983) or Johnson, Kotz and Balakrishnan (1994).

In this paper, we extend the results of Malik (1966) and Huang (1975) by deriving exact explicit expressions for the triple and quadruple moments of Pareto order statistics. Also, we consider the problem of finding confidence intervals for the Pareto parameters θ and σ based on the following pivotal quantities,

$$R_1 = \frac{\theta^* - \theta}{\sigma\sqrt{V_1}}, \quad R_2 = \frac{\sigma^* - \sigma}{\sigma\sqrt{V_2}} \text{ and } R_3 = \frac{\theta^* - \theta}{\sigma^*\sqrt{V_1}}, \tag{14.4}$$

where θ^* and σ^* are the BLUE's of θ and σ with variances $\sigma^2 V_1$ and $\sigma^2 V_2$, respectively. R_1 can be used to draw inference for θ when σ is known, while R_3 can be used to draw inference for θ when σ is unknown. Similarly, R_2 can be used to draw inference for σ when θ is unknown.

Notice that R_1 and R_2 in (14.4) can be rewritten as

$$R_1 = \frac{1}{\sqrt{V_1}}\left(\sum_{i=r+1}^{n-s} a_i Z_{i:n}\right) = \frac{R_1^*}{\sqrt{V_1}} \quad \text{and}$$

$$R_2 = \frac{1}{\sqrt{V_2}}\left(\sum_{i=r+1}^{n-s} b_i Z_{i:n} - 1\right) = \frac{R_2^* - 1}{\sqrt{V_2}}, \tag{14.5}$$

where, $Z_{i:n} = \frac{X_{i:n} - \theta}{\sigma}$, $i = r+1, \ldots, n-s$, is the standardized form of the available Type-II censored sample $X_{i:n}$, $i = r+1, \ldots, n-s$. Thus, they are linear functions of order statistics arising from the standardized Pareto distribution.

Since the distribution of a linear function of order statistics will in general not be known, we consider finding the approximate distribution by using the Edgeworth approximation for a statistic T (with mean 0 and variance 1):

$$G(t) \approx \Phi(t) - \phi(t)$$
$$\times \left\{ \frac{\sqrt{\beta_1}}{6}(t^2 - 1) + \frac{\beta_2 - 3}{24}(t^3 - 3t) + \frac{\beta_1}{72}(t^5 - 10t^3 + 15t) \right\}, \quad (14.6)$$

where $\sqrt{\beta_1}$ and β_2 are the coefficients of skewness and kurtosis of T, respectively, and $\Phi(t)$ is the c.d.f. of the standard normal distribution with corresponding p.d.f. $\phi(t)$.

The coefficients of skewness and kurtosis of linear functions of order statistics require knowledge of the single moments $E(Z_{i:n}^a)$, which we denote by $\mu_{i:n}^{(a)}$, the double moments $E(Z_{i:n}^a Z_{j:n}^b)$, which we denote by $\mu_{i,j:n}^{(a,b)}$, the triple moments $E(Z_{i:n}^a Z_{j:n}^b Z_{k:n}^c)$, which we denote by $\mu_{i,j,k:n}^{(a,b,c)}$, and the quadruple moments $E(Z_{i:n}^a Z_{j:n}^b Z_{k:n}^c Z_{l:n}^d)$, which we denote by $\mu_{i,j,k,l:n}^{(a,b,c,d)}$, of the standardized Pareto order statistics for $1 \leq i < j < k < l \leq n$ and $a, b, c, d \geq 0$, $a+b+c+d \leq 4$.

After presenting the BLUE's of θ and σ in the following section, we derive in Section 14.3 exact explicit expressions for the single, double, triple and quadruple moments of order statistics, which will extend the results of Malik (1966) and Huang (1975). Then, in Section 14.4, we use the differential equation in (14.3) to extend the results of Balakrishnan and Joshi (1982) by deriving recurrence relations for the triple and quadruple moments. These quantities are then used in Section 14.5 to determine the coefficients of skewness and kurtosis of the pivotal quantities R_1 and R_2 based on the BLUE's of θ and σ. We then propose Edgeworth approximations for the distributions of these pivotal quantities and show that this method provides close approximation to the percentage points of the pivotal quantities determined by Monte Carlo simulations. A numerical example to illustrate the method of inference developed in this paper is then presented in Section 14.6. Similar work has been carried out recently for the exponential distribution by Balakrishnan and Gupta (1996), for the Laplace distribution by Balakrishnan et al. (1996), and for the logistic distribution by Childs and Balakrishnan (1996). Finally, in Section 14.7, we generalize the recurrence relations to the case of doubly truncated Pareto distribution, thus extending the results of Balakrishnan and Joshi (1982).

14.2 BLUE's of θ and σ

Let $X_{r+1:n} \leq X_{r+2:n} \leq \ldots \leq X_{n-s:n}$ denote the available doubly Type-II censored sample from the Pareto distribution in (1.1), and let $Z_{i:n} = (X_{i:n} - \theta)/\sigma$,

$i = r+1, r+2, \ldots, n-s$, be the corresponding order statistics from the standard Pareto distribution. Let us denote $E(Z_{i:n})$ by $\mu_{i:n}$, $Var(Z_{i:n})$ by $\sigma_{i,i:n}$, and $Cov(Z_{i:n}, Z_{j:n})$ by $\sigma_{i,j:n}$; further, let

$$\mathbf{X} = (X_{r+1:n}, X_{r+2:n}, \ldots, X_{n-s:n})^T$$
$$\boldsymbol{\mu} = (\mu_{r+1:n}, \mu_{r+2:n}, \ldots, \mu_{n-s:n})^T$$
$$\mathbf{1} = \underbrace{(1, 1, \ldots, 1)^T}_{n-r-s}$$

and $\boldsymbol{\Sigma} = ((\sigma_{i,j:n})), \; r+1 \leq i, \; j \leq n-s$.

Then, the BLUE's of θ and σ are given by [see Balakrishnan and Cohen (1991)]

$$\theta^* = \left\{ \frac{\boldsymbol{\mu}^T \boldsymbol{\Sigma}^{-1} \boldsymbol{\mu} \mathbf{1}^T \boldsymbol{\Sigma}^{-1} - \boldsymbol{\mu}^T \boldsymbol{\Sigma}^{-1} \mathbf{1} \boldsymbol{\mu}^T \boldsymbol{\Sigma}^{-1}}{(\boldsymbol{\mu}^T \boldsymbol{\Sigma}^{-1} \boldsymbol{\mu})(\mathbf{1}^T \boldsymbol{\Sigma}^{-1} \mathbf{1}) - (\boldsymbol{\mu}^T \boldsymbol{\Sigma}^{-1} \mathbf{1})^2} \right\} \mathbf{X} = \sum_{i=r+1}^{n-s} a_i X_{i:n}, \quad (14.7)$$

and

$$\sigma^* = \left\{ \frac{\mathbf{1}^T \boldsymbol{\Sigma}^{-1} \mathbf{1} \boldsymbol{\mu}^T \boldsymbol{\Sigma}^{-1} - \mathbf{1}^T \boldsymbol{\Sigma}^{-1} \boldsymbol{\mu} \mathbf{1}^T \boldsymbol{\Sigma}^{-1}}{(\boldsymbol{\mu}^T \boldsymbol{\Sigma}^{-1} \boldsymbol{\mu})(\mathbf{1}^T \boldsymbol{\Sigma}^{-1} \mathbf{1}) - (\boldsymbol{\mu}^T \boldsymbol{\Sigma}^{-1} \mathbf{1})^2} \right\} \mathbf{X} = \sum_{i=r+1}^{n-s} b_i X_{i:n}. \quad (14.8)$$

Furthermore, the variances and covariance of these BLUE's are given by [see Balakrishnan and Cohen (1991)]

$$Var(\theta^*) = \sigma^2 \left\{ \frac{\boldsymbol{\mu}^T \boldsymbol{\Sigma}^{-1} \boldsymbol{\mu}}{(\boldsymbol{\mu}^T \boldsymbol{\Sigma}^{-1} \boldsymbol{\mu})(\mathbf{1}^T \boldsymbol{\Sigma}^{-1} \mathbf{1}) - (\boldsymbol{\mu}^T \boldsymbol{\Sigma}^{-1} \mathbf{1})^2} \right\} = \sigma^2 V_1, \quad (14.9)$$

$$Var(\sigma^*) = \sigma^2 \left\{ \frac{\mathbf{1}^T \boldsymbol{\Sigma}^{-1} \mathbf{1}}{(\boldsymbol{\mu}^T \boldsymbol{\Sigma}^{-1} \boldsymbol{\mu})(\mathbf{1}^T \boldsymbol{\Sigma}^{-1} \mathbf{1}) - (\boldsymbol{\mu}^T \boldsymbol{\Sigma}^{-1} \mathbf{1})^2} \right\} = \sigma^2 V_2, \quad (14.10)$$

and

$$Cov(\theta^*, \sigma^*) = \sigma^2 \left\{ \frac{-\boldsymbol{\mu}^T \boldsymbol{\Sigma}^{-1} \mathbf{1}}{(\boldsymbol{\mu}^T \boldsymbol{\Sigma}^{-1} \boldsymbol{\mu})(\mathbf{1}^T \boldsymbol{\Sigma}^{-1} \mathbf{1}) - (\boldsymbol{\mu}^T \boldsymbol{\Sigma}^{-1} \mathbf{1})^2} \right\} = \sigma^2 V_3; \quad (14.11)$$

for details, refer to David (1981), Balakrishnan and Cohen (1991), and Arnold, Balakrishnan and Nagaraja (1992).

Exact explicit expressions may be obtained from the above formulas by using the fact that the covariance matrix of the standardized Pareto order statistics $((\sigma_{i,j:n}))$ is of the form $((a_i b_j))$. First, we use the facts that

$$\mu_{i:n} = \frac{\Gamma(n+1)}{\Gamma(n-i+1)} \frac{\Gamma(n-i+1-1/\nu)}{\Gamma(n+1-1/\nu)} \quad (14.12)$$

$$\mu_{i:n}^{(2)} = \frac{\Gamma(n+1)}{\Gamma(n-i+1)} \frac{\Gamma(n-i+1-2/\nu)}{\Gamma(n+1-2/\nu)} \quad (14.13)$$

and

$$\mu_{i,j:n} = \frac{\Gamma(n+1)}{\Gamma(n-j+1)} \frac{\Gamma(n-j+1-1/\nu)}{\Gamma(n-i+1-1/\nu)} \frac{\Gamma(n-i+1-2/\nu)}{\Gamma(n+1-2/\nu)} \quad (14.14)$$

for the standardized Pareto order statistics [Malik (1966) and Huang (1975)], to note that $((\sigma_{i,j:n})) = ((a_i b_j))$ where

$$a_i = \frac{\Gamma(n-i+1-2/\nu)}{\Gamma(n-i+1-1/\nu)\Gamma(n+1-2/\nu)} - \frac{\Gamma(n-i+1-1/\nu)\Gamma(n+1)}{\Gamma(n-i+1)\Gamma(n+1-1/\nu)^2} \quad (14.15)$$

and

$$b_j = \frac{\Gamma(n+1)\Gamma(n-j+1-1/\nu)}{\Gamma(n-j+1)}. \quad (14.16)$$

We are therefore able to invert the covariance matrix $((\sigma_{i,j:n}))$ and obtain explicit expressions for the censored BLUE's θ^* and σ^* of θ and σ respectively, their variances $\text{Var}(\theta^*)$ and $\text{Var}(\sigma^*)$, and their covariance $\text{Cov}(\theta^*, \sigma^*)$, as described, for example, in Arnold, Balakrishnan, and Nagaraja (1992). We have,

$$\begin{aligned}\theta^* &= \left\{\left(\frac{a_{r+2}-a_{r+1}}{a_{r+1}(a_{r+2}b_{r+1}-a_{r+1}b_{r+2})} - \frac{1}{a_{r+1}b_{r+1}}\right) X_{r+1:n}\right. \\ &+ \sum_{i=r+1}^{n-s-2} \frac{a_i(b_{i+1}-b_{i+2}) + a_{i+1}(b_{i+2}-b_i) + a_{i+2}(b_i-b_{i+1})}{(a_{i+1}b_i - a_i b_{i+1})(a_{i+2}b_{i+1} - a_{i+1}b_{i+2})} X_{i+1:n} \\ &+ \left.\frac{b_{n-s-1}-b_{n-s}}{b_{n-s}(a_{n-s}b_{n-s-1} - a_{n-s-1}b_{n-s})} X_{n-s:n}\right\} \left(\sum_{i,j}\sigma^{ij} - \frac{1}{a_{r+1}b_{r+1}}\right)^{-1} ; \\ & \hspace{10cm} (14.17)\end{aligned}$$

$$\begin{aligned}\sigma^* &= \left\{\left(\sum_{i,j}\sigma^{ij} - \frac{a_{r+2}-a_{r+1}}{a_{r+1}(a_{r+2}b_{r+1}-a_{r+1}b_{r+2})}\right) X_{r+1:n}\right. \\ &- \sum_{i=r+1}^{n-s-2} \frac{a_i(b_{i+1}-b_{i+2}) + a_{i+1}(b_{i+2}-b_i) + a_{i+2}(b_i-b_{i+1})}{(a_{i+1}b_i - a_i b_{i+1})(a_{i+2}b_{i+1} - a_{i+1}b_{i+2})} X_{i+1:n} \\ &- \left.\frac{b_{n-s-1}-b_{n-s}}{b_{n-s}(a_{n-s}b_{n-s-1} - a_{n-s-1}b_{n-s})} X_{n-s:n}\right\} \left(\frac{b_{r+1}}{c}\sum_{i,j}\sigma^{ij} - \frac{1}{ca_{r+1}}\right)^{-1} ; \\ & \hspace{10cm} (14.18)\end{aligned}$$

$$\text{Var}(\theta^*) = \sigma^2 \left(\sum_{i,j}\sigma^{ij} - \frac{1}{a_{r+1}b_{r+1}}\right)^{-1} ; \quad (14.19)$$

$$Var(\sigma^*) = \sigma^2 \left(\frac{b_{r+1}}{a_{r+1}c^2} - \frac{1}{c^2 a_{r+1}^2 \sum_{i,j} \sigma^{ij}} \right)^{-1} ; \qquad (14.20)$$

and

$$Cov(\theta^*, \sigma^*) = -\sigma^2 \left(\frac{b_{r+1}}{c} \sum_{i,j} \sigma^{ij} - \frac{1}{ca_{r+1}} \right)^{-1}, \qquad (14.21)$$

where a_i and b_i are given in (14.15) and (14.16) respectively,

$$c = \Gamma(n+1-1/\nu), \qquad (14.22)$$

and $\sum_{i,j} \sigma^{ij}$ is the sum of all of the elements of the inverse matrix of the covariance matrix $((\sigma_{i,j:n}))$ and is given by

$$\sum_{i,j} \sigma^{ij} = \sum_{i=r+2}^{n-s-1} \frac{a_{i-1}(2b_i - b_{i+1}) - 2a_i b_{i-1} + a_{i+1} b_{i-1}}{(a_i b_{i-1} - a_{i-1} b_i)(a_{i+1} b_i - a_i b_{i+1})}$$
$$+ \frac{a_{r+2} - 2a_{r+1}}{a_{r+1}(a_{r+2} b_{r+1} - a_{r+1} b_{r+2})}$$
$$+ \frac{b_{n-s-1}}{b_{n-s}(a_{n-s} b_{n-s-1} - a_{n-s-1} b_{n-s})}. \qquad (14.23)$$

In the full sample case (r=s=0), the above exact explicit expressions for the BLUE's correspond to the results of Kulldorff and Vännman (1973), while exact expressions for the right-censored case ($r = 0$ and $s > 0$) correspond to those of Childs and Balakrishnan (1996).

14.3 Exact Expressions For The Moments of Order Statistics

In this section, we derive exact expressions for the triple and quadruple moments of order statistics from the one-parameter Pareto distribution. For the sake of completeness and better understanding of the results, we also present the corresponding results for the single and double moments which are given by Malik (1966) and Huang (1975). Without loss of generality, we can put $\theta = 0$ in (14.1) to obtain the one-parameter Pareto distribution as

$$f(x) = \nu \sigma^\nu x^{-(\nu+1)}, \qquad x \geq \sigma, \ \sigma, \nu > 0, \qquad (14.24)$$

with c.d.f.

$$F(x) = 1 - \sigma^\nu x^{-\nu}, \qquad x \geq \sigma, \ \sigma, \nu > 0. \qquad (14.25)$$

Notice from the above expression that

$$x = \sigma(1 - F(x))^{-1/\nu}, \qquad \sigma, \nu > 0. \qquad (14.26)$$

14.3.1 Single moments of order statistics

From (14.24) and (14.25), we have

$$\begin{aligned}
\mu_{r:n}^{(a)} &= C_{r:n} \int_\sigma^\infty w^a \left[F(w)\right]^{r-1} \left[1 - F(w)\right]^{n-r} f(w) dw \quad (14.27) \\
&= C_{r:n} \sigma^a \int_0^1 u^{r-1}(1-u)^{n-r-a/\nu} du,
\end{aligned}$$

and hence,

$$\mu_{r:n}^{(a)} = C_{r:n} \sigma^a B(r, n-r+1-a/\nu),$$

where, $B(a,b)$ is the beta function and $C_{r:n} = \frac{n!}{(r-1)!(n-r)!}$. The above equation may be rewritten as:

$$\mu_{r:n}^{(a)} = \sigma^a \frac{\Gamma(n+1)\Gamma(n-r+1-a/\nu)}{\Gamma(n-r+1)\Gamma(n+1-a/\nu)}. \quad (14.28)$$

14.3.2 Double moments of order statistics

The double moments of order statistics are given by

$$\mu_{r,s:n}^{(a,b)} = C_{r,s:n} \int_\sigma^\infty w^a \left[F(w)\right]^{r-1} I(w) f(w) dw, \quad (14.29)$$

where $C_{r,s:n} = \frac{n!}{(r-1)!(s-r-1)!(n-s)!}$ and

$$I(w) = \int_w^\infty x^b (F(x) - F(w))^{s-r-1} (1 - F(x))^{n-s} f(x) dx.$$

We first let $u = \frac{F(x) - F(w)}{1 - F(w)}$ in the above integral and use (14.26) to get

$$I(w) = \sigma^b (1 - F(w))^{n-r-b/\nu} B(s-r, n-s+1-b/\nu). \quad (14.30)$$

Then we substitute (14.30) into (14.29) to get

$$\begin{aligned}
\mu_{r,s:n}^{(a,b)} &= C_{r,s:n} \sigma^{a+b} B(s-r, n-s+1-b/\nu) B(r, n-r+1-a/\nu-b/\nu) \\
&= \sigma^{a+b} \frac{\Gamma(n+1)\Gamma(n-r+1-a/\nu-b/\nu)\Gamma(n-s+1-b/\nu)}{\Gamma(n-s+1)\Gamma(n+1-a/\nu-b/\nu)\Gamma(n-r+1-b/\nu)}.
\end{aligned} \quad (14.31)$$

As a check, we set $b = 0$ in (14.31) and observe from (14.28) that,

$$\mu_{r,s:n}^{(a,0)} = \mu_{r:n}^{(a)}.$$

14.3.3 Triple moments of order statistics

The triple moments of order statistics are given by

$$\mu_{r,s,t:n}^{(a,b,c)} = C_{r,s,t:n} \int_\sigma^\infty \int_w^\infty w^a x^b [F(w)]^{r-1} (F(x) - F(w))^{s-r-1}$$
$$\times I(x) f(w) f(x) dx dw, \qquad (14.32)$$

where $C_{r,s,t:n} = \frac{n!}{(r-1)!(s-r-1)!(t-s-1)!(n-t)!}$ and

$$I(x) = \int_x^\infty y^c (F(y) - F(x))^{t-s-1} (1 - F(y))^{n-t} f(y) dy.$$

We first let $u = \frac{F(y) - F(x)}{1 - F(x)}$ in the above integral and use (14.26) to get

$$I(x) = \sigma^c (1 - F(x))^{n-s-c/\nu} B(t - s, n - t + 1 - c/\nu). \qquad (14.33)$$

Then, we substitute (14.33) into (14.32) to get

$$\mu_{r,s,t:n}^{(a,b,c)} = C_{r,s,t:n} \sigma^c B(t - s, n - t + 1 - c/\nu) \int_\sigma^\infty w^a (F(w))^{r-1} J(w) f(w) dw, \qquad (14.34)$$

where

$$J(w) = \int_w^\infty x^b (F(x) - F(w))^{s-r-1} (1 - F(x))^{n-s-c/\nu} f(x) dx. \qquad (14.35)$$

Next we let $h = \frac{F(x) - F(w)}{1 - F(w)}$ in the above integral and use (14.26) to get

$$J(w) = \sigma^b (1 - F(w))^{n-r-c/\nu - b/\nu} B(s - r, n - s + 1 - b/\nu - c/\nu). \qquad (14.36)$$

Finally, substituting (14.36) into (14.34) gives

$$\mu_{r,s,t:n}^{(a,b,c)} = C_{r,s,t:n} \sigma^{a+b+c} B(t - s, n - t + 1 - c/\nu)$$
$$\times B(s - r, n - s + 1 - b/\nu - c/\nu)$$
$$\times B(r, n - r + 1 - a/\nu - b/\nu - c/\nu)$$
$$= \sigma^{a+b+c} \frac{\Gamma(n+1) \Gamma(n - r + 1 - a/\nu - b/\nu - c/\nu)}{\Gamma(n - t + 1) \Gamma(n + 1 - a/\nu - b/\nu - c/\nu)}$$
$$\times \frac{\Gamma(n - s + 1 - b/\nu - c/\nu) \Gamma(n - t + 1 - c/\nu)}{\Gamma(n - r + 1 - b/\nu - c/\nu) \Gamma(n - s + 1 - c/\nu)}. \qquad (14.37)$$

As a check, we set $c = 0$ in (14.37) and observe from (14.31) that

$$\mu_{r,s,t:n}^{(a,b,0)} = \mu_{r,s:n}^{(a,b)}. \qquad (14.38)$$

14.3.4 Quadruple moments of order statistics

Using the same method as for the single, double and triple moments, the following exact explicit expression for the quadruple moments of order statistics can be obtained:

$$\begin{aligned}\mu_{r,s,t,u:n}^{(a,b,c,d)} &= C_{r,s,t,u:n}\sigma^{a+b+c+d}B(u-t, n-u+1-d/\nu) \\ &\quad \times B(t-s, n-t+1-c/\nu-d/\nu) \\ &\quad \times B(s-r, n-s+1-b/\nu-c/\nu-d/\nu) \\ &\quad \times B(r, n-r+1-a/\nu-b/\nu-c/\nu-d/\nu) \\ &= \sigma^{a+b+c+d}\frac{\Gamma(n+1)\Gamma(n-r+1-a/\nu-b/\nu-c/\nu-d/\nu)}{\Gamma(n-u+1)\Gamma(n+1-a/\nu-b/\nu-c/\nu-d/\nu)} \\ &\quad \times \frac{\Gamma(n-s+1-b/\nu-c/\nu-d/\nu)\Gamma(n-t+1-c/\nu-d/\nu)\Gamma(n-u+1-d/\nu)}{\Gamma(n-r+1-b/\nu-c/\nu-d/\nu)\Gamma(n-s+1-c/\nu-d/\nu)\Gamma(n-t+1-d/\nu)}.\end{aligned}$$
(14.39)

As a check, we set $d = 0$ in (14.39) and observe from (14.37) that

$$\mu_{r,s,t,u:n}^{(a,b,c,0)} = \mu_{r,s,t:n}^{(a,b,c)}. \tag{14.40}$$

14.4 Recurrence Relations For Moments of Order Statistics

In this section, we will use the differential equation in (14.3) to derive some recurrence relations between the single moments, double moments, triple moments and quadruple moments of order statistics. Without loss of generality, we once again take $\theta = 0$.

14.4.1 Relations for single moments

The p.d.f. of $X_{r:n}$ for $1 \leq r \leq n$ is [see David (1981, p. 9)] given by

$$f_{r:n}(w) = C_{r:n}[F(w)]^{r-1}[1-F(w)]^{n-r}f(w), \qquad \sigma \leq w < \infty, \tag{14.41}$$

where $C_{r:n} = \frac{n!}{(r-1)!(n-r)!}$.

From this, we may compute the single moments of $X_{r:n}$ as

$$\mu_{r:n}^{(a)} = \int_\sigma^\infty w^a f_{r:n}(w)dw, \qquad a = 1, 2, \ldots, \ 1 \leq r \leq n. \tag{14.42}$$

Then the single moments $\mu_{r:n}^{(a)}$ for $a = 1, 2, \ldots$, satisfy the following two recurrence relations. It should be mentioned here that the recurrence relations for the single and the double moments of order statistics from Pareto distribution have been derived by Balakrishnan and Joshi (1982) and are presented here for the sake of completeness.

Theorem 14.4.1 *For $n \geq 1$ and $a = 1, 2, \ldots,$*

$$(n\nu - a)\mu_{1:n}^{(a)} = n\nu\sigma^a. \tag{14.43}$$

Theorem 14.4.2 *For $2 \leq r \leq n$ and $a = 1, 2, \ldots,$*

$$\left(n + 1 - r - \frac{a}{\nu}\right)\mu_{r:n}^{(a)} = (n + 1 - r)\mu_{r-1:n}^{(a)}. \tag{14.44}$$

14.4.2 Relations for double moments

The joint p.d.f. of $X_{r:n}$ and $X_{s:n}$ for $1 \leq r < s \leq n$ is given by [see David (1981, p. 10)]

$$f_{r,s:n}(w, x) = C_{r,s:n}[F(w)]^{r-1}[F(x) - F(w)]^{s-r-1}[1 - F(x)]^{n-s}f(w)f(x),$$
$$\sigma \leq w < x < \infty, \tag{14.45}$$

where $C_{r,s:n} = \frac{n!}{(r-1)!(s-r-1)!(n-s)!}$. Using (4.5), we may compute the double moments as

$$\mu_{r,s:n}^{(a,b)} = \int_\sigma^\infty \int_w^\infty w^a x^b f_{r,s:n}(w, x)\,dx\,dw, \quad 1 \leq r < s \leq n,\ a, b \geq 1. \tag{14.46}$$

Then the double moments $\mu_{r,s:n}^{(a,b)}$ for $a, b = 1, 2, \ldots,$ satisfy the recurrence relations presented in the following two theorems.

Theorem 14.4.3 *For $1 \leq r \leq n - 1$ and $a, b = 1, 2, \ldots,$*

$$\left(n - r - \frac{b}{\nu}\right)\mu_{r,r+1:n}^{(a,b)} = (n - r)\mu_{r:n}^{(a+b)}, \tag{14.47}$$

and for $1 \leq r < s \leq n$, $s - r \geq 2$ and $a, b = 1, 2, \ldots,$

$$\left(n - s + 1 - \frac{b}{\nu}\right)\mu_{r,s:n}^{(a,b)} = (n - s + 1)\mu_{r,s-1:n}^{(a,b)}. \tag{14.48}$$

Theorem 14.4.4 *For $n \geq 2$ and $a, b = 1, 2, \ldots,$*

$$\left(1 - \frac{a}{\nu}\right)\mu_{1,2:n}^{(a,b)} = n\left(\sigma^a \mu_{1:n-1}^{(b)} - \mu_{1:n-1}^{(a+b)}\right) + \mu_{2:n}^{(a+b)}; \tag{14.49}$$

for $n \geq 3$, $2 \leq r \leq n - 1$ and $a, b = 1, 2, \ldots,$

$$\left(r - \frac{a}{\nu}\right)\mu_{r,r+1:n}^{(a,b)} = r\mu_{r+1:n}^{(a+b)} + n\left(\mu_{r-1,r:n-1}^{(a,b)} - \mu_{r:n-1}^{(a+b)}\right); \tag{14.50}$$

Higher Order Moments of Order Statistics

for $n \geq 3$, $3 \leq s \leq n$ and $a, b = 1, 2, \ldots$,

$$\left(1 - \frac{a}{\nu}\right)\mu_{1,s:n}^{(a,b)} = \mu_{2,s:n}^{(a,b)} - n\left(\mu_{1,s-1:n-1}^{(a,b)} + \sigma^a \mu_{s-1:n-1}^{(b)}\right), \quad (14.51)$$

and for $n \geq 4$, $2 \leq r < s \leq n$, $s - r \geq 2$ and $a, b = 1, 2, \ldots$,

$$\left(r - \frac{a}{\nu}\right)\mu_{r,s:n}^{(a,b)} = r\mu_{r+1,s:n}^{(a,b)} - n\left(\mu_{r,s-1:n-1}^{(a,b)} - \mu_{r-1,s-1:n-1}^{(a,b)}\right). \quad (14.52)$$

14.4.3 Relations for triple moments

The joint p.d.f. of $X_{r:n}$, $X_{s:n}$ and $X_{t:n}$ for $1 \leq r < s < t \leq n$ is given by

$$f_{r,s,t:n}(w, x, y) = C_{r,s,t:n}[F(w)]^{r-1}[F(x) - F(w)]^{s-r-1}[F(y) - F(x)]^{t-s-1}$$
$$\times [1 - F(y)]^{n-t} f(w)f(x)f(y), \quad \sigma \leq w < x < y < \infty, \quad (14.53)$$

where $C_{r,s,t:n} = \frac{n!}{(r-1)!(s-r-1)!(t-s-1)!(n-t)!}$. Using (14.53), we may compute the triple moments as

$$\mu_{r,s,t:n}^{(a,b,c)} = \int_\sigma^\infty \int_w^\infty \int_x^\infty w^a x^b y^c f_{r,s,t:n}(w, x, y) \, dy \, dx \, dw,$$
$$1 \leq r < s < t \leq n, \ a, b, c \geq 1. \quad (14.54)$$

Then the triple moments $\mu_{r,s,t:n}^{(a,b,c)}$ for $a, b, c = 1, 2, \ldots$, satisfy the recurrence relations presented in the following three theorems, which may be proved by following the same steps as those of Balakrishnan and Gupta (1996).

Theorem 14.4.5 *For $n \geq 3$, $1 \leq r < s \leq n - 1$ and $a, b, c = 1, 2, \ldots$,*

$$\left((n-s) - \frac{c}{\nu}\right)\mu_{r,s,s+1:n}^{(a,b,c)} = (n-s)\mu_{r,s:n}^{(a,b+c)}, \quad (14.55)$$

and for $n \geq 4$, $1 \leq r < s < t \leq n$, $t - s \geq 2$ and $a, b, c = 1, 2, \ldots$,

$$\left((n-t+1) - \frac{c}{\nu}\right)\mu_{r,s,t:n}^{(a,b,c)} = (n-t+1)\mu_{r,s,t-1:n}^{(a,b,c)}. \quad (14.56)$$

Theorem 14.4.6 *For $n \geq 3$, $3 \leq t \leq n$ and $a, b, c = 1, 2, \ldots$,*

$$\left(1 - \frac{a}{\nu}\right)\mu_{1,2,t:n}^{(a,b,c)} = n\left(\sigma^a \mu_{1,t-1:n-1}^{(b,c)} - \mu_{1,t-1:n-1}^{(a+b,c)}\right) + \mu_{2,t:n}^{(a+b,c)}; \quad (14.57)$$

for $n \geq 4$, $2 \leq r < t \leq n$, $t - r \geq 2$ and $a, b, c = 1, 2, \ldots$,

$$\left(r - \frac{a}{\nu}\right)\mu_{r,r+1,t:n}^{(a,b,c)} = n\left(\mu_{r-1,r,t-1:n-1}^{(a,b,c)} - \mu_{r,t-1:n-1}^{(a+b,c)}\right) + r\mu_{r+1,t:n}^{(a+b,c)}; \quad (14.58)$$

for $n \geq 4$, $3 \leq s < t \leq n$, and $a, b, c = 1, 2, \ldots$,

$$\mu_{2,s,t:n}^{(a,b,c)} = \left(1 - \frac{a}{\nu}\right)\mu_{1,s,t:n}^{(a,b,c)} + n\left(\mu_{1,s-1,t-1:n-1}^{(a,b,c)} - \sigma^a \mu_{s-1,t-1:n-1}^{(b,c)}\right); \quad (14.59)$$

and for $n \geq 5$, $2 \leq r < s < t \leq n$, $s - r \geq 2$ and $a, b, c = 1, 2, \ldots$,

$$r\mu_{r+1,s,t:n}^{(a,b,c)} = \left(r - \frac{a}{\nu}\right)\mu_{r,s,t:n}^{(a,b,c)} + n\left(\mu_{r,s-1,t-1:n-1}^{(a,b,c)} - \mu_{r-1,s-1,t-1:n-1}^{(a,b,c)}\right). \quad (14.60)$$

Theorem 14.4.7 For $n \geq 3$, $1 \leq r \leq n-2$ and $a, b, c = 1, 2, \ldots$,

$$\left(1 - \frac{b}{\nu}\right)\mu_{r,r+1,r+2:n}^{(a,b,c)} = \mu_{r,r+2:n}^{(a+b,c)} - (n - r - 1)\left(\mu_{r,r+1:n}^{(a,b+c)} - \mu_{r,r+1:n}^{(a+b,c)}\right); \quad (14.61)$$

for $n \geq 4$, $1 \leq r \leq n$, $t - r \geq 3$ and $a, b, c = 1, 2, \ldots$,

$$\left((t - r - 1) - \frac{b}{\nu}\right)\mu_{r,r+1,t:n}^{(a,b,c)} = (n - t + 1)\left(\mu_{r,t-1:n}^{(a+b,c)} - \mu_{r,r+1,t-1:n}^{(a,b,c)}\right)$$
$$+ (t - r - 1)\mu_{r,t:n}^{(a+b,c)}; \quad (14.62)$$

for $n \geq 4$, $1 \leq r < s \leq n-1$, $s - r \geq 2$ and $a, b, c = 1, 2, \ldots$,

$$\left(1 - \frac{b}{\nu}\right)\mu_{r,s,s+1:n}^{(a,b,c)} = \mu_{r,s-1,s+1:n}^{(a,b,c)} - (n - s)\left(\mu_{r,s:n}^{(a,b+c)} - \mu_{r,s-1,s:n}^{(a,b,c)}\right), \quad (14.63)$$

and for $n \geq 5$, $1 \leq r < s < t \leq n$, $s - r \geq 2$, $t - s \geq 2$ and $a, b, c = 1, 2, \ldots$,

$$\left((t - s) - \frac{b}{\nu}\right)\mu_{r,s,t:n}^{(a,b,c)}$$
$$= (t - s)\mu_{r,s-1,t:n}^{(a,b,c)} - (n - t + 1)\left(\mu_{r,s,t-1:n}^{(a,b,c)} - \mu_{r,s-1,t-1:n}^{(a,b,c)}\right). \quad (14.64)$$

14.4.4 Relations for quadruple moments

The joint p.d.f. of $X_{r:n}$, $X_{s:n}$, $X_{t:n}$ and $X_{u:n}$ for $1 \leq r < s < t < u \leq n$ is given by

$$f_{r,s,t,u:n}(w, x, y, z) = C_{r,s,t,u:n}[F(w)]^{r-1}[F(x) - F(w)]^{s-r-1}$$
$$\times [F(y) - F(x)]^{t-s-1}[F(z) - F(y)]^{u-t-1}$$
$$\times [1 - F(z)]^{n-u}f(w)f(x)f(y)f(z),$$
$$\sigma \leq w < x < y < z < \infty, \quad (14.65)$$

where $C_{r,s,t,u:n} = \frac{n!}{(r-1)!(s-r-1)!(t-s-1)!(u-t-1)!(n-u)!}$. Using (14.65), we may compute the quadruple moments as

$$\mu_{r,s,t,u:n}^{(a,b,c,d)} = \int_\sigma^\infty \int_w^\infty \int_x^\infty \int_y^\infty w^a x^b y^c z^d f_{r,s,t,u:n}(w, x, y, z) dz dy dx dw,$$
$$1 \leq r < s < t < u \leq n, \; a, b, c, d \geq 1. \quad (14.66)$$

Then the quadruple moments $\mu_{r,s,t,u:n}^{(a,b,c,d)}$ for $a, b, c, d = 1, 2, \ldots$, satisfy the recurrence relations presented in the following four theorems, which may once again be established along the lines of Balakrishnan and Gupta (1996).

Higher Order Moments of Order Statistics

Theorem 14.4.8 *For* $n \geq 4$, $1 \leq r < s < t \leq n-1$ *and* $a,b,c,d = 1,2,\ldots,$

$$\left(n-t-\frac{d}{\nu}\right)\mu_{r,s,t,t+1:n}^{(a,b,c,d)} = (n-t)\mu_{r,s,t:n}^{(a,b,c+d)}, \tag{14.67}$$

and for $n \geq 5$, $1 \leq r < s < t < u \leq n$, $u-t \geq 2$ *and* $a,b,c,d = 1,2,\ldots,$

$$\left((n-u+1)-\frac{d}{\nu}\right)\mu_{r,s,t,u:n}^{(a,b,c,d)} = (n-u+1)\mu_{r,s,t,u-1:n}^{(a,b,c,d)}. \tag{14.68}$$

Theorem 14.4.9 *For* $n \geq 4$, $3 \leq t < u \leq n$ *and* $a,b,c,d = 1,2,\ldots,$

$$\left(1-\frac{a}{\nu}\right)\mu_{1,2,t,u:n}^{(a,b,c,d)} = n\left(\sigma^a \mu_{1,t-1,u-1:n-1}^{(b,c,d)} - \mu_{1,t-1,u-1:n-1}^{(a+b,c,d)}\right) + \mu_{2,t,u:n}^{(a+b,c,d)}; \tag{14.69}$$

for $n \geq 5$, $2 \leq r < t < u \leq n$, $t-r \geq 2$ *and* $a,b,c,d = 1,2,\ldots,$

$$\left(r-\frac{a}{\nu}\right)\mu_{r,r+1,t,u:n}^{(a,b,c,d)} = n\left(\mu_{r-1,r,t-1,u-1:n-1}^{(a,b,c,d)} - \mu_{r,t-1,u-1:n-1}^{(a+b,c,d)}\right) + r\mu_{r+1,t,u:n}^{(a+b,c,d)}; \tag{14.70}$$

for $n \geq 5$, $3 \leq s < t < u \leq n$, *and* $a,b,c,d = 1,2,\ldots,$

$$\mu_{2,s,t,u:n}^{(a,b,c,d)} = \left(1-\frac{a}{\nu}\right)\mu_{1,s,t,u:n}^{(a,b,c,d)} + n\left(\mu_{1,s-1,t-1,u-1:n-1}^{(a,b,c,d)} - \sigma^a \mu_{s-1,t-1,u-1:n-1}^{(b,c,d)}\right), \tag{14.71}$$

and for $n \geq 6$, $2 \leq r < s < t < u \leq n$, $s-r \geq 2$ *and* $a,b,c,d = 1,2,\ldots,$

$$r\mu_{r+1,s,t,u:n}^{(a,b,c,d)}$$
$$= \left(r-\frac{a}{\nu}\right)\mu_{r,s,t,u:n}^{(a,b,c,d)} + n\left(\mu_{r,s-1,t-1,u-1:n-1}^{(a,b,c,d)} - \mu_{r-1,s-1,t-1,u-1:n-1}^{(a,b,c,d)}\right). \tag{14.72}$$

Theorem 14.4.10 *For* $n \geq 4$, $1 \leq r < u \leq n$, $u-r \geq 3$ *and* $a,b,c,d = 1,2,\ldots,$

$$\left(1-\frac{b}{\nu}\right)\mu_{r,r+1,r+2,u:n}^{(a,b,c,d)} = \mu_{r,r+2,u:n}^{(a+b,c,d)} - (n-u+1)\left(\mu_{r,r+1,u-1:n}^{(a,b+c,d)} - \mu_{r,r+1,u-1:n}^{(a+b,c,d)}\right)$$
$$- (u-r-2)\left(\mu_{r,r+1,u:n}^{(a,b+c,d)} - \mu_{r,r+1,u:n}^{(a+b,c,d)}\right); \tag{14.73}$$

for $n \geq 5$, $1 \leq r < t < u \leq n$, $t-r \geq 3$ *and* $a,b,c,d = 1,2,\ldots,$

$$\left((t-r-1)-\frac{b}{\nu}\right)\mu_{r,r+1,t,u:n}^{(a,b,c,d)} = (n-u+1)\left(\mu_{r,t-1,u-1:n}^{(a+b,c,d)} - \mu_{r,r+1,t-1,u-1:n}^{(a,b,c,d)}\right)$$
$$- (u-t)\left(\mu_{r,r+1,t-1,u:n}^{(a,b,c,d)} - \mu_{r,t-1,u:n}^{(a+b,c,d)}\right)$$
$$+ (t-r-1)\mu_{r,t,u:n}^{(a+b,c,d)}; \tag{14.74}$$

for $1 \leq r < s < u \leq n$, $s - r \geq 2$, $u - s \geq 2$ and $a, b, c, d = 1, 2, \ldots,$

$$\left(1 - \frac{b}{\nu}\right) \mu_{r,s,s+1,u:n}^{(a,b,c,d)}$$
$$= \mu_{r,s-1,s+1,u:n}^{(a,b,c,d)} - (u - s - 1)\left(\mu_{r,s,u:n}^{(a,b+c,d)} - \mu_{r,s-1,s,u:n}^{(a,b,c,d)}\right)$$
$$+ (n - u + 1)\left(\mu_{r,s-1,s,u-1:n}^{(a,b,c,d)} - \mu_{r,s,u-1:n}^{(a,b+c,d)}\right), \quad (14.75)$$

and for $1 \leq r < s < t < u \leq n$, $s - r \geq 2$, $t - s \geq 2$ and $a, b, c, d = 1, 2, \ldots,$

$$\left((t - s) - \frac{b}{\nu}\right) \mu_{r,s,t,u:n}^{(a,b,c,d)}$$
$$= (t - s)\mu_{r,s-1,t,u:n}^{(a,b,c,d)} - (n - u + 1)\left(\mu_{r,s,t-1,u-1:n}^{(a,b,c,d)} - \mu_{r,s-1,t-1,u-1:n}^{(a,b,c,d)}\right)$$
$$- (u - t)\left(\mu_{r,s,t-1,u:n}^{(a,b,c,d)} - \mu_{r,s-1,t-1,u:n}^{(a,b,c,d)}\right). \quad (14.76)$$

Theorem 14.4.11 For $n \geq 4$, $1 \leq r < s \leq n - 2$, and $a, b, c, d = 1, 2, \ldots,$

$$\left(1 - \frac{c}{\nu}\right) \mu_{r,s,s+1,s+2:n}^{(a,b,c,d)} = \mu_{r,s,s+2:n}^{(a,b+c,d)} - (n - s - 1)\left(\mu_{r,s,s+1:n}^{(a,b,c+d)} - \mu_{r,s,s+1:n}^{(a,b+c,d)}\right); \quad (14.77)$$

for $n \geq 5$, $1 \leq r < s < u \leq n$, $u - s \geq 3$ and $a, b, c, d = 1, 2, \ldots,$

$$\left((u - s - 1) - \frac{c}{\nu}\right) \mu_{r,s,s+1,u:n}^{(a,b,c,d)} = -(n - u + 1)\left(\mu_{r,s,s+1,u-1:n}^{(a,b,c,d)} - \mu_{r,s,u-1:n}^{(a,b+c,d)}\right)$$
$$+ (u - s - 1)\mu_{r,s,u:n}^{(a,b+c,d)}; \quad (14.78)$$

for $n \geq 5$, $1 \leq r < s < t \leq n - 1$, $t - s \geq 2$ and $a, b, c, d = 1, 2, \ldots,$

$$\left(1 - \frac{c}{\nu}\right) \mu_{r,s,t,t+1:n}^{(a,b,c,d)} = \mu_{r,s,t-1,t+1:n}^{(a,b,c,d)} - (n - t)\left(\mu_{r,s,t:n}^{(a,b,c+d)} - \mu_{r,s,t-1,t:n}^{(a,b,c,d)}\right), \quad (14.79)$$

and for $n \geq 6$, $1 \leq r < s < t < u \leq n$, $t - s \geq 2$, $u - t \geq 2$ and $a, b, c, d = 1, 2, \ldots,$

$$\left((u - t) - \frac{c}{\nu}\right) \mu_{r,s,t,u:n}^{(a,b,c,d)} = -(n - u + 1)\left(\mu_{r,s,t,u-1:n}^{(a,b,c,d)} - \mu_{r,s,t-1,u-1:n}^{(a,b,c,d)}\right)$$
$$+ (u - t)\mu_{r,s,t-1,u:n}^{(a,b,c,d)}. \quad (14.80)$$

Remark All the recurrence relations presented in this section are complete in the sense that they will enable one to compute all the single, double, triple and quadruple moments of all order statistics for all sample sizes in a simple recursive manner. This can be done for any choice of the shape parameter ν.

14.5 Approximate Inference

By making use of the recurrence relations in Section 14.4, and the values of a_i and b_i obtained from the exact explicit expressions of the BLUE's given in Section 14.2 for the case of Type-II right-censored samples

$$X_{1:n} \leq X_{2:n} \leq \cdots \leq X_{n-s:n}, \tag{14.81}$$

we determined the values of the mean, variance and the coefficients of skewness and kurtosis ($\sqrt{\beta_1}$ and β_2) of R_1^* and R_2^*, for $n = 5(1)10(5)20$, $\nu = 5(10)25$, and $s = 0(1)(\frac{n}{2}-1)$ for n even and $s = 0(1)[\![\frac{n}{2}]\!]$ for n odd. These values are presented in Tables 14.1–14.3. An examination of the ($\sqrt{\beta_1}, \beta_2$) values in Tables 14.1–14.3 reveals that the distribution of R_1^* (and hence of R_1) is negatively skewed and heavier tailed than normal. For small values of n and ν (most notably $n = 5\text{-}8$ when $\nu = 5$), we see that the values of ($\sqrt{\beta_1}, \beta_2$) are not within the range of the Edgeworth approximation; for details on the possible range for Edgeworth approximation, see Barton and Dennis (1952) and Johnson, Kotz and Balakrishnan (1994). However, for larger values of n when $\nu = 5$, and even for some of the smaller values of n for $\nu = 15$ and 25, we see that the ($\sqrt{\beta_1}, \beta_2$) values lie well within the range of an Edgeworth approximation.

Examination of the ($\sqrt{\beta_1}, \beta_2$) values in Tables 14.1–14.3 for R_2^* shows that they are very similar to those of R_1^*. The β_2 values are almost identical while the $\sqrt{\beta_1}$ values are almost equal, except for the sign. This similarity can be explained by a careful examination of the exact explicit expressions of the BLUE's given in Section 14.2. These expressions reveal that $a_i = -cb_i$ for $i = 2, 3, \ldots n - s$, while a_1 is very close to $-b_1$.

By making use of the entries in Tables 14.1-14.3, we determined the lower and upper 1%, 2.5%, 5% and 10% points of R_1 and R_2 through the Edgeworth approximation in (14.6). These values, for the case of Type-II right-censored samples ($r = 0$) for $s = 0(1)(\frac{n}{2} - 1)$ for n even and $s = 0(1)[\![\frac{n}{2}]\!]$ for n odd and sample size $n = 5(1)10(5)20$, are presented in Tables 14.4, 14.5 and 14.6. For the purpose of comparison, these percentage points were also determined by Monte Carlo simulations (based on 5000 runs) and they are presented along with the Edgeworth percentage points in Tables 14.4, 14.5 and 14.6.

From Tables 14.4–14.6, we see that the Edgeworth approximation of the distribution of R_1 provides quite close agreement with the simulated percentage points. The largest discrepancy occurs at the extreme lower tails of the distribution, but only for small sample sizes, and small values of ν. As the sample size and/or ν increases, the agreement becomes quite close at all levels of censoring, even at the lower extremes of the distribution.

From Tables 14.7–14.9, we see that the Edgeworth approximation of the distribution of R_2 also provides close agreement with the simulated percentage points; in fact, even a bit closer than R_1. This time, however, the largest discrepancies occur at the upper tails of the distribution for small sample sizes, but the discrepancies are not as great as those of R_1. Again, as the sample size and/or ν increases, the discrepancy becomes quite small at all levels of censoring, even at the extremes of the distribution.

In conclusion, we observe that the Edgeworth approximations of the distributions of R_1 and R_2 both work quite satisfactorily even in samples of size as small as 5 (for large ν), and they improve in accuracy as the sample size and/or ν increases. It should also be pointed out here that a similar Edgeworth approximation can not be developed for the percentage points of the pivotal quantities R_3 since it is not a linear function of order statistics. However, as displayed in the next section, approximate inference based on R_1 with σ replaced by σ^* provides quite close results to those based on R_3. For this purpose, we have presented in Tables 14.10, 14.11 and 14.12 some selected percentage points of R_3 determined by Monte Carlo simulations (based on 5000 runs).

Remark Though all the illustrations have been made here with right censored samples, if the available sample is doubly censored, one could use the exact explicit expressions of the BLUE's of θ and σ given in (14.17) and (14.18) and proceed exactly in the same fashion in order to develop Edgeworth approximate inference for the parameters θ and σ.

14.6 Numerical Illustration

In order to illustrate the usefulness of the inference procedures discussed in the previous sections, we consider here a simulated data set of size $n = 20$ (with $\theta = 25$, $\sigma = 5$ and $\nu = 5$):

$$\begin{array}{ccccc} 30.008 & 30.051 & 30.065 & 30.119 & 30.236 \\ 30.358 & 30.366 & 30.473 & 30.492 & 30.544 \\ 30.586 & 30.596 & 30.887 & 31.262 & 31.440 \\ 31.556 & 32.241 & 32.268 & 33.225 & 36.069 \end{array}$$

Using this sample, the BLUE's were calculated based on complete as well as Type-II right-censored samples ($r = 0$ and $s > 0$) by making use of the exact explicit expressions given in Section 14.2. These estimates are presented in the following table along with their standard errors:

n	s	θ^*	σ^*	S.E.(θ^*)	S.E.(σ^*)
20	0	25.43	4.53	1.2529	1.2406
	1	25.48	4.48	1.2488	1.2367
	2	25.53	4.43	1.2506	1.2396
	3	25.31	4.65	1.3356	1.3234

With these estimates and the use of Table 14.1 and Table 14.4, we determined the 90% confidence intervals for θ (when σ is known to be 5 and $\nu = 5$) based on the Edgeworth approximation and on the simulated percentage point using the pivotal quantity R_1 through the formula

$$P\left(\theta^* - \sigma\sqrt{V_1}\,(R_1)_{1-\alpha/2} \leq \theta \leq \theta^* - \sigma\sqrt{V_1}\,(R_1)_{\alpha/2}\right) = 1 - \alpha.$$

These are presented in the table below:

n	s	Edgeworth C.I.	Simulated C.I.
20	0	(23.452,27.947)	(23.439,27.975)
	1	(23.487,28.017)	(23.487,28.003)
	2	(23.511,28.085)	(23.483,28.099)
	3	(23.256,27.924)	(23.271,27.881)

It is clear that the confidence intervals based on the Edgeworth approximation are very close to the confidence intervals determined by simulation, at all levels of censoring.

Similarly, with the use of Table 14.1 and Table 14.7 we determined the 90% confidence intervals for σ, through the formula

$$P\left(\frac{\sigma^*}{1+\sqrt{V_2}\,(R_2)_{1-\alpha/2}} \leq \sigma \leq \frac{\sigma^*}{1+\sqrt{V_2}\,(R_2)_{\alpha/2}}\right) = 1 - \alpha,$$

and they are as follows:

n	s	Edgeworth C.I.	Simulated C.I.
20	0	(3.023,7.446)	(3.007,7.480)
	1	(2.983,7.395)	(2.989,7.395)
	2	(2.943,7.376)	(2.943,7.445)
	3	(3.063,7.841)	(3.081,7.804)

Once again, we observe that the confidence intervals based on the Edgeworth approximation are very close to those based on simulation.

In the case when σ is unknown, the Edgeworth approximation method cannot be used to draw inference for θ using R_3. However, as pointed out in the last section, the Edgeworth approximation for the distribution of pivotal quantity R_1 may be used in this case with σ replaced by σ^* in order to draw approximate inference for θ. By this method, we determined the 90% confidence intervals for θ through the formula

$$P\left(\theta^* - \sigma^*\sqrt{V_1}\,(R_1)_{1-\alpha/2} \leq \theta \leq \theta^* - \sigma^*\sqrt{V_1}\,(R_1)_{\alpha/2}\right) = 1 - \alpha,$$

and these are presented in the following table for the choices of $r = 0$, $s = 0(1)3$. Also presented in this table are the corresponding 90% confidence intervals for

θ based on the simulated percentage points of the pivotal quantity R_3 (given in Table 14.10) through the formula

$$P\left(\theta^* - \sigma^*\sqrt{V_1}\,(R_3)_{1-\alpha/2} \leq \theta \leq \theta^* - \sigma^*\sqrt{V_1}\,(R_3)_{\alpha/2}\right) = 1 - \alpha.$$

n	s	Edgeworth C.I. $R_1(\sigma = \sigma^*)$	Simulated C.I. R_3
20	0	(23.638,27.710)	(22.461,26.971)
	1	(23.694,27.753)	(22.545,26.979)
	2	(23.742,27.794)	(22.491,27.031)
	3	(23.400,27.741)	(22.145,26.886)

It is quite clear that the confidence intervals using the approximate Edgeworth method based on the pivotal quantity R_1 are somewhat close to the confidence intervals determined by simulation of the distribution of the pivotal quantity R_3, at all levels of censoring.

14.7 Recurrence Relations For Moments of Order Statistics in The Doubly Truncated Case

In this section, we will derive some recurrence relations between the single moments, double moments, triple moments and quadruple moments of order statistics in the doubly truncated case.

Let $X_1, X_2, \ldots, X_n$ be a random sample of size n from a one-parameter doubly truncated Pareto population with probability density function (p.d.f.)

$$f(x;\sigma) = \frac{1}{P-Q}\nu\sigma^\nu x^{-(\nu+1)}, \quad Q_1 \leq x \leq P_1,\ \nu > 0,\ \sigma > 0, \qquad (14.82)$$

and cumulative distribution function (c.d.f.)

$$F(x;\sigma) = \frac{1}{P-Q}\left(1 - \sigma^\nu x^{-\nu}\right), \quad Q_1 \leq x \leq P_1,\ \nu > 0,\ \sigma > 0, \qquad (14.83)$$

where

$$P = 1 - \sigma^\nu P_1^{-\nu} \text{ and } Q = 1 - \sigma^\nu Q_1^{-\nu}. \qquad (14.84)$$

From (14.82), and (14.83), we have

$$f(x) = \frac{\nu}{x}\{-Q_2 - F(x)\} = \frac{\nu}{x}\{-P_2 + (1 - F(x))\}, \qquad (14.85)$$

where

$$P_2 = \frac{P-1}{P-Q} \text{ and } Q_2 = \frac{Q-1}{P-Q}. \qquad (14.86)$$

Letting $Q_1 \to \sigma$ and $P_1 \to \infty$ in (14.82)–(14.85), we obtain the untruncated case discussed earlier.

14.7.1 Relations for single moments

The p.d.f. of $X_{r:n}$ for $1 \leq r \leq n$ is given by

$$f_{r:n}(w) = C_{r:n} [F(w)]^{r-1} [1 - F(w)]^{n-r} f(w), \quad Q_1 \leq w \leq P_1, \quad (14.87)$$

where $C_{r:n} = \frac{n!}{(r-1)!(n-r)!}$.

From this, we may compute the single moments of $X_{r:n}$ as

$$\mu_{r:n}^{(a)} = \int_{Q_1}^{P_1} w^a f_{r:n}(w) dw, \quad a = 1, 2, \ldots, 1 \leq r \leq n. \quad (14.88)$$

Then the single moments $\mu_{r:n}^{(a)}$ for $a = 1, 2, \ldots$, satisfy the recurrence relations presented in the following two theorems. Also, it should be mentioned here that the recurrence relations for the single and the double moments of order statistics from the doubly truncated Pareto distribution have been established earlier by Balakrishnan and Joshi (1982), and we are presenting here for the sake of completeness.

Theorem 14.7.1 *For $n \geq 2$ and $a = 1, 2, \ldots$,*

$$(n\nu - a)\mu_{1:n}^{(a)} = n\nu \left[P_2 \mu_{1:n-1}^{(a)} - Q_2 Q_1^a \right]. \quad (14.89)$$

Theorem 14.7.2 *For $2 \leq r \leq n - 1$ and $a = 1, 2, \ldots$,*

$$\left(n + 1 - r - \frac{a}{\nu} \right) \mu_{r:n}^{(a)} = (n + 1 - r)\mu_{r-1:n}^{(a)} + n P_2 \left[\mu_{r:n-1}^{(a)} - \mu_{r-1:n-1}^{(a)} \right]. \quad (14.90)$$

14.7.2 Relations for double moments

The joint p.d.f. of $X_{r:n}$ and $X_{s:n}$ for $1 \leq r < s \leq n$ is given by

$$\begin{aligned} f_{r,s:n}(w, x) &= C_{r,s:n} [F(w)]^{r-1} [F(x) - F(w)]^{s-r-1} \\ &\quad \times [1 - F(x)]^{n-s} f(w) f(x), \quad Q_1 \leq w < x \leq P_1, \quad (14.91) \end{aligned}$$

where $C_{r,s:n} = \frac{n!}{(r-1)!(s-r-1)!(n-s)!}$. Using (14.91), we may compute the double moments as

$$\mu_{r,s:n}^{(a,b)} = \int_{Q_1}^{P_1} \int_w^{P_1} w^a x^b f_{r,s:n}(w, x) dx dw, \quad 1 \leq r < s \leq n, \, a, b \geq 1. \quad (14.92)$$

Then the double moments $\mu_{r,s:n}^{(a,b)}$ for $a, b = 1, 2, \ldots$, satisfy the recurrence relations presented in the following two theorems.

Theorem 14.7.3 *For $n \geq 3$, $1 \leq r \leq n-2$ and $a, b = 1, 2, \ldots$,*

$$\left(n - r - \frac{b}{\nu}\right)\mu^{(a,b)}_{r,r+1:n} = (n-r)\mu^{(a+b)}_{r:n} + nP_2\left(\mu^{(a,b)}_{r,r+1:n-1} - \mu^{(a+b)}_{r:n-1}\right), \quad (14.93)$$

and for $n \geq 4$, $1 \leq r < s \leq n-1$, $s - r \geq 2$ and $a, b = 1, 2, \ldots$,

$$\left(n - s + 1 - \frac{b}{\nu}\right)\mu^{(a,b)}_{r,s:n} = (n-s+1)\mu^{(a,b)}_{r,s-1:n} + nP_2\left(\mu^{(a,b)}_{r,s:n-1} - \mu^{(a,b)}_{r,s-1:n-1}\right). \quad (14.94)$$

Theorem 14.7.4 *For $n \geq 2$ and $a, b = 1, 2, \ldots$,*

$$\left(1 - \frac{a}{\nu}\right)\mu^{(a,b)}_{1,2:n} = nQ_2\left(\mu^{(a+b)}_{1:n-1} - Q_1^a\mu^{(b)}_{1:n-1}\right) + \mu^{(a+b)}_{2:n}; \quad (14.95)$$

for $n \geq 3$, $2 \leq r \leq n-1$ and $a, b = 1, 2, \ldots$,

$$\left(r - \frac{a}{\nu}\right)\mu^{(a,b)}_{r,r+1:n} = r\mu^{(a+b)}_{r+1:n} - nQ_2\left(\mu^{(a,b)}_{r-1,r:n-1} - \mu^{(a+b)}_{r:n-1}\right); \quad (14.96)$$

for $n \geq 3$, $3 \leq s \leq n$ and $a, b = 1, 2, \ldots$,

$$\left(1 - \frac{a}{\nu}\right)\mu^{(a,b)}_{1,s:n} = \mu^{(a,b)}_{2,s:n} + nQ_2\left(\mu^{(a,b)}_{1,s-1:n-1} + Q_1^a\mu^{(b)}_{s-1:n-1}\right), \quad (14.97)$$

and for $n \geq 4$, $2 \leq r < s \leq n$, $s - r \geq 2$ and $a, b = 1, 2, \ldots$,

$$\left(r - \frac{a}{\nu}\right)\mu^{(a,b)}_{r,s:n} = r\mu^{(a,b)}_{r+1,s:n} + nQ_2\left(\mu^{(a,b)}_{r,s-1:n-1} - \mu^{(a,b)}_{r-1,s-1:n-1}\right). \quad (14.98)$$

14.7.3 Relations for triple moments

The joint p.d.f. of $X_{r:n}$, $X_{s:n}$ and $X_{t:n}$ for $1 \leq r < s < t \leq n$ is given by

$$\begin{aligned}f_{r,s,t:n}(w,x,y) &= C_{r,s,t:n}[F(w)]^{r-1}[F(x)-F(w)]^{s-r-1}[F(y)-F(x)]^{t-s-1}\\ &\quad \times [1-F(y)]^{n-t}f(w)f(x)f(y), \quad Q_1 \leq w < x < y \leq P_1,\end{aligned} \quad (14.99)$$

where $C_{r,s,t:n} = \frac{n!}{(r-1)!(s-r-1)!(t-s-1)!(n-t)!}$. Using (14.99), we may compute the triple moments as

$$\mu^{(a,b,c)}_{r,s,t:n} = \int_{Q_1}^{P_1}\int_w^{P_1}\int_x^{P_1} w^a x^b y^c f_{r,s,t:n}(w,x,y)\,dy\,dx\,dw,$$
$$1 \leq r < s < t \leq n, \ a,b,c \geq 1. \quad (14.100)$$

Then the triple moments $\mu^{(a,b,c)}_{r,s,t:n}$ for $a, b, c = 1, 2, \ldots$, satisfy the recurrence relations presented in the following three theorems.

Theorem 14.7.5 *For* $n \geq 3$, $1 \leq r < s \leq n-2$ *and* $a, b, c = 1, 2, \ldots,$

$$\left((n-s) - \frac{c}{\nu}\right) \mu_{r,s,s+1:n}^{(a,b,c)} = (n-s)\mu_{r,s:n}^{(a,b+c)} + nP_2 \left(\mu_{r,s,s+1:n-1}^{(a,b,c)} - \mu_{r,s:n-1}^{(a,b+c)}\right), \tag{14.101}$$

and for $n \geq 4$, $1 \leq r < s < t \leq n-1$, $t-s \geq 2$ *and* $a, b = 1, 2, \ldots,$

$$\left((n-t+1) - \frac{c}{\nu}\right) \mu_{r,s,t:n}^{(a,b,c)}$$
$$= (n-t+1)\mu_{r,s,t-1:n}^{(a,b,c)} + nP_2 \left(\mu_{r,s,t:n-1}^{(a,b,c)} - \mu_{r,s,t-1:n-1}^{(a,b,c)}\right). \tag{14.102}$$

Theorem 14.7.6 *For* $n \geq 3$, $3 \leq t \leq n$ *and* $a, b, c = 1, 2, \ldots,$

$$\left(1 - \frac{a}{\nu}\right) \mu_{1,2,t:n}^{(a,b,c)} = -nQ_2 \left(Q_1^a \mu_{1,t-1:n-1}^{(b,c)} - \mu_{1,t-1:n-1}^{(a+b,c)}\right) + \mu_{2,t:n}^{(a+b,c)}; \tag{14.103}$$

for $n \geq 4$, $2 \leq r < t \leq n$, $t - r \geq 2$ *and* $a, b, c = 1, 2, \ldots,$

$$\left(r - \frac{a}{\nu}\right) \mu_{r,r+1,t:n}^{(a,b,c)} = -nQ_2 \left(\mu_{r-1,r,t-1:n-1}^{(a,b,c)} - \mu_{r,t-1:n-1}^{(a+b,c)}\right) + r\mu_{r+1,t:n}^{(a+b,c)}; \tag{14.104}$$

for $n \geq 4$, $3 \leq s < t \leq n$, *and* $a, b, c = 1, 2, \ldots,$

$$\mu_{2,s,t:n}^{(a,b,c)} = \left(1 - \frac{a}{\nu}\right) \mu_{1,s,t:n}^{(a,b,c)} - nQ_2 \left(\mu_{1,s-1,t-1:n-1}^{(a,b,c)} - Q_1^a \mu_{s-1,t-1:n-1}^{(b,c)}\right), \tag{14.105}$$

and for $n \geq 5$, $2 \leq r < s < t \leq n$, $s - r \geq 2$ *and* $a, b, c = 1, 2, \ldots,$

$$r\mu_{r+1,s,t:n}^{(a,b,c)} = \left(r - \frac{a}{\nu}\right) \mu_{r,s,t:n}^{(a,b,c)} - nQ_2 \left(\mu_{r,s-1,t-1:n-1}^{(a,b,c)} - \mu_{r-1,s-1,t-1:n-1}^{(a,b,c)}\right). \tag{14.106}$$

Theorem 14.7.7 *For* $n \geq 3$, $1 \leq r \leq n-2$ *and* $a, b, c = 1, 2, \ldots,$

$$\left(1 - \frac{b}{\nu}\right) \mu_{r,r+1,r+2:n}^{(a,b,c)} = \mu_{r,r+2:n}^{(a+b,c)} - (n-r-1)\left(\mu_{r,r+1:n}^{(a,b+c)} - \mu_{r,r+1:n}^{(a+b,c)}\right)$$
$$+ nP_2 \left(\mu_{r,r+1:n-1}^{(a,b+c)} - \mu_{r,r+1:n-1}^{(a+b,c)}\right); \tag{14.107}$$

for $n \geq 4$, $1 \leq r < t \leq n$, $t - r \geq 3$ *and* $a, b, c = 1, 2, \ldots,$

$$\left((t-r-1) - \frac{b}{\nu}\right) \mu_{r,r+1,t:n}^{(a,b,c)} = (n-t+1)\left(\mu_{r,t-1:n}^{(a+b,c)} - \mu_{r,r+1,t-1:n}^{(a,b,c)}\right)$$
$$+ nP_2 \left(\mu_{r,r+1,t-1:n-1}^{(a,b,c)} - \mu_{r,t-1:n-1}^{(a+b,c)}\right)$$
$$+ (t-r-1)\mu_{r,t:n}^{(a+b,c)}; \tag{14.108}$$

for $n \geq 4$, $1 \leq r < s \leq n-1$, $s - r \geq 2$ and $a, b, c = 1, 2, \ldots$,

$$\left(1 - \frac{b}{\nu}\right)\mu_{r,s,s+1:n}^{(a,b,c)} = \mu_{r,s-1,s+1:n}^{(a,b,c)} - (n-s)\left(\mu_{r,s:n}^{(a,b+c)} - \mu_{r,s-1,s:n}^{(a,b,c)}\right)$$
$$+ nP_2\left(\mu_{r,s:n-1}^{(a,b+c)} - \mu_{r,s-1,s:n-1}^{(a,b,c)}\right), \quad (14.109)$$

and for $n \geq 5$, $1 \leq r < s < t \leq n$, $s - r \geq 2$, $t - s \geq 2$ and $a, b, c = 1, 2, \ldots$,

$$\left((t-s) - \frac{b}{\nu}\right)\mu_{r,s,t:n}^{(a,b,c)}$$
$$= (t-s)\mu_{r,s-1,t:n}^{(a,b,c)} - (n-t+1)\left(\mu_{r,s,t-1:n}^{(a,b,c)} - \mu_{r,s-1,t-1:n}^{(a,b,c)}\right)$$
$$+ nP_2\left(\mu_{r,s,t-1:n-1}^{(a,b,c)} - \mu_{r,s-1,t-1:n-1}^{(a,b,c)}\right). \quad (14.110)$$

14.7.4 Relations for quadruple moments

The joint p.d.f. of $X_{r:n}$, $X_{s:n}$, $X_{t:n}$ and $X_{u:n}$ for $1 \leq r < s < t < u \leq n$ is given by

$$f_{r,s,t,u:n}(w,x,y,z) = C_{r,s,t,u:n}[F(w)]^{r-1}[F(x) - F(w)]^{s-r-1}$$
$$\times [F(y) - F(x)]^{t-s-1}[F(z) - F(y)]^{u-t-1}$$
$$\times [1 - F(z)]^{n-u} f(w)f(x)f(y)f(z),$$
$$Q_1 \leq w < x < y < z \leq P_1, \quad (14.111)$$

where $C_{r,s,t,u:n} = \frac{n!}{(r-1)!(s-r-1)!(t-s-1)!(u-t-1)!(n-u)!}$. Using (14.111), we may compute the quadruple moments as

$$\mu_{r,s,t,u:n}^{(a,b,c,d)} = \int_{Q_1}^{P_1}\int_{w}^{P_1}\int_{x}^{P_1}\int_{y}^{P_1} w^a x^b y^c z^d f_{r,s,t,u:n}(w,x,y,z)\,dz\,dy\,dx\,dw,$$
$$1 \leq r < s < t < u \leq n, \; a, b, c, d \geq 1. \quad (14.112)$$

Then the quadruple moments $\mu_{r,s,t,u:n}^{(a,b,c,d)}$ for $a, b, c, d = 1, 2, \ldots$, satisfy the recurrence relations presented in the following four theorems.

Theorem 14.7.8 *For* $n \geq 4$, $1 \leq r < s < t \leq n-2$ *and* $a, b, c, d = 1, 2, \ldots$,

$$\left(n - t - \frac{d}{\nu}\right)\mu_{r,s,t,t+1:n}^{(a,b,c,d)} = nP_2\left(\mu_{r,s,t,t+1:n-1}^{(a,b,c,d)} - \mu_{r,s,t:n-1}^{(a,b,c+d)}\right)$$
$$+ (n-t)\mu_{r,s,t:n}^{(a,b,c+d)}, \quad (14.113)$$

and for $n \geq 5$, $1 \leq r < s < t < u \leq n-1$, $u - t \geq 2$ and $a, b, c, d = 1, 2, \ldots$,

$$\left((n-u+1) - \frac{d}{\nu}\right)\mu_{r,s,t,u:n}^{(a,b,c,d)}$$
$$= (n-u+1)\mu_{r,s,t,u-1:n}^{(a,b,c,d)} + nP_2\left(\mu_{r,s,t,u:n-1}^{(a,b,c,d)} - \mu_{r,s,t,u-1:n-1}^{(a,b,c,d)}\right).$$
$$(14.114)$$

Theorem 14.7.9 *For $n \geq 4$, $3 \leq t < u \leq n$ and $a, b, c, d = 1, 2, \ldots,$*

$$\left(1 - \frac{a}{\nu}\right) \mu_{1,2,t,u:n}^{(a,b,c,d)} = -nQ_2 \left(Q_1^a \mu_{1,t-1,u-1:n-1}^{(b,c,d)} - \mu_{1,t-1,u-1:n-1}^{(a+b,c,d)}\right) + \mu_{2,t,u:n}^{(a+b,c,d)};$$

(14.115)

for $n \geq 5$, $2 \leq r < t < u \leq n$, $t - r \geq 2$ and $a, b, c, d = 1, 2, \ldots,$

$$\left(r - \frac{a}{\nu}\right) \mu_{r,r+1,t,u:n}^{(a,b,c,d)}$$
$$= -nQ_2 \left(\mu_{r-1,r,t-1,u-1:n-1}^{(a,b,c,d)} - \mu_{r,t-1,u-1:n-1}^{(a+b,c,d)}\right) + r\mu_{r+1,t,u:n}^{(a+b,c,d)};$$

(14.116)

for $n \geq 5$, $4 \leq t < u \leq n$, and $a, b, c, d = 1, 2, \ldots,$

$$\mu_{2,s,t,u:n}^{(a,b,c,d)} = \left(1 - \frac{a}{\nu}\right) \mu_{1,s,t,u:n}^{(a,b,c,d)}$$
$$- nQ_2 \left(\mu_{1,s-1,t-1,u-1:n-1}^{(a,b,c,d)} - Q_1^a \mu_{s-1,t-1,u-1:n-1}^{(b,c,d)}\right),$$

(14.117)

and for $n \geq 6$, $2 \leq r < s < t < u \leq n$, $s - r \geq 2$ and $a, b, c, d = 1, 2, \ldots,$

$$r\mu_{r+1,s,t,u:n}^{(a,b,c,d)}$$
$$= \left(r - \frac{a}{\nu}\right) \mu_{r,s,t,u:n}^{(a,b,c,d)}$$
$$- nQ_2 \left(\mu_{r,s-1,t-1,u-1:n-1}^{(a,b,c,d)} - \mu_{r-1,s-1,t-1,u-1:n-1}^{(a,b,c,d)}\right) \quad (14.118)$$

Theorem 14.7.10 *For $n \geq 4$, $1 \leq r < u \leq n$, $u - r \geq 3$ and $a, b, c, d = 1, 2, \ldots,$*

$$\left(1 - \frac{b}{\nu}\right) \mu_{r,r+1,r+2,u:n}^{(a,b,c,d)}$$
$$= \mu_{r,r+2,u:n}^{(a+b,c,d)} - (u - r - 2) \left(\mu_{r,r+1,u:n}^{(a,b+c,d)} - \mu_{r,r+1,u:n}^{(a+b,c,d)}\right)$$
$$+ nP_2 \left(\mu_{r,r+1,u-1:n-1}^{(a,b+c,d)} - \mu_{r,r+1,u-1:n-1}^{(a+b,c,d)}\right)$$
$$- (n - u + 1) \left(\mu_{r,r+1,u-1:n}^{(a,b+c,d)} - \mu_{r,r+1,u-1:n}^{(a+b,c,d)}\right);$$

(14.119)

for $n \geq 5$, $1 \leq r < t < u \leq n$, $t - r \geq 3$ and $a, b, c, d = 1, 2, \ldots,$

$$\left((t - r - 1) - \frac{b}{\nu}\right) \mu_{r,r+1,t,u:n}^{(a,b,c,d)}$$
$$= (n - u + 1) \left(\mu_{r,t-1,u-1:n}^{(a+b,c,d)} - \mu_{r,r+1,t-1,u-1:n}^{(a,b,c,d)}\right)$$

$$- (u - t) \left(\mu_{r,r+1,t-1,u:n}^{(a,b,c,d)} - \mu_{r,t-1,u:n}^{(a+b,c,d)} \right)$$
$$+ nP_2 \left(\mu_{r,t-1,u-1:n-1}^{(a+b,c,d)} + \mu_{r,r+1,t-1,u-1:n-1}^{(a,b,c,d)} \right)$$
$$+ (t - r - 1)\mu_{r,t,u:n}^{(a+b,c,d)}; \tag{14.120}$$

for $1 \leq r < s < u \leq n$, $s - r \geq 2$, $u - s \geq 2$ and $a, b, c, d = 1, 2, \ldots$,

$$\left(1 - \frac{b}{\nu}\right) \mu_{r,s,s+1,u:n}^{(a,b,c,d)}$$
$$= \mu_{r,s-1,s+1,u:n}^{(a,b,c,d)} - (u - s - 1) \left(\mu_{r,s,u:n}^{(a,b+c,d)} - \mu_{r,s-1,s,u:n}^{(a,b,c,d)} \right)$$
$$+ (n - u + 1) \left(\mu_{r,s-1,s,u-1:n}^{(a,b,c,d)} - \mu_{r,s,u-1:n}^{(a,b+c,d)} \right)$$
$$+ nP_2 \left(\mu_{r,s,u-1:n-1}^{(a,b+c,d)} - \mu_{r,s-1,s,u-1:n-1}^{(a,b,c,d)} \right), \tag{14.121}$$

and for $1 \leq r < s < t < u \leq n$, $s - r \geq 2$, $t - s \geq 2$ and $a, b, c, d = 1, 2, \ldots$,

$$\left((t - s) - \frac{b}{\nu}\right) \mu_{r,s,t,u:n}^{(a,b,c,d)}$$
$$= (t - s)\mu_{r,s-1,t,u:n}^{(a,b,c,d)} - (n - u + 1) \left(\mu_{r,s,t-1,u-1:n}^{(a,b,c,d)} - \mu_{r,s-1,t-1,u-1:n}^{(a,b,c,d)} \right)$$
$$- (u - t) \left(\mu_{r,s,t-1,u:n}^{(a,b,c,d)} - \mu_{r,s-1,t-1,u:n}^{(a,b,c,d)} \right)$$
$$+ nP_2 \left(\mu_{r,s,t-1,u-1:n-1}^{(a,b,c,d)} - \mu_{r,s-1,t-1,u-1:n-1}^{(a,b,c,d)} \right). \tag{14.122}$$

Theorem 14.7.11 For $n \geq 4$, $1 \leq r < s \leq n - 2$ and $a, b, c, d = 1, 2, \ldots$,

$$\left(1 - \frac{c}{\nu}\right) \mu_{r,s,s+1,s+2:n}^{(a,b,c,d)}$$
$$= \mu_{r,s,s+2:n}^{(a,b+c,d)} - (n - s - 1) \left(\mu_{r,s,s+1:n}^{(a,b,c+d)} - \mu_{r,s,s+1:n}^{(a,b+c,d)} \right)$$
$$+ nP_2 \left(\mu_{r,s,s+1:n-1}^{(a,b,c+d)} - \mu_{r,s,s+1:n-1}^{(a,b+c,d)} \right); \tag{14.123}$$

for $n \geq 5$, $1 \leq r < s < u \leq n$, $u - s \geq 3$ and $a, b, c, d = 1, 2, \ldots$,

$$\left((u - s - 1) - \frac{c}{\nu}\right) \mu_{r,s,s+1,u:n}^{(a,b,c,d)}$$
$$= -(n - u + 1) \left(\mu_{r,s,s+1,u-1:n}^{(a,b,c,d)} - \mu_{r,s,u-1:n}^{(a,b+c,d)} \right)$$
$$+ nP_2 \left(\mu_{r,s,s+1,u-1:n-1}^{(a,b,c,d)} - \mu_{r,s,u-1:n-1}^{(a,b+c,d)} \right)$$
$$+ (u - s - 1)\mu_{r,s,u:n}^{(a,b+c,d)}; \tag{14.124}$$

for $n \geq 5$, $1 \leq r < s < t \leq n - 1$, $t - s \geq 2$ and $a, b, c, d = 1, 2, \ldots$,

$$\left(1 - \frac{c}{\nu}\right) \mu_{r,s,t,t+1:n}^{(a,b,c,d)}$$
$$= \mu_{r,s,t-1,t+1:n}^{(a,b,c,d)} - (n - t) \left(\mu_{r,s,t:n}^{(a,b,c+d)} - \mu_{r,s,t-1,t:n}^{(a,b,c,d)} \right)$$
$$+ = nP_2 \left(\mu_{r,s,t:n-1}^{(a,b,c+d)} - \mu_{r,s,t-1,t:n-1}^{(a,b,c,d)} \right), \tag{14.125}$$

and for $n \geq 6$, $1 \leq r < s < t < u \leq n$, $t - s \geq 2$, $u - t \geq 2$ and $a, b, c, d = 1, 2, \ldots$,

$$\left((u-t) - \frac{c}{\nu}\right) \mu_{r,s,t,u:n}^{(a,b,c,d)} = -(n-u+1)\left(\mu_{r,s,t,u-1:n}^{(a,b,c,d)} - \mu_{r,s,t-1,u-1:n}^{(a,b,c,d)}\right)$$
$$+ nP_2 \left(\mu_{r,s,t,u-1:n-1}^{(a,b,c,d)} - \mu_{r,s,t,t-1,u-1:n-1}^{(a,b,c,d)}\right)$$
$$+ (u-t)\mu_{r,s,t-1,u:n}^{(a,b,c,d)}. \tag{14.126}$$

Acknowledgments. The authors thank the Natural Sciences and Engineering Research Council of Canada for funding this research.

References

1. Arnold, B. C. (1983). *Pareto Distributions*, International Cooperative Publishing House, Fairland, Maryland.

2. Arnold, B. C., Balakrishnan, N. and Nagaraja, H. N. (1992). *A First Course in Order Statistics*, New York: John Wiley & Sons.

3. Balakrishnan, N. and Cohen, A. C. (1991). *Order Statistics and Inference: Estimation Methods*, San Diego: Academic Press.

4. Balakrishnan, N. and Gupta, S. S. (1996). Higher order moments of order statistics from exponential and right-truncated exponential distributions and applications to life-testing problems, In *Handbook of Statistics-16: Order Statistics and Their Applications* (Eds., C.R. Rao and N. Balakrishnan), Amsterdam: North-Holland (to appear).

5. Balakrishnan, N. and Joshi, P. C. (1982). Moments of order statistics from doubly truncated Pareto distribution, *Journal of Indian Statistical Association*, **20**, 109–117.

6. Balakrishnan, N., Childs, A., Govindarajulu, Z. and Chandramouleeswaran, M. P. (1996). Inference on parameters of the Laplace distribution based on Type-II censored samples using Edgeworth approximation, *submitted for publication*.

7. Barton, D. E. and Dennis, K. E. R. (1952). The conditions under which Gram-Charlier and Edgeworth curves are positive definite and unimodal, *Biometrika*, **39**, 425–427.

8. Childs, A. and Balakrishnan, N. (1996). Generalized recurrence relations for moments of order statistics from non-identical Pareto and truncated

Pareto random variables with applications to robustness, In *Handbook of Statistics-16: Order Statistics and Their Applications* (Eds., C. R. Rao and N. Balakrishnan), Amsterdam: North-Holland (to appear).

9. David, H. A. (1981). *Order Statistics*, Second edition, New York: John Wiley & Sons.

10. Huang, J. S. (1975). A note on order statistics from the Pareto distribution, *Scandinavian Actuarial Journal*, **2**, 187–190.

11. Johnson, N. L., Kotz, S. and Balakrishnan, N. (1994). *Continuous Univariate Distributions*, Vol. 1, Second edition, New York: John Wiley & Sons.

12. Kulldorff, G. and Vännman, K. (1973). Estimation of the location and scale parameters of a Pareto distribution by linear functions of order statistics, *Journal of the American Statistical Association*, **68**, 218–227.

13. Malik, H. J. (1966). Exact moments of order statistics from the Pareto distribution, *Skandinavisk Aktuarietidskrift*, **1966**, 144–157.

Table 14.1: Mean, variance and coefficients of skewness and kurtosis of $R1^*$ and $R2^*$ when $v = 5$.

n	s	R_1^* Mean	Variance	$\sqrt{\beta_1}$	β_2	n	s	R_2^* Mean	Variance	$\sqrt{\beta_1}$	β_2
5	0	0.0000	0.4058	-1.7002	9.2492	5	0	1.0000	0.3757	1.7068	9.2603
	1	0.0000	0.4718	-1.6706	8.1331		1	1.0000	0.4365	1.6776	8.1548
	2	0.0000	0.6470	-1.8267	8.8398		2	1.0000	0.5980	1.8332	8.8633
6	0	0.0000	0.3166	-1.4798	7.5171	6	0	1.0000	0.2970	1.4850	7.5266
	1	0.0000	0.3525	-1.4534	6.8523		1	1.0000	0.3306	1.4589	6.8669
	2	0.0000	0.4346	-1.5264	7.0890		2	1.0000	0.4073	1.5316	7.1047
7	0	0.0000	0.2592	-1.3262	6.5139	7	0	1.0000	0.2455	1.3304	6.5216
	1	0.0000	0.2813	-1.3040	6.0819		1	1.0000	0.2663	1.3084	6.0924
	2	0.0000	0.3279	-1.3437	6.1721		2	1.0000	0.3103	1.3480	6.1833
	3	0.0000	0.4135	-1.4465	6.5736		3	1.0000	0.3911	1.4504	6.5851
8	0	0.0000	0.2193	-1.2116	5.8667	8	0	1.0000	0.2091	1.2151	5.8729
	1	0.0000	0.2340	-1.1931	5.5676		1	1.0000	0.2231	1.1967	5.5755
	2	0.0000	0.2636	-1.2166	5.6011		2	1.0000	0.2513	1.2202	5.6096
	3	0.0000	0.3132	-1.2780	5.8041		3	1.0000	0.2984	1.2814	5.8128
9	0	0.0000	0.1899	-1.1220	5.4171	9	0	1.0000	0.1821	1.1250	5.4222
	1	0.0000	0.2004	-1.1064	5.1999		1	1.0000	0.1921	1.1095	5.2061
	2	0.0000	0.2206	-1.1211	5.2089		2	1.0000	0.2115	1.1242	5.2154
	3	0.0000	0.2526	-1.1611	5.3226		3	1.0000	0.2420	1.1641	5.3293
	4	0.0000	0.3033	-1.2338	5.5704		4	1.0000	0.2905	1.2366	5.5772
10	0	0.0000	0.1674	-1.0496	5.0876	10	0	1.0000	0.1612	1.0522	5.0919
	1	0.0000	0.1752	-1.0363	4.9241		1	1.0000	0.1687	1.0390	4.9291
	2	0.0000	0.1898	-1.0458	4.9217		2	1.0000	0.1827	1.0484	4.9270
	3	0.0000	0.2120	-1.0734	4.9895		3	1.0000	0.2040	1.0760	4.9949
	4	0.0000	0.2449	-1.1224	5.1379		4	1.0000	0.2356	1.1248	5.1434
15	0	0.0000	0.1050	-0.8229	4.2379	15	0	1.0000	0.1024	0.8243	4.2399
	1	0.0000	0.1076	-0.8160	4.1816		1	1.0000	0.1049	0.8174	4.1838
	2	0.0000	0.1121	-0.8170	4.1709		2	1.0000	0.1093	0.8184	4.1731
	3	0.0000	0.1184	-0.8237	4.1778		3	1.0000	0.1154	0.8251	4.1801
	4	0.0000	0.1265	-0.8359	4.2008		4	1.0000	0.1233	0.8374	4.2031
	5	0.0000	0.1369	-0.8544	4.2414		5	1.0000	0.1335	0.8557	4.2438
	6	0.0000	0.1504	-0.8800	4.3033		6	1.0000	0.1466	0.8814	4.3057
	7	0.0000	0.1681	-0.9149	4.3934		7	1.0000	0.1639	0.9162	4.3957
20	0	0.0000	0.0765	-0.6989	3.8789	20	0	1.0000	0.0750	0.6998	3.8801
	1	0.0000	0.0777	-0.6946	3.8521		1	1.0000	0.0762	0.6956	3.8533
	2	0.0000	0.0797	-0.6942	3.8441		2	1.0000	0.0783	0.6951	3.8453
	3	0.0000	0.0825	-0.6964	3.8433		3	1.0000	0.0810	0.6973	3.8446
	4	0.0000	0.0859	-0.7011	3.8486		4	1.0000	0.0843	0.7020	3.8499
	5	0.0000	0.0901	-0.7083	3.8601		5	1.0000	0.0884	0.7093	3.8613
	6	0.0000	0.0951	-0.7183	3.8779		6	1.0000	0.0933	0.7192	3.8792
	7	0.0000	0.1011	-0.7312	3.9031		7	1.0000	0.0992	0.7321	3.9044
	8	0.0000	0.1084	-0.7476	3.9368		8	1.0000	0.1064	0.7485	3.9381
	9	0.0000	0.1173	-0.7681	3.9812		9	1.0000	0.1151	0.7689	3.9825

Table 14.2: Mean, variance and coefficients of skewness and kurtosis of $R1^*$ and $R2^*$ when $v = 15$.

n	s	\multicolumn{4}{c}{R_1^*}	n	s	\multicolumn{4}{c}{R_2^*}
		Mean	Variance	$\sqrt{\beta_1}$	β_2			Mean	Variance	$\sqrt{\beta_1}$	β_2
5	0	0.0000	0.2936	-1.2087	5.4478	5	0	1.0000	0.2860	1.2097	5.4500
	1	0.0000	0.3733	-1.3093	5.7756		1	1.0000	0.3636	1.3102	5.7779
	2	0.0000	0.5435	-1.5378	6.7421		2	1.0000	0.5293	1.5386	6.7444
6	0	0.0000	0.2335	-1.0776	4.9386	6	0	1.0000	0.2285	1.0784	4.9401
	1	0.0000	0.2800	-1.1404	5.1093		1	1.0000	0.2740	1.1411	5.1109
	2	0.0000	0.3635	-1.2691	5.5618		2	1.0000	0.3556	1.2697	5.5634
7	0	0.0000	0.1938	-0.9814	4.6036	7	0	1.0000	0.1902	0.9820	4.6047
	1	0.0000	0.2241	-1.0243	4.7039		1	1.0000	0.2199	1.0249	4.7050
	2	0.0000	0.2734	-1.1079	4.9596		2	1.0000	0.2683	1.1084	4.9608
	3	0.0000	0.3578	-1.2459	5.4425		3	1.0000	0.3511	1.2464	5.4436
8	0	0.0000	0.1656	-0.9070	4.3669	8	0	1.0000	0.1629	0.9075	4.3677
	1	0.0000	0.1868	-0.9382	4.4305		1	1.0000	0.1838	0.9387	4.4314
	2	0.0000	0.2192	-0.9972	4.5919		2	1.0000	0.2157	0.9976	4.5927
	3	0.0000	0.2693	-1.0881	4.8711		3	1.0000	0.2649	1.0885	4.8720
9	0	0.0000	0.1446	-0.8473	4.1907	9	0	1.0000	0.1425	0.8477	4.1914
	1	0.0000	0.1602	-0.8709	4.2335		1	1.0000	0.1579	0.8713	4.2342
	2	0.0000	0.1831	-0.9149	4.3429		2	1.0000	0.1804	0.9153	4.3436
	3	0.0000	0.2161	-0.9799	4.5226		3	1.0000	0.2130	0.9803	4.5233
	4	0.0000	0.2665	-1.0746	4.8118		4	1.0000	0.2626	1.0749	4.8125
10	0	0.0000	0.1282	-0.7979	4.0546	10	0	1.0000	0.1266	0.7982	4.0552
	1	0.0000	0.1402	-0.8164	4.0846		1	1.0000	0.1384	0.8168	4.0851
	2	0.0000	0.1572	-0.8506	4.1628		2	1.0000	0.1551	0.8510	4.1633
	3	0.0000	0.1806	-0.8997	4.2869		3	1.0000	0.1782	0.9000	4.2874
	4	0.0000	0.2139	-0.9678	4.4749		4	1.0000	0.2111	0.9681	4.4754
15	0	0.0000	0.0820	-0.6375	3.6705	15	0	1.0000	0.0813	0.6377	3.6707
	1	0.0000	0.0864	-0.6449	3.6777		1	1.0000	0.0857	0.6451	3.6780
	2	0.0000	0.0923	-0.6585	3.7005		2	1.0000	0.0915	0.6587	3.7007
	3	0.0000	0.0994	-0.6769	3.7345		3	1.0000	0.0986	0.6770	3.7347
	4	0.0000	0.1083	-0.7000	3.7801		4	1.0000	0.1073	0.7002	3.7804
	5	0.0000	0.1192	-0.7286	3.8397		5	1.0000	0.1182	0.7288	3.8400
	6	0.0000	0.1330	-0.7639	3.9175		6	1.0000	0.1318	0.7641	3.9177
	7	0.0000	0.1508	-0.8080	4.0204		7	1.0000	0.1495	0.8081	4.0206
20	0	0.0000	0.0602	-0.5462	3.4912	20	0	1.0000	0.0598	0.5463	3.4913
	1	0.0000	0.0625	-0.5501	3.4935		1	1.0000	0.0621	0.5502	3.4936
	2	0.0000	0.0654	-0.5574	3.5032		2	1.0000	0.0650	0.5575	3.5034
	3	0.0000	0.0688	-0.5671	3.5177		3	1.0000	0.0683	0.5672	3.5179
	4	0.0000	0.0727	-0.5789	3.5366		4	1.0000	0.0722	0.5790	3.5368
	5	0.0000	0.0773	-0.5929	3.5601		5	1.0000	0.0768	0.5930	3.5602
	6	0.0000	0.0827	-0.6094	3.5887		6	1.0000	0.0821	0.6095	3.5889
	7	0.0000	0.0890	-0.6285	3.6235		7	1.0000	0.0884	0.6287	3.6236
	8	0.0000	0.0965	-0.6510	3.6657		8	1.0000	0.0958	0.6511	3.6659
	9	0.0000	0.1055	-0.6773	3.7176		9	1.0000	0.1048	0.6774	3.7177

Table 14.3: Mean, variance and coefficients of skewness and kurtosis of $R1^*$ and $R2^*$ when $v = 25$.

		R_1^*						R_2^*			
n	s	Mean	Variance	$\sqrt{\beta_1}$	β_2	n	s	Mean	Variance	$\sqrt{\beta_1}$	β_2
5	0	0.0000	0.2753	-1.1230	5.0302	5	0	1.0000	0.2709	1.1234	5.0309
	1	0.0000	0.3566	-1.2457	5.4413		1	1.0000	0.3510	1.2460	5.4421
	2	0.0000	0.5255	-1.4869	6.4261		2	1.0000	0.5172	1.4872	6.4269
6	0	0.0000	0.2195	-1.0033	4.6179	6	0	1.0000	0.2166	1.0035	4.6185
	1	0.0000	0.2676	-1.0831	4.8489		1	1.0000	0.2640	1.0834	4.8495
	2	0.0000	0.3511	-1.2223	5.3240		2	1.0000	0.3464	1.2226	5.3245
7	0	0.0000	0.1825	-0.9151	4.3445	7	0	1.0000	0.1805	0.9153	4.3449
	1	0.0000	0.2141	-0.9717	4.4900		1	1.0000	0.2117	0.9719	4.4904
	2	0.0000	0.2638	-1.0640	4.7659		2	1.0000	0.2608	1.0642	4.7663
	3	0.0000	0.3477	-1.2088	5.2571		3	1.0000	0.3438	1.2089	5.2575
8	0	0.0000	0.1562	-0.8466	4.1499	8	0	1.0000	0.1547	0.8468	4.1502
	1	0.0000	0.1785	-0.8891	4.2487		1	1.0000	0.1767	0.8893	4.2490
	2	0.0000	0.2113	-0.9556	4.4273		2	1.0000	0.2092	0.9557	4.4276
	3	0.0000	0.2614	-1.0524	4.7160		3	1.0000	0.2588	1.0525	4.7163
9	0	0.0000	0.1365	-0.7915	4.0044	9	0	1.0000	0.1353	0.7917	4.0047
	1	0.0000	0.1531	-0.8247	4.0752		1	1.0000	0.1517	0.8248	4.0755
	2	0.0000	0.1763	-0.8752	4.1993		2	1.0000	0.1748	0.8754	4.1996
	3	0.0000	0.2095	-0.9453	4.3881		3	1.0000	0.2077	0.9455	4.3883
	4	0.0000	0.2597	-1.0444	4.6823		4	1.0000	0.2575	1.0445	4.6825
10	0	0.0000	0.1212	-0.7459	3.8916	10	0	1.0000	0.1202	0.7460	3.8918
	1	0.0000	0.1340	-0.7726	3.9444		1	1.0000	0.1329	0.7728	3.9446
	2	0.0000	0.1513	-0.8125	4.0351		2	1.0000	0.1501	0.8126	4.0353
	3	0.0000	0.1749	-0.8661	4.1674		3	1.0000	0.1735	0.8662	4.1676
	4	0.0000	0.2082	-0.9382	4.3608		4	1.0000	0.2066	0.9383	4.3610
15	0	0.0000	0.0777	-0.5972	3.5706	15	0	1.0000	0.0773	0.5973	3.5707
	1	0.0000	0.0826	-0.6092	3.5882		1	1.0000	0.0821	0.6092	3.5883
	2	0.0000	0.0886	-0.6262	3.6175		2	1.0000	0.0882	0.6263	3.6176
	3	0.0000	0.0960	-0.6474	3.6563		3	1.0000	0.0955	0.6474	3.6564
	4	0.0000	0.1049	-0.6729	3.7057		4	1.0000	0.1043	0.6730	3.7058
	5	0.0000	0.1159	-0.7037	3.7684		5	1.0000	0.1153	0.7038	3.7685
	6	0.0000	0.1297	-0.7411	3.8486		6	1.0000	0.1291	0.7412	3.8486
	7	0.0000	0.1476	-0.7870	3.9533		7	1.0000	0.1468	0.7871	3.9534
20	0	0.0000	0.0571	-0.5123	3.4195	20	0	1.0000	0.0569	0.5123	3.4195
	1	0.0000	0.0597	-0.5191	3.4277		1	1.0000	0.0594	0.5192	3.4277
	2	0.0000	0.0627	-0.5288	3.4414		2	1.0000	0.0625	0.5288	3.4414
	3	0.0000	0.0662	-0.5404	3.4590		3	1.0000	0.0660	0.5405	3.4590
	4	0.0000	0.0703	-0.5540	3.4804		4	1.0000	0.0700	0.5540	3.4805
	5	0.0000	0.0749	-0.5696	3.5060		5	1.0000	0.0746	0.5697	3.5061
	6	0.0000	0.0803	-0.5875	3.5365		6	1.0000	0.0800	0.5876	3.5365
	7	0.0000	0.0867	-0.6081	3.5728		7	1.0000	0.0863	0.6081	3.5728
	8	0.0000	0.0942	-0.6318	3.6164		8	1.0000	0.0938	0.6318	3.6165
	9	0.0000	0.1033	-0.6593	3.6694		9	1.0000	0.1029	0.6594	3.6695

Table 14.4: Percentage points of the distribution of R_1 when $v = 5$.

n	s	Edgeworth								Simulated							
		1%	2.5%	5%	10%	90%	95%	97.5%	99%	1%	2.5%	5%	10%	90%	95%	97.5%	99%
5	0	-3.71	-3.24	-2.65	-0.94	0.98	1.11	1.19	2.67	-3.33	-2.54	-1.91	-1.26	1.02	1.16	1.27	1.38
	1	-3.66	-3.17	-2.40	-1.08	1.03	1.16	1.24	2.86	-3.35	-2.37	-1.84	-1.28	1.01	1.16	1.26	1.34
	2	-3.73	-3.25	-2.58	-1.05	1.02	1.14	1.21	2.79	-3.42	-2.59	-1.9	-1.28	0.98	1.08	1.16	1.22
6	0	-3.58	-3.07	-2.25	-1.08	1.03	1.18	1.27	2.99	-3.1	-2.34	-1.8	-1.26	1.07	1.21	1.33	1.44
	1	-3.54	-3.00	-2.04	-1.16	1.06	1.21	1.31	1.38	-3.32	-2.37	-1.85	-1.23	1.04	1.19	1.33	1.43
	2	-3.58	-3.05	-2.06	-1.16	1.06	1.21	1.30	1.36	-3.27	-2.42	-1.85	-1.29	1.04	1.18	1.29	1.39
7	0	-3.49	-2.93	-2.03	-1.17	1.07	1.23	1.34	1.42	-3.34	-2.52	-1.92	-1.34	1.09	1.25	1.39	1.55
	1	-3.45	-2.87	-1.96	-1.21	1.09	1.25	1.36	1.44	-3.17	-2.37	-1.86	-1.28	1.09	1.25	1.38	1.53
	2	-3.47	-2.89	-1.96	-1.21	1.09	1.25	1.35	1.43	-3.28	-2.51	-1.93	-1.36	1.05	1.22	1.36	1.46
	3	-3.53	-2.97	-1.98	-1.20	1.08	1.23	1.33	1.39	-3.15	-2.47	-1.92	-1.3	1.05	1.21	1.3	1.39
8	0	-3.40	-2.81	-1.95	-1.21	1.09	1.27	1.39	1.48	-3.2	-2.44	-1.88	-1.33	1.09	1.27	1.42	1.57
	1	-3.37	-2.76	-1.93	-1.24	1.10	1.28	1.40	1.50	-3.08	-2.41	-1.9	-1.32	1.05	1.27	1.4	1.53
	2	-3.38	-2.77	-1.93	-1.25	1.10	1.28	1.40	1.49	-3.17	-2.43	-1.9	-1.3	1.09	1.27	1.4	1.54
	3	-3.42	-2.82	-1.93	-1.24	1.10	1.27	1.38	1.46	-3.24	-2.43	-1.88	-1.29	1.08	1.25	1.37	1.48
9	0	-3.33	-2.72	-1.92	-1.24	1.11	1.30	1.42	1.53	-3.03	-2.43	-1.91	-1.31	1.1	1.3	1.45	1.6
	1	-3.31	-2.68	-1.90	-1.26	1.12	1.31	1.44	1.54	-3.19	-2.35	-1.83	-1.27	1.12	1.32	1.47	1.59
	2	-3.31	-2.68	-1.90	-1.26	1.12	1.31	1.43	1.54	-3.16	-2.39	-1.84	-1.31	1.1	1.31	1.44	1.62
	3	-3.34	-2.71	-1.91	-1.26	1.12	1.30	1.42	1.52	-3.06	-2.43	-1.92	-1.33	1.11	1.32	1.45	1.59
	4	-3.39	-2.77	-1.92	-1.25	1.11	1.28	1.40	1.49	-3.03	-2.45	-1.9	-1.32	1.08	1.26	1.39	1.49
10	0	-3.27	-2.64	-1.90	-1.26	1.12	1.32	1.46	1.57	-3.06	-2.38	-1.92	-1.4	1.12	1.33	1.49	1.65
	1	-3.25	-2.61	-1.89	-1.28	1.13	1.33	1.47	1.58	-3.18	-2.39	-1.93	-1.35	1.12	1.32	1.49	1.66
	2	-3.25	-2.61	-1.89	-1.28	1.13	1.33	1.46	1.58	-3.05	-2.38	-1.85	-1.3	1.12	1.32	1.47	1.6
	3	-3.27	-2.63	-1.89	-1.28	1.13	1.32	1.45	1.56	-3.2	-2.4	-1.86	-1.33	1.11	1.3	1.45	1.58
	4	-3.31	-2.67	-1.90	-1.27	1.12	1.31	1.44	1.54	-3.18	-2.4	-1.86	-1.3	1.11	1.29	1.43	1.56
15	0	-3.06	-2.42	-1.84	-1.30	1.16	1.39	1.56	1.71	-2.89	-2.29	-1.8	-1.35	1.16	1.37	1.59	1.78
	1	-3.05	-2.41	-1.84	-1.30	1.17	1.39	1.56	1.72	-2.98	-2.34	-1.81	-1.28	1.17	1.42	1.59	1.78
	2	-3.05	-2.41	-1.84	-1.30	1.17	1.39	1.56	1.72	-3.07	-2.29	-1.82	-1.32	1.14	1.39	1.59	1.77
	3	-3.05	-2.41	-1.84	-1.31	1.17	1.39	1.56	1.71	-2.99	-2.28	-1.82	-1.34	1.18	1.39	1.6	1.78
	4	-3.06	-2.42	-1.84	-1.31	1.17	1.39	1.56	1.70	-2.95	-2.27	-1.81	-1.32	1.17	1.39	1.56	1.75
	5	-3.08	-2.43	-1.85	-1.31	1.16	1.39	1.55	1.69	-3.02	-2.32	-1.85	-1.37	1.15	1.37	1.53	1.7
	6	-3.10	-2.45	-1.85	-1.30	1.16	1.38	1.54	1.68	-2.98	-2.34	-1.75	-1.3	1.15	1.37	1.54	1.69
	7	-3.13	-2.47	-1.86	-1.30	1.16	1.37	1.52	1.65	-2.94	-2.35	-1.81	-1.29	1.15	1.37	1.52	1.68
20	0	-2.94	-2.33	-1.82	-1.31	1.18	1.43	1.62	1.80	-2.84	-2.34	-1.84	-1.37	1.19	1.44	1.6	1.82
	1	-2.93	-2.32	-1.82	-1.31	1.19	1.43	1.62	1.80	-2.81	-2.19	-1.81	-1.32	1.19	1.43	1.62	1.8
	2	-2.93	-2.32	-1.81	-1.31	1.19	1.43	1.62	1.80	-2.74	-2.26	-1.82	-1.34	1.21	1.45	1.64	1.84
	3	-2.93	-2.32	-1.82	-1.31	1.19	1.43	1.62	1.80	-2.78	-2.25	-1.79	-1.34	1.17	1.42	1.62	1.85
	4	-2.93	-2.32	-1.82	-1.31	1.19	1.43	1.62	1.80	-2.83	-2.21	-1.79	-1.28	1.2	1.44	1.62	1.81
	5	-2.94	-2.33	-1.82	-1.31	1.18	1.43	1.61	1.79	-2.83	-2.24	-1.8	-1.32	1.14	1.4	1.6	1.81
	6	-2.94	-2.33	-1.82	-1.31	1.18	1.43	1.61	1.78	-2.89	-2.17	-1.73	-1.27	1.17	1.42	1.62	1.81
	7	-2.96	-2.34	-1.82	-1.31	1.18	1.42	1.60	1.77	-2.69	-2.23	-1.75	-1.29	1.2	1.4	1.59	1.84
	8	-2.97	-2.35	-1.82	-1.31	1.18	1.42	1.60	1.76	-2.9	-2.24	-1.81	-1.28	1.18	1.41	1.6	1.81
	9	-2.99	-2.36	-1.83	-1.31	1.18	1.41	1.59	1.75	-2.91	-2.35	-1.85	-1.35	1.15	1.4	1.57	1.74

Table 14.5: Percentage points of the distribution of R_1 when $v = 15$.

		Edgeworth								Simulated							
n	s	1%	2.5%	5%	10%	90%	95%	97.5%	99%	1%	2.5%	5%	10%	90%	95%	97.5%	99%
5	0	-3.37	-2.75	-1.91	-1.26	1.11	1.29	1.41	1.50	-3.11	-2.45	-1.89	-1.27	1.08	1.27	1.40	1.54
	1	-3.43	-2.83	-1.93	-1.25	1.10	1.27	1.38	1.46	-3.16	-2.33	-1.85	-1.31	1.07	1.23	1.34	1.45
	2	-3.57	-3.02	-1.96	-1.21	1.08	1.22	1.31	1.37	-3.29	-2.54	-1.92	-1.32	1.03	1.14	1.21	1.27
6	0	-3.27	-2.62	-1.89	-1.28	1.13	1.32	1.46	1.56	-2.90	-2.30	-1.81	-1.29	1.13	1.32	1.45	1.59
	1	-3.31	-2.67	-1.90	-1.28	1.13	1.31	1.44	1.53	-3.13	-2.34	-1.83	-1.27	1.09	1.27	1.41	1.54
	2	-3.40	-2.78	-1.91	-1.26	1.11	1.28	1.39	1.48	-3.08	-2.35	-1.85	-1.31	1.08	1.23	1.36	1.45
7	0	-3.19	-2.53	-1.87	-1.30	1.15	1.35	1.49	1.62	-3.05	-2.44	-1.89	-1.35	1.15	1.34	1.50	1.68
	1	-3.22	-2.56	-1.88	-1.30	1.14	1.34	1.48	1.59	-3.02	-2.32	-1.83	-1.31	1.14	1.31	1.47	1.62
	2	-3.28	-2.63	-1.89	-1.29	1.13	1.32	1.45	1.55	-3.13	-2.46	-1.94	-1.38	1.09	1.28	1.41	1.55
	3	-3.38	-2.75	-1.91	-1.27	1.12	1.29	1.40	1.49	-3.13	-2.45	-1.94	-1.32	1.09	1.25	1.35	1.44
8	0	-3.12	-2.47	-1.86	-1.30	1.16	1.37	1.53	1.66	-3.02	-2.39	-1.89	-1.34	1.14	1.35	1.51	1.69
	1	-3.14	-2.48	-1.86	-1.30	1.15	1.36	1.51	1.64	-2.95	-2.36	-1.89	-1.36	1.10	1.32	1.47	1.64
	2	-3.19	-2.53	-1.87	-1.30	1.15	1.35	1.49	1.61	-3.08	-2.37	-1.87	-1.33	1.13	1.34	1.47	1.63
	3	-3.26	-2.61	-1.88	-1.29	1.14	1.33	1.46	1.56	-3.09	-2.38	-1.86	-1.32	1.11	1.30	1.42	1.53
9	0	-3.06	-2.42	-1.84	-1.31	1.17	1.39	1.55	1.70	-2.90	-2.30	-1.87	-1.32	1.15	1.37	1.53	1.72
	1	-3.08	-2.43	-1.85	-1.31	1.16	1.38	1.54	1.68	-3.01	-2.30	-1.80	-1.27	1.17	1.38	1.55	1.69
	2	-3.12	-2.46	-1.85	-1.31	1.16	1.37	1.52	1.66	-3.00	-2.38	-1.83	-1.32	1.13	1.37	1.50	1.70
	3	-3.17	-2.51	-1.87	-1.30	1.15	1.36	1.50	1.62	-2.93	-2.40	-1.90	-1.34	1.15	1.36	1.51	1.67
	4	-3.25	-2.59	-1.88	-1.30	1.14	1.33	1.46	1.57	-2.93	-2.40	-1.88	-1.34	1.11	1.29	1.44	1.54
10	0	-3.02	-2.38	-1.83	-1.31	1.17	1.40	1.57	1.73	-2.91	-2.28	-1.89	-1.41	1.15	1.40	1.58	1.74
	1	-3.03	-2.39	-1.84	-1.31	1.17	1.40	1.57	1.72	-2.99	-2.34	-1.88	-1.36	1.16	1.38	1.57	1.77
	2	-3.06	-2.41	-1.84	-1.31	1.17	1.39	1.55	1.70	-2.96	-2.35	-1.81	-1.32	1.15	1.37	1.53	1.70
	3	-3.10	-2.45	-1.85	-1.31	1.16	1.38	1.53	1.67	-3.03	-2.35	-1.84	-1.35	1.15	1.35	1.50	1.65
	4	-3.16	-2.50	-1.86	-1.31	1.15	1.36	1.50	1.63	-3.10	-2.36	-1.84	-1.31	1.14	1.34	1.48	1.61
15	0	-2.86	-2.28	-1.80	-1.32	1.20	1.45	1.65	1.84	-2.77	-2.21	-1.79	-1.34	1.18	1.44	1.68	1.89
	1	-2.86	-2.28	-1.80	-1.32	1.20	1.45	1.65	1.84	-2.83	-2.29	-1.79	-1.28	1.21	1.47	1.65	1.88
	2	-2.88	-2.29	-1.81	-1.32	1.20	1.45	1.64	1.83	-2.91	-2.25	-1.80	-1.32	1.17	1.44	1.64	1.85
	3	-2.89	-2.30	-1.81	-1.32	1.19	1.44	1.63	1.81	-2.88	-2.23	-1.80	-1.34	1.19	1.44	1.65	1.85
	4	-2.91	-2.31	-1.81	-1.32	1.19	1.43	1.62	1.80	-2.86	-2.22	-1.79	-1.32	1.20	1.43	1.62	1.82
	5	-2.94	-2.33	-1.82	-1.32	1.19	1.43	1.61	1.78	-2.89	-2.31	-1.86	-1.37	1.17	1.40	1.57	1.77
	6	-2.97	-2.35	-1.83	-1.32	1.18	1.42	1.59	1.75	-2.90	-2.30	-1.75	-1.31	1.16	1.39	1.58	1.75
	7	-3.01	-2.38	-1.83	-1.32	1.18	1.40	1.57	1.72	-2.86	-2.31	-1.77	-1.30	1.17	1.40	1.56	1.73
20	0	-2.77	-2.22	-1.78	-1.32	1.21	1.48	1.69	1.91	-2.71	-2.25	-1.84	-1.37	1.22	1.47	1.68	1.90
	1	-2.77	-2.23	-1.78	-1.32	1.21	1.48	1.69	1.91	-2.69	-2.18	-1.77	-1.32	1.21	1.48	1.67	1.88
	2	-2.78	-2.23	-1.78	-1.32	1.21	1.48	1.69	1.90	-2.68	-2.19	-1.78	-1.35	1.23	1.49	1.72	1.92
	3	-2.79	-2.23	-1.79	-1.32	1.21	1.47	1.68	1.89	-2.70	-2.21	-1.76	-1.33	1.20	1.45	1.68	1.90
	4	-2.80	-2.24	-1.79	-1.32	1.21	1.47	1.68	1.88	-2.74	-2.17	-1.76	-1.29	1.23	1.48	1.68	1.88
	5	-2.81	-2.25	-1.79	-1.32	1.21	1.47	1.67	1.87	-2.75	-2.20	-1.78	-1.31	1.15	1.43	1.65	1.87
	6	-2.83	-2.26	-1.80	-1.32	1.20	1.46	1.66	1.86	-2.76	-2.14	-1.72	-1.28	1.19	1.45	1.65	1.87
	7	-2.84	-2.27	-1.80	-1.32	1.20	1.46	1.65	1.85	-2.65	-2.19	-1.75	-1.29	1.22	1.43	1.64	1.88
	8	-2.86	-2.28	-1.80	-1.32	1.20	1.45	1.64	1.83	-2.83	-2.21	-1.79	-1.29	1.20	1.44	1.64	1.86
	9	-2.89	-2.29	-1.81	-1.32	1.19	1.44	1.63	1.81	-2.82	-2.31	-1.83	-1.35	1.17	1.43	1.60	1.79

Table 14.6: Percentage points of the distribution of R_1 when $v = 25$.

n	s	Edgeworth								Simulated							
		1%	2.5%	5%	10%	90%	95%	97.5%	99%	1%	2.5%	5%	10%	90%	95%	97.5%	99%
5	0	-3.30	-2.65	-1.89	-1.28	1.13	1.32	1.44	1.54	-3.04	-2.44	-1.89	-1.29	1.10	1.29	1.43	1.58
	1	-3.38	-2.75	-1.91	-1.27	1.12	1.29	1.40	1.49	-3.13	-2.31	-1.86	-1.31	1.08	1.25	1.36	1.47
	2	-3.53	-2.97	-1.94	-1.23	1.09	1.24	1.33	1.40	-3.26	-2.53	-1.92	-1.33	1.03	1.15	1.22	1.28
6	0	-3.20	-2.54	-1.87	-1.30	1.15	1.35	1.49	1.61	-2.86	-2.29	-1.80	-1.30	1.15	1.34	1.47	1.62
	1	-3.26	-2.60	-1.88	-1.29	1.14	1.33	1.46	1.56	-3.10	-2.34	-1.83	-1.28	1.10	1.29	1.43	1.57
	2	-3.36	-2.73	-1.90	-1.28	1.12	1.30	1.41	1.50	-3.05	-2.34	-1.85	-1.31	1.08	1.25	1.37	1.47
7	0	-3.12	-2.46	-1.85	-1.31	1.16	1.37	1.52	1.66	-3.01	-2.42	-1.89	-1.35	1.15	1.35	1.52	1.70
	1	-3.17	-2.50	-1.86	-1.30	1.15	1.36	1.50	1.62	-3.00	-2.31	-1.83	-1.32	1.15	1.33	1.49	1.65
	2	-3.24	-2.58	-1.88	-1.30	1.14	1.33	1.47	1.57	-3.10	-2.45	-1.95	-1.38	1.10	1.29	1.45	1.56
	3	-3.35	-2.71	-1.90	-1.28	1.12	1.30	1.42	1.51	-3.12	-2.44	-1.94	-1.33	1.09	1.26	1.36	1.45
8	0	-3.06	-2.41	-1.84	-1.31	1.17	1.39	1.55	1.70	-2.96	-2.38	-1.88	-1.35	1.15	1.37	1.53	1.72
	1	-3.09	-2.44	-1.85	-1.31	1.16	1.38	1.54	1.67	-2.93	-2.33	-1.89	-1.36	1.12	1.34	1.49	1.66
	2	-3.15	-2.49	-1.86	-1.31	1.16	1.36	1.51	1.63	-3.06	-2.35	-1.87	-1.33	1.14	1.35	1.48	1.65
	3	-3.23	-2.57	-1.88	-1.30	1.14	1.34	1.47	1.58	-3.07	-2.37	-1.87	-1.32	1.11	1.31	1.43	1.55
9	0	-3.00	-2.37	-1.83	-1.32	1.18	1.41	1.58	1.73	-2.87	-2.30	-1.87	-1.32	1.16	1.39	1.55	1.76
	1	-3.03	-2.39	-1.84	-1.31	1.17	1.40	1.56	1.71	-2.96	-2.28	-1.78	-1.28	1.18	1.39	1.57	1.70
	2	-3.08	-2.43	-1.85	-1.31	1.17	1.38	1.54	1.68	-2.98	-2.36	-1.83	-1.31	1.14	1.37	1.51	1.72
	3	-3.14	-2.48	-1.86	-1.31	1.16	1.37	1.51	1.64	-2.92	-2.38	-1.90	-1.35	1.16	1.37	1.53	1.69
	4	-3.22	-2.56	-1.87	-1.30	1.14	1.34	1.48	1.59	-2.91	-2.39	-1.88	-1.34	1.12	1.29	1.45	1.55
10	0	-2.96	-2.34	-1.82	-1.32	1.18	1.42	1.60	1.76	-2.90	-2.27	-1.87	-1.42	1.16	1.42	1.60	1.77
	1	-2.98	-2.36	-1.83	-1.32	1.18	1.41	1.59	1.75	-2.96	-2.34	-1.87	-1.36	1.17	1.39	1.59	1.79
	2	-3.02	-2.38	-1.84	-1.32	1.17	1.40	1.57	1.72	-2.94	-2.33	-1.80	-1.32	1.16	1.38	1.55	1.71
	3	-3.07	-2.42	-1.84	-1.31	1.17	1.39	1.55	1.69	-3.00	-2.34	-1.83	-1.35	1.15	1.36	1.51	1.67
	4	-3.13	-2.47	-1.86	-1.31	1.16	1.37	1.52	1.64	-3.09	-2.35	-1.83	-1.31	1.14	1.34	1.50	1.62
15	0	-2.82	-2.25	-1.79	-1.32	1.20	1.47	1.67	1.87	-2.74	-2.19	-1.78	-1.35	1.20	1.46	1.69	1.92
	1	-2.83	-2.26	-1.80	-1.32	1.20	1.46	1.66	1.86	-2.81	-2.28	-1.77	-1.27	1.21	1.48	1.67	1.90
	2	-2.84	-2.27	-1.80	-1.32	1.20	1.46	1.65	1.85	-2.89	-2.23	-1.80	-1.32	1.18	1.45	1.66	1.87
	3	-2.86	-2.28	-1.80	-1.32	1.20	1.45	1.64	1.83	-2.85	-2.22	-1.80	-1.34	1.20	1.45	1.67	1.87
	4	-2.88	-2.29	-1.81	-1.32	1.19	1.44	1.63	1.81	-2.83	-2.21	-1.79	-1.32	1.21	1.44	1.63	1.83
	5	-2.91	-2.31	-1.81	-1.32	1.19	1.43	1.62	1.79	-2.86	-2.30	-1.85	-1.36	1.18	1.40	1.57	1.78
	6	-2.95	-2.33	-1.82	-1.32	1.19	1.42	1.60	1.77	-2.89	-2.29	-1.75	-1.31	1.17	1.40	1.59	1.76
	7	-2.99	-2.36	-1.83	-1.32	1.18	1.41	1.58	1.74	-2.84	-2.30	-1.77	-1.31	1.17	1.40	1.56	1.74
20	0	-2.74	-2.20	-1.77	-1.32	1.22	1.49	1.71	1.94	-2.68	-2.24	-1.84	-1.36	1.23	1.48	1.69	1.91
	1	-2.74	-2.21	-1.78	-1.32	1.22	1.49	1.71	1.93	-2.67	-2.17	-1.76	-1.32	1.22	1.48	1.68	1.91
	2	-2.75	-2.21	-1.78	-1.32	1.21	1.49	1.70	1.92	-2.67	-2.17	-1.77	-1.35	1.23	1.50	1.73	1.95
	3	-2.76	-2.22	-1.78	-1.32	1.21	1.48	1.70	1.91	-2.67	-2.20	-1.76	-1.33	1.21	1.46	1.69	1.92
	4	-2.77	-2.23	-1.78	-1.32	1.21	1.48	1.69	1.90	-2.72	-2.15	-1.76	-1.29	1.23	1.49	1.68	1.89
	5	-2.79	-2.23	-1.79	-1.32	1.21	1.47	1.68	1.89	-2.73	-2.19	-1.78	-1.31	1.16	1.43	1.66	1.89
	6	-2.80	-2.24	-1.79	-1.32	1.21	1.47	1.67	1.88	-2.74	-2.14	-1.71	-1.28	1.20	1.45	1.66	1.89
	7	-2.82	-2.25	-1.80	-1.32	1.20	1.46	1.66	1.86	-2.63	-2.18	-1.75	-1.30	1.22	1.44	1.65	1.90
	8	-2.84	-2.27	-1.80	-1.32	1.20	1.46	1.65	1.84	-2.82	-2.21	-1.79	-1.29	1.20	1.45	1.65	1.87
	9	-2.87	-2.28	-1.81	-1.32	1.20	1.45	1.64	1.82	-2.81	-2.30	-1.83	-1.35	1.17	1.43	1.61	1.80

Table 14.7: Percentage points of the distribution of R_2 when $v = 5$.

n	s	Edgeworth									Simulated							
		1%	2.5%	5%	10%	90%	95%	97.5%	99%	1%	2.5%	5%	10%	90%	95%	97.5%	99%
5	0	-1.24	-1.19	-1.11	-0.98	0.94	2.65	3.24	3.71	-1.38	-1.27	-1.16	-1.01	1.27	1.90	2.54	3.36
	1	-1.30	-1.24	-1.16	-1.03	1.08	2.40	3.17	3.67	-1.34	-1.25	-1.15	-1.01	1.29	1.84	2.37	3.38
	2	-1.26	-1.21	-1.14	-1.02	1.05	2.59	3.25	3.73	-1.20	-1.15	-1.08	-0.98	1.27	1.90	2.59	3.43
6	0	-1.34	-1.27	-1.18	-1.03	1.09	2.25	3.07	3.59	-1.44	-1.33	-1.22	-1.07	1.25	1.78	2.34	3.12
	1	-1.38	-1.31	-1.21	-1.06	1.16	2.04	3.01	3.55	-1.43	-1.32	-1.20	-1.04	1.22	1.85	2.40	3.34
	2	-1.36	-1.29	-1.20	-1.06	1.16	2.06	3.05	3.58	-1.37	-1.28	-1.18	-1.03	1.29	1.84	2.41	3.27
7	0	-1.42	-1.34	-1.23	-1.07	1.17	2.03	2.93	3.49	-1.52	-1.39	-1.25	-1.09	1.34	1.92	2.51	3.35
	1	-1.44	-1.36	-1.25	1.09	1.21	1.96	2.87	3.45	-1.52	-1.38	-1.25	-1.08	1.27	1.86	2.37	3.14
	2	-1.43	-1.35	-1.25	-1.09	1.21	1.96	2.90	3.47	-1.46	-1.37	-1.22	-1.05	1.37	1.93	2.51	3.31
	3	-1.39	-1.32	-1.23	-1.08	1.20	1.98	2.97	3.53	-1.37	-1.29	-1.20	-1.05	1.31	1.91	2.47	3.15
8	0	-1.48	-1.38	-1.27	-1.09	1.21	1.95	2.82	3.41	-1.55	-1.41	-1.27	-1.09	1.33	1.89	2.46	3.17
	1	-1.50	-1.40	-1.28	-1.10	1.24	1.93	2.76	3.38	-1.54	-1.39	-1.25	-1.06	1.34	1.92	2.43	3.09
	2	-1.49	-1.40	-1.28	-1.10	1.25	1.93	2.78	3.39	-1.53	-1.39	-1.28	-1.09	1.29	1.91	2.43	3.17
	3	-1.46	-1.38	-1.27	-1.10	1.24	1.94	2.82	3.42	-1.47	-1.36	-1.25	-1.08	1.29	1.87	2.44	3.24
9	0	-1.53	-1.42	-1.30	-1.11	1.24	1.92	2.72	3.34	-1.61	-1.44	-1.30	-1.11	1.33	1.92	2.41	3.06
	1	-1.54	-1.44	-1.31	-1.12	1.26	1.90	2.68	3.31	-1.58	-1.47	-1.32	-1.13	1.27	1.83	2.36	3.17
	2	-1.54	-1.43	-1.31	-1.12	1.26	1.90	2.68	3.32	-1.61	-1.43	-1.31	-1.09	1.31	1.85	2.37	3.18
	3	-1.52	-1.42	-1.30	-1.12	1.26	1.91	2.71	3.34	-1.61	-1.45	-1.31	-1.12	1.31	1.92	2.45	3.07
	4	-1.48	-1.40	-1.28	-1.11	1.25	1.92	2.78	3.39	-1.49	-1.38	-1.24	-1.08	1.33	1.91	2.46	3.04
10	0	-1.57	-1.45	-1.32	-1.12	1.26	1.90	2.64	3.27	-1.62	-1.50	-1.34	-1.12	1.41	1.92	2.38	3.07
	1	-1.58	-1.47	-1.33	-1.13	1.28	1.89	2.61	3.25	-1.67	-1.49	-1.32	-1.12	1.35	1.94	2.42	3.22
	2	-1.58	-1.46	-1.33	-1.13	1.28	1.89	2.61	3.25	-1.62	-1.47	-1.31	-1.11	1.30	1.83	2.34	3.07
	3	-1.56	-1.45	-1.32	-1.13	1.28	1.89	2.63	3.27	-1.57	-1.44	-1.31	-1.12	1.34	1.86	2.39	3.19
	4	-1.54	-1.44	-1.31	-1.12	1.27	1.90	2.67	3.31	-1.55	-1.44	-1.29	-1.11	1.31	1.87	2.40	3.18
15	0	-1.71	-1.56	-1.39	-1.16	1.30	1.84	2.42	3.06	-1.79	-1.60	-1.38	-1.16	1.35	1.81	2.28	2.88
	1	-1.72	-1.56	-1.39	-1.17	1.30	1.84	2.41	3.05	-1.78	-1.59	-1.42	-1.18	1.29	1.81	2.35	2.99
	2	-1.72	-1.56	-1.39	-1.17	1.30	1.84	2.41	3.05	-1.77	-1.59	-1.38	-1.14	1.32	1.83	2.30	3.07
	3	-1.71	-1.56	-1.39	-1.17	1.31	1.84	2.41	3.05	-1.77	-1.60	-1.39	-1.17	1.34	1.81	2.31	2.98
	4	-1.70	-1.55	-1.39	-1.17	1.31	1.84	2.42	3.06	-1.75	-1.56	-1.40	-1.17	1.33	1.79	2.27	2.93
	5	-1.69	-1.55	-1.39	-1.16	1.31	1.85	2.43	3.08	-1.71	-1.52	-1.37	-1.14	1.37	1.85	2.33	3.00
	6	-1.68	-1.54	-1.38	-1.16	1.30	1.85	2.45	3.10	-1.70	-1.54	-1.37	-1.13	1.30	1.75	2.34	2.96
	7	-1.65	-1.52	-1.37	-1.16	1.30	1.86	2.47	3.13	-1.68	-1.52	-1.37	-1.15	1.29	1.81	2.34	2.93
20	0	-1.80	-1.62	-1.43	-1.18	1.31	1.82	2.33	2.94	-1.82	-1.60	-1.44	-1.19	1.37	1.85	2.33	2.84
	1	-1.80	-1.62	-1.43	-1.19	1.31	1.82	2.32	2.93	-1.80	-1.61	-1.43	-1.18	1.31	1.81	2.19	2.79
	2	-1.80	-1.62	-1.43	-1.19	1.31	1.81	2.32	2.93	-1.83	-1.64	-1.45	-1.21	1.34	1.81	2.28	2.72
	3	-1.80	-1.62	-1.43	-1.19	1.31	1.82	2.32	2.93	-1.84	-1.61	-1.42	-1.18	1.34	1.79	2.24	2.82
	4	-1.80	-1.62	-1.43	-1.19	1.31	1.82	2.32	2.93	-1.83	-1.62	-1.45	-1.20	1.28	1.80	2.22	2.83
	5	-1.79	-1.61	-1.43	-1.18	1.31	1.82	2.33	2.94	-1.82	-1.61	-1.40	-1.14	1.32	1.79	2.28	2.82
	6	-1.78	-1.61	-1.43	-1.18	1.31	1.82	2.33	2.94	-1.82	-1.62	-1.41	-1.18	1.27	1.74	2.18	2.87
	7	-1.77	-1.60	-1.42	-1.18	1.31	1.82	2.34	2.96	-1.84	-1.59	-1.40	-1.20	1.29	1.74	2.23	2.69
	8	-1.76	-1.60	-1.42	-1.18	1.31	1.82	2.35	2.97	-1.81	-1.60	-1.41	-1.18	1.28	1.81	2.23	2.89
	9	-1.75	-1.59	-1.41	-1.18	1.31	1.83	2.36	2.99	-1.74	-1.57	-1.40	-1.15	1.34	1.85	2.34	2.91

Table 14.8: Percentage points of the distribution of R_2 when $v = 15$.

n	s	Edgeworth								Simulated						
		1%	2.5%	5%	10%	90%	95%	97.5%	99%	1%	2.5%	5%	10%	90%	95%	97.5%	99%
5	0	-1.50	-1.41	-1.29	-1.11	1.26	1.91	2.75	3.37	-1.55	-1.40	-1.27	-1.09	1.27	1.89	2.45	3.13
	1	-1.46	-1.38	-1.27	-1.10	1.25	1.93	2.83	3.43	-1.45	-1.34	-1.23	-1.07	1.31	1.85	2.33	3.18
	2	-1.37	-1.31	-1.22	-1.08	1.21	1.96	3.02	3.57	-1.27	-1.21	-1.14	-1.02	1.33	1.92	2.54	3.30
6	0	-1.56	-1.46	-1.32	-1.13	1.28	1.89	2.62	3.27	-1.59	-1.45	-1.31	-1.14	1.29	1.81	2.29	2.91
	1	-1.53	-1.44	-1.31	-1.13	1.28	1.90	2.67	3.31	-1.55	-1.41	-1.28	-1.09	1.27	1.84	2.34	3.14
	2	-1.48	-1.39	-1.28	-1.11	1.26	1.91	2.78	3.40	-1.45	-1.35	-1.24	-1.07	1.31	1.85	2.36	3.08
7	0	-1.62	-1.49	-1.35	-1.15	1.30	1.87	2.53	3.19	-1.66	-1.49	-1.33	-1.15	1.36	1.89	2.43	3.05
	1	-1.59	-1.48	-1.34	-1.14	1.30	1.88	2.56	3.22	-1.62	-1.47	-1.32	-1.13	1.32	1.82	2.32	3.01
	2	-1.55	-1.45	-1.32	-1.13	1.29	1.89	2.63	3.28	-1.54	-1.45	-1.28	-1.09	1.38	1.94	2.46	3.14
	3	-1.49	-1.40	-1.29	-1.12	1.27	1.91	2.76	3.38	-1.44	-1.35	-1.24	-1.09	1.33	1.94	2.44	3.11
8	0	-1.66	-1.53	-1.37	-1.16	1.30	1.86	2.47	3.12	-1.69	-1.51	-1.35	-1.15	1.34	1.89	2.38	3.01
	1	-1.64	-1.51	-1.36	-1.15	1.30	1.86	2.49	3.14	-1.63	-1.47	-1.32	-1.11	1.36	1.89	2.36	2.94
	2	-1.61	-1.49	-1.35	-1.15	1.30	1.87	2.53	3.19	-1.63	-1.47	-1.34	-1.13	1.33	1.88	2.38	3.07
	3	-1.56	-1.46	-1.33	-1.14	1.29	1.88	2.61	3.26	-1.54	-1.42	-1.30	-1.11	1.32	1.87	2.37	3.09
9	0	-1.70	-1.55	-1.39	-1.17	1.31	1.84	2.42	3.06	-1.73	-1.53	-1.37	-1.15	1.32	1.87	2.29	2.90
	1	-1.68	-1.54	-1.38	-1.16	1.31	1.85	2.43	3.08	-1.68	-1.55	-1.38	-1.17	1.28	1.79	2.30	2.99
	2	-1.66	-1.52	-1.37	-1.16	1.31	1.85	2.46	3.12	-1.70	-1.50	-1.36	-1.13	1.32	1.84	2.38	3.00
	3	-1.62	-1.50	-1.36	-1.15	1.30	1.87	2.51	3.17	-1.68	-1.52	-1.37	-1.16	1.33	1.90	2.39	2.94
	4	-1.57	-1.46	-1.33	-1.14	1.29	1.88	2.59	3.25	-1.54	-1.43	-1.29	-1.11	1.35	1.89	2.41	2.93
10	0	-1.73	-1.57	-1.40	-1.17	1.31	1.83	2.38	3.02	-1.74	-1.59	-1.40	-1.15	1.41	1.90	2.28	2.91
	1	-1.72	-1.57	-1.40	-1.17	1.31	1.84	2.39	3.03	-1.77	-1.56	-1.38	-1.16	1.36	1.90	2.35	2.98
	2	-1.69	-1.55	-1.39	-1.17	1.31	1.84	2.41	3.06	-1.70	-1.54	-1.36	-1.15	1.31	1.81	2.34	2.96
	3	-1.66	-1.53	-1.38	-1.16	1.31	1.85	2.45	3.10	-1.65	-1.50	-1.35	-1.15	1.35	1.83	2.35	3.04
	4	-1.63	-1.50	-1.36	-1.15	1.31	1.86	2.50	3.16	-1.61	-1.49	-1.33	-1.14	1.31	1.83	2.37	3.10
15	0	-1.84	-1.65	-1.45	-1.20	1.32	1.80	2.28	2.86	-1.89	-1.68	-1.45	-1.19	1.35	1.79	2.21	2.77
	1	-1.84	-1.65	-1.45	-1.20	1.32	1.80	2.28	2.86	-1.88	-1.65	-1.47	-1.21	1.28	1.78	2.29	2.84
	2	-1.83	-1.64	-1.45	-1.20	1.32	1.81	2.29	2.88	-1.85	-1.65	-1.43	-1.17	1.33	1.80	2.25	2.93
	3	-1.81	-1.63	-1.44	-1.19	1.32	1.81	2.30	2.89	-1.85	-1.65	-1.44	-1.20	1.34	1.79	2.24	2.88
	4	-1.80	-1.62	-1.43	-1.19	1.32	1.81	2.31	2.91	-1.82	-1.62	-1.43	-1.20	1.32	1.79	2.22	2.86
	5	-1.78	-1.61	-1.43	-1.19	1.32	1.82	2.33	2.94	-1.78	-1.57	-1.40	-1.17	1.36	1.86	2.30	2.90
	6	-1.75	-1.59	-1.42	-1.18	1.32	1.83	2.35	2.97	-1.75	-1.58	-1.39	-1.16	1.30	1.75	2.30	2.90
	7	-1.72	-1.57	-1.40	-1.18	1.32	1.83	2.38	3.01	-1.73	-1.56	-1.40	-1.16	1.30	1.78	2.31	2.86
20	0	-1.91	-1.69	-1.48	-1.21	1.32	1.78	2.23	2.77	-1.90	-1.67	-1.47	-1.22	1.37	1.83	2.25	2.72
	1	-1.91	-1.69	-1.48	-1.21	1.32	1.78	2.23	2.77	-1.89	-1.67	-1.48	-1.21	1.32	1.77	2.18	2.68
	2	-1.90	-1.69	-1.48	-1.21	1.32	1.78	2.23	2.78	-1.93	-1.72	-1.49	-1.23	1.35	1.78	2.19	2.68
	3	-1.89	-1.68	-1.47	-1.21	1.32	1.79	2.23	2.79	-1.90	-1.68	-1.45	-1.20	1.34	1.76	2.22	2.69
	4	-1.88	-1.68	-1.47	-1.21	1.32	1.79	2.24	2.80	-1.88	-1.68	-1.48	-1.23	1.29	1.76	2.17	2.73
	5	-1.87	-1.67	-1.47	-1.21	1.32	1.79	2.25	2.81	-1.88	-1.66	-1.43	-1.15	1.31	1.78	2.21	2.76
	6	-1.86	-1.66	-1.46	-1.20	1.32	1.80	2.26	2.83	-1.87	-1.66	-1.44	-1.20	1.28	1.72	2.15	2.75
	7	-1.85	-1.65	-1.46	-1.20	1.32	1.80	2.27	2.84	-1.89	-1.64	-1.43	-1.22	1.29	1.75	2.20	2.64
	8	-1.83	-1.64	-1.45	-1.20	1.32	1.80	2.28	2.86	-1.87	-1.64	-1.44	-1.20	1.28	1.80	2.21	2.83
	9	-1.81	-1.63	-1.44	-1.19	1.32	1.81	2.29	2.89	-1.80	-1.60	-1.43	-1.17	1.35	1.83	2.31	2.82

Table 14.9: Percentage points of the distribution of R_2 when $v = 25$.

		Edgeworth								Simulated							
n	s	1%	2.5%	5%	10%	90%	95%	97.5%	99%	1%	2.5%	5%	10%	90%	95%	97.5%	99%
5	0	-1.54	-1.44	-1.32	-1.13	1.28	1.89	2.65	3.30	-1.58	-1.43	-1.30	-1.10	1.28	1.89	2.43	3.06
	1	-1.49	-1.40	-1.29	-1.12	1.27	1.91	2.75	3.38	-1.47	-1.36	-1.25	-1.08	1.31	1.86	2.32	3.13
	2	-1.40	-1.33	-1.24	-1.09	1.23	1.94	2.97	3.53	-1.28	-1.22	-1.15	-1.03	1.33	1.93	2.53	3.26
6	0	-1.61	-1.49	-1.35	-1.15	1.30	1.87	2.54	3.20	-1.62	-1.47	-1.33	-1.15	1.30	1.81	2.29	2.87
	1	-1.56	-1.46	-1.33	-1.14	1.29	1.88	2.60	3.26	-1.57	-1.43	-1.29	-1.10	1.28	1.83	2.33	3.10
	2	-1.50	-1.41	-1.30	-1.12	1.28	1.90	2.73	3.36	-1.47	-1.37	-1.25	-1.08	1.31	1.86	2.34	3.05
7	0	-1.66	-1.52	-1.37	-1.16	1.31	1.85	2.46	3.12	-1.69	-1.53	-1.36	-1.16	1.36	1.88	2.42	3.00
	1	-1.62	-1.50	-1.36	-1.15	1.30	1.86	2.50	3.17	-1.65	-1.49	-1.33	-1.14	1.32	1.83	2.31	2.99
	2	-1.57	-1.47	-1.33	-1.14	1.30	1.88	2.58	3.24	-1.56	-1.46	-1.29	-1.10	1.38	1.94	2.44	3.10
	3	-1.51	-1.42	-1.30	-1.12	1.28	1.90	2.71	3.35	-1.45	-1.36	-1.25	-1.09	1.33	1.94	2.43	3.12
8	0	-1.70	-1.55	-1.39	-1.17	1.31	1.84	2.41	3.06	-1.72	-1.53	-1.37	-1.16	1.36	1.88	2.38	2.97
	1	-1.67	-1.54	-1.38	-1.16	1.31	1.85	2.44	3.09	-1.66	-1.49	-1.33	-1.12	1.35	1.89	2.33	2.93
	2	-1.63	-1.51	-1.36	-1.16	1.31	1.86	2.49	3.15	-1.64	-1.48	-1.35	-1.14	1.34	1.87	2.36	3.05
	3	-1.58	-1.47	-1.34	-1.14	1.30	1.88	2.57	3.23	-1.55	-1.43	-1.31	-1.12	1.32	1.86	2.37	3.07
9	0	-1.73	-1.58	-1.41	-1.18	1.32	1.83	2.37	3.00	-1.76	-1.55	-1.39	-1.16	1.32	1.86	2.29	2.86
	1	-1.71	-1.56	-1.40	-1.17	1.31	1.84	2.39	3.03	-1.70	-1.57	-1.39	-1.18	1.28	1.78	2.28	2.95
	2	-1.68	-1.54	-1.38	-1.17	1.31	1.85	2.43	3.08	-1.72	-1.51	-1.37	-1.14	1.32	1.83	2.36	2.98
	3	-1.64	-1.51	-1.37	-1.16	1.31	1.86	2.48	3.14	-1.70	-1.53	-1.38	-1.16	1.34	1.90	2.37	2.92
	4	-1.59	-1.48	-1.34	-1.14	1.30	1.87	2.56	3.22	-1.55	-1.44	-1.30	-1.12	1.35	1.88	2.40	2.91
10	0	-1.76	-1.60	-1.42	-1.18	1.32	1.82	2.34	2.96	-1.77	-1.60	-1.41	-1.17	1.42	1.87	2.27	2.89
	1	-1.75	-1.59	-1.41	-1.18	1.32	1.83	2.36	2.98	-1.79	-1.58	-1.40	-1.17	1.36	1.88	2.34	2.96
	2	-1.72	-1.57	-1.40	-1.17	1.32	1.84	2.38	3.02	-1.72	-1.55	-1.37	-1.15	1.32	1.80	2.34	2.93
	3	-1.69	-1.55	-1.39	-1.17	1.31	1.84	2.42	3.07	-1.67	-1.51	-1.36	-1.15	1.35	1.83	2.34	3.00
	4	-1.64	-1.52	-1.37	-1.16	1.31	1.86	2.47	3.13	-1.62	-1.50	-1.34	-1.14	1.31	1.83	2.36	3.09
15	0	-1.87	-1.67	-1.46	-1.20	1.32	1.79	2.25	2.82	-1.92	-1.70	-1.46	-1.20	1.35	1.78	2.19	2.74
	1	-1.86	-1.66	-1.46	-1.20	1.32	1.80	2.26	2.83	-1.90	-1.67	-1.48	-1.21	1.27	1.78	2.27	2.81
	2	-1.85	-1.65	-1.46	-1.20	1.32	1.80	2.27	2.84	-1.87	-1.67	-1.45	-1.17	1.32	1.79	2.23	2.90
	3	-1.83	-1.64	-1.45	-1.20	1.32	1.80	2.28	2.86	-1.87	-1.67	-1.45	-1.20	1.34	1.79	2.23	2.86
	4	-1.81	-1.63	-1.44	-1.19	1.32	1.81	2.29	2.88	-1.83	-1.63	-1.44	-1.21	1.32	1.79	2.21	2.83
	5	-1.79	-1.62	-1.43	-1.19	1.32	1.81	2.31	2.91	-1.79	-1.57	-1.40	-1.18	1.36	1.85	2.29	2.87
	6	-1.77	-1.60	-1.42	-1.19	1.32	1.82	2.33	2.95	-1.76	-1.59	-1.40	-1.17	1.31	1.75	2.29	2.88
	7	-1.74	-1.58	-1.41	-1.18	1.32	1.83	2.36	2.99	-1.74	-1.56	-1.40	-1.17	1.31	1.77	2.30	2.84
20	0	-1.94	-1.71	-1.49	-1.22	1.32	1.77	2.21	2.74	-1.92	-1.70	-1.48	-1.22	1.36	1.84	2.23	2.69
	1	-1.93	-1.71	-1.49	-1.22	1.32	1.78	2.21	2.74	-1.91	-1.68	-1.48	-1.22	1.32	1.76	2.17	2.66
	2	-1.92	-1.70	-1.49	-1.21	1.32	1.78	2.21	2.75	-1.94	-1.73	-1.50	-1.23	1.35	1.77	2.17	2.67
	3	-1.91	-1.70	-1.48	-1.21	1.32	1.78	2.22	2.76	-1.91	-1.69	-1.46	-1.21	1.34	1.77	2.19	2.67
	4	-1.90	-1.69	-1.48	-1.21	1.32	1.78	2.23	2.77	-1.89	-1.69	-1.49	-1.24	1.29	1.76	2.15	2.72
	5	-1.89	-1.68	-1.47	-1.21	1.32	1.79	2.23	2.79	-1.89	-1.66	-1.43	-1.16	1.31	1.78	2.19	2.74
	6	-1.88	-1.67	-1.47	-1.21	1.32	1.79	2.24	2.80	-1.89	-1.66	-1.45	-1.20	1.28	1.71	2.14	2.73
	7	-1.86	-1.66	-1.46	-1.20	1.32	1.80	2.25	2.82	-1.90	-1.65	-1.44	-1.22	1.29	1.75	2.19	2.63
	8	-1.84	-1.65	-1.46	-1.20	1.32	1.80	2.27	2.84	-1.87	-1.65	-1.45	-1.20	1.28	1.80	2.21	2.82
	9	-1.82	-1.64	-1.45	-1.20	1.32	1.81	2.28	2.87	-1.81	-1.61	-1.43	-1.17	1.35	1.82	2.30	2.81

Table 14.10: Simulated percentage points of the distribution of R_3 when $v = 5$.

n	s	1%	2.5%	5%	10%	90%	95%	97.5%	99%
5	0	-1.10	-0.99	-0.88	-0.71	2.70	3.97	5.71	9.15
	1	-1.05	-0.92	-0.83	-0.69	3.06	4.83	7.05	11.66
	2	-0.94	-0.86	-0.77	-0.64	4.01	6.58	10.13	16.68
6	0	-1.15	-1.03	-0.91	-0.75	2.57	3.63	4.86	6.90
	1	-1.14	-1.01	-0.90	-0.72	2.58	3.84	5.46	8.18
	2	-1.06	-0.95	-0.85	-0.70	2.99	4.71	7.10	10.94
7	0	-1.26	-1.12	-0.99	-0.81	2.35	3.35	4.47	6.39
	1	-1.20	-1.06	-0.95	-0.77	2.44	3.50	4.87	7.11
	2	-1.16	-1.05	-0.93	-0.77	2.54	3.86	5.74	7.71
	3	-1.07	-0.97	-0.88	-0.72	3.07	4.76	6.74	9.79
8	0	-1.29	-1.16	-1.02	-0.83	2.18	3.03	4.01	5.35
	1	-1.26	-1.12	-1.00	-0.82	2.10	3.09	4.10	5.55
	2	-1.22	-1.10	-0.97	-0.79	2.42	3.56	4.62	6.80
	3	-1.17	-1.05	-0.93	-0.76	2.62	3.90	5.27	7.33
9	0	-1.32	-1.19	-1.06	-0.84	2.09	2.91	3.75	5.12
	1	-1.33	-1.16	-1.01	-0.82	2.24	3.13	4.13	5.11
	2	-1.28	-1.14	-1.00	-0.82	2.19	3.27	4.20	6.15
	3	-1.22	-1.11	-0.99	-0.81	2.48	3.68	5.03	7.58
	4	-1.15	-1.06	-0.94	-0.77	2.60	3.77	5.42	7.51
10	0	-1.37	-1.22	-1.08	-0.90	2.02	2.87	3.74	4.67
	1	-1.37	-1.21	-1.08	-0.87	2.07	2.88	3.86	5.31
	2	-1.33	-1.18	-1.03	-0.84	2.14	3.03	3.93	5.23
	3	-1.30	-1.15	-1.01	-0.83	2.25	3.17	4.14	5.36
	4	-1.26	-1.12	-0.97	-0.79	2.41	3.48	4.71	6.20
15	0	-1.51	-1.32	-1.14	-0.94	1.85	2.45	3.23	4.14
	1	-1.52	-1.34	-1.14	-0.91	1.89	2.62	3.30	4.18
	2	-1.52	-1.30	-1.14	-0.92	1.85	2.55	3.36	4.19
	3	-1.49	-1.28	-1.13	-0.92	1.95	2.64	3.49	4.43
	4	-1.45	-1.26	-1.11	-0.90	1.98	2.73	3.48	4.56
	5	-1.43	-1.26	-1.11	-0.91	1.96	2.74	3.44	4.54
	6	-1.39	-1.24	-1.05	-0.87	2.02	2.86	3.76	4.80
	7	-1.34	-1.20	-1.04	-0.85	2.14	3.09	3.93	5.23
20	0	-1.60	-1.43	-1.23	-1.00	1.76	2.37	2.85	3.60
	1	-1.59	-1.37	-1.20	-0.97	1.77	2.35	2.90	3.60
	2	-1.56	-1.38	-1.20	-0.97	1.82	2.43	3.04	3.76
	3	-1.55	-1.37	-1.18	-0.97	1.77	2.37	2.99	3.90
	4	-1.56	-1.34	-1.18	-0.93	1.84	2.50	3.06	3.87
	5	-1.53	-1.34	-1.18	-0.95	1.73	2.39	3.08	3.94
	6	-1.54	-1.30	-1.13	-0.92	1.83	2.48	3.21	4.04
	7	-1.46	-1.31	-1.13	-0.92	1.94	2.50	3.19	4.38
	8	-1.49	-1.29	-1.14	-0.91	1.92	2.62	3.36	4.46
	9	-1.46	-1.31	-1.14	-0.93	1.89	2.67	3.37	4.23

Table 14.11: Simulated percentage points of the distribution of R_3 when $v = 15$.

n	s	1%	2.5%	5%	10%	90%	95%	97.5%	99%
5	0	-1.17	-1.06	-0.94	-0.76	2.59	3.97	5.63	8.99
	1	-1.09	-0.97	-0.88	-0.73	3.03	4.78	7.09	11.57
	2	-0.97	-0.89	-0.80	-0.67	4.01	6.56	10.16	16.72
6	0	-1.22	-1.10	-0.97	-0.80	2.48	3.56	4.69	6.65
	1	-1.18	-1.05	-0.94	-0.77	2.55	3.85	5.37	8.20
	2	-1.08	-0.98	-0.88	-0.73	2.97	4.67	7.08	10.94
7	0	-1.31	-1.19	-1.04	-0.85	2.30	3.21	4.27	6.11
	1	-1.26	-1.11	-0.98	-0.81	2.42	3.45	4.73	6.77
	2	-1.19	-1.08	-0.97	-0.80	2.52	3.79	5.71	7.77
	3	-1.10	-1.00	-0.90	-0.74	3.05	4.76	6.77	9.84
8	0	-1.36	-1.22	-1.07	-0.87	2.12	2.97	3.85	5.29
	1	-1.30	-1.17	-1.05	-0.86	2.10	3.03	4.00	5.47
	2	-1.27	-1.13	-1.00	-0.82	2.38	3.56	4.62	6.66
	3	-1.20	-1.07	-0.95	-0.79	2.59	3.88	5.21	7.33
9	0	-1.39	-1.23	-1.10	-0.88	2.04	2.84	3.64	4.97
	1	-1.37	-1.20	-1.05	-0.85	2.19	3.04	4.01	5.04
	2	-1.32	-1.18	-1.03	-0.84	2.18	3.26	4.16	6.08
	3	-1.24	-1.14	-1.01	-0.83	2.48	3.68	5.05	7.50
	4	-1.17	-1.08	-0.96	-0.80	2.58	3.78	5.41	7.35
10	0	-1.43	-1.26	-1.13	-0.94	1.96	2.80	3.62	4.58
	1	-1.42	-1.25	-1.11	-0.90	2.05	2.84	3.75	5.22
	2	-1.36	-1.22	-1.06	-0.87	2.10	2.96	3.88	5.10
	3	-1.33	-1.18	-1.04	-0.86	2.22	3.15	4.09	5.47
	4	-1.28	-1.13	-1.00	-0.82	2.36	3.43	4.67	6.16
15	0	-1.55	-1.36	-1.19	-0.97	1.79	2.46	3.22	4.11
	1	-1.55	-1.37	-1.17	-0.93	1.86	2.57	3.21	4.19
	2	-1.55	-1.34	-1.16	-0.94	1.81	2.54	3.27	4.18
	3	-1.51	-1.31	-1.15	-0.94	1.91	2.63	3.45	4.42
	4	-1.48	-1.29	-1.13	-0.92	1.98	2.69	3.45	4.51
	5	-1.45	-1.29	-1.13	-0.93	1.96	2.69	3.39	4.55
	6	-1.41	-1.25	-1.07	-0.89	2.02	2.82	3.72	4.79
	7	-1.36	-1.22	-1.05	-0.87	2.12	3.05	3.91	5.24
20	0	-1.63	-1.45	-1.27	-1.02	1.73	2.29	2.85	3.57
	1	-1.61	-1.42	-1.22	-0.99	1.73	2.33	2.87	3.55
	2	-1.59	-1.40	-1.23	-1.00	1.78	2.41	3.06	3.77
	3	-1.58	-1.40	-1.21	-0.99	1.75	2.32	2.99	3.81
	4	-1.58	-1.37	-1.19	-0.96	1.83	2.46	3.05	3.80
	5	-1.56	-1.37	-1.19	-0.96	1.70	2.36	3.06	3.92
	6	-1.54	-1.33	-1.15	-0.93	1.82	2.47	3.14	4.06
	7	-1.48	-1.33	-1.15	-0.93	1.91	2.49	3.20	4.30
	8	-1.51	-1.31	-1.15	-0.92	1.91	2.61	3.33	4.41
	9	-1.48	-1.32	-1.15	-0.94	1.88	2.64	3.33	4.29

Table 14.12: Simulated percentage points of the distribution of R_3 when $v = 25$.

n	s	1%	2.5%	5%	10%	90%	95%	97.5%	99%
5	0	-1.18	-1.07	-0.95	-0.77	2.57	4.01	5.56	8.93
	1	-1.10	-0.98	-0.89	-0.74	3.01	4.79	7.09	11.53
	2	-0.98	-0.90	-0.81	-0.68	4.00	6.55	10.21	16.71
6	0	-1.23	-1.11	-0.98	-0.81	2.47	3.51	4.70	6.59
	1	-1.19	-1.06	-0.94	-0.77	2.53	3.80	5.41	8.12
	2	-1.09	-0.98	-0.89	-0.74	2.96	4.70	7.10	10.98
7	0	-1.32	-1.20	-1.05	-0.86	2.26	3.20	4.34	6.01
	1	-1.26	-1.11	-0.99	-0.82	2.42	3.43	4.72	6.80
	2	-1.20	-1.09	-0.98	-0.81	2.50	3.77	5.69	7.85
	3	-1.10	-1.00	-0.91	-0.74	3.04	4.77	6.79	9.83
8	0	-1.37	-1.23	-1.08	-0.88	2.12	2.94	3.80	5.32
	1	-1.31	-1.18	-1.05	-0.86	2.10	3.03	4.01	5.47
	2	-1.28	-1.13	-1.01	-0.83	2.39	3.55	4.58	6.62
	3	-1.20	-1.07	-0.96	-0.79	2.58	3.88	5.22	7.34
9	0	-1.40	-1.24	-1.11	-0.89	2.04	2.83	3.63	4.98
	1	-1.38	-1.21	-1.05	-0.86	2.17	3.06	4.02	5.06
	2	-1.33	-1.19	-1.04	-0.85	2.17	3.21	4.13	6.08
	3	-1.25	-1.14	-1.02	-0.84	2.48	3.69	5.03	7.55
	4	-1.18	-1.08	-0.96	-0.80	2.58	3.78	5.41	7.33
10	0	-1.44	-1.27	-1.14	-0.95	1.95	2.78	3.63	4.61
	1	-1.42	-1.26	-1.11	-0.91	2.03	2.83	3.74	5.18
	2	-1.37	-1.22	-1.06	-0.87	2.09	2.96	3.89	5.12
	3	-1.33	-1.18	-1.04	-0.86	2.22	3.15	4.08	5.49
	4	-1.29	-1.14	-1.00	-0.82	2.36	3.42	4.68	6.17
15	0	-1.56	-1.36	-1.19	-0.98	1.79	2.46	3.21	4.13
	1	-1.55	-1.38	-1.17	-0.93	1.86	2.58	3.21	4.18
	2	-1.55	-1.34	-1.17	-0.95	1.80	2.53	3.26	4.20
	3	-1.52	-1.32	-1.16	-0.95	1.91	2.62	3.46	4.42
	4	-1.48	-1.29	-1.13	-0.93	1.98	2.69	3.44	4.49
	5	-1.45	-1.29	-1.14	-0.93	1.96	2.68	3.38	4.55
	6	-1.42	-1.26	-1.07	-0.89	2.01	2.82	3.72	4.78
	7	-1.36	-1.22	-1.05	-0.87	2.11	3.04	3.89	5.23
20	0	-1.64	-1.46	-1.28	-1.03	1.73	2.29	2.85	3.53
	1	-1.62	-1.42	-1.23	-1.00	1.73	2.32	2.85	3.57
	2	-1.60	-1.41	-1.23	-1.01	1.77	2.40	3.05	3.78
	3	-1.58	-1.40	-1.21	-0.99	1.75	2.33	2.99	3.77
	4	-1.58	-1.37	-1.20	-0.96	1.84	2.46	3.03	3.79
	5	-1.56	-1.37	-1.20	-0.97	1.70	2.35	3.04	3.92
	6	-1.54	-1.33	-1.15	-0.94	1.82	2.46	3.14	4.07
	7	-1.48	-1.33	-1.15	-0.94	1.90	2.49	3.20	4.29
	8	-1.51	-1.31	-1.16	-0.92	1.90	2.62	3.33	4.40
	9	-1.48	-1.32	-1.15	-0.94	1.87	2.64	3.33	4.29

15

Higher Order Moments of Order Statistics From the Power Function Distribution and Edgeworth Approximate Inference

K. S. Sultan, Aaron Childs and N. Balakrishnan

Al-Azhar University, Nasr City, Cairo, Egypt
McMaster University, Hamilton, ON, Canada
McMaster University, Hamilton, ON, Canada

Abstract: In this paper, we first derive exact explicit expressions for the triple and quadruple moments of order statistics from the power function distribution. Also, we present recurrence relations for single, double, triple and quadruple moments of order statistics from the power function distribution. These relations will enable one to find all moments (of order up to four) of order statistics for all sample sizes in a simple recursive manner. We then use these results to determine the mean, variance, and coefficients of skewness and kurtosis of certain linear functions of order statistics. We then derive approximate confidence intervals for the parameters of the power function distribution using the Edgeworth approximation. Finally, we extend the recurrence relations to the case of the doubly truncated power function distribution.

Keywords and phrases: Order statistics, exact moments, single moments, double moments, triple moments, quadruple moments, power function distribution, doubly truncated distribution, recurrence relations, Edgeworth approximation, coefficients of skewness and kurtosis, approximate confidence interval, pivotal quantity

15.1 Introduction

Let $X_1, X_2, \ldots, X_n$ be a random sample of size n from the power function population with probability density function (p.d.f.)

$$f(x;\theta,\sigma) = \frac{\nu}{\sigma}\left(\frac{x-\theta}{\sigma}\right)^{\nu-1}, \qquad \theta \le x \le \sigma+\theta,\ \nu > 0,\ \sigma > 0, \qquad (15.1)$$

and cumulative distribution function (c.d.f.)

$$F(x;\theta,\sigma) = \sigma^{-\nu}(x-\theta)^{\nu}, \qquad \theta \leq x \leq \sigma+\theta,\ \nu > 0,\ \sigma > 0. \qquad (15.2)$$

Let $X_{1:n} \leq X_{2:n} \leq \ldots \leq X_{n:n}$ be the order statistics obtained by arranging the above sample in increasing order of magnitude. Notice from (15.1) and (15.2) that

$$(x-\theta)f(x) = \nu F(x), \qquad \theta \leq x \leq \sigma+\theta,\ \nu > 0,\ \sigma > 0. \qquad (15.3)$$

Balakrishnan and Joshi (1981) used the above differential equation to derive some recurrence relations satisfied by the single and the product moments of order statistics from the power function distribution. These recurrence relations allow one to compute means, variances and covariances of all order statistics from the power function distribution for all sample sizes n in a simple recursive manner. Exact explicit expressions for the single and the product moments of order statistics have been derived by Malik (1967); see also Arnold, Balakrishnan and Nagaraja (1992).

In this paper, we extend the results of Malik (1967) by deriving exact explicit expressions for the triple and quadruple moments of order statistics from the power function distribution. Also, we consider the problem of finding confidence intervals for the parameters θ and σ in (15.1) based on the following pivotal quantities,

$$R_1 = \frac{\theta^* - \theta}{\sigma\sqrt{V_1}}, \quad R_2 = \frac{\sigma^* - \sigma}{\sigma\sqrt{V_2}} \text{ and } R_3 = \frac{\theta^* - \theta}{\sigma^*\sqrt{V_1}}, \qquad (15.4)$$

where θ^* and σ^* are the BLUE's of θ and σ with variances $\sigma^2 V_1$ and $\sigma^2 V_2$, respectively. R_1 can be used to draw inference for θ when σ is known, while R_3 can be used to draw inference for θ when σ is unknown. Similarly, R_2 can be used to draw inference for σ when θ is unknown.

Notice that R_1 and R_2 in (15.4) can be rewritten as

$$R_1 = \frac{1}{\sqrt{V_1}}\left(\sum_{i=r+1}^{n-s} a_i Z_{i:n}\right) = \frac{R_1^*}{\sqrt{V_1}} \quad \text{and}$$

$$R_2 = \frac{1}{\sqrt{V_2}}\left(\sum_{i=r+1}^{n-s} b_i Z_{i:n} - 1\right) = \frac{R_2^* - 1}{\sqrt{V_2}}, \qquad (15.5)$$

where, $Z_{i:n} = \frac{X_{i:n} - \theta}{\sigma}$, $i = r+1, \ldots, n-s$, is the standardized form of the available Type-II censored sample $X_{i:n}$, $i = r+1, \ldots, n-s$. Thus, they are linear functions of order statistics arising from the standardized power function distribution. Since the distribution of a linear function of order statistics will in general not be known, we consider finding the approximate distribution by using the Edgeworth approximation for a statistic T (with mean 0 and variance 1):

$$G(t) \approx \Phi(t) - \phi(t)\left\{\frac{\sqrt{\beta_1}}{6}(t^2 - 1) + \frac{\beta_2 - 3}{24}(t^3 - 3t) + \frac{\beta_1}{72}(t^5 - 10t^3 + 15t)\right\}, \tag{15.6}$$

where $\sqrt{\beta_1}$ and β_2 are the coefficients of skewness and kurtosis of T, respectively, and $\Phi(t)$ is the c.d.f. of the standard normal distribution with corresponding p.d.f. $\phi(t)$.

The coefficients of skewness and kurtosis of linear functions of order statistics require knowledge of the single moments $E(Z_{i:n}^a)$ which we denote by $\mu_{i:n}^{(a)}$, the double moments $E(Z_{i:n}^a Z_{j:n}^b)$ which we denote by $\mu_{i,j:n}^{(a,b)}$, the triple moments $E(Z_{i:n}^a Z_{j:n}^b Z_{k:n}^c)$ which we denote by $\mu_{i,j,k:n}^{(a,b,c)}$, and the quadruple moments $E(Z_{i:n}^a Z_{j:n}^b Z_{k:n}^c Z_{l:n}^d)$ which we denote by $\mu_{i,j,k,l:n}^{(a,b,c,d)}$, of order statistics from the standardized power function distribution for $1 \leq i < j < k < l \leq n$ and $a,b,c,d \geq 0$, $a+b+c+d \leq 4$.

After presenting the BLUE's of θ and σ in the following section, we derive in Section 15.3, exact explicit expressions for the single, double, triple and quadruple moments of order statistics, which will extend the results of Malik (1967). Then, in Section 15.4, we use the differential equation in (15.3) to extend the results of Balakrishnan and Joshi (1981) by deriving recurrence relations for the triple and quadruple moments. These quantities are then used in Section 15.5 to determine the coefficients of skewness and kurtosis of the pivotal quantities R_1 and R_2 based on the BLUE's of θ and σ. We then propose Edgeworth approximations for the distributions of these pivotal quantities and show that this method provides close approximations to the percentage points of the pivotal quantities determined by Monte Carlo simulations. A numerical example to illustrate the method of inference developed in this paper is then presented in Section 15.6. Similar work has been carried out recently for the exponential distribution by Balakrishnan and Gupta (1996), for the Laplace distribution by Balakrishnan et al. (1996), for the logistic distribution by Childs and Balakrishnan (1996), and for the Pareto distribution by Childs, Sultan and Balakrishnan (1996). Finally, in Section 15.7, we generalize the recurrence relations to the case of doubly truncated power function distribution, thus extending the results of Balakrishnan and Joshi (1981).

15.2 BLUE's of θ and σ

Let $X_{r+1:n} \leq X_{r+2:n} \leq \ldots \leq X_{n-s:n}$ denote the available doubly Type-II censored sample from the power function distribution in (15.1), and let $Z_{i:n} =$

$(X_{i:n} - \theta)/\sigma$, $i = r+1, r+2, \ldots, n-s$, be the corresponding order statistics from the standard power function distribution. Let us denote $E(Z_{i:n})$ by $\mu_{i:n}$, $\text{Var}(Z_{i:n})$ by $\sigma_{i,i:n}$, and $\text{Cov}(Z_{i:n}, Z_{j:n})$ by $\sigma_{i,j:n}$; further, let

$$\boldsymbol{X} = (X_{r+1:n}, X_{r+2:n}, \ldots, X_{n-s:n})^T$$
$$\boldsymbol{\mu} = (\mu_{r+1:n}, \mu_{r+2:n}, \ldots, \mu_{n-s:n})^T$$
$$\boldsymbol{1} = \underbrace{(1, 1, \ldots, 1)^T}_{n-r-s}$$

and $\boldsymbol{\Sigma} = ((\sigma_{i,j:n}))$, $r+1 \leq i, j \leq n-s$.

Then, the BLUE's of θ and σ are given by [see Balakrishnan and Cohen (1991)]

$$\theta^* = \left\{ \frac{\boldsymbol{\mu}^T \boldsymbol{\Sigma}^{-1} \boldsymbol{\mu} \boldsymbol{1}^T \boldsymbol{\Sigma}^{-1} - \boldsymbol{\mu}^T \boldsymbol{\Sigma}^{-1} \boldsymbol{1} \boldsymbol{\mu}^T \boldsymbol{\Sigma}^{-1}}{(\boldsymbol{\mu}^T \boldsymbol{\Sigma}^{-1} \boldsymbol{\mu})(\boldsymbol{1}^T \boldsymbol{\Sigma}^{-1} \boldsymbol{1}) - (\boldsymbol{\mu}^T \boldsymbol{\Sigma}^{-1} \boldsymbol{1})^2} \right\} \boldsymbol{X} = \sum_{i=r+1}^{n-s} a_i X_{i:n}, \quad (15.7)$$

and

$$\sigma^* = \left\{ \frac{\boldsymbol{1}^T \boldsymbol{\Sigma}^{-1} \boldsymbol{1} \boldsymbol{\mu}^T \boldsymbol{\Sigma}^{-1} - \boldsymbol{1}^T \boldsymbol{\Sigma}^{-1} \boldsymbol{\mu} \boldsymbol{1}^T \boldsymbol{\Sigma}^{-1}}{(\boldsymbol{\mu}^T \boldsymbol{\Sigma}^{-1} \boldsymbol{\mu})(\boldsymbol{1}^T \boldsymbol{\Sigma}^{-1} \boldsymbol{1}) - (\boldsymbol{\mu}^T \boldsymbol{\Sigma}^{-1} \boldsymbol{1})^2} \right\} \boldsymbol{X} = \sum_{i=r+1}^{n-s} b_i X_{i:n}. \quad (15.8)$$

Furthermore, the variances and covariance of these BLUE's are given by [see Balakrishnan and Cohen (1991)]

$$\text{Var}(\theta^*) = \sigma^2 \left\{ \frac{\boldsymbol{\mu}^T \boldsymbol{\Sigma}^{-1} \boldsymbol{\mu}}{(\boldsymbol{\mu}^T \boldsymbol{\Sigma}^{-1} \boldsymbol{\mu})(\boldsymbol{1}^T \boldsymbol{\Sigma}^{-1} \boldsymbol{1}) - (\boldsymbol{\mu}^T \boldsymbol{\Sigma}^{-1} \boldsymbol{1})^2} \right\} = \sigma^2 V_1, \quad (15.9)$$

$$\text{Var}(\sigma^*) = \sigma^2 \left\{ \frac{\boldsymbol{1}^T \boldsymbol{\Sigma}^{-1} \boldsymbol{1}}{(\boldsymbol{\mu}^T \boldsymbol{\Sigma}^{-1} \boldsymbol{\mu})(\boldsymbol{1}^T \boldsymbol{\Sigma}^{-1} \boldsymbol{1}) - (\boldsymbol{\mu}^T \boldsymbol{\Sigma}^{-1} \boldsymbol{1})^2} \right\} = \sigma^2 V_2, \quad (15.10)$$

and

$$\text{Cov}(\theta^*, \sigma^*) = \sigma^2 \left\{ \frac{-\boldsymbol{\mu}^T \boldsymbol{\Sigma}^{-1} \boldsymbol{1}}{(\boldsymbol{\mu}^T \boldsymbol{\Sigma}^{-1} \boldsymbol{\mu})(\boldsymbol{1}^T \boldsymbol{\Sigma}^{-1} \boldsymbol{1}) - (\boldsymbol{\mu}^T \boldsymbol{\Sigma}^{-1} \boldsymbol{1})^2} \right\} = \sigma^2 V_3; \quad (15.11)$$

for details, refer to David (1981), Balakrishnan and Cohen (1991), and Arnold, Balakrishnan and Nagaraja (1992).

Exact explicit expressions may be obtained from the above formulae by using the fact that the covariance matrix of the standardized power function order statistics $((\sigma_{i,j:n}))$ is of the form $((a_i b_j))$. First, we use the fact that

$$\mu_{i:n} = \frac{\Gamma(n+1)}{\Gamma(i)} \frac{\Gamma(i+1/\nu)}{\Gamma(n+1+1/\nu)} \quad (15.12)$$

$$\mu_{i:n}^{(2)} = \frac{\Gamma(n+1)}{\Gamma(i)} \frac{\Gamma(i+2/\nu)}{\Gamma(n+1+2/\nu)} \quad (15.13)$$

and
$$\mu_{i,j:n} = \frac{\Gamma(n+1)}{\Gamma(n+1+2/\nu)} \frac{\Gamma(i+1/\nu)}{\Gamma(i)} \frac{\Gamma(j+2/\nu)}{\Gamma(j+1/\nu)} \quad (15.14)$$

[Arnold, Balakrishnan and Nagaraja (1992)], to note that $((\sigma_{i,j:n})) = ((a_i b_j))$, where

$$a_i = \frac{\Gamma(n+1)\Gamma(i+1/\nu)}{\Gamma(i)} \quad (15.15)$$

and

$$b_j = \frac{\Gamma(j+2/\nu)}{\Gamma(j+1/\nu)\Gamma(n+1+2/\nu)} - \frac{\Gamma(j+1/\nu)\Gamma(n+1)}{\Gamma(j)\Gamma(n+1+1/\nu)^2}. \quad (15.16)$$

We are therefore able to invert the covariance matrix $((\sigma_{i,j:n}))$ and obtain explicit expressions for the BLUE's θ^* and σ^* of θ and σ respectively, their variances $\mathrm{Var}(\theta^*)$ and $\mathrm{Var}(\sigma^*)$, and their covariance $\mathrm{Cov}(\theta^*, \sigma^*)$, as described, for example, in Arnold, Balakrishnan and Nagaraja (1992). We have

$$\theta^* = \left\{ \frac{a_{r+2} - a_{r+1}}{a_{r+1}(a_{r+2}b_{r+1} - a_{r+1}b_{r+2})} X_{r+1:n} \right.$$
$$+ \sum_{i=r+1}^{n-s-2} \frac{a_i(b_{i+1} - b_{i+2}) + a_{i+1}(b_{i+2} - b_i) + a_{i+2}(b_i - b_{i+1})}{(a_{i+1}b_i - a_i b_{i+1})(a_{i+2}b_{i+1} - a_{i+1}b_{i+2})} X_{i+1:n}$$
$$\left. + \left(\frac{b_{n-s-1} - b_{n-s}}{b_{n-s}(a_{n-s}b_{n-s-1} - a_{n-s-1}b_{n-s})} - \frac{1}{b_{n-s}a_{n-s}} \right) X_{n-s:n} \right\}$$
$$\times \left(\sum_{i,j} \sigma^{ij} - \frac{1}{a_{n-s}b_{n-s}} \right)^{-1}; \quad (15.17)$$

$$\sigma^* = \left\{ \frac{a_{r+1} - a_{r+2}}{a_{r+1}(a_{r+2}b_{r+1} - a_{r+1}b_{r+2})} X_{r+1:n} \right.$$
$$- \sum_{i=r+1}^{n-s-2} \frac{a_i(b_{i+1} - b_{i+2}) + a_{i+1}(b_{i+2} - b_i) + a_{i+2}(b_i - b_{i+1})}{(a_{i+1}b_i - a_i b_{i+1})(a_{i+2}b_{i+1} - a_{i+1}b_{i+2})} X_{i+1:n}$$
$$\left. + \left(\sum_{i,j} \sigma^{ij} - \frac{b_{n-s-1} - b_{n-s}}{b_{n-s}(a_{n-s}b_{n-s-1} - a_{n-s-1}b_{n-s})} \right) X_{n-s:n} \right\}$$
$$\times \left(\frac{a_{n-s}}{c} \sum_{i,j} \sigma^{ij} - \frac{1}{c b_{n-s}} \right)^{-1}; \quad (15.18)$$

$$\mathrm{Var}(\theta^*) = \sigma^2 \left(\sum_{i,j} \sigma^{ij} - \frac{1}{a_{n-s}b_{n-s}} \right)^{-1}; \quad (15.19)$$

$$\operatorname{Var}(\sigma^*) = \sigma^2 \left(\frac{a_{n-s}}{b_{n-s}c^2} - \frac{1}{c^2 b_{n-s}^2 \sum_{i,j} \sigma^{ij}} \right)^{-1}; \qquad (15.20)$$

and

$$\operatorname{Cov}(\theta^* \sigma^*) = -\sigma^2 \left(\frac{a_{n-s}}{c} \sum_{i,j} \sigma^{ij} - \frac{1}{cb_{n-s}} \right)^{-1}, \qquad (15.21)$$

where a_i and b_i are as given in (15.15) and (15.16), respectively,

$$c = \Gamma(n+1+1/\nu), \qquad (15.22)$$

and $\sum_{i,j} \sigma^{ij}$ is the sum of all of the elements of the inverse matrix of the covariance matrix $((\sigma_{i,j:n}))$ and is given by

$$\begin{aligned}
\sum_{i,j} \sigma^{ij} &= \sum_{i=r+2}^{n-s-1} \frac{a_{i-1}(2b_i - b_{i+1}) - 2a_i b_{i-1} + a_{i+1} b_{i-1}}{(a_i b_{i-1} - a_{i-1} b_i)(a_{i+1} b_i - a_i b_{i+1})} \\
&+ \frac{a_{r+2} - 2a_{r+1}}{a_{r+1}(a_{r+2} b_{r+1} - a_{r+1} b_{r+2})} + \frac{b_{n-s-1}}{b_{n-s}(a_{n-s} b_{n-s-1} - a_{n-s-1} b_{n-s})}.
\end{aligned} \qquad (15.23)$$

Similar expressions have been derived by Childs, Sultan and Balakrishnan (1996) for the Pareto distribution.

15.3 Exact Expressions For The Moments of Order Statistics

In this section, we derive exact expressions for the triple and quadruple moments of order statistics from the one-parameter power function distribution. For the sake of completeness and better understanding of the results, we also present the corresponding results for the single and double moments which are given in Malik (1967). Without loss of generality, we can put $\theta = 0$ in (15.1) to obtain the one-parameter power function distribution as

$$f(x) = \frac{\nu}{\sigma} \left(\frac{x}{\sigma} \right)^{\nu-1}, \qquad 0 \le x \le \sigma, \ \sigma, \nu > 0, \qquad (15.24)$$

and with c.d.f.

$$F(x) = \sigma^{-\nu} x^\nu, \qquad 0 \le x \le \sigma, \ \sigma, \nu > 0. \qquad (15.25)$$

Note from the above expression that

$$x = \sigma \{F(x)\}^{1/\nu}, \qquad 0 \le x \le \sigma, \ \sigma, \nu > 0. \qquad (15.26)$$

15.3.1 Single moments of order statistics

From (15.24) and (15.25), we have

$$\mu_{r:n}^{(a)} = C_{r:n} \int_0^\sigma w^a \{F(w)\}^{r-1} \{1 - F(w)\}^{n-r} f(w) dw \quad (15.27)$$

$$= C_{r:n} \sigma^a \int_0^1 u^{r+a/\nu-1} (1-u)^{n-r} du,$$

and hence,

$$\mu_{r:n}^{(a)} = C_{r:n} \sigma^a B(r + a/\nu, n - r + 1),$$

where, $B(a, b)$ is the beta function and $C_{r:n} = \frac{n!}{(r-1)!(n-r)!}$. The above equation may be rewritten as

$$\mu_{r:n}^{(a)} = \sigma^a \frac{\Gamma(n+1)\Gamma(r+a/\nu)}{\Gamma(r)\Gamma(n+1+a/\nu)}. \quad (15.28)$$

15.3.2 Double moments of order statistics

The double moments of order statistics are given by

$$\mu_{r,s:n}^{(a,b)} = C_{r,s:n} \int_0^\sigma x^b \{1 - F(x)\}^{n-s} I(x) f(x) dx, \quad (15.29)$$

where $C_{r,s:n} = \frac{n!}{(r-1)!(s-r-1)!(n-s)!}$ and

$$I(x) = \int_0^x w^a \{F(w)\}^{r-1} \{F(x) - F(w)\}^{s-r-1} f(w) dw.$$

We first let $u = \frac{F(w)}{F(x)}$ in the above integral and use (15.26) to get

$$I(x) = \sigma^a \{F(x)\}^{s+a/\nu-1} B(r + a/\nu, s - r). \quad (15.30)$$

Then we substitute (15.30) into (15.29) to get

$$\mu_{r,s:n}^{(a,b)} = C_{r,s:n} \sigma^{a+b} B(r + a/\nu, s - r) B(s + a/\nu + b/\nu, n - s + 1)$$

$$= \sigma^{a+b} \frac{\Gamma(n+1)\Gamma(r+a/\nu)\Gamma(s+a/\nu+b/\nu)}{\Gamma(r)\Gamma(s+a/\nu)\Gamma(n+1+a/\nu+b/\nu)}. \quad (15.31)$$

As a check, we set $b = 0$ in (15.31) and observe from (15.28) that $\mu_{r,s:n}^{(a,0)} = \mu_{r:n}^{(a)}$.

15.3.3 Triple moments of order statistics

The triple moments of order statistics are given by

$$\mu_{r,s,t:n}^{(a,b,c)} = C_{r,s,t:n} \int_0^\sigma \int_0^y x^b y^c \{F(y)-F(x)\}^{t-s-1}\{1-F(y)\}^{n-t} I(x) f(y) f(x) dx dy, \tag{15.32}$$

where $C_{r,s,t:n} = \frac{n!}{(r-1)!(s-r-1)!(t-s-1)!(n-t)!}$ and

$$I(x) = \int_0^x w^a \{F(w)\}^{r-1}\{F(x)-F(w)\}^{s-r-1} f(w) dw.$$

We first let $u = \frac{F(w)}{F(x)}$ in the above integral and use (15.26) to get

$$I(x) = \sigma^a \{F(x)\}^{s+a/\nu-1} B(r+a/\nu, s-r). \tag{15.33}$$

Then, we substitute (15.33) into (15.32) to get

$$\mu_{r,s,t:n}^{(a,b,c)} = C_{r,s,t:n} \sigma^a B(r+a/\nu, s-r) \int_0^\sigma y^c \{1-F(y)\}^{n-t} J(y) f(y) dy, \tag{15.34}$$

where

$$J(y) = \int_0^y x^b \{F(x)\}^{s+a/\nu+1}\{F(y)-F(x)\}^{t-s-1} f(x) dx. \tag{15.35}$$

Next we let $h = \frac{F(x)}{F(y)}$ in the above integral and use (15.26) to get

$$J(y) = \sigma^b \{F(y)\}^{t+a/\nu+b/\nu-1} B(s+a/\nu+b/\nu, t-s). \tag{15.36}$$

Finally, substituting (15.36) into (15.34) gives

$$\begin{aligned}\mu_{r,s,t:n}^{(a,b,c)} &= C_{r,s,t:n} \sigma^{a+b+c} B(r+a/\nu, s-r) B(s+a/\nu+b/\nu, t-s) \\ &\quad \times B(t+a/\nu+b/\nu+c/\nu, n-t+1) \\ &= \sigma^{a+b+c} \frac{\Gamma(n+1)\Gamma(r+a/\nu)}{\Gamma(r)\Gamma(s+a/\nu)} \\ &\quad \times \frac{\Gamma(s+a/\nu+b/\nu)\Gamma(t+a/\nu+b/\nu+c/\nu)}{\Gamma(t+a/\nu+b/\nu)\Gamma(n+1+a/\nu+b/\nu+c/\nu)}. \end{aligned} \tag{15.37}$$

As a check, we set $c=0$ in (15.37) and observe from (15.31) that $\mu_{r,s,t:n}^{(a,b,0)} = \mu_{r,s:n}^{(a,b)}$.

15.3.4 Quadruple moments of order statistics

Using the same method as for the single, product and triple moments, the following exact explicit expression for the quadruple moments of order statistics can be obtained:

$$\begin{aligned}
\mu^{(a,b,c,d)}_{r,s,t,u:n} &= C_{r,s,t,u:n} \sigma^{a+b+c+d} B(r+a/\nu, s-r) B(s+a/\nu+b/\nu, t-s) \\
&\quad \times B(t+a/nu+b/nu+c/nu, u-t) \\
&\quad \times B(u+a/\nu+b/\nu+c/\nu+d/\nu, n-u+1) \\
&= \sigma^{a+b+c+d} \frac{\Gamma(n+1)\Gamma(r+a/\nu)}{\Gamma(r)\Gamma(s+a/\nu)} \\
&\quad \times \frac{\Gamma(s+a/\nu+b/\nu)\Gamma(t+a/\nu+b/\nu+c/\nu)\Gamma(u+a/\nu+b/\nu+c/\nu+d/\nu)}{\Gamma(t+a/\nu+b/\nu)\Gamma(u+a/\nu+b/\nu+c/\nu)\Gamma(n+1+a/\nu+b/\nu+c/\nu+d/\nu)}.
\end{aligned} \quad (15.38)$$

As a check, we set $d = 0$ in (15.38) and observe from (15.37) that $\mu^{(a,b,c,0)}_{r,s,t,u:n} = \mu^{(a,b,c)}_{r,s,t:n}$.

15.4 Recurrence Relations For Moments of Order Statistics

In this section, we will use the differential equation in (15.3) to derive some recurrence relations between the single moments, double moments, triple moments and quadruple moments of order statistics. Without loss of generality, we once again take $\theta = 0$ and $\sigma = 1$.

15.4.1 Relations for single moments

The p.d.f. of $X_{r:n}$ for $1 \leq r \leq n$ is [see David (1981, p. 9)] given by

$$f_{r:n}(w) = C_{r:n}\{F(w)\}^{r-1}\{1-F(w)\}^{n-r}f(w), \quad 0 \leq w \leq 1, \quad (15.39)$$

where $C_{r:n} = \frac{n!}{(r-1)!(n-r)!}$.

From this, we may compute the single moments of $X_{r:n}$ as

$$\mu^{(a)}_{r:n} = \int_0^1 w^a f_{r:n}(w)dw, \quad a = 1, 2, \ldots, 1 \leq r \leq n. \quad (15.40)$$

Then the single moments $\mu^{(a)}_{r:n}$ for $a = 1, 2, \ldots$, satisfy the following recurrence relations. It should be mentioned here that the recurrence relations for the single and the product moments of order statistics from power function distribution have been derived by Balakrishnan and Joshi (1981) and are presented here for the sake of completeness.

Theorem 15.4.1 *For $n \geq 1$ and $a = 1, 2, \ldots,$*

$$\mu_{n:n}^{(a)} = \frac{n\nu}{n\nu + a} ; \qquad (15.41)$$

for $1 \leq r \leq n-1$ and $a = 1, 2, \ldots,$

$$\mu_{r+1:n}^{(a)} = \frac{r\nu + a}{r\nu} \mu_{r:n}^{(a)}; \qquad (15.42)$$

for $n \geq 2$ and $a = 1, 2, \ldots,$

$$\mu_{1:n}^{(a)} = \frac{n\nu}{n\nu + a} \mu_{1:n-1}^{(a)}; \qquad (15.43)$$

and for $a = 1, 2, \ldots,$

$$\mu_{1:1}^{(a)} = \frac{\nu}{\nu + a}. \qquad (15.44)$$

15.4.2 Relations for double moments

The joint p.d.f. of $X_{r:n}$ and $X_{s:n}$ for $1 \leq r < s \leq n$ is given by [see David (1981, p. 10)]

$$f_{r,s:n}(w,x) = C_{r,s:n}\{F(w)\}^{r-1}\{F(x)-F(w)\}^{s-r-1}\{1-F(x)\}^{n-s}f(w)f(x),$$
$$0 \leq w < x \leq 1, \qquad (15.45)$$

where $C_{r,s:n} = \frac{n!}{(r-1)!(s-r-1)!(n-s)!}$. Using (15.45), we may compute the double moments as

$$\mu_{r,s:n}^{(a,b)} = \int_0^1 \int_w^1 w^a x^b f_{r,s:n}(w,x) dx dw, \quad 1 \leq r < s \leq n, \ a, b \geq 1. \qquad (15.46)$$

Then the double moments $\mu_{r,s:n}^{(a,b)}$ for $a, b = 1, 2, \ldots,$ satisfy the recurrence relations presented in the following two theorems.

Theorem 15.4.2 *For $1 \leq r \leq n-1$ and $a, b = 1, 2, \ldots,$*

$$\mu_{r,r+1:n}^{(a,b)} = \frac{r\nu}{r\nu + a} \mu_{r+1:n}^{(a+b)}; \qquad (15.47)$$

and for $1 \leq r < s \leq n$, $s - r \geq 2$ and $a, b = 1, 2, \ldots,$

$$\mu_{r,s:n}^{(a,b)} = \frac{r\nu}{r\nu + a} \mu_{r+1,s:n}^{(a,b)}. \qquad (15.48)$$

Theorem 15.4.3 *For $a, b = 1, 2, \ldots,$*

$$\mu_{1,2:2}^{(a,b)} = \frac{\nu}{\nu + b}\left[\mu_{1:2}^{(a+b)} + 2\left\{\mu_{1:1}^{(a)} - \mu_{1:1}^{(a+b)}\right\}\right]; \qquad (15.49)$$

for $1 \leq r \leq n-2$ and $a, b = 1, 2, \ldots,$

$$\mu_{r,r+1:n}^{(a,b)} = \frac{\nu}{(n-r)\nu + b}\left[(n-r)\mu_{r:n}^{(a+b)} + n\left\{\mu_{r,r+1:n-1}^{(a,b)} - \mu_{r:n-1}^{(a+b)}\right\}\right]; \quad (15.50)$$

for $n \geq 2$ and $a, b = 1, 2, \ldots,$

$$\mu_{n-1,n:n}^{(a,b)} = \frac{\nu}{\nu + b}\left[\mu_{n-1:n}^{(a+b)} + n\left\{\mu_{n-1:n-1}^{(a)} - \mu_{n-1:n-1}^{(a+b)}\right\}\right]; \quad (15.51)$$

for $1 \leq r < s \leq n-1$, $s - r \geq 2$ and $a, b = 1, 2, \ldots,$

$$\mu_{r,s:n}^{(a,b)} = \frac{\nu}{(n-s+1)\nu + b}\left[(n-s+1)\mu_{r,s-1:n}^{(a,b)} + n\left\{\mu_{r,s:n-1}^{(a,b)} - \mu_{r,s-1:n-1}^{(a,b)}\right\}\right]; \quad (15.52)$$

and for $1 \leq r \leq n-1$ and $a, b = 1, 2, \ldots,$

$$\mu_{r,n:n}^{(a,b)} = \frac{\nu}{\nu + b}\left[\mu_{r,n-1:n}^{(a,b)} + n\left\{\mu_{r:n-1}^{(a)} - \mu_{r,n-1:n-1}^{(a,b)}\right\}\right]. \quad (15.53)$$

15.4.3 Relations for triple moments

The joint p.d.f. of $X_{r:n}$, $X_{s:n}$ and $X_{t:n}$ for $1 \leq r < s < t \leq n$ is given by

$$\begin{aligned}
f_{r,s,t:n}(w, x, y) &= C_{r,s,t:n}\{F(w)\}^{r-1}\{F(x) - F(w)\}^{s-r-1}\{F(y) - F(x)\}^{t-s-1} \\
&\quad \times \{1 - F(y)\}^{n-t} f(w)f(x)f(y), \quad 0 \leq w < x < y \leq 1,
\end{aligned} \quad (15.54)$$

where $C_{r,s,t:n} = \frac{n!}{(r-1)!(s-r-1)!(t-s-1)!(n-t)!}$. Using (15.55), we may compute the triple moments as

$$\begin{aligned}
\mu_{r,s,t:n}^{(a,b,c)} &= \int_0^1 \int_w^1 \int_x^1 w^a x^b y^c f_{r,s,t:n}(w, x, y) dy dx dw, \\
&\quad 1 \leq r < s < t \leq n, \; a, b, c \geq 1.
\end{aligned} \quad (15.55)$$

Then the triple moments $\mu_{r,s,t:n}^{(a,b,c)}$ for $a, b, c = 1, 2, \ldots,$ satisfy the recurrence relations presented in the following three theorems, which may be proved by following the same steps as those of Balakrishnan and Gupta (1996) and Childs, Sultan and Balakrishnan (1996).

Theorem 15.4.4 For $n \geq 3$, $1 \leq r < t \leq n$, $t - r \geq 2$ and $a, b, c = 1, 2, \ldots,$

$$\mu_{r,r+1,t:n}^{(a,b,c)} = \frac{r\nu}{r\nu + a}\mu_{r+1,t:n}^{(a+b,c)}; \quad (15.56)$$

and for $n \geq 3$, $1 \leq r < s < t \leq n$, $s - r \geq 2$ and $a, b, c = 1, 2, \ldots,$

$$\mu_{r,s,t:n}^{(a,b,c)} = \frac{r\nu}{r\nu + a}\mu_{r+1,s,t:n}^{(a,b,c)}. \quad (15.57)$$

Theorem 15.4.5 *For* $n \geq 3$, $1 \leq r \leq n$ *and* $a, b, c = 1, 2, \ldots,$

$$\mu_{r,r+1,r+2:n}^{(a,b,c)} = \frac{\nu}{\nu+b} \left[\mu_{r,r+2:n}^{(a,b+c)} + r \left\{ \mu_{r+1,r+2:n}^{(a,b+c)} - \mu_{r+1,r+2:n}^{(a+b,c)} \right\} \right]; \qquad (15.58)$$

for $n \geq 4$, $1 \leq r < s \leq n$, $s - r \geq 2$ *and* $a, b, c = 1, 2, \ldots,$

$$\mu_{r,s,s+1:n}^{(a,b,c)} = \frac{\nu}{(s-r)\nu+b} \left[(s-r)\mu_{r,s+1,:n}^{(a,b+c)} + r \left\{ \mu_{r+1,s+1:n}^{(a,b+c)} - \mu_{r+1,s,s+1:n}^{(a,b,c)} \right\} \right];$$

$$(15.59)$$

for $n \geq 4$, $1 \leq r < t \leq n$, $t - r \geq 3$ *and* $a, b, c = 1, 2, \ldots,$

$$\mu_{r,r+1,t:n}^{(a,b,c)} = \frac{\nu}{\nu+b} \left[\mu_{r,r+2,t:n}^{(a,b,c)} + r \left\{ \mu_{r+1,r+2,t:n}^{(a,b,c)} - \mu_{r+1,t:n}^{(a+b,c)} \right\} \right]; \qquad (15.60)$$

and for $n \geq 4$, $1 \leq r < s < t \leq n$, $s - r \geq 2$, $t - s \geq 2$ *and* $a, b, c = 1, 2, \ldots,$

$$\mu_{r,s,t:n}^{(a,b,c)} = \frac{\nu}{(s-r)\nu+b} \left[(s-r)\mu_{r,s+1,t:n}^{(a,b,c)} + r \left\{ \mu_{r+1,s+1,t:n}^{(a,b,c)} - \mu_{r+1,s,t:n}^{(a,b,c)} \right\} \right].$$

$$(15.61)$$

Theorem 15.4.6 *For* $1 \leq r \leq n-1$ *and* $a, b, c = 1, 2, \ldots,$

$$\mu_{r,n-1,n:n}^{(a,b,c)} = \frac{\nu}{\nu+c} \left[\mu_{r,n-1:n}^{(a,b+c)} + n \left\{ \mu_{r,n-1:n-1}^{(a,b)} - \mu_{r,n-1:n-1}^{(a,b+c)} \right\} \right]; \qquad (15.62)$$

for $1 \leq r < s \leq n-2$, *and* $a, b, c = 1, 2, \ldots,$

$$\mu_{r,s,s+1:n}^{(a,b,c)} = \frac{\nu}{(n-s)\nu+c} \left[(n-s)\mu_{r,s:n}^{(a,b+c)} + n \left\{ \mu_{r,s,s+1:n-1}^{(a,b,c)} - \mu_{r,s:n-1}^{(a,b+c)} \right\} \right];$$

$$(15.63)$$

for $1 \leq r < s < t \leq n-1$, $t - s \geq 2$ *and* $a, b, c = 1, 2, \ldots,$

$$\mu_{r,s,t:n}^{(a,b,c)}$$
$$= \frac{\nu}{(n-t+1)\nu+c} \left[(n-t+1)\mu_{r,s,t-1:n}^{(a,b,c)} + n \left\{ \mu_{r,s,t:n-1}^{(a,b,c)} - \mu_{r,s,t-1:n-1}^{(a,b,c)} \right\} \right];$$

$$(15.64)$$

and for $1 \leq r < s < t \leq n$ *and* $a, b, c = 1, 2, \ldots,$

$$\mu_{r,s,n:n}^{(a,b,c)} = \frac{\nu}{\nu+c} \left[\mu_{r,s,n-1:n}^{(a,b,c)} + n \left\{ \mu_{r,s:n-1}^{(a,b)} - \mu_{r,s,t-1:n-1}^{(a,b,c)} \right\} \right]. \qquad (15.65)$$

15.4.4 Relations for quadruple moments

The joint p.d.f. of $X_{r:n}$, $X_{s:n}$, $X_{t:n}$ and $X_{u:n}$ for $1 \leq r < s < t < u \leq n$ is given by

$$\begin{aligned}
f_{r,s,t,u:n}&(w,x,y,z) \\
&= C_{r,s,t,u:n}\{F(w)\}^{r-1}\{F(x)-F(w)\}^{s-r-1}\{F(y)-F(x)\}^{t-s-1} \\
&\quad \times \{F(z)-F(y)\}^{u-t-1}\{1-F(z)\}^{n-u}f(w)f(x)f(y)f(z), \\
&\qquad 0 \leq w < x < y < z \leq 1,
\end{aligned} \quad (15.66)$$

where $C_{r,s,t,u:n} = \frac{n!}{(r-1)!(s-r-1)!(t-s-1)!(u-t-1)!(n-u)!}$. Using (15.67), we may compute the quadruple moments as

$$\mu_{r,s,t,u:n}^{(a,b,c,d)} = \int_0^1 \int_w^1 \int_x^1 \int_y^1 w^a x^b y^c z^d f_{r,s,t,u:n}(w,x,y,z)dz\,dy\,dx\,dw,$$
$$1 \leq r < s < t < u \leq n,\ a,b,c,d \geq 1. \quad (15.67)$$

Then the quadruple moments $\mu_{r,s,t,u:n}^{(a,b,c,d)}$ for $a,b,c,d = 1,2,\ldots$, satisfy the recurrence relations presented in the following four theorems, which may once again be established along the lines of Balakrishnan and Gupta (1996) and Childs, Sultan and Balakrishnan (1996).

Theorem 15.4.7 *For $1 \leq r < t < u \leq n$, $t - r \geq 2$ and $a,b,c,d = 1,2,\ldots$,*

$$\mu_{r,r+1,t,u:n}^{(a,b,c,d)} = \frac{r\nu}{r\nu + a}\mu_{r+1,t,u:n}^{(a+b,c,d)}; \quad (15.68)$$

and for $1 \leq r < s < t < u \leq n$, $s - r \geq 2$ and $a,b,c,d = 1,2,\ldots$,

$$\mu_{r,s,t,u:n}^{(a,b,c,d)} = \frac{r\nu}{r\nu + a}\mu_{r+1,s,t,u:n}^{(a,b,c,d)}. \quad (15.69)$$

Theorem 15.4.8 *For $1 \leq r < u \leq n$, $u - r \geq 3$ and $a,b,c,d = 1,2,\ldots$,*

$$\mu_{r,r+1,r+2,u:n}^{(a,b,c,d)} = \frac{\nu}{\nu + b}\left[\mu_{r,r+2,u:n}^{(a,b+c,d)} + r\left\{\mu_{r+1,r+2,u:n}^{(a,b+c,d)} - \mu_{r+1,r+2,u:n}^{(a+b,c,d)}\right\}\right]; \quad (15.70)$$

for $1 \leq r < s < u \leq n$, $s - r \geq 2$, $u - s \geq 2$ and $a,b,c,d = 1,2,\ldots$,

$$\begin{aligned}
&\mu_{r,s,s+1,u:n}^{(a,b,c,d)} \\
&= \frac{\nu}{(s-r)\nu + b}\left[(s-r)\mu_{r,s+1,u:n}^{(a,b+c,d)} + r\left\{\mu_{r+1,s+1,u:n}^{(a,b+c,d)} - \mu_{r+1,s,s+1,u:n}^{(a,b,c,d)}\right\}\right];
\end{aligned} \quad (15.71)$$

for $1 \leq r < t < u \leq n$, $t - r \geq 3$ and $a,b,c,d = 1,2,\ldots$,

$$\mu_{r,r+1,t,u:n}^{(a,b,c,d)} = \frac{\nu}{\nu + b}\left[\mu_{r,r+2,t,u:n}^{(a,b,c,d)} + r\left\{\mu_{r+1,r+2,t,u:n}^{(a,b,c,d)} - \mu_{r+1,t,u:n}^{(a+b,c,d)}\right\}\right]; \quad (15.72)$$

and for $1 \leq r < s < t < u \leq n$, $s - r \geq 2$, $t - s \geq 2$ and $a, b, c, d = 1, 2, \ldots,$

$$\mu_{r,s,t,u:n}^{(a,b,c,d)} = \frac{\nu}{(s-r)\nu + b} \left[(s-r)\mu_{r,s+1,t,u:n}^{(a,b,c,d)} + r \left\{ \mu_{r+1,s+1,t,u:n}^{(a,b,c,d)} - \mu_{r+1,s,t,u:n}^{(a,b,c,d)} \right\} \right]. \tag{15.73}$$

Theorem 15.4.9 For $1 \leq r < s \leq n - 2$, $s - r \geq 2$ and $a, b, c, d = 1, 2, \ldots,$

$$\mu_{r,s,s+1,s+2:n}^{(a,b,c,d)} = \frac{\nu}{\nu + c} \left[\mu_{r,s,s+2:n}^{(a,b,c+d)} + r \left\{ \mu_{r+1,s+1,s+2:n}^{(a,b,c+d)} - \mu_{r+1,s+1,s+2:n}^{(a,b+c,d)} \right\} \right.$$
$$\left. + (s-r) \left\{ \mu_{r,s+1,s+2:n}^{(a,b,c+d)} - \mu_{r,s+1,s+2:n}^{(a,b+c,d)} \right\} \right]; \tag{15.74}$$

for $1 \leq r < s < t \leq n - 1$, $s - r \geq 2$, $t - s \geq 2$ and $a, b, c, d = 1, 2, \ldots,$

$$\mu_{r,s,t,t+1:n}^{(a,b,c,d)}$$
$$= \frac{\nu}{(t-s)\nu + c} \left[(t-s)\mu_{r,s,t+1:n}^{(a,b,c+d)} + r \left\{ \mu_{r+1,s+1,t+1:n}^{(a,b,c+d)} - \mu_{r+1,s+1,t,t+1:n}^{(a,b,c,d)} \right\} \right.$$
$$\left. + (s-r) \left\{ \mu_{r,s+1,t+1:n}^{(a,b,c+d)} - \mu_{r,s+1,t,t+1:n}^{(a,b,c,d)} \right\} \right]; \tag{15.75}$$

for $1 \leq r < s < u \leq n$, $s - r \geq 2$, $u - s \geq 2$ and $a, b, c, d = 1, 2, \ldots,$

$$\mu_{r,s,s+1,u:n}^{(a,b,c,d)} = \frac{\nu}{\nu + c} \left[\mu_{r,s,s+2,u:n}^{(a,b,c,d)} + r \left\{ \mu_{r+1,s+1,s+2,u:n}^{(a,b,c,d)} - \mu_{r+1,s+1,u:n}^{(a,b+c,d)} \right\} \right.$$
$$\left. + (s-r) \left\{ \mu_{r,s+1,s+2,u:n}^{(a,b,c,d)} - \mu_{r,s+1,u:n}^{(a,b+c,d)} \right\} \right]; \tag{15.76}$$

and for $1 \leq r < s < t < u \leq n$, $s - r \geq 2$, $t - s \geq 2$, $u - t \geq 2$ and $a, b, c, d = 1, 2, \ldots,$

$$\mu_{r,s,t,u:n}^{(a,b,c,d)}$$
$$= \frac{\nu}{(t-s)\nu + c} \left[(t-s)\mu_{r,s,t+1,u:n}^{(a,b,c,d)} + r \left\{ \mu_{r+1,s+1,t+1,u:n}^{(a,b,c,d)} - \mu_{r+1,s+1,t,u:n}^{(a,b,c,d)} \right\} \right.$$
$$\left. + (s-r) \left\{ \mu_{r,s+1,t+1,u:n}^{(a,b,c,d)} - \mu_{r,s+1,t,u:n}^{(a,b,c,d)} \right\} \right]. \tag{15.77}$$

Theorem 15.4.10 For $1 \leq r < s \leq n - 2$, and $a, b, c, d = 1, 2, \ldots,$

$$\mu_{r,s,n-1,n:n}^{(a,b,c,d)} = \frac{\nu}{\nu + d} \left[\mu_{r,s,n-1:n}^{(a,b,c+d)} + n \left\{ \mu_{r,s,n-1:n-1}^{(a,b,c)} - \mu_{r,s,n-1:n-1}^{(a,b,c+d)} \right\} \right]; \tag{15.78}$$

for $1 \leq r < s < t \leq n - 2$, and $a, b, c, d = 1, 2, \ldots,$

$$\mu_{r,s,t,t+1:n}^{(a,b,c,d)}$$
$$= \frac{\nu}{(n-t)\nu + d} \left[(n-t)\mu_{r,s,t:n}^{(a,b,c+d)} + n \left\{ \mu_{r,s,t,t+1:n-1}^{(a,b,c,d)} - \mu_{r,s,t:n-1}^{(a,b,c+d)} \right\} \right]; \tag{15.79}$$

Higher Order Moments of Order Statistics

for $1 \leq r < s < t < u \leq n-1$, $u - t \geq 2$ and $a, b, c, d = 1, 2, \ldots$,

$$\mu_{r,s,t,u:n}^{(a,b,c,d)} = \frac{\nu}{(n-u+1)\nu + d}$$
$$\times \left[(n-u+1)\mu_{r,s,t,u-1:n}^{(a,b,c,d)} + n \left\{ \mu_{r,s,t,u:n-1}^{(a,b,c,d)} - \mu_{r,s,t,u-1:n-1}^{(a,b,c,d)} \right\} \right]; \quad (15.80)$$

and for $1 \leq r < s < t \leq n-2$ and $a, b, c, d = 1, 2, \ldots$,

$$\mu_{r,s,t,n:n}^{(a,b,c,d)} = \frac{\nu}{\nu + d} \left[\mu_{r,s,t,n-1:n}^{(a,b,c,d)} + n \left\{ \mu_{r,s,t:n-1}^{(a,b,c)} - \mu_{r,s,t,n-1:n-1}^{(a,b,c,d)} \right\} \right]. \quad (15.81)$$

Remark All the recurrence relations presented in this section are complete in the sense that they will enable one to compute all the single, double, triple and quadruple moments of all order statistics for all sample sizes in a simple recursive manner. This can be done for any choice of the shape parameter ν.

15.5 Approximate Inference

By making use of the recurrence relations in Section 15.4, and the values of a_i and b_i obtained from the exact explicit expressions for the BLUE's given in Section 15.2 for the case of Type-II right-censored samples

$$X_{1:n} \leq X_{2:n} \leq \cdots \leq X_{n-s:n},$$

we determined the values of the mean, variance and the coefficients of skewness and kurtosis ($\sqrt{\beta_1}$ and β_2) of R_1^* and R_2^*, for $n = 5(1)10(5)20$, $\nu = 3/2$, 3 and 6, and $s = 0(1)(\frac{n}{2} - 1)$ for n even and $s = 0(1)[\![\frac{n}{2}]\!]$ for n odd. These values are presented in Tables 15.1–15.3. An examination of the ($\sqrt{\beta_1}, \beta_2$) values in Table 15.1 reveals that when $\nu = 3/2$ the distribution of R_1^* (and hence of R_1) is positively skewed and heavier tailed than normal except for small sample sizes where it is positively skewed and lighter tailed than normal. When $\nu = 3$, we see from Table 15.2 that the distribution of R_1^* is lighter tailed than normal for all sample sizes, and negatively skewed for small samples. From Table 15.3, we see that when $\nu = 6$, the distribution of R_1^* is negatively skewed for all sizes, and slightly heavier tailed than normal for small sample sizes. In each case, the ($\sqrt{\beta_1}, \beta_2$) values lie well within the range of the Edgeworth approximation. Similar observation may be made from the ($\sqrt{\beta_1}, \beta_2$) values for R_2^* in Tables 15.1–15.3, except that the $\sqrt{\beta_1}$ values for R_2^* usually are opposite in sign to those of R_1^*; for details on the possible range for Edgeworth approximation, see Barton and Dennis (1952) and Johnson, Kotz and Balakrishnan (1994). This similarity can be explained by a careful examination of the exact explicit expressions for the BLUE's given in Section 15.2. These expressions reveal that $a_i = -cb_i$ for $i = r+1, r+2, \ldots, n-s-1$, while a_{n-s} is very close to $-b_{n-s}$.

By making use of the entries in Tables 15.1–15.3, we determined the lower and upper 1%, 2.5%, 5% and 10% points of R_1 and R_2 through the Edgeworth approximation in (15.6). These values, for the case of Type-II right-censored samples $(r = 0)$ for $s = 0(1)(\frac{n}{2} - 1)$ for n even and $s = 0(1)[\![\frac{n}{2}]\!]$ for n odd and sample size $n = 5(1)10(5)20$, are presented in Tables 15.4, 15.5 and 15.6. For the purpose of comparison, these percentage points were also determined by Monte Carlo simulations (based on 5000 runs) and they are presented along with the Edgeworth percentage points in Tables 15.4, 15.5 and 15.6.

From Tables 15.4–15.6, we see that the Edgeworth approximation of the distribution of R_1 provides quite close agreement with the simulated percentage points. The largest discrepancy occurs at the extreme lower tails of the distribution, but only for small sample sizes, and small values of ν. As the sample size and/or ν increases, the agreement becomes quite close at all levels of censoring, even at the lower extremes of the distribution.

From Tables 15.7–15.9, we see that the Edgeworth approximation of the distribution of R_2 also provides close agreement with the simulated percentage points; in fact, even a bit closer than R_1. This time, however, the largest discrepancies occur at the upper tails of the distribution for small sample sizes, but the discrepancies are not as great as those of R_1. Again, as the sample size and/or ν increases, the discrepancy becomes quite small at all levels of censoring, even at the extremes of the distribution.

In conclusion, we observe that the Edgeworth approximations of the distributions of R_1 and R_2 both work quite satisfactorily even in samples of size as small as 5 (for large ν), and they improve in accuracy as the sample size and/or ν increases. It should also be pointed out here that a similar Edgeworth approximation can not be developed for the percentage points of the pivotal quantities R_3 since it is not a linear function of order statistics. However, as displayed in the next section, approximate inference based on R_1 with σ replaced by σ^* provides quite close results to those based on R_3. For this purpose, we have presented in Tables 15.10, 15.11 and 15.12 some selected percentage points of R_3 determined by Monte Carlo simulations (based on 5000 runs).

Remark. Though all the illustrations have been made here with right censored samples, if the available sample is doubly censored, one could use the exact explicit expressions of the BLUE's of θ and σ given in (2.11) and (2.12) and proceed exactly in the same fashion in order to develop Edgeworth approximate inference for the parameters θ and σ.

15.6 Numerical Illustration

In order to illustrate the usefulness of the inference procedures discussed in the previous sections, we consider here a simulated data set of size $n = 20$ (with $\theta = 15$, $\sigma = 10$ and $\nu = 3/2$):

$$\begin{array}{ccccc}
16.704 & 16.869 & 17.586 & 17.941 & 17.994 \\
18.303 & 19.013 & 19.085 & 19.348 & 20.520 \\
21.152 & 21.500 & 21.627 & 21.743 & 22.466 \\
22.795 & 23.842 & 24.345 & 24.359 & 24.650
\end{array}$$

Using this sample, the BLUE's were calculated based on complete as well as Type-II right-censored samples ($r = 0$ and $s > 0$) by making use of the exact explicit expressions given in Section 15.2. These estimates are presented in the following table:

n	s	θ^*	σ^*	S.E.(θ^*)	S.E.(σ^*)
20	0	15.36	9.60	0.837	0.921
	1	15.35	9.64	0.846	0.011
	2	15.27	10.07	0.884	1.148
	3	15.31	9.83	0.868	1.216

With these estimates and the use of Table 15.1 and Table 15.4, we determined the 90% confidence intervals for θ (when σ is known to be 10 and $\nu = 3/2$) based on the Edgeworth approximation and on the simulated percentage points using the pivotal quantity R_1 through the formula

$$P\left(\theta^* - \sigma\sqrt{V_1}\,(R_1)_{1-\alpha/2} \leq \theta \leq \theta^* - \sigma\sqrt{V_1}\,(R_1)_{\alpha/2}\right) = 1 - \alpha.$$

These are presented in the table below:

n	s	Edgeworth C.I.	Simulated C.I.
20	0	(14.558,15.970)	(14.541,15.918)
	1	(14.443,15.864)	(14.412,15.820)
	2	(14.463,15.884)	(14.463,15.845)
	3	(14.502,15.928)	(14.449,15.888)

It is clear that the confidence intervals based on the Edgeworth approximation are very close to the confidence interval determined by simulation, at all levels of censoring.

Similarly, with the use of Table 15.1 and Table 15.7 we determined the 90% confidence intervals for σ, through the formula

$$P\left(\frac{\sigma^*}{1+\sqrt{V_2}\,(R_2)_{1-\alpha/2}} \leq \sigma \leq \frac{\sigma^*}{1+\sqrt{V_2}\,(R_2)_{\alpha/2}}\right) = 1-\alpha,\ .$$

and they are presented below:

n	s	Edgeworth C.I.	Simulated C.I.
20	0	(8.442,11.630)	(8.499,11.685)
	1	(8.360,11.883)	(8.398,11.899)
	2	(8.608,12.634)	(8.616,12.725)
	3	(8.283,12.566)	(8.283,12.605)

Once again, we observe that the confidence intervals based on the Edgeworth approximation are very close to those based on simulation.

In the case when σ is unknown, the Edgeworth approximation method cannot be used to draw inference for θ using R_3. However, as pointed out in the last section, the Edgeworth approximation for the distribution of the pivotal quantity R_1 may be used in this case with σ replaced by σ^* in order to draw approximate inference for θ. By this method, we determined the 90% confidence intervals for θ through the formula

$$P\left(\theta^* - \sigma^*\sqrt{V_1}\,(R_1)_{1-\alpha/2} \leq \theta \leq \theta^* - \sigma^*\sqrt{V_1}\,(R_1)_{\alpha/2}\right) = 1-\alpha,$$

and these are presented in the following table for the choices of $r = 0$, $s = 0(1)3$. Also presented in this table are the corresponding 90% confidence intervals for θ based on the simulated percentage points of the pivotal quantity R_3 (given in Table 15.10) through the formula

$$P\left(\theta^* - \sigma^*\sqrt{V_1}\,(R_3)_{1-\alpha/2} \leq \theta \leq \theta^* - \sigma^*\sqrt{V_1}\,(R_3)_{\alpha/2}\right) = 1-\alpha,$$

		Edgeworth C.I.	Simulated C.I.
n	s	$R_1(\sigma = \sigma^*)$	R_3
20	0	(13.820,16.532)	(13.469,16.322)
	1	(13.694,16.434)	(13.279,16.231)
	2	(13.644,16.507)	(13.255,16.295)
	3	(13.721,16.525)	(13.183,16.308)

It is quite clear that the confidence intervals using the approximate Edgeworth method based on the pivotal quantity R_1 are somewhat close to the confidence intervals determined by simulation of the distribution of pivotal quantity R_3 at all levels of censoring.

15.7 Recurrence Relations For Moments of Order Statistics in the Doubly Truncated Case

In this section, we present some recurrence relations between the single moments, double moments, triple moments and quadruple moments of order statistics in the doubly truncated case.

Let $X_1, X_2, \ldots, X_n$ be a random sample of size n from a one-parameter doubly truncated power function population with probability density function (p.d.f.)

$$f(x) = \frac{\nu}{P-Q} x^{\nu-1}, \qquad Q_1 \leq x \leq P_1,\ \nu > 0, \tag{15.82}$$

and cumulative distribution function (c.d.f.)

$$F(x) = \frac{x^\nu}{P-Q} - Q_2, \qquad Q_1 \leq x \leq P_1,\ \nu > 0, \tag{15.83}$$

where

$$P = P_1^\nu \text{ and } Q = Q_1^\nu. \tag{15.84}$$

From (15.82) and (15.83), we have

$$f(x) = \frac{\nu}{x}\{F(x) + Q_2\} = \frac{\nu}{x}\left[P_2 - \{1 - F(x)\}\right], \tag{15.85}$$

where

$$P_2 = \frac{P}{P-Q} \quad \text{and} \tag{15.86}$$

$$quad Q_2 = \frac{Q}{P-Q}. \tag{15.87}$$

Letting $Q_1 \to 0$ and $P_1 \to 1$ in (15.82)–(15.85), we obtain the standard untruncated case discussed earlier.

15.7.1 Relations for single moments

The p.d.f. of $X_{r:n}$ for $1 \leq r \leq n$ is given by

$$f_{r:n}(w) = C_{r:n}\{F(w)\}^{r-1}\{1-F(w)\}^{n-r} f(w), \qquad Q_1 \leq w \leq P_1, \tag{15.88}$$

where $C_{r:n} = \frac{n!}{(r-1)!(n-r)!}$.

From this, we may compute the single moments of $X_{r:n}$ as

$$\mu_{r:n}^{(a)} = \int_{Q_1}^{P_1} w^a f_{r:n}(w)\, dw, \qquad a = 1, 2, \ldots, 1 \leq r \leq n. \tag{15.89}$$

Then the single moments $\mu_{r:n}^{(a)}$ for $a = 1, 2, \ldots$, satisfy the recurrence relations presented in the following two theorems. Also, it should be mentioned here that the recurrence relations for the single and the product moments of order statistics from the doubly truncated power function distribution have been proved by Balakrishnan and Joshi (1981), and we are presenting here for the sake of completeness.

Theorem 15.7.1 *For $n \geq 2$ and $a = 1, 2, \ldots$,*

$$\left(n + \frac{a}{\nu}\right) \mu_{n:n}^{(a)} = nP_1^a + nQ_2 \left(P_1^a - \mu_{n-1:n-1}^{(a)}\right); \qquad (15.90)$$

for $2 \leq r \leq n - 1$ and $a = 1, 2, \ldots$,

$$\mu_{r:n}^{(a)} = \frac{\nu}{r\nu + a} \left[r\mu_{r+1:n}^{(a)} + nQ_2 \left\{\mu_{r:n-1}^{(a)} - \mu_{r-1:n-1}^{(a)}\right\}\right]; \qquad (15.91)$$

for $n \geq 2$ and $a = 1, 2, \ldots$,

$$\mu_{1:n}^{(a)} = \frac{n\nu}{n\nu + a} \left[\mu_{1:n-1}^{(a)} + Q_2 \left\{\mu_{1:n-1}^{(a)} - Q_1^a\right\}\right]; \qquad (15.92)$$

and for $a = 1, 2, \ldots$,

$$\mu_{1:1}^{(a)} = \frac{\nu}{\nu + a} \left[P_2 P_1^a - Q_2 Q_1^a\right]. \qquad (15.93)$$

15.7.2 Relations for double moments

The joint p.d.f. of $X_{r:n}$ and $X_{s:n}$ for $1 \leq r < s \leq n$ is given by

$$f_{r,s:n}(w, x) = C_{r,s:n} \{F(w)\}^{r-1} \{F(x) - F(w)\}^{s-r-1} \{1 - F(x)\}^{n-s} f(w) f(x),$$
$$Q_1 \leq w < x \leq P_1, \qquad (15.94)$$

where $C_{r,s:n} = \frac{n!}{(r-1)!(s-r-1)!(n-s)!}$. Using (15.94), we may compute the double moments as

$$\mu_{r,s:n}^{(a,b)} = \int_{Q_1}^{P_1} \int_w^{P_1} w^a x^b f_{r,s:n}(w, x) dx dw, \ 1 \leq r < s \leq n, \ a, b \geq 1. \qquad (15.95)$$

Then the double moments $\mu_{r,s:n}^{(a,b)}$ for $a, b = 1, 2, \ldots$, satisfy the recurrence relations presented in the following two theorems.

Theorem 15.7.2 *For $2 \leq r \leq n - 1$ and $a, b = 1, 2, \ldots$,*

$$\mu_{r,r+1:n}^{(a,b)} = \frac{\nu}{r\nu + a} \left[r\mu_{r+1:n}^{(a+b)} + nQ_2 \left\{\mu_{r,n-1}^{(a+b)} - \mu_{r-1,r:n-1}^{(a,b)}\right\}\right]; \qquad (15.96)$$

and for $2 \leq r < s \leq n$, $s - r \geq 2$ and $a, b = 1, 2, \ldots$,

$$\mu_{r,s:n}^{(a,b)} = \frac{\nu}{r\nu + a} \left[r\mu_{r+1,s:n}^{(a,b)} + nQ_2 \left\{\mu_{r,s-r:n-1}^{(a,b)} - \mu_{r-1,s-1:n-1}^{(a,b)}\right\}\right]. \qquad (15.97)$$

Theorem 15.7.3 *For* $a, b = 1, 2, \ldots,$

$$\mu_{1,2:2}^{(a,b)} = \frac{\nu}{\nu + b} \left[\mu_{1:2}^{(a+b)} + 2P_2 \left\{ P_1^b \mu_{1:1}^{(a)} - \mu_{1:1}^{(a+b)} \right\} \right]; \qquad (15.98)$$

for $1 \leq r \leq n - 2$ *and* $a, b = 1, 2, \ldots,$

$$\mu_{r,r+1:n}^{(a,b)}$$
$$= \frac{\nu}{(n-r)\nu + b} \left[(n-r)\mu_{r:n}^{(a+b)} + nP_2 \left\{ \mu_{r,r+1:n-1}^{(a,b)} - \mu_{r:n-1}^{(a+b)} \right\} \right]; \qquad (15.99)$$

for $n \geq 2$ *and* $a, b = 1, 2, \ldots,$

$$\mu_{n-1,n:n}^{(a,b)} = \frac{\nu}{\nu + b} \left[\mu_{n-1:n}^{(a+b)} + nP_2 \left\{ P_1^b \mu_{n-1:n-1}^{(a)} - \mu_{n-1:n-1}^{(a+b)} \right\} \right]; \qquad (15.100)$$

for $1 \leq r < s \leq n - 1$, $s - r \geq 2$ *and* $a, b = 1, 2, \ldots,$

$$\mu_{r,s:n}^{(a,b)}$$
$$= \frac{n\nu}{(n-s+1)\nu + b} \left[(n-s+1)\mu_{r,s-1:n}^{(a,b)} + nP_2 \left\{ \mu_{r,s:n-1}^{(a,b)} - \mu_{r,s-1:n-1}^{(a,b)} \right\} \right]; \qquad (15.101)$$

and for $1 \leq r \leq n - 1$ *and* $a, b = 1, 2, \ldots,$

$$\mu_{r,n:n}^{(a,b)} = \frac{\nu}{\nu + b} \left[\mu_{r,n-1:n}^{(a,b)} + nP_2 \left\{ P_1^b \mu_{r:n-1}^{(a)} - \mu_{r,n-1:n-1}^{(a,b)} \right\} \right]. \qquad (15.102)$$

15.7.3 Relations for triple moments

The joint p.d.f of $X_{r:n}$, $X_{s:n}$ and $X_{t:n}$ for $1 \leq r < s < t \leq n$ is given by

$$\begin{aligned} f_{r,s,t:n}(w,x,y) &= C_{r,s,t:n} \{F(w)\}^{r-1} \{F(x) - F(w)\}^{s-r-1} \{F(y) - F(x)\}^{t-s-1} \\ &\quad \times \{1 - F(y)\}^{n-t} f(w)f(x)f(y), \\ &\quad Q_1 \leq w < x < y \leq P_1, \end{aligned} \qquad (15.103)$$

where $C_{r,s,t:n} = \frac{n!}{(r-1)!(s-r-1)!(t-s-1)!(n-t)!}$. Using (7.22), we may compute the triple moments as

$$\mu_{r,s,t:n}^{(a,b,c)} = \int_{Q_1}^{P_1} \int_w^{P_1} \int_x^{P_1} w^a x^b y^c f_{r,s,t:n}(w,x,y) \, dy \, dx \, dw,$$
$$1 \leq r < s < t \leq n, \ a, b, c \geq 1. \qquad (15.104)$$

Then the triple moments $\mu_{r,s,t:n}^{(a,b,c)}$ for $a, b, c = 1, 2, \ldots,$ satisfy the recurrence relations presented in the following two theorems.

Theorem 15.7.4 *For* $n \geq 3$, $2 \leq r < t \leq n$, $t - r \geq 2$ *and* $a, b, c = 1, 2, \ldots,$

$$\mu_{r,r+1,t:n}^{(a,b,c)} = \frac{\nu}{r\nu + a} \left[r\mu_{r+1,t:n}^{(a+b,c)} + nQ_2 \left\{ \mu_{r,t-t:n-1}^{(a+b,c)} - \mu_{r-1,r,t-1:n-1}^{(a,b,c)} \right\} \right]; \quad (15.105)$$

and for $n \geq 4$, $2 \leq r < s < t \leq n$, $s - r > 2$ *and* $a, b, c = 1, 2, \ldots,$

$$\mu_{r,s,t:n}^{(a,b,c)} = \frac{\nu}{r\nu + a} \left[r\mu_{r+1,s,t:n}^{(a,b,c)} + nQ_2 \left\{ \mu_{r,s-1,t-1:n-1}^{(a,b,c)} - \mu_{r-1,s-1,t-1:n-1}^{(a,b,c)} \right\} \right]. \quad (15.106)$$

Theorem 15.7.5 *For* $n \geq 3$, $1 \leq r \leq n$ *and* $a, b, c = 1, 2, \ldots,$

$$\mu_{r,r+1,r+2:n}^{(a,b,c)} = \frac{\nu}{\nu + b} \left[\mu_{r,r+2:n}^{(a,b+c)} + r \left\{ \mu_{r+1,r+2:n}^{(a,b+c)} - \mu_{r+1,r+2:n}^{(a+b,c)} \right\} \right.$$
$$\left. + nQ_2 \left\{ \mu_{r,r+1:n-1}^{(a,b+c)} - \mu_{r,r+1:n-1}^{(a+b,c)} \right\} \right]; \quad (15.107)$$

for $n \geq 3$, $1 \leq r < s \leq n$, $s - r \geq 2$ *and* $a, b, c = 1, 2, \ldots,$

$$\mu_{r,s,s+1:n}^{(a,b,c)} = \frac{\nu}{(s-r)\nu + b} \left[(s-r)\mu_{r,s+1,:n}^{(a,b+c)} + r \left\{ \mu_{r+1,s+1:n}^{(a,b+c)} - \mu_{r+1,s,s+1:n}^{(a,b,c)} \right\} \right.$$
$$\left. + nQ_2 \left\{ \mu_{r,s:n-1}^{(a,b+c)} - \mu_{r,s-1,s:n-1}^{(a,b,c)} \right\} \right]; \quad (15.108)$$

for $n \geq 4$, $1 \leq r < s < t \leq n$, $t - r \geq 3$ *and* $a, b, c = 1, 2, \ldots,$

$$\mu_{r,r+1,t:n}^{(a,b,c)} = \frac{\nu}{\nu + b} \left[\mu_{r,r+2,t:n}^{(a,b,c)} + r \left\{ \mu_{r+1,r+2,t:n}^{(a,b,c)} - \mu_{r+1,t:n}^{(a+b,c)} \right\} \right.$$
$$\left. + nQ_2 \left\{ \mu_{r,r+1,t-1:n-1}^{(a,b,c)} - \mu_{r,t-1:n-1}^{(a+b,c)} \right\} \right]; \quad (15.109)$$

and for $n \geq 4$, $1 \leq r < s < t \leq n$, $s - r \geq 2$, $t - s \geq 2$ *and* $a, b, c = 1, 2, \ldots,$

$$\mu_{r,s,t:n}^{(a,b,c)} = \frac{\nu}{(s-r)\nu + b} \left[(s-r)\mu_{r,s+1,t:n}^{(a,b,c)} + r \left\{ \mu_{r+1,s+1,t:n}^{(a,b,c)} - \mu_{r+1,s,t:n}^{(a,b,c)} \right\} \right.$$
$$\left. + nQ_2 \left\{ \mu_{r,s,t-1:n-1}^{(a,b,c)} - \mu_{r,s-1,t-1:n-1}^{(a,b,c)} \right\} \right]. \quad (15.110)$$

Theorem 15.7.6 *For* $1 \leq r \leq n - 1$ *and* $a, b, c = 1, 2, \ldots,$

$$\mu_{r,n-1,n:n}^{(a,b,c)} = \frac{\nu}{\nu + c} \left[\mu_{r,n-1:n}^{(a,b+c)} + nP_2 \left\{ P_1^c \mu_{r,n-1:n-1}^{(a,b)} - \mu_{r,n-1:n-1}^{(a,b+c)} \right\} \right]; \quad (15.111)$$

for $1 \leq r < s < t \leq n - 2$, *and* $a, b, c = 1, 2, \ldots,$

$$\mu_{r,s,s+1:n}^{(a,b,c)}$$
$$= \frac{\nu}{(n-s)\nu + c} \left[(n-s)\mu_{r,s:n}^{(a,b+c)} + nP_2 \left\{ \mu_{r,s,s+1:n-1}^{(a,b,c)} - \mu_{r,s:n-1}^{(a,b+c)} \right\} \right]; \quad (15.112)$$

Higher Order Moments of Order Statistics

for $1 \leq r < s \leq n-1$, $t-s \geq 2$ and $a, b, c = 1, 2, \ldots$,

$$\mu_{r,s,t:n}^{(a,b,c)} = \frac{\nu}{(n-t+1)\nu + c}\Big[(n-t+1)\mu_{r,s,t-1:n}^{(a,b,c)} \\ + nP_2\left\{\mu_{r,s,t:n-1}^{(a,b,c)} - \mu_{r,s,t-1:n-1}^{(a,b,c)}\right\}\Big]; \quad (15.113)$$

and for $1 \leq r < s < t \leq n$ and $a, b, c = 1, 2, \ldots$,

$$\mu_{r,s,n:n}^{(a,b,c)} = \frac{\nu}{\nu + c}\left[\mu_{r,s,n-1:n}^{(a,b,c)} + nP_2\left\{P_1^c \mu_{r,s:n-1}^{(a,b)} - \mu_{r,s,t-1:n-1}^{(a,b,c)}\right\}\right]. \quad (15.114)$$

15.7.4 Relations for quadruple moments

The joint p.d.f. of $X_{r:n}$, $X_{s:n}$, $X_{t:n}$ and $X_{u:n}$ for $1 \leq r < s < t < u \leq n$ is given by

$$f_{r,s,t,u:n}(w, x, y, z) \\ = C_{r,s,t,u:n}\{F(w)\}^{r-1}\{F(x) - F(w)\}^{s-r-1}\{F(y) - F(x)\}^{t-s-1} \\ \times \{F(z) - F(y)\}^{u-t-1}\{1 - F(z)\}^{n-u} f(w)f(x)f(y)f(z), \\ Q_1 \leq w < x < y < z \leq P_1, \quad (15.115)$$

where $C_{r,s,t,u:n} = \frac{n!}{(r-1)!(s-r-1)!(t-s-1)!(u-t-1)!(n-u)!}$. Using (7.34), we may compute the quadruple moments as

$$\mu_{r,s,t,u:n}^{(a,b,c,d)} = \int_{Q_1}^{P_1}\int_{w}^{P_1}\int_{x}^{P_1}\int_{y}^{P_1} w^a x^b y^c z^d f_{r,s,t,u:n}(w, x, y, z)\,dz\,dy\,dx\,dw, \\ 1 \leq r < s < t < u \leq n,\ a, b, c, d \geq 1. \quad (15.116)$$

Then the quadruple moments $\mu_{r,s,t,u:n}^{(a,b,c,d)}$ for $a, b, c, d = 1, 2, \ldots$, satisfy the recurrence relations presented in the following four theorems.

Theorem 15.7.7 *For $1 < r < t < u \leq n$, $t - r \geq 2$ and $a, b, c, d = 1, 2, \ldots$,*

$$\mu_{r,r+1,t,u:n}^{(a,b,c,d)} \\ = \frac{\nu}{r\nu + a}\left[r\mu_{r+1,t,u:n}^{(a+b,c,d)} + nQ_2\left\{\mu_{r,r-1,u-1:n-1}^{(a+b,c,d)} - \mu_{r-1,r,t,u:n-1}^{(a,b,c,d)}\right\}\right]; \quad (15.117)$$

and for $2 \leq r < s < t < u \leq n$, $s - r \geq 2$ and $a, b, c, d = 1, 2, \ldots$,

$$\mu_{r,s,t,u:n}^{(a,b,c,d)} = \frac{\nu}{r\nu + a}\Big[r\mu_{r+1,s,t,u:n}^{(a,b,c,d)} + nQ_2\Big\{\mu_{r,s-1,t-1,u-1:n-1}^{(a,b,c,d)} \\ - \mu_{r-1,s-1,t-1,u-1:n-1}^{(a,b,c,d)}\Big\}\Big]. \quad (15.118)$$

Theorem 15.7.8 *For $1 \leq r \leq u \leq n$, $u - r \geq 3$ and $a, b, c, d = 1, 2, \ldots,$*

$$\mu_{r,r+1,r+2,u:n}^{(a,b,c,d)} = \frac{\nu}{\nu + b}\left[\mu_{r,r+2,u:n}^{(a,b+c,d)} + r\left\{\mu_{r+1,r+2,u:n}^{(a,b+c,d)} - \mu_{r+1,r+2,u:n}^{(a+b,c,d)}\right\}\right.$$
$$\left. + nQ_2\left\{\mu_{r,r+1,u-1:n-1}^{(a,b+c,d)} - \mu_{r,r+1,u-1:n-1}^{(a+b,c,d)}\right\}\right]; \quad (15.119)$$

for $1 \leq r < s < u \leq n$, $s - r \geq 2$, $u - s \geq 2$ and $a, b, c, d = 1, 2, \ldots,$

$$\mu_{r,s,s+1,u:n}^{(a,b,c,d)} = \frac{\nu}{(s-r)\nu + b}\left[(s-r)\mu_{r,s+1,u:n}^{(a,b+c,d)}\right.$$
$$+ r\left\{\mu_{r+1,s+1,u:n}^{(a,b+c,d)} - \mu_{r+1,s,s+1,u:n}^{(a,b,c,d)}\right\}$$
$$\left. + nQ_2\left\{\mu_{r,s,u-1:n-1}^{(a,b+c,d)} - \mu_{r,s-1,s,u-1:n-1}^{(a,b,c,d)}\right\}\right]; \quad (15.120)$$

for $1 \leq r < t < u \leq n$, $t - r \geq 3$ and $a, b, c, d = 1, 2, \ldots,$

$$\mu_{r,r+1,t,u:n}^{(a,b,c,d)} = \frac{\nu}{\nu + b}\left[\mu_{r,r+2,t,u:n}^{(a,b,c,d)} + r\left\{\mu_{r+1,r+2,t,u:n}^{(a,b,c,d)} - \mu_{r+1,t,u:n}^{(a+b,c,d)}\right\}\right.$$
$$\left. + nQ_2\left\{\mu_{r,r+1,t-1,u-1:n-1}^{(a,b,c,d)} - \mu_{r,t-1,u-1:n-1}^{(a+b,c,d)}\right\}\right]; \quad (15.121)$$

and for $1 \leq r < s < t < u \leq n$, $s - r \geq 2$, $t - s \geq 2$ and $a, b, c, d = 1, 2, \ldots,$

$$\mu_{r,s,t,u:n}^{(a,b,c,d)}$$
$$= \frac{\nu}{(s-r)\nu + b}\left[(s-r)\mu_{r,s+1,t,u:n}^{(a,b,c,d)} + r\left\{\mu_{r+1,s+1,t,u:n}^{(a,b,c,d)} - \mu_{r+1,s,t,u:n}^{(a,b,c,d)}\right\}\right.$$
$$\left. + nQ_2\left\{\mu_{r,s,t-1,u-1:n-1}^{(a,b,c,d)} - \mu_{r,s-1,t-1,u-1:n-1}^{(a,b,c,d)}\right\}\right]. \quad (15.122)$$

Theorem 15.7.9 *For $1 \leq r < s \leq n - 2$, $s - r \geq 2$ and $a, b, c, d = 1, 2, \ldots,$*

$$\mu_{r,s,s+1,s+2:n}^{(a,b,c,d)} = \frac{\nu}{\nu + c}\left[\mu_{r,s,s+2:n}^{(a,b,c+d)} + r\left\{\mu_{r+1,s+1,s+2:n}^{(a,b,c+d)} - \mu_{r+1,s+1,s+2:n}^{(a,b+c,d)}\right\}\right.$$
$$+ (s-r)\left\{\mu_{r,s+1,s+2:n}^{(a,b,c+d)} - \mu_{r,s+1,s+2:n}^{(a,b+c,d)}\right\}$$
$$\left. + nQ_2\left\{\mu_{r,s,s+1:n-1}^{(a,b,c+d)} - \mu_{r,s,s+1:n-1}^{(a,b+c,d)}\right\}\right]; \quad (15.123)$$

for $1 \leq r < s < t \leq n - 1$, $s - r \geq 2$, $t - s \geq 2$ and $a, b, c, d = 1, 2, \ldots,$

$$\mu_{r,s,t,t+1:n}^{(a,b,c,d)}$$
$$= \frac{\nu}{(t-s)\nu + c}\left[(t-s)\mu_{r,s,t+1:n}^{(a,b,c+d)} + r\left\{\mu_{r+1,s+1,t+1:n}^{(a,b,c+d)} - \mu_{r+1,s+1,t,t+1:n}^{(a,b,c,d)}\right\}\right.$$
$$+ (s-r)\left\{\mu_{r,s+1,t+1:n}^{(a,b,c+d)} - \mu_{r,s+1,t,t+1:n}^{(a,b,c,d)}\right\}$$
$$\left. + nQ_2\left\{\mu_{r,s,t:n-1}^{(a,b,c+d)} - \mu_{r,s-1,t-1,t:n-1}^{(a,b,c,d)}\right\}\right]; \quad (15.124)$$

for $1 \leq r < s < u \leq n$, $s - r \geq 2$, $u - r \geq 3$ and $a, b, c, d = 1, 2, \ldots,$

$$\mu_{r,s,s+1,u:n}^{(a,b,c,d)} = \frac{\nu}{\nu + c} \left[\mu_{r,s,s+2,u:n}^{(a,b,c,d)} + r \left\{ \mu_{r+1,s+1,s+2,u:n}^{(a,b,c,d)} - \mu_{r+1,s+1,u:n}^{(a,b+c,d)} \right\} \right.$$
$$+ (s - r) \left\{ \mu_{r,s+1,s+2,u:n}^{(a,b,c,d)} - \mu_{r,s+1,u:n}^{(a,b+c,d)} \right\}$$
$$\left. + nQ_2 \left\{ \mu_{r,s,s+1,u-1:n-1}^{(a,b,c,d)} - \mu_{r,s,u-1:n-1}^{(a,b+c,d)} \right\} \right]; \qquad (15.125)$$

and for $1 \leq r < s < t < u \leq n$, $s - r \geq 2$, $t - s \geq 2$, $u - t \geq 2$ and $a, b, c, d = 1, 2, \ldots,$

$$\mu_{r,s,t,u:n}^{(a,b,c,d)}$$
$$= \frac{\nu}{(t-s)\nu + c} \left[(t-s)\mu_{r,s,t+1,u:n}^{(a,b,c,d)} + r \left\{ \mu_{r+1,s+1,t+1,u:n}^{(a,b,c,d)} - \mu_{r+1,s+1,t,u:n}^{(a,b,c,d)} \right\} \right.$$
$$+ (s-r) \left\{ \mu_{r,s+1,t+1,u:n}^{(a,b,c,d)} - \mu_{r,s+1,t,u:n}^{(a,b,c,d)} \right\}$$
$$\left. + nQ_2 \left\{ \mu_{r,s,t,u-1:n-1}^{(a,b,c,d)} - \mu_{r,s,t-1,u-1:n-1}^{(a,b,c,d)} \right\} \right]. \qquad (15.126)$$

Theorem 15.7.10 For $1 \leq r < s \leq n - 2$, and $a, b, c, d = 1, 2, \ldots,$

$$\mu_{r,s,n-1,n:n}^{(a,b,c,d)} = \frac{\nu}{\nu + d} \left[\mu_{r,s,n-1:n}^{(a,b,c+d)} + nP_2 \left\{ P_1^d \mu_{r,s,n-1:n-1}^{(a,b,c)} - \mu_{r,s,n-1:n-1}^{(a,b,c+d)} \right\} \right]; \qquad (15.127)$$

for $1 \leq r < s < t \leq n - 2$, and $a, b, c, d = 1, 2, \ldots,$

$$\mu_{r,s,t,t+1:n}^{(a,b,c,d)}$$
$$= \frac{\nu}{(n-t)\nu + d} \left[(n-t)\mu_{r,s,t:n}^{(a,b,c+d)} + nP_2 \left\{ \mu_{r,s,t,t+1:n-1}^{(a,b,c,d)} - \mu_{r,s,t:n-1}^{(a,b,c+d)} \right\} \right]; \qquad (15.128)$$

for $1 \leq r < s < t < u \leq n - 1$, $u - t \geq 2$ and $a, b, c, d = 1, 2, \ldots,$

$$\mu_{r,s,t,u:n}^{(a,b,c,d)} = \frac{\nu}{(n-u+1)\nu + d} \left[(n-u+1)\mu_{r,s,t,u-1:n}^{(a,b,c,d)} \right.$$
$$\left. + nP_2 \left\{ \mu_{r,s,t,u:n-1}^{(a,b,c,d)} - \mu_{r,s,t,u-1:n-1}^{(a,b,c,d)} \right\} \right]; \qquad (15.129)$$

and for $1 \leq r < s < t \leq n - 2$ and $a, b, c, d = 1, 2, \ldots,$

$$\mu_{r,s,t,n:n}^{(a,b,c,d)} = \frac{\nu}{\nu + d} \left[\mu_{r,s,t,n-1:n}^{(a,b,c,d)} + nP_2 \left\{ P_1^d \mu_{r,s,t:n-1}^{(a,b,c)} - \mu_{r,s,t,n-1:n-1}^{(a,b,c,d)} \right\} \right]. \qquad (15.130)$$

Acknowledgments. The authors thank the Natural Sciences and Engineering Research Council of Canada for funding this research.

15.8 References

1. Arnold, B. C., Balakrishnan, N. and Nagaraja, H. N. (1992). *A First Course in Order Statistics*, New York: John Wiley & Sons.

2. Balakrishnan, N. and Cohen, A. C. (1991). *Order Statistics and Inference: Estimation Methods*, San Diego: Academic Press.

3. Balakrishnan, N. and Gupta, S. S. (1996). Higher order moments of order statistics from exponential and right-truncated exponential distributions and applications to life-testing problems, In *Handbook of Statistics-16: Order Statistics and Their Applications* (Eds., C. R. Rao and N. Balakrishnan), Amsterdam: North-Holland (to appear).

4. Balakrishnan, N. and Joshi, P. C. (1981). Moments of order statistics from doubly truncated power function distribution, *Aligarh Journal of Statistics*, **1**, 98–105.

5. Balakrishnan, N., Childs, A., Govindarajulu, Z. and Chandramouleeswaran, M. P. (1996). Inference on parameters of the Laplace distribution based on Type-II censored samples using Edgeworth approximation, *submitted for publication*.

6. Barton, D. E. and Dennis, K. E. R. (1952). The conditions under which Gram-Charlier and Edgeworth curves are positive definite and unimodal, *Biometrika*, **39**, 425–427.

7. Childs, A. and Balakrishnan, N. (1996). Generalized recurrence relations for moments of order statistics from non-identical Pareto and truncated Pareto random variables with applications to robustness, In *Handbook of Statistics-16: Order Statistics and Their Applications* (Eds., C. R. Rao and N. Balakrishnan), Amsterdam: North-Holland (to appear).

8. Childs, A., Sultan, K. S. and Balakrishnan, N. (1996). Higher order momentss of order statistics from the Pareto distribution and Edgeworth approximate inference, *submitted for publication*.

9. David, H. A. (1981). *Order Statistics*, Second edition, New York: John Wiley & Sons.

10. Johnson, N. L., Kotz, S. and Balakrishnan, N. (1994). *Continuous Univariate Distributions*, Vol. 1, Second edition, New York: John Wiley & Sons.

11. Malik, H. J. (1967). Exact moments of order statistics from a power function distribution, *Skandinavisk Aktuarietidskrift*, 64–69.

Table 15.1: Mean, variance and coefficients of skewness and kurtosis of $R1^*$ and $R2^*$ when $v = 3/2$.

		R_1^*						R_2^*			
n	s	Mean	Variance	$\sqrt{\beta_1}$	β_2	n	s	Mean	Variance	$\sqrt{\beta_1}$	β_2
5	0	0.0000	0.0566	0.5086	2.7528	5	0	1.0000	0.0867	-0.2641	2.4795
	1	0.0000	0.0625	0.3814	2.6630		1	1.0000	0.1445	0.0415	2.3832
	2	0.0000	0.0738	0.1521	2.6080		2	1.0000	0.2633	0.4028	2.5738
6	0	0.0000	0.0432	0.5975	2.9299	6	0	1.0000	0.0634	-0.3688	2.6284
	1	0.0000	0.0463	0.5154	2.8469		1	1.0000	0.0977	-0.1169	2.4645
	2	0.0000	0.0512	0.3865	2.7540		2	1.0000	0.1560	0.1453	2.4622
7	0	0.0000	0.0345	0.6608	3.0775	7	0	1.0000	0.0489	-0.4455	2.7617
	1	0.0000	0.0363	0.6036	3.0069		1	1.0000	0.0715	-0.2266	2.5675
	2	0.0000	0.0389	0.5207	2.9217		2	1.0000	0.1060	-0.0145	2.4921
	3	0.0000	0.0430	0.3905	2.8264		3	1.0000	0.1648	0.2209	2.5417
8	0	0.0000	0.0284	0.7083	3.2008	8	0	1.0000	0.0392	-0.5043	2.8784
	1	0.0000	0.0295	0.6662	3.1416		1	1.0000	0.0552	-0.3080	2.6694
	2	0.0000	0.0311	0.6085	3.0695		2	1.0000	0.0779	-0.1260	2.5597
	3	0.0000	0.0333	0.5250	2.9826		3	1.0000	0.1126	0.0613	2.5334
9	0	0.0000	0.0239	0.7452	3.3047	9	0	1.0000	0.0324	-0.5509	2.9802
	1	0.0000	0.0247	0.7130	3.2550		1	1.0000	0.0442	-0.3710	2.7639
	2	0.0000	0.0257	0.6706	3.1948		2	1.0000	0.0602	-0.2091	2.6361
	3	0.0000	0.0271	0.6125	3.1215		3	1.0000	0.0830	-0.0503	2.5738
	4	0.0000	0.0290	0.5284	3.0331		4	1.0000	0.1179	0.1200	2.5779
10	0	0.0000	0.0205	0.7746	3.3931	10	0	1.0000	0.0273	-0.5888	3.0693
	1	0.0000	0.0211	0.7493	3.3511		1	1.0000	0.0364	-0.4215	2.8495
	2	0.0000	0.0218	0.7169	3.3007		2	1.0000	0.0483	-0.2739	2.7114
	3	0.0000	0.0227	0.6742	3.2397		3	1.0000	0.0644	-0.1340	2.6298
	4	0.0000	0.0239	0.6158	3.1653		4	1.0000	0.0873	0.0088	2.5972
15	0	0.0000	0.0115	0.8617	3.6882	15	0	1.0000	0.0143	-0.7063	3.3829
	1	0.0000	0.0116	0.8516	3.6676		1	1.0000	0.0178	-0.5749	3.1663
	2	0.0000	0.0118	0.8398	3.6441		2	1.0000	0.0219	-0.4638	3.0142
	3	0.0000	0.0120	0.8258	3.6170		3	1.0000	0.0268	-0.3657	2.9042
	4	0.0000	0.0123	0.8088	3.5855		4	1.0000	0.0327	-0.2755	2.8237
	5	0.0000	0.0125	0.7882	3.5484		5	1.0000	0.0400	-0.1896	2.7657
	6	0.0000	0.0129	0.7623	3.5045		6	1.0000	0.0493	-0.1047	2.7265
	7	0.0000	0.0133	0.7294	3.4518		7	1.0000	0.0613	-0.0175	2.7051
20	0	0.0000	0.0076	0.9034	3.8527	20	0	1.0000	0.0092	-0.7672	3.5702
	1	0.0000	0.0077	0.8983	3.8410		1	1.0000	0.0110	-0.6539	3.3644
	2	0.0000	0.0077	0.8926	3.8280		2	1.0000	0.0130	-0.5589	3.2136
	3	0.0000	0.0078	0.8860	3.8135		3	1.0000	0.0153	-0.4765	3.0997
	4	0.0000	0.0079	0.8784	3.7971		4	1.0000	0.0179	-0.4030	3.0118
	5	0.0000	0.0080	0.8697	3.7786		5	1.0000	0.0209	-0.3358	2.9427
	6	0.0000	0.0081	0.8596	3.7576		6	1.0000	0.0244	-0.2729	2.8882
	7	0.0000	0.0082	0.8476	3.7335		7	1.0000	0.0286	-0.2127	2.8452
	8	0.0000	0.0084	0.8335	3.7057		8	1.0000	0.0335	-0.1538	2.8120
	9	0.0000	0.0086	0.8164	3.6734		9	1.0000	0.0394	-0.0950	2.7874

Table 15.2: Mean, variance and coefficients of skewness and kurtosis of $R1^*$ and $R2^*$ when $v = 3$.

		R_1^*						R_2^*			
n	s	Mean	Variance	$\sqrt{\beta_1}$	β_2	n	s	Mean	Variance	$\sqrt{\beta_1}$	β_2
5	0	0.0000	0.1152	-0.1523	2.5686	5	0	1.0000	0.1349	0.1962	2.5818
	1	0.0000	0.1358	-0.2579	2.6016		1	1.0000	0.1913	0.3471	2.6519
	2	0.0000	0.1740	-0.4484	2.7506		2	1.0000	0.3070	0.5830	2.8834
6	0	0.0000	0.0917	-0.0853	2.5940	6	0	1.0000	0.1049	0.1223	2.6004
	1	0.0000	0.1036	-0.1529	2.5946		1	1.0000	0.1381	0.2280	2.6213
	2	0.0000	0.1222	-0.2589	2.6279		2	1.0000	0.1946	0.3736	2.6982
7	0	0.0000	0.0760	-0.0387	2.6238	7	0	1.0000	0.0854	0.0705	2.6264
	1	0.0000	0.0837	-0.0856	2.6137		1	1.0000	0.1072	0.1501	2.6280
	2	0.0000	0.0946	-0.1533	2.6143		2	1.0000	0.1404	0.2518	2.6537
	3	0.0000	0.1116	-0.2597	2.6479		3	1.0000	0.1971	0.3934	2.7350
8	0	0.0000	0.0647	-0.0046	2.6527	8	0	1.0000	0.0718	0.0323	2.6532
	1	0.0000	0.0701	-0.0388	2.6394		1	1.0000	0.0872	0.0951	2.6465
	2	0.0000	0.0773	-0.0857	2.6293		2	1.0000	0.1090	0.1716	2.6515
	3	0.0000	0.0873	-0.1537	2.6299		3	1.0000	0.1423	0.2703	2.6804
9	0	0.0000	0.0563	0.0215	2.6792	9	0	1.0000	0.0618	0.0030	2.6785
	1	0.0000	0.0603	-0.0046	2.6654		1	1.0000	0.0732	0.0542	2.6682
	2	0.0000	0.0653	-0.0389	2.6520		2	1.0000	0.0886	0.1146	2.6641
	3	0.0000	0.0719	-0.0859	2.6419		3	1.0000	0.1104	0.1887	2.6716
	4	0.0000	0.0813	-0.1539	2.6425		4	1.0000	0.1437	0.2851	2.7027
10	0	0.0000	0.0498	0.0420	2.7031	10	0	1.0000	0.0542	-0.0201	2.7016
	1	0.0000	0.0528	0.0215	2.6897		1	1.0000	0.0630	0.0228	2.6899
	2	0.0000	0.0565	-0.0046	2.6758		2	1.0000	0.0743	0.0721	2.6816
	3	0.0000	0.0612	-0.0389	2.6624		3	1.0000	0.0897	0.1305	2.6795
	4	0.0000	0.0674	-0.0860	2.6522		4	1.0000	0.1116	0.2027	2.6888
15	0	0.0000	0.0313	0.1003	2.7898	15	0	1.0000	0.0331	-0.0863	2.7873
	1	0.0000	0.0324	0.0924	2.7810		1	1.0000	0.0365	-0.0642	2.7769
	2	0.0000	0.0336	0.0831	2.7714		2	1.0000	0.0404	-0.0403	2.7668
	3	0.0000	0.0351	0.0720	2.7608		3	1.0000	0.0450	-0.0143	2.7571
	4	0.0000	0.0367	0.0586	2.7492		4	1.0000	0.0507	0.0144	2.7481
	5	0.0000	0.0387	0.0422	2.7366		5	1.0000	0.0576	0.0467	2.7401
	6	0.0000	0.0410	0.0216	2.7231		6	1.0000	0.0664	0.0837	2.7339
	7	0.0000	0.0439	-0.0046	2.7089		7	1.0000	0.0778	0.1271	2.7309
20	0	0.0000	0.0226	0.1265	2.8422	20	0	1.0000	0.0237	-0.1164	2.8398
	1	0.0000	0.0232	0.1226	2.8364		1	1.0000	0.0254	-0.1024	2.8321
	2	0.0000	0.0238	0.1182	2.8302		2	1.0000	0.0273	-0.0876	2.8244
	3	0.0000	0.0245	0.1131	2.8234		3	1.0000	0.0295	-0.0720	2.8167
	4	0.0000	0.0252	0.1073	2.8161		4	1.0000	0.0320	-0.0555	2.8090
	5	0.0000	0.0260	0.1005	2.8080		5	1.0000	0.0349	-0.0377	2.8013
	6	0.0000	0.0270	0.0926	2.7991		6	1.0000	0.0382	-0.0186	2.7936
	7	0.0000	0.0280	0.0833	2.7894		7	1.0000	0.0421	0.0024	2.7860
	8	0.0000	0.0292	0.0722	2.7788		8	1.0000	0.0468	0.0255	2.7788
	9	0.0000	0.0306	0.0587	2.7671		9	1.0000	0.0524	0.0514	2.7720

Table 15.3: Mean, variance and coefficients of skewness and kurtosis of $R1^*$ and $R2^*$ when $v = 6$.

		R_1^*						R_2^*			
n	s	Mean	Variance	$\sqrt{\beta_1}$	β_2	n	s	Mean	Variance	$\sqrt{\beta_1}$	β_2
5	0	0.0000	0.1685	-0.5431	3.1248	5	0	1.0000	0.1810	0.5518	3.1346
	1	0.0000	0.2092	-0.6543	3.2672		1	1.0000	0.2451	0.6731	3.2930
	2	0.0000	0.2861	-0.8480	3.6029		2	1.0000	0.3748	0.8784	3.6571
6	0	0.0000	0.1353	-0.4702	3.0597	6	0	1.0000	0.1437	0.4771	3.0666
	1	0.0000	0.1596	-0.5437	3.1335		1	1.0000	0.1818	0.5586	3.1507
	2	0.0000	0.1982	-0.6550	3.2763		2	1.0000	0.2460	0.6791	3.3100
7	0	0.0000	0.1130	-0.4178	3.0233	7	0	1.0000	0.1190	0.4235	3.0284
	1	0.0000	0.1292	-0.4705	3.0661		1	1.0000	0.1443	0.4827	3.0784
	2	0.0000	0.1524	-0.5441	3.1400		2	1.0000	0.1825	0.5636	3.1630
	3	0.0000	0.1892	-0.6555	3.2831		3	1.0000	0.2467	0.6835	3.3229
8	0	0.0000	0.0970	-0.3780	3.0014	8	0	1.0000	0.1015	0.3828	3.0053
	1	0.0000	0.1085	-0.4180	3.0281		1	1.0000	0.1195	0.4282	3.0374
	2	0.0000	0.1240	-0.4708	3.0710		2	1.0000	0.1448	0.4871	3.0877
	3	0.0000	0.1463	-0.5445	3.1450		3	1.0000	0.1830	0.5676	3.1726
9	0	0.0000	0.0850	-0.3466	2.9875	9	0	1.0000	0.0885	0.3507	2.9907
	1	0.0000	0.0935	-0.3782	3.0052		1	1.0000	0.1019	0.3869	3.0124
	2	0.0000	0.1046	-0.4182	3.0320		2	1.0000	0.1198	0.4320	3.0447
	3	0.0000	0.1196	-0.4711	3.0749		3	1.0000	0.1451	0.4905	3.0952
	4	0.0000	0.1411	-0.5447	3.1490		4	1.0000	0.1833	0.5707	3.1804
10	0	0.0000	0.0756	-0.3211	2.9786	10	0	1.0000	0.0784	0.3247	2.9811
	1	0.0000	0.0822	-0.3467	2.9907		1	1.0000	0.0888	0.3543	2.9964
	2	0.0000	0.0905	-0.3783	3.0083		2	1.0000	0.1022	0.3902	3.0183
	3	0.0000	0.1012	-0.4184	3.0351		3	1.0000	0.1201	0.4351	3.0508
	4	0.0000	0.1157	-0.4712	3.0781		4	1.0000	0.1454	0.4934	3.1015
15	0	0.0000	0.0486	-0.2413	2.9632	15	0	1.0000	0.0498	0.2434	2.9644
	1	0.0000	0.0512	-0.2531	2.9657		1	1.0000	0.0539	0.2574	2.9682
	2	0.0000	0.0542	-0.2666	2.9691		2	1.0000	0.0586	0.2733	2.9731
	3	0.0000	0.0576	-0.2821	2.9735		3	1.0000	0.0642	0.2913	2.9795
	4	0.0000	0.0617	-0.3002	2.9797		4	1.0000	0.0710	0.3121	2.9880
	5	0.0000	0.0665	-0.3215	2.9882		5	1.0000	0.0793	0.3364	2.9994
	6	0.0000	0.0723	-0.3471	3.0003		6	1.0000	0.0897	0.3654	3.0150
	7	0.0000	0.0796	-0.3788	3.0181		7	1.0000	0.1031	0.4006	3.0373
20	0	0.0000	0.0358	-0.1983	2.9626	20	0	1.0000	0.0364	0.1997	2.9633
	1	0.0000	0.0372	-0.2053	2.9632		1	1.0000	0.0386	0.2082	2.9645
	2	0.0000	0.0387	-0.2130	2.9639		2	1.0000	0.0410	0.2174	2.9661
	3	0.0000	0.0403	-0.2215	2.9650		3	1.0000	0.0437	0.2276	2.9681
	4	0.0000	0.0422	-0.2309	2.9664		4	1.0000	0.0467	0.2387	2.9706
	5	0.0000	0.0443	-0.2414	2.9682		5	1.0000	0.0503	0.2510	2.9737
	6	0.0000	0.0467	-0.2533	2.9707		6	1.0000	0.0543	0.2648	2.9776
	7	0.0000	0.0494	-0.2668	2.9741		7	1.0000	0.0591	0.2804	2.9827
	8	0.0000	0.0525	-0.2823	2.9786		8	1.0000	0.0647	0.2981	2.9892
	9	0.0000	0.0562	-0.3003	2.9847		9	1.0000	0.0714	0.3186	2.9978

Table 15.4: Percentage points of the distribution of R_1 when $v = 3/2$.

n	s	Edgeworth								Simulated							
		1%	2.5%	5%	10%	90%	95%	97.5%	99%	1%	2.5%	5%	10%	90%	95%	97.5%	99%
5	0	-1.90	-1.71	-1.52	-1.26	1.37	1.77	2.12	2.55	-1.72	-1.60	-1.45	-1.23	1.39	1.78	2.13	2.52
	1	-1.98	-1.77	-1.55	-1.27	1.35	1.74	2.08	2.47	-1.84	-1.67	-1.49	-1.25	1.41	1.83	2.16	2.52
	2	-2.14	-1.87	-1.61	-1.29	1.33	1.69	2.00	2.34	-2.07	-1.83	-1.59	-1.28	1.35	1.72	2.05	2.35
6	0	-1.84	-1.68	-1.49	-1.24	1.37	1.78	2.16	2.64	-1.69	-1.55	-1.38	-1.19	1.38	1.89	2.23	2.66
	1	-1.90	-1.71	-1.51	-1.25	1.36	1.77	2.13	2.58	-1.70	-1.54	-1.39	-1.17	1.43	1.83	2.17	2.57
	2	-1.98	-1.77	-1.55	-1.26	1.35	1.74	2.09	2.50	-1.86	-1.65	-1.47	-1.23	1.44	1.87	2.22	2.57
7	0	-1.81	-1.65	-1.47	-1.23	1.37	1.79	2.19	2.71	-1.65	-1.51	-1.39	-1.20	1.40	1.86	2.22	2.75
	1	-1.84	-1.67	-1.49	-1.24	1.36	1.79	2.17	2.67	-1.69	-1.55	-1.40	-1.19	1.40	1.85	2.22	2.70
	2	-1.90	-1.71	-1.51	-1.25	1.36	1.77	2.14	2.60	-1.81	-1.61	-1.45	-1.23	1.33	1.81	2.19	2.59
	3	-1.99	-1.77	-1.54	-1.26	1.34	1.75	2.10	2.52	-1.97	-1.74	-1.52	-1.22	1.35	1.82	2.15	2.55
8	0	-1.78	-1.63	-1.46	-1.22	1.36	1.80	2.22	2.77	-1.64	-1.51	-1.37	-1.19	1.38	1.88	2.31	2.78
	1	-1.81	-1.65	-1.47	-1.23	1.36	1.80	2.20	2.74	-1.66	-1.52	-1.39	-1.20	1.38	1.85	2.23	2.72
	2	-1.84	-1.67	-1.48	-1.23	1.36	1.79	2.18	2.69	-1.69	-1.54	-1.40	-1.18	1.37	1.84	2.27	2.62
	3	-1.90	-1.71	-1.51	-1.24	1.35	1.77	2.15	2.62	-1.83	-1.61	-1.45	-1.20	1.40	1.85	2.19	2.63
9	0	-1.76	-1.61	-1.45	-1.22	1.36	1.81	2.24	2.82	-1.65	-1.49	-1.34	-1.15	1.35	1.86	2.30	2.72
	1	-1.78	-1.63	-1.46	-1.22	1.36	1.80	2.23	2.79	-1.63	-1.51	-1.37	-1.18	1.39	1.87	2.35	2.89
	2	-1.80	-1.64	-1.47	-1.23	1.36	1.80	2.21	2.75	-1.69	-1.54	-1.39	-1.20	1.40	1.90	2.31	2.76
	3	-1.84	-1.67	-1.48	-1.23	1.36	1.79	2.19	2.70	-1.76	-1.58	-1.41	-1.19	1.38	1.84	2.20	2.65
	4	-1.90	-1.70	-1.50	-1.24	1.35	1.77	2.16	2.64	-1.82	-1.63	-1.44	-1.20	1.35	1.79	2.22	2.67
10	0	-1.74	-1.60	-1.44	-1.21	1.36	1.81	2.26	2.86	-1.60	-1.46	-1.34	-1.16	1.38	1.92	2.34	2.89
	1	-1.76	-1.61	-1.45	-1.22	1.36	1.81	2.25	2.83	-1.61	-1.47	-1.34	-1.15	1.40	1.86	2.36	2.85
	2	-1.78	-1.62	-1.45	-1.22	1.36	1.81	2.23	2.80	-1.66	-1.52	-1.35	-1.16	1.34	1.82	2.21	2.65
	3	-1.80	-1.64	-1.46	-1.22	1.36	1.80	2.22	2.77	-1.71	-1.54	-1.39	-1.20	1.43	1.88	2.32	2.82
	4	-1.84	-1.67	-1.48	-1.23	1.35	1.79	2.19	2.72	-1.78	-1.60	-1.41	-1.19	1.41	1.86	2.24	2.76
15	0	-1.69	-1.56	-1.41	-1.20	1.35	1.83	2.32	2.97	-1.58	-1.46	-1.33	-1.15	1.39	1.94	2.38	2.93
	1	-1.70	-1.57	-1.41	-1.20	1.35	1.83	2.32	2.96	-1.56	-1.44	-1.31	-1.13	1.35	1.85	2.32	2.85
	2	-1.71	-1.57	-1.42	-1.20	1.35	1.83	2.31	2.95	-1.57	-1.46	-1.31	-1.14	1.41	1.96	2.44	2.98
	3	-1.71	-1.58	-1.42	-1.20	1.35	1.82	2.30	2.94	-1.58	-1.48	-1.35	-1.16	1.39	1.90	2.37	2.85
	4	-1.72	-1.58	-1.42	-1.20	1.35	1.82	2.30	2.92	-1.62	-1.49	-1.35	-1.17	1.36	1.84	2.27	2.93
	5	-1.74	-1.59	-1.43	-1.20	1.35	1.82	2.29	2.90	-1.68	-1.53	-1.37	-1.17	1.35	1.84	2.28	2.76
	6	-1.75	-1.60	-1.44	-1.21	1.35	1.82	2.27	2.88	-1.64	-1.49	-1.32	-1.14	1.38	1.87	2.28	2.82
	7	-1.77	-1.62	-1.44	-1.21	1.35	1.81	2.26	2.85	-1.72	-1.55	-1.38	-1.17	1.37	1.82	2.20	2.67
20	0	-1.67	-1.55	-1.40	-1.19	1.35	1.84	2.36	3.03	-1.56	-1.41	-1.28	-1.11	1.37	1.88	2.31	2.86
	1	-1.67	-1.55	-1.40	-1.19	1.35	1.84	2.36	3.02	-1.57	-1.43	-1.30	-1.11	1.37	1.91	2.35	2.91
	2	-1.68	-1.55	-1.40	-1.19	1.35	1.84	2.35	3.02	-1.60	-1.46	-1.31	-1.14	1.31	1.84	2.34	2.93
	3	-1.68	-1.55	-1.40	-1.19	1.35	1.83	2.35	3.01	-1.60	-1.47	-1.31	-1.13	1.42	1.95	2.39	2.97
	4	-1.68	-1.56	-1.40	-1.19	1.35	1.83	2.35	3.00	-1.57	-1.44	-1.30	-1.11	1.35	1.83	2.30	2.82
	5	-1.69	-1.56	-1.41	-1.19	1.35	1.83	2.34	2.99	-1.60	-1.46	-1.33	-1.14	1.37	1.93	2.40	2.98
	6	-1.69	-1.56	-1.41	-1.19	1.35	1.83	2.34	2.99	-1.62	-1.45	-1.32	-1.14	1.33	1.90	2.41	2.98
	7	-1.70	-1.57	-1.41	-1.19	1.35	1.83	2.33	2.97	-1.62	-1.47	-1.33	-1.13	1.37	1.87	2.38	2.85
	8	-1.71	-1.57	-1.42	-1.19	1.35	1.83	2.32	2.96	-1.62	-1.47	-1.35	-1.15	1.34	1.85	2.34	2.83
	9	-1.72	-1.58	-1.42	-1.20	1.35	1.83	2.31	2.95	-1.66	-1.49	-1.33	-1.17	1.41	1.87	2.31	2.82

Table 15.5: Percentage points of the distribution of R_1 when $v = 3$.

		Edgeworth								Simulated							
n	s	1%	2.5%	5%	10%	90%	95%	97.5%	99%	1%	2.5%	5%	10%	90%	95%	97.5%	99%
5	0	-2.33	-1.99	-1.69	-1.33	1.30	1.61	1.87	2.13	-2.35	-2.01	-1.74	-1.36	1.28	1.57	1.78	2.03
	1	-2.39	-2.03	-1.72	-1.34	1.28	1.58	1.82	2.06	-2.36	-2.03	-1.78	-1.35	1.29	1.58	1.77	1.99
	2	-2.52	-2.10	-1.76	-1.36	1.26	1.53	1.74	1.94	-2.50	-2.14	-1.78	-1.37	1.25	1.48	1.66	1.81
6	0	-2.30	-1.97	-1.67	-1.32	1.30	1.63	1.90	2.18	-2.30	-1.97	-1.65	-1.32	1.28	1.62	1.88	2.13
	1	-2.34	-2.00	-1.69	-1.33	1.29	1.61	1.87	2.14	-2.26	-1.94	-1.61	-1.26	1.31	1.59	1.83	2.07
	2	-2.40	-2.04	-1.72	-1.34	1.28	1.58	1.82	2.07	-2.35	-1.97	-1.71	-1.35	1.32	1.60	1.81	2.01
7	0	-2.27	-1.95	-1.66	-1.31	1.30	1.64	1.92	2.22	-2.27	-1.97	-1.70	-1.32	1.31	1.63	1.89	2.17
	1	-2.30	-1.97	-1.67	-1.32	1.30	1.63	1.90	2.19	-2.24	-1.93	-1.70	-1.32	1.34	1.62	1.86	2.14
	2	-2.34	-2.00	-1.69	-1.33	1.29	1.61	1.87	2.14	-2.37	-2.02	-1.72	-1.35	1.27	1.57	1.82	2.04
	3	-2.41	-2.04	-1.72	-1.34	1.28	1.58	1.82	2.07	-2.44	-2.09	-1.78	-1.35	1.27	1.55	1.77	2.00
8	0	-2.26	-1.94	-1.65	-1.31	1.31	1.65	1.94	2.25	-2.24	-1.97	-1.66	-1.35	1.28	1.65	1.94	2.24
	1	-2.28	-1.95	-1.66	-1.31	1.30	1.64	1.92	2.22	-2.29	-1.97	-1.68	-1.32	1.28	1.61	1.88	2.17
	2	-2.30	-1.97	-1.67	-1.32	1.30	1.63	1.90	2.19	-2.24	-1.91	-1.64	-1.32	1.29	1.62	1.88	2.15
	3	-2.34	-2.00	-1.69	-1.33	1.29	1.61	1.87	2.14	-2.35	-2.03	-1.72	-1.32	1.30	1.62	1.84	2.10
9	0	-2.24	-1.93	-1.65	-1.30	1.31	1.66	1.95	2.27	-2.35	-1.98	-1.66	-1.29	1.27	1.63	1.91	2.26
	1	-2.26	-1.94	-1.65	-1.31	1.30	1.65	1.94	2.25	-2.22	-1.90	-1.65	-1.34	1.30	1.64	1.97	2.26
	2	-2.28	-1.95	-1.66	-1.31	1.30	1.64	1.92	2.23	-2.24	-1.94	-1.67	-1.33	1.33	1.64	1.91	2.23
	3	-2.31	-1.97	-1.67	-1.32	1.30	1.63	1.90	2.19	-2.22	-1.90	-1.65	-1.32	1.29	1.60	1.85	2.18
	4	-2.35	-2.00	-1.69	-1.32	1.29	1.61	1.87	2.14	-2.33	-2.00	-1.66	-1.31	1.27	1.58	1.84	2.08
10	0	-2.23	-1.92	-1.64	-1.30	1.31	1.66	1.96	2.29	-2.20	-1.90	-1.62	-1.29	1.30	1.68	1.99	2.30
	1	-2.25	-1.93	-1.64	-1.30	1.31	1.66	1.95	2.27	-2.21	-1.88	-1.61	-1.29	1.33	1.67	1.97	2.30
	2	-2.26	-1.94	-1.65	-1.30	1.30	1.65	1.94	2.25	-2.22	-1.93	-1.64	-1.31	1.28	1.60	1.87	2.13
	3	-2.28	-1.96	-1.66	-1.31	1.30	1.64	1.92	2.23	-2.30	-1.98	-1.67	-1.35	1.33	1.65	1.92	2.23
	4	-2.31	-1.97	-1.67	-1.32	1.30	1.63	1.90	2.19	-2.32	-1.98	-1.68	-1.32	1.32	1.63	1.90	2.15
15	0	-2.21	-1.90	-1.62	-1.29	1.31	1.68	1.99	2.35	-2.21	-1.88	-1.63	-1.30	1.31	1.70	2.00	2.37
	1	-2.21	-1.90	-1.62	-1.29	1.31	1.67	1.99	2.34	-2.15	-1.84	-1.61	-1.30	1.31	1.66	1.99	2.35
	2	-2.22	-1.91	-1.63	-1.29	1.31	1.67	1.98	2.33	-2.21	-1.95	-1.64	-1.30	1.31	1.66	1.98	2.32
	3	-2.22	-1.91	-1.63	-1.29	1.31	1.67	1.98	2.32	-2.19	-1.89	-1.62	-1.32	1.31	1.68	1.99	2.32
	4	-2.23	-1.92	-1.63	-1.29	1.31	1.67	1.97	2.31	-2.16	-1.87	-1.63	-1.28	1.31	1.65	1.97	2.27
	5	-2.24	-1.92	-1.64	-1.30	1.30	1.66	1.96	2.30	-2.21	-1.88	-1.64	-1.31	1.30	1.69	2.01	2.29
	6	-2.25	-1.93	-1.64	-1.30	1.30	1.66	1.95	2.28	-2.22	-1.94	-1.64	-1.34	1.28	1.65	1.98	2.29
	7	-2.27	-1.94	-1.65	-1.30	1.30	1.65	1.94	2.26	-2.26	-1.97	-1.67	-1.33	1.28	1.62	1.89	2.19
20	0	-2.20	-1.89	-1.61	-1.28	1.31	1.68	2.00	2.37	-2.19	-1.83	-1.58	-1.27	1.32	1.70	1.97	2.40
	1	-2.20	-1.89	-1.61	-1.28	1.31	1.68	2.00	2.37	-2.24	-1.91	-1.60	-1.30	1.33	1.71	2.00	2.36
	2	-2.20	-1.89	-1.62	-1.28	1.31	1.68	2.00	2.37	-2.16	-1.83	-1.58	-1.30	1.34	1.74	2.04	2.36
	3	-2.20	-1.90	-1.62	-1.28	1.31	1.68	2.00	2.36	-2.20	-1.88	-1.62	-1.28	1.34	1.74	2.06	2.43
	4	-2.21	-1.90	-1.62	-1.28	1.31	1.68	1.99	2.36	-2.18	-1.88	-1.60	-1.26	1.29	1.67	2.02	2.49
	5	-2.21	-1.90	-1.62	-1.28	1.31	1.68	1.99	2.35	-2.29	-1.96	-1.61	-1.30	1.28	1.64	1.92	2.25
	6	-2.22	-1.90	-1.62	-1.29	1.31	1.67	1.99	2.34	-2.16	-1.84	-1.57	-1.24	1.31	1.67	2.01	2.33
	7	-2.22	-1.91	-1.63	-1.29	1.31	1.67	1.98	2.34	-2.22	-1.93	-1.63	-1.31	1.33	1.69	2.02	2.37
	8	-2.23	-1.91	-1.63	-1.29	1.31	1.67	1.98	2.33	-2.17	-1.87	-1.62	-1.28	1.34	1.66	1.96	2.27
	9	-2.23	-1.92	-1.63	-1.29	1.30	1.67	1.97	2.31	-2.23	-1.91	-1.65	-1.31	1.32	1.66	1.96	2.25

Table 15.6: Percentage points of the distribution of R_1 when $v = 6$.

n	s	Edgeworth									Simulated						
		1%	2.5%	5%	10%	90%	95%	97.5%	99%	1%	2.5%	5%	10%	90%	95%	97.5%	99%
5	0	-2.67	-2.17	-1.78	-1.34	1.23	1.50	1.70	1.89	-2.67	-2.25	-1.86	-1.38	1.19	1.44	1.61	1.79
	1	-2.76	-2.22	-1.80	-1.35	1.22	1.47	1.65	1.82	-2.73	-2.25	-1.85	-1.39	1.20	1.43	1.57	1.73
	2	-2.95	-2.31	-1.83	-1.36	1.20	1.42	1.57	1.70	-2.87	-2.33	-1.88	-1.37	1.16	1.32	1.44	1.54
6	0	-2.62	-2.15	-1.76	-1.34	1.24	1.52	1.73	1.94	-2.66	-2.16	-1.77	-1.35	1.22	1.47	1.70	1.90
	1	-2.67	-2.17	-1.78	-1.34	1.23	1.50	1.70	1.89	-2.60	-2.14	-1.71	-1.28	1.23	1.45	1.64	1.83
	2	-2.77	-2.22	-1.80	-1.35	1.22	1.47	1.65	1.82	-2.65	-2.18	-1.82	-1.36	1.23	1.44	1.60	1.75
7	0	-2.58	-2.13	-1.75	-1.33	1.25	1.53	1.75	1.98	-2.62	-2.20	-1.82	-1.36	1.23	1.49	1.72	1.93
	1	-2.62	-2.15	-1.76	-1.34	1.24	1.52	1.73	1.94	-2.55	-2.19	-1.79	-1.32	1.25	1.49	1.67	1.89
	2	-2.68	-2.17	-1.78	-1.34	1.23	1.50	1.70	1.89	-2.71	-2.23	-1.84	-1.40	1.19	1.43	1.64	1.77
	3	-2.77	-2.22	-1.80	-1.35	1.22	1.47	1.65	1.82	-2.83	-2.30	-1.88	-1.38	1.19	1.40	1.57	1.71
8	0	-2.56	-2.11	-1.74	-1.33	1.25	1.54	1.77	2.01	-2.53	-2.19	-1.83	-1.39	1.22	1.52	1.76	1.98
	1	-2.58	-2.13	-1.75	-1.33	1.25	1.53	1.75	1.98	-2.65	-2.18	-1.80	-1.35	1.23	1.50	1.68	1.94
	2	-2.62	-2.15	-1.76	-1.34	1.24	1.52	1.73	1.94	-2.57	-2.12	-1.76	-1.34	1.22	1.50	1.67	1.91
	3	-2.68	-2.18	-1.78	-1.34	1.23	1.50	1.70	1.89	-2.68	-2.23	-1.85	-1.37	1.22	1.47	1.64	1.84
9	0	-2.54	-2.10	-1.74	-1.33	1.25	1.55	1.79	2.04	-2.70	-2.22	-1.79	-1.33	1.22	1.52	1.74	2.05
	1	-2.56	-2.11	-1.74	-1.33	1.25	1.54	1.77	2.01	-2.52	-2.09	-1.79	-1.38	1.23	1.52	1.79	1.99
	2	-2.59	-2.13	-1.75	-1.33	1.25	1.53	1.75	1.98	-2.53	-2.12	-1.78	-1.36	1.25	1.52	1.72	1.96
	3	-2.62	-2.15	-1.76	-1.34	1.24	1.52	1.73	1.94	-2.49	-2.09	-1.73	-1.36	1.22	1.48	1.67	1.92
	4	-2.68	-2.18	-1.78	-1.34	1.23	1.50	1.70	1.89	-2.66	-2.21	-1.77	-1.31	1.21	1.45	1.65	1.83
10	0	-2.52	-2.09	-1.73	-1.32	1.26	1.56	1.80	2.06	-2.50	-2.07	-1.74	-1.33	1.25	1.56	1.81	2.05
	1	-2.54	-2.10	-1.74	-1.33	1.25	1.55	1.79	2.04	-2.50	-2.08	-1.72	-1.33	1.25	1.57	1.76	2.04
	2	-2.56	-2.11	-1.74	-1.33	1.25	1.54	1.77	2.01	-2.56	-2.12	-1.75	-1.35	1.23	1.49	1.68	1.90
	3	-2.59	-2.13	-1.75	-1.33	1.25	1.53	1.75	1.98	-2.68	-2.22	-1.81	-1.39	1.26	1.51	1.72	1.95
	4	-2.62	-2.15	-1.76	-1.34	1.24	1.51	1.73	1.94	-2.62	-2.16	-1.77	-1.34	1.24	1.50	1.70	1.89
15	0	-2.47	-2.06	-1.71	-1.31	1.26	1.58	1.84	2.12	-2.51	-2.06	-1.72	-1.32	1.26	1.58	1.83	2.12
	1	-2.48	-2.06	-1.71	-1.31	1.26	1.57	1.83	2.12	-2.44	-2.02	-1.72	-1.34	1.25	1.55	1.83	2.15
	2	-2.49	-2.07	-1.72	-1.32	1.26	1.57	1.83	2.10	-2.50	-2.13	-1.75	-1.33	1.26	1.55	1.80	2.12
	3	-2.50	-2.08	-1.72	-1.32	1.26	1.57	1.82	2.09	-2.46	-2.11	-1.75	-1.33	1.25	1.56	1.80	2.05
	4	-2.51	-2.08	-1.73	-1.32	1.26	1.56	1.81	2.08	-2.47	-2.05	-1.74	-1.31	1.25	1.54	1.80	2.04
	5	-2.52	-2.09	-1.73	-1.32	1.25	1.56	1.80	2.06	-2.50	-2.08	-1.74	-1.34	1.25	1.58	1.78	2.04
	6	-2.54	-2.10	-1.74	-1.33	1.25	1.55	1.79	2.04	-2.54	-2.10	-1.78	-1.38	1.23	1.53	1.79	2.04
	7	-2.56	-2.11	-1.74	-1.33	1.25	1.54	1.77	2.01	-2.57	-2.18	-1.79	-1.37	1.24	1.50	1.71	1.93
20	0	-2.45	-2.04	-1.70	-1.31	1.27	1.59	1.86	2.16	-2.46	-1.99	-1.66	-1.29	1.27	1.60	1.83	2.15
	1	-2.45	-2.05	-1.70	-1.31	1.26	1.59	1.86	2.16	-2.49	-2.06	-1.71	-1.33	1.29	1.58	1.82	2.13
	2	-2.46	-2.05	-1.70	-1.31	1.26	1.59	1.85	2.15	-2.39	-2.01	-1.68	-1.32	1.29	1.59	1.89	2.15
	3	-2.46	-2.05	-1.71	-1.31	1.26	1.58	1.85	2.14	-2.44	-2.04	-1.72	-1.31	1.27	1.61	1.91	2.18
	4	-2.47	-2.06	-1.71	-1.31	1.26	1.58	1.84	2.13	-2.48	-2.06	-1.69	-1.30	1.26	1.56	1.83	2.22
	5	-2.47	-2.06	-1.71	-1.31	1.26	1.58	1.84	2.13	-2.62	-2.14	-1.72	-1.34	1.25	1.53	1.79	2.07
	6	-2.48	-2.07	-1.71	-1.31	1.26	1.57	1.83	2.12	-2.42	-2.00	-1.66	-1.27	1.25	1.56	1.81	2.11
	7	-2.49	-2.07	-1.72	-1.32	1.26	1.57	1.83	2.10	-2.55	-2.12	-1.75	-1.33	1.27	1.59	1.83	2.14
	8	-2.50	-2.08	-1.72	-1.32	1.26	1.57	1.82	2.09	-2.46	-2.04	-1.73	-1.29	1.27	1.55	1.78	2.10
	9	-2.51	-2.08	-1.73	-1.32	1.26	1.56	1.81	2.08	-2.54	-2.09	-1.75	-1.35	1.25	1.54	1.79	2.02

Higher Order Moments of Order Statistics

Table 15.7: Percentage points of the distribution of R_2 when $v = 3/2$.

n	s	Edgeworth 1%	2.5%	5%	10%	90%	95%	97.5%	99%	Simulated 1%	2.5%	5%	10%	90%	95%	97.5%	99%
5	0	-2.37	-2.03	-1.72	-1.35	1.29	1.59	1.82	2.05	-2.29	-2.02	-1.74	-1.35	1.30	1.55	1.74	1.90
	1	-2.18	-1.91	-1.65	-1.32	1.33	1.67	1.94	2.23	-2.12	-1.89	-1.65	-1.35	1.33	1.67	1.94	2.15
	2	-1.96	-1.76	-1.55	-1.27	1.36	1.75	2.07	2.46	-1.76	-1.61	-1.48	-1.27	1.35	1.78	2.11	2.48
6	0	-2.46	-2.07	-1.74	-1.35	1.27	1.56	1.77	1.99	-2.45	-2.13	-1.81	-1.39	1.25	1.49	1.69	1.88
	1	-2.29	-1.97	-1.68	-1.33	1.31	1.62	1.88	2.14	-2.25	-1.99	-1.70	-1.37	1.25	1.52	1.75	1.98
	2	-2.12	-1.87	-1.62	-1.30	1.34	1.69	1.98	2.30	-2.09	-1.86	-1.62	-1.33	1.30	1.63	1.91	2.19
7	0	-2.53	-2.10	-1.76	-1.36	1.26	1.53	1.74	1.94	-2.54	-2.16	-1.77	-1.39	1.24	1.47	1.65	1.81
	1	-2.37	-2.02	-1.71	-1.34	1.29	1.59	1.84	2.08	-2.37	-2.04	-1.72	-1.34	1.27	1.55	1.76	1.98
	2	-2.23	-1.94	-1.66	-1.32	1.31	1.65	1.92	2.21	-2.13	-1.86	-1.62	-1.32	1.37	1.66	1.93	2.17
	3	-2.08	-1.84	-1.60	-1.29	1.34	1.71	2.02	2.36	-1.97	-1.79	-1.59	-1.32	1.37	1.73	2.07	2.43
8	0	-2.59	-2.13	-1.77	-1.36	1.25	1.51	1.72	1.91	-2.62	-2.22	-1.81	-1.34	1.25	1.48	1.65	1.79
	1	-2.44	-2.06	-1.73	-1.34	1.28	1.57	1.80	2.03	-2.47	-2.07	-1.72	-1.33	1.29	1.54	1.75	1.92
	2	-2.31	-1.98	-1.69	-1.33	1.30	1.62	1.88	2.15	-2.36	-1.98	-1.68	-1.35	1.30	1.60	1.79	2.02
	3	-2.19	-1.91	-1.64	-1.31	1.32	1.67	1.96	2.27	-2.13	-1.89	-1.62	-1.31	1.33	1.65	1.89	2.22
9	0	-2.64	-2.16	-1.78	-1.36	1.24	1.50	1.70	1.88	-2.59	-2.21	-1.79	-1.30	1.21	1.42	1.61	1.77
	1	-2.49	-2.08	-1.74	-1.35	1.27	1.55	1.77	2.00	-2.52	-2.13	-1.82	-1.40	1.25	1.52	1.71	1.93
	2	-2.38	-2.02	-1.70	-1.33	1.29	1.60	1.84	2.10	-2.39	-2.06	-1.74	-1.36	1.26	1.56	1.79	2.07
	3	-2.27	-1.95	-1.67	-1.32	1.31	1.64	1.91	2.21	-2.38	-1.99	-1.68	-1.35	1.34	1.66	1.94	2.24
	4	-2.16	-1.88	-1.62	-1.30	1.32	1.68	1.98	2.31	-2.07	-1.85	-1.60	-1.30	1.33	1.69	2.03	2.25
10	0	-2.68	-2.18	-1.78	-1.36	1.24	1.49	1.68	1.86	-2.72	-2.24	-1.85	-1.39	1.19	1.42	1.59	1.75
	1	-2.54	-2.11	-1.75	-1.35	1.26	1.54	1.75	1.97	-2.51	-2.17	-1.78	-1.36	1.22	1.48	1.66	1.84
	2	-2.43	-2.05	-1.72	-1.34	1.28	1.58	1.82	2.06	-2.37	-2.03	-1.69	-1.34	1.25	1.55	1.77	2.00
	3	-2.33	-1.99	-1.69	-1.32	1.29	1.62	1.88	2.16	-2.31	-2.00	-1.72	-1.35	1.31	1.60	1.86	2.13
	4	-2.24	-1.93	-1.65	-1.31	1.31	1.65	1.94	2.25	-2.14	-1.92	-1.66	-1.35	1.32	1.66	1.94	2.28
15	0	-2.82	-2.25	-1.81	-1.35	1.21	1.45	1.63	1.78	-2.82	-2.30	-1.88	-1.39	1.20	1.41	1.54	1.67
	1	-2.70	-2.19	-1.78	-1.35	1.23	1.49	1.68	1.87	-2.67	-2.22	-1.84	-1.35	1.18	1.41	1.58	1.79
	2	-2.60	-2.14	-1.76	-1.34	1.24	1.52	1.73	1.95	-2.67	-2.27	-1.83	-1.38	1.23	1.49	1.66	1.84
	3	-2.52	-2.10	-1.74	-1.33	1.26	1.55	1.78	2.01	-2.47	-2.10	-1.72	-1.31	1.26	1.54	1.75	1.98
	4	-2.46	-2.06	-1.72	-1.33	1.27	1.57	1.82	2.08	-2.45	-2.04	-1.71	-1.34	1.30	1.59	1.81	2.03
	5	-2.39	-2.02	-1.70	-1.32	1.28	1.60	1.86	2.14	-2.38	-2.03	-1.70	-1.30	1.29	1.61	1.84	2.08
	6	-2.34	-1.99	-1.68	-1.31	1.29	1.62	1.90	2.19	-2.36	-2.01	-1.68	-1.30	1.27	1.59	1.86	2.17
	7	-2.27	-1.95	-1.66	-1.30	1.30	1.65	1.93	2.25	-2.26	-1.95	-1.66	-1.29	1.31	1.68	1.94	2.24
20	0	-2.90	-2.29	-1.82	-1.35	1.20	1.43	1.60	1.75	-2.75	-2.28	-1.86	-1.35	1.14	1.35	1.50	1.67
	1	-2.79	-2.23	-1.80	-1.34	1.22	1.46	1.65	1.82	-2.79	-2.22	-1.81	-1.34	1.17	1.41	1.57	1.78
	2	-2.70	-2.19	-1.78	-1.34	1.23	1.49	1.69	1.88	-2.68	-2.20	-1.83	-1.35	1.24	1.48	1.66	1.84
	3	-2.63	-2.15	-1.76	-1.34	1.24	1.51	1.73	1.94	-2.61	-2.19	-1.78	-1.35	1.25	1.51	1.70	1.87
	4	-2.57	-2.12	-1.75	-1.33	1.25	1.53	1.76	1.99	-2.45	-2.06	-1.72	-1.32	1.21	1.50	1.70	1.94
	5	-2.52	-2.09	-1.73	-1.33	1.26	1.55	1.79	2.04	-2.56	-2.10	-1.73	-1.31	1.25	1.54	1.76	2.00
	6	-2.47	-2.06	-1.72	-1.32	1.26	1.57	1.82	2.09	-2.50	-2.06	-1.70	-1.29	1.26	1.54	1.84	2.12
	7	-2.43	-2.04	-1.70	-1.32	1.27	1.59	1.85	2.13	-2.44	-2.04	-1.71	-1.29	1.27	1.58	1.89	2.15
	8	-2.38	-2.01	-1.69	-1.31	1.28	1.61	1.88	2.17	-2.40	-2.00	-1.69	-1.29	1.26	1.56	1.83	2.09
	9	-2.34	-1.99	-1.67	-1.31	1.29	1.62	1.90	2.21	-2.27	-1.97	-1.66	-1.30	1.28	1.62	1.94	2.20

Table 15.8: Percentage points of the distribution of R_2 when $v = 3$.

n	s	\multicolumn{7}{c	}{Edgeworth}	\multicolumn{7}{c	}{Simulated}
		1%	2.5%	5%	10%	90%	95%	97.5%	99%	1%	2.5%	5%	10%	90%	95%	97.5%	99%
5	0	-2.11	-1.85	-1.60	-1.29	1.33	1.70	2.01	2.36	-1.96	-1.78	-1.57	-1.26	1.36	1.75	2.04	2.40
	1	-2.00	-1.78	-1.56	-1.27	1.35	1.74	2.07	2.45	-1.90	-1.72	-1.53	-1.27	1.37	1.77	2.08	2.44
	2	-1.85	-1.68	-1.50	-1.25	1.37	1.78	2.15	2.62	-1.65	-1.53	-1.40	-1.24	1.38	1.84	2.22	2.58
6	0	-2.16	-1.88	-1.62	-1.30	1.32	1.68	1.98	2.32	-2.10	-1.86	-1.64	-1.31	1.32	1.66	1.99	2.36
	1	-2.09	-1.84	-1.59	-1.29	1.34	1.71	2.02	2.38	-2.01	-1.78	-1.57	-1.28	1.29	1.65	1.95	2.30
	2	-1.99	-1.77	-1.55	-1.27	1.35	1.74	2.08	2.48	-1.92	-1.74	-1.54	-1.28	1.35	1.73	2.01	2.41
7	0	-2.20	-1.91	-1.63	-1.30	1.32	1.67	1.97	2.29	-2.16	-1.86	-1.61	-1.29	1.34	1.70	2.01	2.29
	1	-2.14	-1.87	-1.61	-1.29	1.32	1.69	2.00	2.34	-2.05	-1.83	-1.60	-1.30	1.33	1.68	1.99	2.31
	2	-2.07	-1.83	-1.58	-1.28	1.34	1.71	2.04	2.40	-1.93	-1.74	-1.53	-1.26	1.39	1.75	2.06	2.45
	3	-1.98	-1.76	-1.55	-1.27	1.35	1.75	2.09	2.49	-1.85	-1.69	-1.49	-1.27	1.36	1.83	2.17	2.56
8	0	-2.23	-1.92	-1.64	-1.30	1.31	1.66	1.95	2.27	-2.19	-1.92	-1.64	-1.26	1.33	1.70	1.99	2.30
	1	-2.19	-1.90	-1.63	-1.30	1.32	1.68	1.98	2.31	-2.14	-1.85	-1.61	-1.28	1.35	1.71	2.01	2.29
	2	-2.13	-1.86	-1.61	-1.29	1.33	1.70	2.01	2.36	-2.09	-1.81	-1.60	-1.30	1.32	1.67	1.97	2.29
	3	-2.06	-1.82	-1.58	-1.28	1.34	1.72	2.04	2.42	-2.00	-1.79	-1.56	-1.29	1.32	1.74	2.11	2.46
9	0	-2.26	-1.94	-1.65	-1.30	1.30	1.65	1.94	2.26	-2.26	-1.90	-1.64	-1.26	1.30	1.67	1.97	2.38
	1	-2.22	-1.92	-1.64	-1.30	1.31	1.67	1.96	2.29	-2.20	-1.93	-1.66	-1.30	1.33	1.68	1.96	2.24
	2	-2.18	-1.89	-1.62	-1.29	1.32	1.68	1.99	2.33	-2.13	-1.87	-1.63	-1.31	1.34	1.69	1.99	2.28
	3	-2.12	-1.86	-1.60	-1.29	1.33	1.70	2.01	2.37	-2.16	-1.84	-1.58	-1.29	1.33	1.70	2.00	2.34
	4	-2.05	-1.81	-1.57	-1.28	1.34	1.72	2.05	2.43	-1.96	-1.77	-1.54	-1.27	1.33	1.73	2.05	2.44
10	0	-2.28	-1.95	-1.66	-1.30	1.30	1.65	1.93	2.25	-2.28	-1.97	-1.65	-1.32	1.29	1.63	1.93	2.20
	1	-2.24	-1.93	-1.64	-1.30	1.31	1.66	1.95	2.28	-2.20	-1.94	-1.68	-1.32	1.28	1.62	1.93	2.23
	2	-2.21	-1.91	-1.63	-1.30	1.31	1.67	1.97	2.31	-2.08	-1.84	-1.58	-1.28	1.32	1.66	1.96	2.29
	3	-2.17	-1.88	-1.62	-1.29	1.32	1.68	1.99	2.34	-2.13	-1.87	-1.63	-1.31	1.35	1.70	2.02	2.36
	4	-2.11	-1.85	-1.60	-1.28	1.33	1.70	2.02	2.38	-2.03	-1.82	-1.60	-1.30	1.34	1.70	2.05	2.43
15	0	-2.34	-1.98	-1.67	-1.31	1.29	1.63	1.91	2.22	-2.37	-1.99	-1.70	-1.29	1.29	1.63	1.90	2.24
	1	-2.32	-1.97	-1.67	-1.30	1.29	1.63	1.92	2.23	-2.29	-1.95	-1.65	-1.30	1.30	1.60	1.86	2.17
	2	-2.30	-1.96	-1.66	-1.30	1.29	1.64	1.93	2.25	-2.27	-1.94	-1.65	-1.29	1.32	1.67	1.95	2.28
	3	-2.28	-1.95	-1.65	-1.30	1.30	1.65	1.94	2.26	-2.27	-1.93	-1.65	-1.28	1.30	1.67	1.93	2.24
	4	-2.26	-1.94	-1.65	-1.30	1.30	1.65	1.95	2.28	-2.21	-1.92	-1.66	-1.31	1.33	1.67	1.91	2.22
	5	-2.24	-1.92	-1.64	-1.29	1.31	1.66	1.96	2.30	-2.26	-1.94	-1.63	-1.28	1.33	1.67	1.94	2.26
	6	-2.21	-1.91	-1.63	-1.29	1.31	1.67	1.98	2.32	-2.21	-1.90	-1.63	-1.29	1.33	1.71	2.02	2.32
	7	-2.18	-1.89	-1.62	-1.29	1.32	1.68	2.00	2.35	-2.09	-1.80	-1.58	-1.29	1.35	1.70	2.00	2.35
20	0	-2.37	-2.00	-1.68	-1.31	1.28	1.62	1.90	2.20	-2.39	-1.98	-1.66	-1.31	1.28	1.59	1.83	2.16
	1	-2.36	-1.99	-1.68	-1.31	1.28	1.62	1.90	2.21	-2.33	-1.98	-1.70	-1.31	1.30	1.59	1.89	2.25
	2	-2.35	-1.99	-1.67	-1.30	1.28	1.62	1.91	2.22	-2.33	-2.00	-1.69	-1.33	1.29	1.60	1.85	2.18
	3	-2.33	-1.98	-1.67	-1.30	1.29	1.63	1.92	2.23	-2.39	-2.02	-1.70	-1.33	1.29	1.62	1.91	2.22
	4	-2.32	-1.97	-1.66	-1.30	1.29	1.63	1.92	2.24	-2.35	-1.99	-1.67	-1.28	1.27	1.63	1.88	2.18
	5	-2.31	-1.96	-1.66	-1.30	1.29	1.64	1.93	2.26	-2.17	-1.88	-1.64	-1.29	1.30	1.66	1.97	2.26
	6	-2.29	-1.95	-1.65	-1.30	1.29	1.64	1.94	2.27	-2.31	-1.94	-1.65	-1.31	1.25	1.59	1.85	2.21
	7	-2.28	-1.95	-1.65	-1.30	1.30	1.65	1.95	2.28	-2.31	-1.98	-1.69	-1.31	1.31	1.68	1.95	2.30
	8	-2.26	-1.93	-1.64	-1.29	1.30	1.66	1.96	2.30	-2.22	-1.92	-1.68	-1.32	1.25	1.61	2.00	2.29
	9	-2.24	-1.92	-1.64	-1.29	1.30	1.66	1.97	2.31	-2.17	-1.92	-1.63	-1.29	1.31	1.67	1.96	2.32

Table 15.9: Percentage points of the distribution of R_2 when $v = 6$.

		Edgeworth								Simulated							
n	s	1%	2.5%	5%	10%	90%	95%	97.5%	99%	1%	2.5%	5%	10%	90%	95%	97.5%	99%
5	0	-1.89	-1.69	-1.49	-1.23	1.35	1.78	2.18	2.68	-1.76	-1.62	-1.43	-1.19	1.38	1.88	2.24	2.70
	1	-1.80	-1.64	-1.46	-1.22	1.35	1.80	2.22	2.78	-1.71	-1.56	-1.42	-1.20	1.37	1.87	2.28	2.73
	2	-1.69	-1.56	-1.41	-1.20	1.36	1.83	2.32	2.97	-1.51	-1.41	-1.30	-1.15	1.26	1.89	2.33	2.88
6	0	-1.94	-1.73	-1.51	-1.24	1.34	1.76	2.15	2.62	-1.89	-1.69	-1.48	-1.22	1.35	1.76	2.17	2.67
	1	-1.88	-1.69	-1.49	-1.23	1.35	1.78	2.18	2.69	-1.82	-1.63	-1.46	-1.22	1.28	1.72	2.14	2.60
	2	-1.80	-1.64	-1.46	-1.22	1.35	1.80	2.23	2.79	-1.74	-1.59	-1.43	-1.22	1.36	1.80	2.18	2.68
7	0	-1.98	-1.75	-1.53	-1.25	1.33	1.75	2.13	2.59	-1.93	-1.72	-1.48	-1.22	1.37	1.83	2.21	2.63
	1	-1.94	-1.72	-1.51	-1.24	1.34	1.77	2.15	2.63	-1.89	-1.66	-1.49	-1.24	1.33	1.80	2.17	2.58
	2	-1.88	-1.69	-1.49	-1.23	1.35	1.78	2.18	2.69	-1.77	-1.61	-1.43	-1.19	1.39	1.85	2.24	2.72
	3	-1.80	-1.64	-1.46	-1.22	1.35	1.80	2.23	2.79	-1.68	-1.55	-1.39	-1.21	1.39	1.92	2.32	2.83
8	0	-2.01	-1.77	-1.54	-1.25	1.33	1.74	2.11	2.56	-1.96	-1.77	-1.52	-1.21	1.38	1.80	2.19	2.55
	1	-1.98	-1.75	-1.53	-1.24	1.33	1.75	2.13	2.59	-1.91	-1.68	-1.49	-1.23	1.35	1.81	2.18	2.68
	2	-1.93	-1.72	-1.51	-1.24	1.34	1.77	2.15	2.63	-1.88	-1.66	-1.49	-1.22	1.35	1.76	2.12	2.59
	3	-1.88	-1.69	-1.49	-1.23	1.34	1.78	2.18	2.70	-1.82	-1.64	-1.45	-1.22	1.34	1.84	2.25	2.74
9	0	-2.04	-1.79	-1.55	-1.25	1.33	1.74	2.10	2.54	-2.04	-1.74	-1.52	-1.21	1.34	1.79	2.22	2.72
	1	-2.01	-1.77	-1.54	-1.25	1.33	1.75	2.12	2.56	-1.98	-1.76	-1.55	-1.23	1.37	1.79	2.11	2.54
	2	-1.97	-1.75	-1.53	-1.24	1.33	1.76	2.13	2.60	-1.94	-1.72	-1.52	-1.25	1.35	1.78	2.14	2.55
	3	-1.93	-1.72	-1.51	-1.24	1.34	1.77	2.16	2.64	-1.94	-1.67	-1.48	-1.22	1.37	1.75	2.08	2.52
	4	-1.87	-1.69	-1.49	-1.23	1.34	1.78	2.19	2.70	-1.78	-1.62	-1.44	-1.21	1.32	1.74	2.23	2.68
10	0	-2.06	-1.80	-1.56	-1.25	1.32	1.73	2.09	2.52	-2.05	-1.81	-1.55	-1.25	1.33	1.76	2.06	2.48
	1	-2.03	-1.78	-1.55	-1.25	1.33	1.74	2.10	2.54	-2.00	-1.76	-1.56	-1.26	1.32	1.70	2.08	2.53
	2	-2.01	-1.77	-1.54	-1.25	1.33	1.75	2.12	2.57	-1.89	-1.69	-1.49	-1.22	1.35	1.75	2.14	2.57
	3	-1.97	-1.75	-1.52	-1.24	1.33	1.76	2.13	2.60	-1.94	-1.73	-1.52	-1.26	1.39	1.81	2.22	2.67
	4	-1.93	-1.72	-1.51	-1.24	1.34	1.77	2.16	2.64	-1.85	-1.68	-1.50	-1.24	1.35	1.79	2.23	2.65
15	0	-2.12	-1.84	-1.58	-1.26	1.31	1.71	2.06	2.47	-2.13	-1.84	-1.57	-1.25	1.33	1.71	2.05	2.52
	1	-2.11	-1.83	-1.57	-1.26	1.32	1.71	2.07	2.48	-2.13	-1.80	-1.56	-1.25	1.35	1.71	2.03	2.43
	2	-2.10	-1.82	-1.57	-1.26	1.32	1.72	2.07	2.49	-2.09	-1.81	-1.53	-1.25	1.34	1.75	2.18	2.54
	3	-2.08	-1.82	-1.56	-1.26	1.32	1.72	2.08	2.51	-2.04	-1.80	-1.54	-1.25	1.33	1.76	2.09	2.48
	4	-2.07	-1.80	-1.56	-1.26	1.32	1.73	2.09	2.52	-2.02	-1.80	-1.54	-1.26	1.35	1.75	2.06	2.48
	5	-2.05	-1.79	-1.55	-1.25	1.32	1.73	2.10	2.53	-2.06	-1.77	-1.57	-1.24	1.36	1.75	2.08	2.52
	6	-2.03	-1.78	-1.54	-1.25	1.33	1.74	2.11	2.55	-2.02	-1.79	-1.53	-1.22	1.38	1.78	2.12	2.56
	7	-2.00	-1.76	-1.53	-1.25	1.33	1.75	2.12	2.58	-1.93	-1.70	-1.48	-1.25	1.37	1.81	2.19	2.63
20	0	-2.16	-1.86	-1.59	-1.27	1.31	1.70	2.04	2.45	-2.13	-1.83	-1.58	-1.27	1.29	1.66	2.00	2.46
	1	-2.15	-1.86	-1.59	-1.26	1.31	1.70	2.05	2.45	-2.11	-1.83	-1.58	-1.28	1.34	1.71	2.06	2.51
	2	-2.15	-1.85	-1.58	-1.26	1.31	1.70	2.05	2.46	-2.14	-1.86	-1.59	-1.30	1.31	1.67	2.01	2.40
	3	-2.14	-1.85	-1.58	-1.26	1.31	1.71	2.06	2.47	-2.14	-1.90	-1.60	-1.27	1.31	1.72	2.04	2.44
	4	-2.13	-1.84	-1.58	-1.26	1.31	1.71	2.06	2.47	-2.19	-1.83	-1.57	-1.25	1.31	1.71	2.06	2.48
	5	-2.12	-1.83	-1.58	-1.26	1.31	1.71	2.06	2.48	-2.03	-1.76	-1.54	-1.25	1.33	1.74	2.14	2.63
	6	-2.11	-1.83	-1.57	-1.26	1.32	1.72	2.07	2.49	-2.09	-1.79	-1.55	-1.25	1.27	1.63	2.00	2.40
	7	-2.09	-1.82	-1.57	-1.26	1.32	1.72	2.08	2.50	-2.10	-1.83	-1.58	-1.26	1.34	1.76	2.11	2.53
	8	-2.08	-1.81	-1.56	-1.26	1.32	1.72	2.08	2.51	-2.08	-1.77	-1.56	-1.27	1.29	1.72	2.08	2.46
	9	-2.06	-1.80	-1.56	-1.25	1.32	1.73	2.09	2.52	-2.01	-1.78	-1.52	-1.25	1.36	1.75	2.10	2.54

Table 15.10: Simulated percentage points of the distribution of R_3 when $v = 3/2$.

n	s	1%	2.5%	5%	10%	90%	95%	97.5%	99%
5	0	-1.14	-1.08	-1.02	-0.93	2.17	3.50	5.11	6.98
	1	-1.08	-1.03	-0.98	-0.89	2.64	4.44	6.55	10.66
	2	-0.99	-0.95	-0.90	-0.82	3.40	6.11	9.47	18.26
6	0	-1.18	-1.11	-1.04	-0.94	2.05	3.33	4.53	6.58
	1	-1.12	-1.07	-1.00	-0.89	2.25	3.55	5.11	7.61
	2	-1.07	-1.02	-0.97	-0.88	2.79	4.53	7.09	11.84
7	0	-1.21	-1.15	-1.08	-0.96	1.94	3.00	4.06	6.21
	1	-1.17	-1.11	-1.05	-0.94	2.12	3.25	4.46	6.38
	2	-1.13	-1.08	-1.02	-0.92	2.12	3.40	4.98	7.05
	3	-1.07	-1.03	-0.97	-0.87	2.60	4.00	6.56	9.81
8	0	-1.23	-1.17	-1.09	-0.98	1.82	2.82	3.97	5.54
	1	-1.21	-1.14	-1.07	-0.95	1.90	3.03	4.12	5.74
	2	-1.16	-1.10	-1.03	-0.93	2.01	3.20	4.46	6.56
	3	-1.12	-1.07	-1.01	-0.92	2.27	3.62	5.17	7.71
9	0	-1.27	-1.19	-1.10	-0.97	1.73	2.68	3.72	5.04
	1	-1.22	-1.15	-1.09	-0.97	1.88	2.87	3.97	5.74
	2	-1.19	-1.14	-1.07	-0.95	1.99	3.07	4.22	5.69
	3	-1.15	-1.10	-1.03	-0.93	2.11	3.11	4.44	7.14
	4	-1.12	-1.06	-1.00	-0.90	2.20	3.44	5.08	7.79
10	0	-1.26	-1.19	-1.10	-0.98	1.73	2.73	3.61	4.99
	1	-1.23	-1.16	-1.09	-0.97	1.88	2.75	3.74	5.10
	2	-1.22	-1.15	-1.08	-0.96	1.78	2.64	3.69	5.02
	3	-1.20	-1.13	-1.06	-0.96	1.98	3.03	4.16	5.91
	4	-1.16	-1.10	-1.03	-0.93	2.06	3.30	4.52	6.29
15	0	-1.34	-1.26	-1.16	-1.02	1.64	2.48	3.21	4.41
	1	-1.31	-1.22	-1.13	-1.00	1.60	2.35	3.23	4.21
	2	-1.28	-1.21	-1.13	-1.00	1.71	2.54	3.55	4.74
	3	-1.28	-1.21	-1.13	-1.01	1.68	2.55	3.36	4.43
	4	-1.27	-1.20	-1.12	-1.00	1.71	2.48	3.35	4.77
	5	-1.25	-1.20	-1.12	-0.99	1.72	2.56	3.48	4.55
	6	-1.25	-1.15	-1.07	-0.96	1.82	2.69	3.69	4.94
	7	-1.20	-1.15	-1.07	-0.96	1.85	2.76	3.64	5.22
20	0	-1.36	-1.25	-1.15	-1.01	1.56	2.26	2.93	3.90
	1	-1.36	-1.26	-1.16	-1.01	1.57	2.33	3.04	3.96
	2	-1.36	-1.26	-1.16	-1.03	1.52	2.28	3.00	4.10
	3	-1.34	-1.26	-1.15	-1.02	1.65	2.45	3.14	4.06
	4	-1.32	-1.23	-1.13	-1.00	1.58	2.24	3.03	3.94
	5	-1.32	-1.24	-1.14	-1.01	1.62	2.47	3.28	4.30
	6	-1.30	-1.22	-1.14	-1.00	1.56	2.46	3.32	4.38
	7	-1.29	-1.22	-1.13	-0.99	1.68	2.48	3.34	4.41
	8	-1.27	-1.20	-1.12	-1.01	1.64	2.45	3.35	4.35
	9	-1.28	-1.20	-1.11	-0.99	1.78	2.49	3.42	4.66

Table 15.11: Simulated percentage points of the distribution of R_3 when $v = 3$.

n	s	1%	2.5%	5%	10%	90%	95%	97.5%	99%
5	0	-1.26	-1.16	-1.06	-0.92	2.34	3.65	5.19	6.97
	1	-1.18	-1.09	-1.00	-0.87	2.86	4.67	6.82	11.18
	2	-1.05	-0.98	-0.91	-0.80	3.76	6.47	10.05	19.31
6	0	-1.31	-1.20	-1.09	-0.93	2.20	3.34	4.67	6.65
	1	-1.24	-1.14	-1.02	-0.86	2.44	3.72	5.40	7.78
	2	-1.15	-1.07	-0.99	-0.87	3.00	4.80	7.38	12.21
7	0	-1.36	-1.25	-1.14	-0.96	2.10	3.06	4.14	5.86
	1	-1.30	-1.20	-1.10	-0.93	2.33	3.38	4.52	6.22
	2	-1.24	-1.16	-1.06	-0.91	2.34	3.57	5.23	7.11
	3	-1.18	-1.09	-1.00	-0.85	2.85	4.30	6.76	9.84
8	0	-1.39	-1.29	-1.17	-1.00	1.93	2.89	4.04	5.40
	1	-1.37	-1.25	-1.13	-0.96	2.04	3.08	4.12	5.62
	2	-1.30	-1.19	-1.08	-0.92	2.22	3.41	4.52	6.68
	3	-1.25	-1.15	-1.06	-0.90	2.51	3.79	5.41	8.09
9	0	-1.47	-1.33	-1.18	-0.98	1.86	2.72	3.63	5.23
	1	-1.38	-1.26	-1.15	-0.99	1.97	2.98	4.12	5.49
	2	-1.35	-1.24	-1.12	-0.96	2.11	3.17	4.24	5.80
	3	-1.27	-1.19	-1.06	-0.93	2.20	3.32	4.59	7.30
	4	-1.24	-1.15	-1.03	-0.89	2.41	3.69	5.30	7.42
10	0	-1.44	-1.32	-1.18	-0.99	1.86	2.72	3.63	4.96
	1	-1.42	-1.28	-1.15	-0.99	1.96	2.86	3.78	5.02
	2	-1.39	-1.27	-1.15	-0.98	1.97	2.76	3.70	4.85
	3	-1.39	-1.25	-1.13	-0.97	2.16	3.18	4.26	5.96
	4	-1.30	-1.19	-1.08	-0.93	2.22	3.41	4.63	6.47
15	0	-1.58	-1.41	-1.26	-1.04	1.70	2.46	3.13	4.11
	1	-1.53	-1.36	-1.23	-1.05	1.74	2.41	3.16	4.18
	2	-1.53	-1.40	-1.24	-1.04	1.75	2.47	3.21	4.14
	3	-1.51	-1.36	-1.21	-1.03	1.78	2.58	3.31	4.40
	4	-1.47	-1.32	-1.19	-1.00	1.84	2.61	3.41	4.41
	5	-1.47	-1.32	-1.19	-1.00	1.88	2.72	3.69	4.83
	6	-1.43	-1.29	-1.17	-1.00	1.88	2.80	3.68	5.16
	7	-1.41	-1.28	-1.15	-0.98	1.99	2.83	3.58	5.22
20	0	-1.65	-1.42	-1.27	-1.07	1.65	2.29	2.82	3.84
	1	-1.64	-1.46	-1.27	-1.08	1.69	2.32	2.93	3.69
	2	-1.61	1.42	1.26	-1.07	1.71	2.40	3.03	3.78
	3	-1.59	-1.43	-1.27	-1.05	1.71	2.44	3.12	4.10
	4	-1.58	-1.43	-1.25	-1.04	1.66	2.38	3.10	4.30
	5	-1.61	-1.44	-1.25	-1.05	1.68	2.33	2.93	3.70
	6	-1.53	-1.36	-1.22	-1.01	1.73	2.44	3.23	4.20
	7	-1.55	-1.38	-1.23	-1.04	1.80	2.53	3.33	4.54
	8	-1.50	-1.35	-1.22	-1.01	1.84	2.55	3.26	4.37
	9	-1.49	-1.34	-1.21	-1.02	1.81	2.61	3.45	4.29

Table 15.12: Simulated percentage points of the distribution of R_3 when $v = 6$.

n	s	1%	2.5%	5%	10%	90%	95%	97.5%	99%
5	0	-1.26	-1.15	-1.04	-0.88	2.41	3.65	5.21	7.07
	1	-1.18	-1.08	-0.97	-0.82	2.97	4.76	6.71	11.59
	2	-1.04	-0.96	-0.88	-0.75	3.91	6.51	10.39	19.27
6	0	-1.33	-1.18	-1.06	-0.90	2.27	3.33	4.69	6.74
	1	-1.23	-1.12	-0.99	-0.83	2.53	3.84	5.26	7.77
	2	-1.15	-1.06	-0.96	-0.82	3.10	4.88	7.30	12.30
7	0	-1.37	-1.25	-1.12	-0.92	2.13	3.03	4.22	5.83
	1	-1.31	-1.20	-1.06	-0.88	2.35	3.39	4.46	6.33
	2	-1.25	-1.15	-1.04	-0.87	2.40	3.65	5.25	7.02
	3	-1.17	-1.07	-0.97	-0.82	2.95	4.48	6.75	9.62
8	0	-1.41	-1.29	-1.16	-0.96	1.99	2.93	3.98	5.28
	1	-1.38	-1.24	-1.11	-0.91	2.14	3.10	3.98	5.58
	2	-1.30	-1.18	-1.06	-0.89	2.30	3.46	4.53	6.66
	3	-1.25	-1.14	-1.04	-0.86	2.57	3.88	5.54	8.07
9	0	-1.50	-1.34	-1.17	-0.96	1.91	2.78	3.57	5.25
	1	-1.40	-1.27	-1.14	-0.96	2.02	3.05	4.08	5.46
	2	-1.35	-1.22	-1.10	-0.92	2.19	3.20	4.24	5.86
	3	-1.28	-1.17	-1.04	-0.90	2.28	3.36	4.51	7.37
	4	-1.23	-1.14	-1.00	-0.84	2.49	3.83	5.40	7.78
10	0	-1.47	-1.31	-1.17	-0.96	1.93	2.76	3.70	4.80
	1	-1.43	-1.28	-1.14	-0.96	2.00	2.90	3.66	5.08
	2	-1.41	-1.27	-1.13	-0.95	2.02	2.83	3.63	4.76
	3	-1.40	-1.26	-1.11	-0.95	2.22	3.18	4.21	5.86
	4	-1.31	-1.18	-1.07	-0.89	2.31	3.51	4.71	6.41
15	0	-1.60	-1.41	-1.24	-1.02	1.75	2.43	3.09	4.00
	1	-1.56	-1.37	-1.23	-1.03	1.76	2.41	3.16	4.24
	2	-1.56	-1.41	-1.23	-1.00	1.81	2.46	3.23	4.35
	3	-1.52	-1.37	-1.21	-0.99	1.82	2.55	3.28	4.21
	4	-1.48	-1.32	-1.19	-0.97	1.87	2.63	3.44	4.37
	5	-1.48	-1.31	-1.17	-0.97	1.91	2.82	3.59	4.87
	6	-1.45	-1.31	-1.17	-0.98	1.93	2.86	3.80	5.09
	7	-1.44	-1.28	-1.14	-0.95	2.07	2.81	3.68	5.00
20	0	-1.68	-1.44	-1.27	-1.03	1.67	2.28	2.79	3.62
	1	-1.67	-1.47	-1.28	-1.06	1.73	2.30	2.83	3.63
	2	-1.63	-1.43	-1.25	-1.04	1.74	2.34	3.03	3.81
	3	-1.63	-1.44	-1.26	-1.03	1.74	2.41	3.16	3.96
	4	-1.61	-1.43	-1.24	-1.02	1.71	2.35	3.04	4.23
	5	-1.65	-1.45	-1.24	-1.04	1.73	2.33	2.95	3.76
	6	-1.54	-1.37	-1.20	-0.98	1.77	2.44	3.13	4.15
	7	-1.58	-1.40	-1.23	-1.01	1.83	2.59	3.26	4.34
	8	-1.52	-1.35	-1.20	-0.98	1.87	2.57	3.26	4.51
	9	-1.53	-1.36	-1.20	-0.99	1.86	2.60	3.37	4.33

16

Selecting from Normal Populations the One with the Largest Absolute Mean: Common Unknown Variance Case

S. Jeyaratnam and S. Panchapakesan

Southern Illinois University, Carbondale, IL, USA

Abstract: Rizvi (1971) studied a single-stage procedure for selecting from several normal populations the one with the largest absolute mean using the indifference-zone formulation of Bechhofer (1954), assuming that the populations have a common known variance. When the common variance is unknown, a single-stage procedure that guarantees a minimum probability of a correct selection does not exist. In this paper, a non-eliminating two-stage procedure is proposed and studied for this situation.

Keywords and phrases: Indifference-zone, ranking absolute means, two-stage procedure, expected total sample size

16.1 Introduction

Let $\Pi_1, \Pi_2, \ldots, \Pi_k$ denote k independent normal populations with *unknown* means $\mu_1, \ldots, \mu_k$, respectively, and common *unknown* variance σ^2. Let $\theta_i = |\mu_i|$, $i = 1, 2, \ldots, k$. The populations are ranked according to the absolute values of their means. Let $\theta_{[1]} \leq \theta_{[2]} \leq \ldots \leq \theta_{[k]}$ denote the ordered θ_i. It is assumed that there is no prior knowledge about the correspondence between the ordered and the unordered θ_i. Our goal is to select the population associated with the largest θ_i. Since the populations have a common variance σ^2, the problem is equivalent to selecting the population associated with the largest $\theta_i/\sigma = |\mu_i|/\sigma$, which is the signal-to-noise ratio well-known in communications theory. In terms of comparing k different electronic devices, our goal is to select the device having the largest signal-to-noise ratio. It should also be noted that θ_i/σ is the

Mahalanobis distance between the population Π_i and $N(0, \sigma^2)$ population. In this context, we are interested in selecting the population that is farthest from $N(0, \sigma^2)$.

We now formulate our problem using the indifference-zone approach of Bechhofer (1954). Based on the sample data, we want to select one of the k populations and claim it to be the one associated with $\theta_{[k]}$, which is called the *best* population. A *correct selection* (CS) is said to occur if the selected population is indeed a best population. It is required that the probability of a correct selection (PCS) is at least $P^*(1/k < P^* < 1)$ whenever $\theta_{[k]} - \theta_{[k-1]} \geq \delta^* > 0$, where δ^* and P^* are specified in advance by the experimenter. Let $\Omega = \{\theta = (\theta_1, \ldots, \theta_k) : \theta_i \geq 0, i = 1, 2, \ldots, k\}$ and $\Omega(\delta^*) = \{\theta : \theta_{[k]} - \theta_{[k-1]} \geq \delta^*\}$. The subset $\Omega(\delta^*)$ of the parameter space Ω is called the *preference-zone* and its complement w.r.t. Ω is the so-called *indifference-zone*. Letting $P(CS|R)$ denote the PCS for any prescribed selection rule R, the requirement is:

$$P(CS|R) \geq P^* \text{ whenever } \theta \in \Omega(\delta^*). \tag{16.1}$$

We refer to (16.1) as the probability requirement or the *P^*-condition*. Obviously, P^* is the minimum guaranteed PCS. It is specified such that $1/k < P^* < 1$; otherwise, (16.1) can be trivially satisfied by choosing one of the k populations randomly.

Rizvi (1971) considered the above selection problem assuming σ^2 to be *known*. In this case, he proposed a single-stage procedure based on samples of same size n from the k populations. His procedure R_S selects the population that yields the largest $|\overline{X}_i|$, where the $\overline{X}_i$ are the means of the samples drawn from the k populations. He showed that the PCS attains its infimum over $\Omega(\delta^*)$ for a configuration of the parameters of the form $(\theta, \ldots, \theta, \theta + \delta^*)$ and further showed that the PCS for such a configuration is monotonically increasing in $\theta \geq 0$. This yields the so-called least favorable configuration (LFC) $\theta = (0, \ldots, 0, \delta^*)$ where the PCS attains its infimum over $\Omega(\delta^*)$. Thus the smallest sample size n necessary to satisfy the P^*-condition is the smallest n for which the PCS evaluated at the LFC is at least P^*. This is given by [see Rizvi (1971)]

$$n = \left\langle \left(\frac{\lambda \sigma}{\delta^*}\right)^2 \right\rangle \tag{16.2}$$

where $\langle s \rangle$ denotes the smallest integer greater than or equal to s, and λ is the solution of

$$2(k-1) \int_0^\infty \left\{2\Phi(u) - 1\right\}^{k-2} \left\{\Phi(-u+\lambda) + \Phi(-u-\lambda)\right\} \varphi(u)\, du = P^*. \tag{16.3}$$

Here and in the sequel, Φ and φ denote, respectively, the standard normal distribution function and its density.

Selecting from Normal Populations

Now, when σ is *unknown*, we cannot decide the sample size n necessary without the knowledge of σ in order to satisfy the P^*-condition. So, in this case, we propose and investigate a two-stage procedure analogous to that of Bechhofer, Dunnett and Sobel (1954) for selecting the normal population having the largest mean μ_i. It should be pointed out that, when σ is unknown, a single-stage procedure can be defined if we adopt the subset selection approach of Gupta (1956); this has been done by Rizvi (1971) and will not be discussed here.

16.2 Proposed Procedure R_T

We first take a random sample of n_0 observations from each of the k populations. Let these be X_{ij}, $j = 1, 2, \ldots, n_0; i = 1, 2, \ldots, k$. Define

$$\bar{X}_i = \sum_{j=1}^{n_0} X_{ij}/n_0, \; i = 1, 2, \ldots, k,$$

$$S^2 = \sum_{i=1}^{k} \sum_{j=1}^{n_0} \left(X_{ij} - \bar{X}_i\right)^2 / k(n_0 - 1).$$

We then take a second sample of $(N - n_0)$ observations from each population, where

$$N \equiv N(S^2) = \max\{n_0, \langle (hS/\delta^*)^2 \rangle\}, \tag{16.4}$$

where h is a positive constant to be defined later in (16.8). Let $\bar{Y}_i$ denote the over-all mean of the N observations from Π_i and let $W_i = |\bar{Y}_i|$, $i = 1, 2, \ldots, k$. The proposed procedure R_T is:

"Select as the best the population that yields the largest W_i."

We shall show that the procedure R_T, by a proper choice of h, satisfies the probability requirement (16.1). First, for any fixed N, let $F_N(w, \theta_i)$ and $f_N(w, \theta_i)$ denote the distribution function and the density function of W_i, $i = 1, 2, \ldots, k$. Then it is easy to see that

$$F_N(w, \theta_i) = \begin{cases} \Phi\left\{\frac{\sqrt{N}}{\sigma}(w - \theta_i)\right\} - \Phi\left\{\frac{\sqrt{N}}{\sigma}(-w - \theta_i)\right\} &, w \geq 0, \\ 0 &, w < 0, \end{cases}$$

and

$$f_N(w, \theta_i) = \frac{\sqrt{N}}{\sigma}\left[\varphi\left\{\frac{\sqrt{N}}{\sigma}(w - \theta_i)\right\} + \varphi\left\{\frac{\sqrt{N}}{\sigma}(w + \theta_i)\right\}\right], \; w \geq 0.$$

The family $\{F_N(w, \theta)\}$ is stochastically increasing in θ; in other words, for $0 \leq \theta_1 < \theta_2$, $F_N(w, \theta_2) \leq F_N(w, \theta_1)$. Finally, it is also known that $\nu S^2/\sigma^2$ has a chi-square distribution with $\nu = k(n_0 - 1)$ degrees of freedom.

The PCS for procedure R_T is given by

$$P(CS|R_T) = \sum_{m=n_0}^{\infty} Pr\{N(S^2) = m\} P(CS|R_T; N(S^2) = m), \quad (16.5)$$

where $P(CS|R_T; N(S^2) = m)$ is the conditional PCS for R_T given that $N(S^2) = m$. For this conditional PCS, it follows from Rizvi (1971) that the infimum over $\Omega(\delta^*)$ occurs when $\theta_{[1]} = \theta_{[2]} = \ldots = \theta_{[k-1]} = 0$ and $\theta_{[k]} = \delta^*$. In other words, $\theta = (0, \ldots, 0, \delta^*)$ is the LFC for $P(CS|R_T; N(S^2) = m)$. Since this LFC does not depend on the fixed value of m, it follows that it is the LFC for $P(CS|R_T)$. It is shown by Rizvi (1971) that $P(CS|R_T; N(S^2) = m)$ at the LFC is given by

$$P_{LFC}(CS|R_T; N(S^2) = m) = T\left(\frac{\sqrt{m}\,\delta^*}{\sigma}\right), \quad (16.6)$$

where $T(\lambda)$ is the left-hand side expression in (16.3). Using (16.6) in (16.5), we have

$$P_{LFC}(CS|R_T) = \sum_{m=n_0}^{\infty} Pr\{N(S^2) = m\} T\left(\frac{\sqrt{m}\,\delta^*}{\sigma}\right)$$

$$\geq \sum_{m=n_0}^{\infty} Pr\{N(S^2) = m\} T\left(\frac{hS}{\sigma}\right), \text{ using (16.4).}$$

Let A_m denote the set in the space of S^2 for which $N(S^2) = m$, $m \geq n_0$. Then $[0, \infty) = \bigcup_{m=n_0}^{\infty} A_m$ is a partition of the space of S^2. Thus

$$P_{LFC}(CS|R_T) \geq \int_0^{\infty} T(hS/\sigma) q(S^2)\, dS^2, \quad (16.7)$$

where $q(S^2)$ is the density function of S^2. Letting $U = \nu S^2 \sigma^2$, (16.7) can be written as

$$P_{LFC}(CS|R_T) \geq \int_0^{\infty} T(h\sqrt{u/\nu}) g_\nu(u)\, du,$$

where $g_\nu(u)$ is the chi-square density with $\nu = k(n_0 - 1)$ degrees of freedom. Thus the probability requirement (16.1) is satisfied if, for a given choice of k, P^* and n_0, we choose h such that

$$\int_0^{\infty} T(h\sqrt{u/\nu}) g_\nu(u)\, du = P^*. \quad (16.8)$$

The values of the constant h are given in Tables 16.1 and 16.2 for $P^* = 0.90$ and 0.95, respectively, for $k = 2(1)10$, and selected values of n_0 for each k.

Table 16.1: Value of h satisfying equation (16.8).

$P^* = 0.90$

k	n_0	h	k	n_0	h
2	5	2.5547	4	15	2.8723
	10	2.4057		20	2.8622
	15	2.3657	5	5	3.1416
	20	2.3473		10	3.0463
	25	2.3367		15	3.0202
	30	2.3298	6	5	3.1882
	35	2.3249		10	3.1070
	40	2.3304		15	3.1550
	45	2.5478	7	5	3.2348
3	5	2.6414		10	3.1632
	10	2.5382	8	5	3.2918
	15	2.5100		10	3.2271
	20	2.4975	9	5	3.3504
	25	2.4900		10	3.3162
	30	2.6259	10	5	3.4020
4	5	3.0093		10	3.7243
	10	2.9018			

Table 16.2: Value of h satisfying equation (16.8).

$P^* = 0.95$

k	n_0	h	k	n_0	h
2	5	3.2156	4	15	3.3424
	10	2.9483		20	3.3270
	15	2.8785	5	5	3.6947
	20	2.8476		10	3.5423
	25	2.8293		15	3.5010
	30	2.8188	6	5	3.6992
	35	2.8101		10	3.5722
	40	2.8201		15	3.6675
	45	3.2917	7	5	3.7218
3	5	3.0836		10	3.6117
	10	2.9264	8	5	3.7733
	15	2.8839		10	3.6743
	20	2.8635	9	5	3.8350
	25	2.8527		10	3.7894
	30	3.0709	10	5	3.8901
4	5	3.5611		10	5.0743
	10	3.3896			

16.3 Approximation for h

Let U be a random variable having a chi-square distribution with $\nu = k(n_0 - 1)$ degrees of freedom. For large ν, U is approximately distributed as U^* which is normal with mean ν and variance 2ν. Since $T(\lambda)$ is defined only for $\lambda \geq 0$, we use the distribution of U^* left-truncated at zero and approximate the left-hand side of (16.8) by

$$\int_0^\infty T\left[h\sqrt{\frac{u^*}{\nu}}\right] \frac{1}{2\sqrt{\pi\nu}} e^{-\frac{(u^*-\nu)^2}{4\nu}} du^*$$

$$= \int_{-\sqrt{\nu}/2}^\infty T\left[h\sqrt{1+\frac{2t}{\sqrt{\nu}}}\right] \frac{1}{\sqrt{\pi}} e^{-t^2} dt \qquad (16.9)$$

by making the transformation $t = \frac{u^*-\nu}{\sqrt{4\nu}}$. Thus, for large ν, an approximation for h can be obtained by equating the right-hand side integral in (16.9) to P^* and solving for h. For $k = 2(1)4$ and $n_0 = 20$, and for $k = 2$ and $n_0 = 40$, both exact and approximate values of h are presented in Table 16.3 in order to indicate the closeness of the approximation.

Table 16.3: Exact and approximate value of h
$P^* = 0.90$ (top entry) and 0.95 (bottom entry).

k	n_0	exact	approximate
2	20	2.3473	2.3521
		2.8476	2.8588
	40	2.3304	2.3234
		2.8201	2.8082
3	20	2.4975	2.4996
		2.8635	2.8682
4	20	2.8622	2.8612
		3.3270	3.3260

16.4 Expected Sample Size

As mentioned in Section 16.1, our procedure is analogous to that of Bechhofer, Dunnett and Sobel (1954) for selecting the normal population with the largest mean when the common variance is unknown. Their expression for $E(N)$, the

Selecting from Normal Populations

expected value of the total sample size N, is applicable to our procedure R_T. Thus, $E(N)$ is given (to within a quantity less than unity) by the equation

$$E(N) = n_0 Pr\left\{\chi_\nu^2 < \nu n_0 \left(\frac{\delta^*}{h\sigma}\right)^2\right\} + \left(\frac{h\sigma}{\delta^*}\right)^2 Pr\left\{\chi_{\nu+2}^2 > \nu n_0 \left(\frac{\delta^*}{h\sigma}\right)^2\right\}$$
$$+ \tau Pr\left\{\chi_\nu^2 > \nu n_0 \left(\frac{\delta^*}{h\sigma}\right)^2\right\}, \tag{16.10}$$

where $0 \leq \tau \leq 1$ and χ_ν^2 is a chi-square variable with ν degrees of freedom. Note that h in (16.10) is given by (16.8).

In order to assess the gain due to the knowledge of σ, one would assume that σ is known and compare $E(N)$ in (16.10) with the sample size N_S needed for the single-stage procedure R_S of Rizvi (1971) defined in Section 16.1. Since the first-stage sample size n_0 for procedure R_T is arbitrarily chosen, we should consider the behavior of $E(N)/N_S$, for given k, P^* and δ^*/σ, as a function of n_0. This requires an extensive computational analysis. We consider a limited comparison here by assuming that δ^*/σ is known and taking n_0 to be equal to N_S. We then look at $E(N)/N_S$. Tables 16.4 and 16.5 give, respectively, for $P^* = 0.90$ and 0.95, the values of N_S and $E(N)/N_S$ for $k = 2(1)10$ and selected values of δ^*/σ. In the table values, our $E(N)$ is slightly larger than the exact value because we have taken τ to be equal to unity in (16.10). It can be seen that the expected excess of the total sample size for each population in the case of the two-stage procedure R_T is marginal compared to the sample size per population for the single-stage procedure R_S (knowing σ). Only in the case of $k = 10$, $\delta^* = 1.0384$, $P^* = 0.95$, the ratio $E(N)/N_S$ is close to 2. In all other cases tabulated, the ratio varies from 1.0302 to 1.5949.

16.5 Concluding Remarks

Our two-stage procedure is a non-eliminating procedure in the sense that no population is eliminated at the first stage before sampling stops. We can define a two-stage procedure which (depending on data) eliminates a random number $m(0 \leq m \leq k-1)$ of the k populations after the first-stage sampling, using the subset selection procedure of Gupta (1956). When $m = k - 1$, the one population that remains will be selected without any additional sampling. Establishing the LFC for the PCS for such a procedure is a challenging problem.

In Section 16.1, we mentioned that there is no single-stage procedure that satisfies the probability requirement (16.1) when σ is unknown because the necessary sample size cannot be determined. If we modify the preference-zone to be $\Omega(\delta^*) = \{\theta : \theta_{[k]} - \theta_{[k-1]} \geq \delta^* \sigma\}$ using standardized distance between populations, we can use the single-stage procedure R_S of Rizvi (1971).

Table 16.4: Values of N_S and $E(N)/N_S$ with $n_0 = N_S$.

$$P^* = 0.90$$

k	δ^*/σ	N_S	$E(N)/N_S$
2	1.0236	5	1.4900
	0.7238	10	1.2565
	0.5910	15	1.1857
	0.5118	20	1.1502
3	1.1773	5	1.2551
	0.8325	10	1.1073
	0.6797	15	1.0641
	0.5887	20	1.0438
4	1.2608	5	1.3511
	0.8915	10	1.1863
	0.7279	15	1.1344
	0.6304	20	1.1100
5	1.3177	5	1.3383
	0.9318	10	1.1859
	0.7608	15	1.1373
6	1.3606	5	1.2946
	0.9621	10	1.1581
	0.7855	15	1.1532
7	1.3948	5	1.2675
	0.9863	10	1.1413
8	1.4232	5	1.2570
	1.0064	10	1.1371
9	1.4475	5	1.2546
	1.0235	10	1.1526
10	1.4685	5	1.2531
	1.0384	10	1.3843

Table 16.5: Values of N_S and $E(N)/N_S$ with $n_0 = N_S$.

$P^* = 0.95$

k	δ^*/σ	N_S	$E(N)/N_S$
2	1.2325	5	1.5949
	0.8715	10	1.2888
	0.7116	15	1.2029
	0.6163	20	1.1618
3	1.37743	5	1.2517
	0.9739	10	1.0914
	0.7952	15	1.0490
	0.6887	20	1.0302
4	1.4571	5	1.4020
	1.0303	10	1.2051
	0.8413	15	1.1454
	0.7286	20	1.1189
5	1.5117	5	1.3931
	1.0689	10	1.2110
	0.8728	15	1.1553
6	1.5529	5	1.3293
	1.0981	10	1.1710
	0.8966	15	1.1886
7	1.5859	5	1.2920
	1.1214	10	1.1486
8	1.6133	5	1.2803
	1.1408	10	1.1449
9	1.6366	5	1.2807
	1.1573	10	1.1729
10	1.6570	5	1.2823
	1.1717	10	1.9757

Finally, even when σ is known and we use $\Omega(\delta^*)$ as defined in Section 16.1, we can define a two-stage procedure for selecting the best by using Gupta's subset selection procedure at the first stage to screen inferior populations and then take further samples from the remaining populations, if necessary. Such an eliminating procedure was studied by Tamhane and Bechhofer (1977, 1979) for the problem of ranking in terms of the population means. The problems mentioned above are under investigation by the authors.

References

1. Bechhofer, R. E. (1954). A single-sample multiple decision procedure for ranking means of normal populations with known variances, *Annals of Mathematics and Statistics*, **25**, 16–39.

2. Bechhofer, R. E., Dunnett, C. W. and Sobel, M. (1954). A two-sample multiple decision procedure for ranking means of normal populations with a common unknown variance, *Biometrika*, **41**, 170–176.

3. Gupta, S. S. (1956). On a decision rule for a problem in ranking means, *Ph.D. Thesis*, Also, Mimeo. Ser. No. 150, Inst. of Statistics, Univ. of North Carolina, Chapel Hill, NC.

4. Rizvi, M .H. (1971). Some selection problems involving folded normal distribution, *Technometrics*, **13**, 355–369.

5. Tamhane, A .C. and Bechhofer, R. E. (1977). A two-stage minimax procedure with screening for selecting the largest normal mean, *Communications In Statistics—Theory and Methods*, **A6**, 1003–1033.

6. Tamhane, A. C. and Bechhofer, R. E. (1979). A two-stage minimax procedure with screening for selecting the largest normal mean (II): An improved PCS lower bound and associated tables *Communications In Statistics—Theory and Methods*, **A8**, 337–358.

17

Conditional Inference for the Parameters of Pareto Distributions when Observed Samples are Progressively Censored

Rita Aggarwala and Aaron Childs

University of Calgary, Calgary, AB, Canada
McMaster University, Hamilton, ON, Canada

Abstract: In this paper we develop procedures for obtaining confidence intervals for the location and scale parameters of a Pareto distribution as well as upper and lower γ probability tolerance intervals for a proportion β when the observed samples are progressively censored. The intervals are exact, and are obtained by conditioning on the observed values of the ancillary statistics. Since the procedures assume that the shape parameter ν is known, a sensitivity analysis is also carried out to see how the procedures are affected by changes in ν.

Keywords and phrases: Ancillary statistics, tolerance intervals, confidence intervals, order statistics, Pareto distribution, progressive censoring, type-II right censoring, best linear unbiased estimators, conditional inference

17.1 Introduction

Progressive Type-II right censoring arises in life-testing and reliability studies when live units are removed or lost from experimentation at points other than the final termination point of an experiment. This loss may occur unintentionally, or it may be designed into the study. Unintentional loss may occur, for example, in the case of accidental breakage of an experimental unit, or if the experimenter loses contact with an individual under study, or if, due to some unforeseen circumstance, such as depletion of funds, experimentation must cease. Removal may be desirable in the case of studies of wear, in which the study of the actual aging process requires items to be fully disassembled at various stages in the experiment.

A request for statistical methodology for the analysis of progressively censored data came from an industrial source dealing with life-testing of electrical components, and appeared in a query in *Technometrics* (1966, p. 539). Although progressive censoring allows for greater flexibility in life-testing experimentation, this article demonstrates that for the Pareto distribution the level of difficulty of the statistical analysis is very similar to that of conventional Type-II right censoring [see Childs and Balakrishnan (1997)] provided that conditional inferences are sought.

Consider the following Type-II censoring scheme: Experimentation begins at time 0 with n units placed on life test. Immediately following the first observed failure, R_1 surviving items are removed from the test at random. Similarly, following the second observed failure, R_2 surviving items are removed from the test at random. This process continues until, immediately following the mth observed failure, the remaining $R_m = n - R_1 - R_2 - \cdots - R_{m-1} - m$ items are all removed from the experiment. In this scheme, $R_1, R_2, \ldots, R_m$ are pre-determined. The resulting m ordered failure times, which we will denote $X_{1:m:n}, X_{2:m:n}, \ldots, X_{m:m:n}$, are referred to as progressive Type-II right censored order statistics. The special case where $R_1 = R_2 = \cdots = R_{m-1} = 0$ so that $R_m = n - m$ is the case of conventional Type-II right censored sampling.

Balakrishnan and Sandhu (1996) and Aggarwala and Balakrishnan (1998) have discussed some properties of progressive Type-II right censored order statistics from exponential and uniform distributions, including parameter estimation using BLUE's and MLE's. A number of authors, including Mann (1969) and Thomas and Wilson (1972) have discussed linear inference for progressive Type-II right censored samples from the Weibull and exponential distributions. Work on conditional inference based on progressively censored data for the exponential and Weibull distributions has been carried out by Viveros and Balakrishnan (1994). For the Pareto distribution, Childs and Balakrishnan (1997) considered the case of conventional Type-II right censored samples ($R_1 = R_2 = \cdots = R_{m-1} = 0$, $R_m = n - m$). A thorough discussion of developments in the area of progressive censoring for specific and general distributions is given in Aggarwala and Balakrishnan (1999).

We show in this paper how to derive confidence intervals for the location and scale parameters, μ and σ, of the Pareto(ν) distribution, as well as tolerance intervals, based on a progressive Type-II right censored sample, using conditional inference. Tolerance intervals are useful in setting quality control limits and designing warranties in the manufacturing industry, which is one place in which progressive censoring can be very useful. A good example of the use of progressive censoring in the reliability industry is given in Montanari and Cacciari (1988). They conducted an aging experiment on insulated cable in which they employed progressive censoring in collecting their life-time data and analyzed the resulting values using inferential techniques based on progressively censored order statistics from the Weibull distribution.

If the failure times of the n items on test are independent and from a continuous population with cumulative distribution function $F(x)$ and probability density function $f(x)$, the joint probability density function of the m progressive Type-II right censored order statistics is given by

$$f_{X_{1:m:n}, X_{2:m:n}, \ldots, X_{m:m:n}}(x_1, x_2, \cdots, x_m) = c \prod_{i=1}^{m} f(x_i) \left[1 - F(x_i)\right]^{R_i}, \quad (17.1)$$

where $c = n(n - R_1 - 1)(n - R_1 - R_2 - 2) \cdots (n - R_1 - R_2 - \cdots - R_{m-1} - m + 1)$.

In this paper we consider progressive Type-II right censored order statistics from the location-scale shifted Pareto(ν) distribution, which has probability density function

$$f(x) = \nu \sigma^{\nu}(x - \mu)^{-(\nu+1)}, \, x \geq \sigma + \mu, \, \nu > 0, \, -\infty < \mu < \infty, \, \sigma > 0 \quad (17.2)$$

and cumulative distribution function

$$F(x) = 1 - \left(\frac{x - \mu}{\sigma}\right)^{-\nu}, \, x \geq \sigma + \mu. \quad (17.3)$$

For a detailed discussion of various aspects of the Pareto distribution one may refer to Arnold (1983). A more concise discussion of the Pareto distribution, including more recent developments, may be found in Chapter 20 of Johnson, Kotz, and Balakrishnan (1994)

In Section 17.2 we first present exact explicit expressions for the best linear unbiased estimators (BLUE's), μ^* and σ^*, of μ and σ derived recently by Aggarwala and Balakrishnan (1997). We then use the BLUE's in Section 17.3 to derive the joint conditional distribution of the pivotal quantities $v = \frac{\mu^* - \mu}{\sigma^*}$ and $w = \frac{\sigma^*}{\sigma}$ given the ancillary statistics

$$u_i = \frac{X_{i:m:n} - \mu^*}{\sigma^*}, \, i = 1, 2, \ldots, m. \quad (17.4)$$

The marginal conditional distribution of the pivotal quantity v is then derived and used to obtain conditional confidence intervals for μ. Conditional confidence intervals for σ are obtained numerically from the joint distribution using the pivotal quantity w.

In Section 17.4 we similarly derive conditional tolerance intervals. We show how to find q such that the intervals $(\mu^* - q\sigma^*, \infty)$ and $(-\infty, \mu^* - q\sigma^*)$ are lower and upper γ probability conditional tolerance intervals for proportion β, respectively. By definition, this is equivalent to finding q such that

$$\Pr\left\{\int_{\mu^* - q\sigma^*}^{\infty} f(x; \mu, \sigma) \, dx \geq \beta \, | \, u_1, u_2, \ldots, u_r\right\} = \gamma \quad (17.5)$$

and
$$\Pr\left\{\int_{-\infty}^{\mu^*-q\sigma^*} f(x;\mu,\sigma)\,dx \geq \beta\,|\,u_1, u_2, \ldots, u_r\right\} = \gamma \tag{17.6}$$
respectively.

In Section 17.5 we give an illustrative example. Since all of the procedures assume that the value of the shape parameter ν is known, in Section 17.6 we carry out a sensitivity analysis to see how the procedures are affected by small changes in ν.

17.2 Best Linear Unbiased Estimation

Let $X_{1:m:n} \leq X_{2:m:n} \leq \cdots \leq X_{m:m:n}$ denote a progressive Type-II right censored sample from the distribution in (17.2), and let $Z_{i:m:n} = (X_{i:m:n}-\mu)/\sigma$, $i = 1, 2, \ldots, m$, be the corresponding order statistics from the standardized distribution. Let us denote $\mathrm{E}(Z_{i:m:n})$ by μ_i, $\mathrm{Var}(Z_{i:m:n})$ by σ_i, and $\mathrm{Cov}(Z_{i:m:n}, Z_{j:m:n})$ by $\sigma_{i,j}$; further, let

$$\mathbf{X} = (X_{1:m:n}, X_{2:m:n}, \ldots, X_{m:m:n})^T,$$

$$\boldsymbol{\mu} = (\mu_1, \mu_2, \ldots, \mu_m)^T,$$

$$\mathbf{1} = \underbrace{(1, 1, \ldots, 1)}_{m}{}^T,$$

and

$$\boldsymbol{\Sigma} = ((\sigma_{i,j})),\, 1 \leq i, j \leq m.$$

Then, the BLUE's of μ and σ are given by

$$\mu^* = \left\{\frac{\boldsymbol{\mu}^T\boldsymbol{\Sigma}^{-1}\boldsymbol{\mu}\mathbf{1}^T\boldsymbol{\Sigma}^{-1} - \boldsymbol{\mu}^T\boldsymbol{\Sigma}^{-1}\mathbf{1}\boldsymbol{\mu}^T\boldsymbol{\Sigma}^{-1}}{(\boldsymbol{\mu}^T\boldsymbol{\Sigma}^{-1}\boldsymbol{\mu})(\mathbf{1}^T\boldsymbol{\Sigma}^{-1}\mathbf{1}) - (\boldsymbol{\mu}^T\boldsymbol{\Sigma}^{-1}\mathbf{1})^2}\right\}\mathbf{X} = \sum_{j=1}^{m} a_j X_{j:m:n}$$

and

$$\sigma^* = \left\{\frac{\mathbf{1}^T\boldsymbol{\Sigma}^{-1}\mathbf{1}\boldsymbol{\mu}^T\boldsymbol{\Sigma}^{-1} - \mathbf{1}^T\boldsymbol{\Sigma}^{-1}\boldsymbol{\mu}\mathbf{1}^T\boldsymbol{\Sigma}^{-1}}{(\boldsymbol{\mu}^T\boldsymbol{\Sigma}^{-1}\boldsymbol{\mu})(\mathbf{1}^T\boldsymbol{\Sigma}^{-1}\mathbf{1}) - (\boldsymbol{\mu}^T\boldsymbol{\Sigma}^{-1}\mathbf{1})^2}\right\}\mathbf{X} = \sum_{j=1}^{m} b_j X_{j:m:n}.$$

For details, refer to David (1981), or Arnold, Balakrishnan and Nagaraja (1992). By inverting the variance-covariance matrix $\boldsymbol{\Sigma}$, Aggarwala and Balakrishnan

(1997) recently derived the following exact explicit expressions for the coefficients a_j and b_j of the BLUE's,

$$a_j = -\frac{1}{\Delta} \sum_{i=1}^{m} \sum_{l=1}^{m} \sum_{k=1}^{m} \mu_i c_{i,k}(\mu_l - \mu_k) c_{l,j} \quad (17.7)$$

and

$$b_j = \frac{1}{\Delta} \sum_{i=1}^{m} \sum_{l=1}^{m} \sum_{k=1}^{m} c_{i,k}(\mu_l - \mu_k) c_{l,j}, \quad (17.8)$$

where

$$\Delta = \left(\sum_{i=1}^{m}\sum_{j=1}^{m} c_{i,j}\right)\left(\sum_{i=1}^{m}\sum_{j=1}^{m} \mu_i \mu_j c_{i,j}\right) - \left(\sum_{i=1}^{m}\sum_{j=1}^{m} \mu_i c_{i,j}\right)^2. \quad (17.9)$$

The elements of the symmetric tri-diagonal inverted variance-covariance matrix $\Sigma^{-1} = (c_{i,j})$ are given by:

$$c_{i,i} = \frac{\gamma_i \gamma_{i+1} - \beta_i^2 \beta_{i+1}^2}{\left(\prod_{k=1}^{i} \gamma_k\right)\left(\gamma_i - \beta_i^2\right)\left(\gamma_{i+1} - \beta_{i+1}^2\right)}, i = 1, 2, ..., m-1$$

$$c_{m,m} = \frac{1}{\left(\prod_{k=1}^{m-1} \gamma_k\right)\left(\gamma_m - \beta_m^2\right)}$$

$$c_{i,i+1} = \frac{\beta_{i+1}}{\left(\prod_{k=1}^{i} \gamma_k\right)\left(\beta_{i+1}^2 - \gamma_{i+1}\right)}, i = 1, 2, ..., m-1.$$

$$c_{i,j} = 0 \text{ otherwise}$$

and the elements of the mean vector (μ_i) are given by $\mu_i = \prod_{k=1}^{i} \beta_k$, $i = 1, 2, \ldots, m$, where

$$\alpha_1 = \nu n$$
$$\alpha_i = \nu(n - R_1 - R_2 - \cdots - R_{i-1} - i + 1), i = 2, 3, \ldots, m$$
$$\beta_i = \frac{\alpha_i}{\alpha_i - 1}, i = 1, 2, \ldots, m, \alpha_i > 1$$
$$\gamma_i = \frac{\alpha_i}{\alpha_i - 2}, i = 1, 2, \ldots, m, \alpha_i > 2.$$

17.3 Conditional Confidence Intervals

Based on a progressive Type-II right censored sample, $X_{1:m:n} \leq X_{2:m:n} \leq \cdots \leq X_{m:m:n}$, for any member of a general location-scale family $f(x) = \frac{1}{\sigma} g\left(\frac{x-\mu}{\sigma}\right)$, the joint conditional probability density function of the pivotal quantities, v and w, given the ancillary statistics, $u_1, u_2, \ldots, u_m$, is given by

$$kw^{m-1} \prod_{i=1}^{m} g\left((u_i + v)w\right) \left[1 - G\left((u_i + v)w\right)\right]^{R_i}, \quad -\infty < v < \infty, w > 0, \quad (17.10)$$

where k is the constant of proportionality; for example, see Lawless (1982) for details. For the Pareto distribution (where $g(x) = \nu \left(\frac{1}{x}\right)^{\nu+1}$), using (17.2) and (17.3), (17.10) gives the joint conditional density function of v and w, given $u_1, u_2, \ldots, u_m$, to be

$$f(v, w \mid u_1, u_2, \ldots, u_m) = \frac{k\nu^m}{w^{\nu n+1}} \prod_{i=1}^{m} \frac{1}{(u_i + v)^{\nu(R_i+1)+1}},$$
$$u_1 + v \geq 0, \ (u_1 + v)w \geq 1. \quad (17.11)$$

To obtain the marginal conditional distribution of v given $u_1, u_2 \ldots u_m$, we integrate (17.11) with respect to w to get

$$f(v \mid u_1, u_2, \ldots, u_m) = \frac{k\nu^{m-1}(u_1 + v)^{\nu n}}{n} \prod_{i=1}^{m} \frac{1}{(u_i + v)^{\nu(R_i+1)+1}}, \ v \geq -u_1.$$
$$(17.12)$$

Once the ancillary statistics have been observed, conditional $100(1-\alpha)\%$ confidence intervals for the parameter μ can be produced by first integrating (17.12) from $-u_1$ to ∞ to find the value of the constant k, then finding two constants d and e such that

$$\Pr\{v > e \mid u_1, u_2, \ldots, u_m\} = \int_{e}^{\infty} f(v \mid u_1, u_2, \ldots, u_m) dv = \alpha/2 \quad (17.13)$$

and

$$\Pr\{v < d \mid u_1, u_2, \ldots, u_m\} = \int_{-u_1}^{d} f(v \mid u_1, u_2, \ldots, u_m) dv = \alpha/2. \quad (17.14)$$

The $100(1-\alpha)\%$ conditional confidence interval for μ is then $(\mu^* - e\sigma^*, \mu^* - d\sigma^*)$ with μ^* and σ^* given Section 17.2.

Conditional $100(1-\alpha)\%$ confidence intervals for the parameter σ can be produced by using the joint density function in (17.11) to find two constants d and e such that

$$\Pr\{w > e \mid u_1, u_2, \ldots, u_m\} = \alpha/2 \quad (17.15)$$

and

$$\Pr\{w < d \mid u_1, u_2, \ldots, u_m\} = \alpha/2. \quad (17.16)$$

This is accomplished by solving the following equations,

Conditional Inference Under Progressive Censoring

$$\int_e^\infty \int_{\frac{1}{w}-u_1}^\infty f(v,w\,|\,u_1,u_2,\ldots,u_m)\,dv\,dw = \alpha/2 \qquad (17.17)$$

and

$$\int_0^d \int_{\frac{1}{w}-u_1}^\infty f(v,w\,|\,u_1,u_2,\ldots,u_m)\,dv\,dw = \alpha/2 \qquad (17.18)$$

for d and e, with $f(v,w\,|\,u_1,u_2,\ldots,u_m)$ given in (17.11). The $100\,(1-\alpha)\,\%$ conditional confidence interval for σ is then $(\sigma^*/e, \sigma^*/d)$.

17.4 Conditional Tolerance Intervals

To find a lower γ probability conditional tolerance interval for proportion β of the form $(\mu^* - q\sigma^*, \infty)$, we first evaluate the integral in (17.5) to get

$$\Pr\left\{w \le \frac{1}{(v-q)\beta^{1/\nu}}\,\bigg|\,u_1,u_2,\ldots,u_m\right\} = \gamma \qquad (17.19)$$

where, as before, $v = \frac{\mu^*-\mu}{\sigma^*}$ and $w = \frac{\sigma^*}{\sigma}$. We now write (17.19) as

$$\int_{-u_1}^\infty \int_{\frac{1}{u_1+v}}^{\max\left\{\frac{1}{u_1+v},\frac{1}{(v-q)\beta^{1/\nu}}\right\}} f(v,w\,|\,u_1,u_2,\ldots,u_m)\,dw\,dv = \gamma \qquad (17.20)$$

which gives

$$\int_{-u_1}^\infty k\nu^{m-1}\,\frac{(u_1+v)^{\nu n} - \left[\min\left\{u_1+v,(v-q)\beta^{1/\nu}\right\}\right]^{\nu n}}{n}$$
$$\times \prod_{i=1}^m \frac{1}{(u_i+v)^{\nu(R_i+1)+1}}\,dv = \gamma. \qquad (17.21)$$

Once the ancillary statistics have been observed, we find the value of q by simply solving the above equation as we will see in the following section.

An upper γ probability conditional tolerance interval for proportion β of the form $(-\infty, \mu^* - q\sigma^*)$ is similarly obtained by solving for q in the equation

$$\int_{-u_1}^\infty k\nu^{m-1}\,\frac{\left[\min\left\{(v-q)(1-\beta)^{1/\nu}, u_1+v\right\}\right]^{\nu n}}{n}\prod_{i=1}^m \frac{1}{(u_i+v)^{\nu(R_i+1)+1}}\,dv = \gamma. \qquad (17.22)$$

17.5 An Example

Aggarwala and Balakrishnan (1997) considered a sample of size $n = 15$, and the following progressive Type-II right censoring scheme: $R_1 = 5$, $R_2 = 0$, $R_3 = 2$, $R_4 = 0$, $R_5 = 3$ (so that $m = 5$). The following progressive Type-II right censored sample from the Pareto(3) distribution with location parameter $\mu = 0$ and scale parameter $\sigma = 5$ was generated,

$$5.11073,\ 5.34932,\ 5.36434,\ 5.70137,\ 5.90067,$$

and the following BLUE's were obtained using the expressions given in Section 17.2: $\mu^* = 1.87680$, $\sigma^* = 3.16207$.

To obtain a 90% confidence interval for μ based on the above progressively censored sample using the conditional approach, we first integrate (17.12) from $-u_1\ (= -1.0227)$ to ∞ to get $k = 13.7598$. Then we numerically solve the equation

$$\int_e^\infty f(v \mid u_1, u_2, \ldots, u_m) dv = .05$$

to get $e = 2.2215$, and we solve $\int_{-u_1}^d f(v \mid u_1, u_2, \ldots, u_m) dv = .05$ to get $d = -.5233$. So the conditional confidence interval is

$$(\mu^* - e\sigma^*, \mu^* - d\sigma^*) = (-5.148, 3.531).$$

A 90% confidence interval for σ, using the conditional approach, is obtained by numerically solving equations (17.17) and (17.18) to get $e = 2.049$ and $d = .3155$. The confidence interval is then $(\sigma^*/e, \sigma^*/d) = (1.543, 10.021)$.

To find a 90% conditional lower tolerance interval for proportion 0.8 based on the censored sample, we take $\beta = 0.8$, $\gamma = 0.9$ in the procedure described in Section 17.4 and solve for q in (17.21) to get $q = -1.043$. We conclude that a lower 0.9 probability tolerance interval for proportion 0.8 is $(\mu^* - q\sigma^*, \infty) = (5.175, \infty)$. Similarly, solving for q in (17.22) to get $q = -2.725$ gives an upper 0.9 probability tolerance interval for proportion 0.8 to be $x < \mu^* - q\sigma^* = 10.493$.

17.6 Sensitivity Analysis

The procedures in Sections 17.3 and 17.4 assume that the shape parameter ν is known. In the example in the previous section, it was assumed that ν was known to be 3. Since ν is normally not known, it must be first estimated. To see how the confidence intervals are affected by changing the value of ν, we calculated, using the data in the previous section, 90% confidence intervals for μ and σ for various values of ν. These intervals are given in the following table.

ν	90% CI for μ	90% CI for σ
2.6	(-3.791, 3.771)	(1.304, 8.674)
2.8	(-4.471, 3.652)	(1.423, 9.350)
3	(-5.148, 3.531)	(1.543, 10.021)
3.2	(-5.821, 3.411)	(1.663, 10.688)
3.4	(-6.490, 3.291)	(1.784, 11.349)

From the table, we see that for this set of data, for all of the values of ν chosen in the sensitivity analysis, the confidence intervals still contain the true values of μ and σ.

Acknowledgments. The authors thank the Natural Sciences and Engineering Research Council of Canada for supporting this research.

References

1. Aggarwala, R. and Balakrishnan, N. (1999). *Progressive Censored Samples: Theory and Application*, New York: Birkhäuser (to appear).

2. Aggarwala, R. and Balakrishnan, N. (1998). Some properties of progressive censored order statistics from arbitrary and uniform distributions, with applications to inference and simulation, *Journal of Statistical Planning and Inference*, **70**, 35–49.

3. Aggarwala, R. and Balakrishnan, N. (1997). Some properties of progressive censored order statistics from the Pareto distribution with applications to inference, *Statistics and Probability Letters*, submitted for publication.

4. Arnold, B.C. (1983). *Pareto Distributions*, Fairland, MD: International Cooperative Publishing House.

5. Arnold, B.C., Balakrishnan, N., and Nagaraja, H.N. (1992). *A First Course in Order Statistics*, New York: John Wiley & Sons.

6. Balakrishnan, N. and Sandhu, R. A. (1996). Best linear unbiased and maximum likelihood estimation for exponential distributions under general progressive Type-II censored samples, *Sankhyā, Series B*, **58**, 1–9.

7. Childs, A. and Balakrishnan, N. (1997). Conditional inference procedures for the Pareto distribution based on Type-II right censored samples, submitted for publication.

8. David, H. A. (1981). *Order Statistics*, second edition, New York: John Wiley & Sons.

9. Johnson N. L., Kotz S., and Balakrishnan N. (1994). *Continuous Univariate Distributions*, Vol. 1, second edition, New York: John Wiley & Sons.

10. Lawless, J. F. (1982). *Statistical Models and Methods for Lifetime Data*, New York: John Wiley & Sons.

11. Mann, N. R. (1969). Exact three-order-statistic confidence bounds on reliable life for a Weibull model with progressive censoring, *Journal of the American Statistical Association*, **64**, 306–315.

12. Montanari, G. C. and Cacciari, M. (1988). Progressively-censored aging tests on XLPE-insulated cable models, *IEEE Transactions on Electrical Insulation*, **23**, 365–372.

13. Thomas, D. R. and Wilson, W. M. (1972). Linear order statistic estimation for the two-parameter Weibull and Extreme-value distributions from Type-II progressively censored samples, *Technometrics*, **14**, 679–691.

14. Viveros, R. and Balakrishnan, N. (1994). Interval estimation of parameters of life characteristics from progressively censored data, *Technometrics*, **36**, 84–91.

PART IV
APPLIED STATISTICS AND RELATED TOPICS

18

On Randomizing Estimators in Linear Regression Models

S. Ermakov and R. Schwabe

St. Petersburg State University, St. Petersburg, Russia
Free University, Berlin, Germany

Abstract: In this work we consider a special kind of randomization in the analysis of linear regression. This randomization is connected to the Δ^2-distribution which was first introduced by Ermakov and Zolotukhin (1960) for decreasing the variance in the Monte Carlo calculation of integrals. The resulting resampling procedure allows for separating the systematic and the random components of the variance. Further, some problems are discussed in designing regression experiments.

Keywords and phrases: Randomization, least squares estimator, linear regression, bootstrap, experimental design

In the article by Ermakov and Zolotukhin (1960) a special distribution was introduced for the nodes in interpolation-cubature formulae which will be called the Δ^2-distribution. This distribution appeared to be efficient in the theory of Monte Carlo methods. Subsequently, Ermakov(1963, 1989) studied interpolation procedures where the nodes were chosen at random according to the Δ^2-distribution. That allowed for extending the application to the field of experimental design and to prove a number of existence theorems for interpolation-cubature formulae [Ermakov(1993)]. Apparently, there is a close connection between those results and the least squares estimators in the analysis of linear models though this question has not been studied, yet, properly. In general there were only common considerations discussed.

Here we will show this connection in detail and obtain some new results concerning resampling procedures as well as connections between D-optimal designs and the average of random interpolation nodes.

18.1 Some Properties of the Δ^2 Distribution

First, we will make use of the following known results [Ermakov (1975)]:

Let $\mathcal{X}$ be a set, μ a probability measure on $\mathcal{X}$ and $f \in L^2(\mu)$ a function of interest on $\mathcal{X}$ which is square integrable with respect to μ.

Instead of f a function ξ, $\xi(x_i) = \xi_i$, $i = 1, 2, ..., m$ may be calculated (observed) at m points $x_1, ..., x_m$ from $\mathcal{X}$. Moreover, we assume that ξ is a random perturbation of f:

$$\xi_i = f(x_i) + \varepsilon_i, \qquad \mathbf{E}\varepsilon_i = 0, \qquad \mathbf{E}(\varepsilon_i \varepsilon_j) = \begin{cases} \sigma^2 &, i = j \\ 0 &, i \neq j \end{cases}$$

$i, j = 1, ..., m$.

A system of m functions $\{\phi_l\}_{l=1}^m$ on $\mathcal{X}$ is selected which is orthonormal with respect to μ. Now the problem is considered of approximating the function f by means of a linear combination $\sum_{l=1}^m c_l \phi_l$ of $\phi_1, ..., \phi_l$.

We will choose the points $x_1, ..., x_m$ at random according to the density

$$p(x_1, ..., x_m) = (m!)^{-1} \left(\det \|\phi_i(x_j)\|_{i,j=1}^m \right)^2$$

(Δ^2-distribution).

Then we solve the problem of interpolation for each set $x_1^{(k)}, ..., x_m^{(k)}$ with observations $\xi_1^{(k)}, ..., \xi_m^{(k)}$ where k is the replication number, i.e. we determine the coefficients $c_l^{(k)}$ from the system of equations

$$\sum_{l=1}^m c_l^{(k)} \phi_l(x_j^{(k)}) = \xi_j^{(k)}. \tag{18.1}$$

Theorem 18.1.1 *The expectation of $c_l^{(k)}$ is equal to the lth Fourier coefficient of the function f with respect to the system $\{\phi_l\}_{l=1}^m$:*

$$\mathbf{E} c_l^{(k)} = \int_{\mathcal{X}} f(x) \phi_l(x) \mu(dx). \tag{18.2}$$

Further, we introduce the definition of D-regularity of a system of functions which weakens standard regularity conditions [Ermakov and Kurotschka (1995)].

Definition A system of functions is D-regular if

$$\mu^m \Big(\{(x_1, ..., x_m) : \det \|\phi_i(x_j)\|_{i,j=1}^m = 0\} \setminus \{(x_1, ..., x_m) : \\ \det \|\phi_1(x_j), ..., \phi_{l-1}(x_j), f(x_j), \phi_{l+1}(x_j), ..., \phi_m(x_j)\|_{j=1}^m = 0\} \Big) = 0$$

for $l = 1, ..., m$ and every function f on $\mathcal{X}$.

This definition is, in particular, suitable for discrete measures μ as it requires regularity only on the support of μ.

Theorem 18.1.2 *If $\phi_1, ..., \phi_m$ is a D-regular system, then*

$$Cov(c_l^{(k)}, c_r^{(k)}) = \begin{cases} \int (f(x) - \sum_{i=1}^m \langle f, \phi_i \rangle \phi_i(x))^2 \mu(dx) + \sigma^2, & l = r \\ 0, & l \neq r \end{cases} \quad (18.3)$$

where $\langle f, \phi_i \rangle = \int f(x)\phi_i(x)\mu(dx)$.

18.2 Least Squares Estimators and Their Randomization

In this section we consider the problem of estimating the coefficients in linear regression in the simplest formulation corresponding to the previous section. Only a finite set of points $x_1, ..., x_N$ is of interest and we use the uniform measure μ which is concentrated on the points $x_1, ..., x_N$ with equal weights $\frac{1}{N}$. It is supposed that data $\xi_1, ..., \xi_N$ can be made available, $\xi_i = f(x_i) + \varepsilon_i$, with $\mathbf{E}\varepsilon_i = 0$, $Var(\varepsilon_i) = \sigma^2$, and $Cov(\varepsilon_i, \varepsilon_j) = 0$ for $i \neq j$.

If we approximate f by a linear combination $\sum_{l=1}^m c_l \phi_l$ where the ϕ_l are linearly independent functions on the set $x_1, ..., x_N$ then the least squares estimators $\hat{c}_l$ of the coefficients c_l are expressed by

$$\hat{c}_l = \frac{\det \|\langle \phi_1, \phi_k \rangle, ..., \langle \phi_{l-1}, \phi_k \rangle, \langle \xi, \phi_k \rangle, \langle \phi_{l+1}, \phi_k \rangle, ..., \langle \phi_m, \phi_k \rangle\|_{k=1}^m}{\det \|\langle \phi_k, \phi_r \rangle\|_{k,r=1}^m} \quad (18.4)$$

where, here, $\langle \phi_k, \phi_r \rangle = \frac{1}{N}\sum_{j=1}^N \phi_k(x_j)\phi_r(x_j)$ and $\langle \xi, \phi_k \rangle = \frac{1}{N}\sum_{j=1}^N \xi_j \phi_k(x_j)$. If the functions ϕ_l are orthonormal on the set of points $x_1, ..., x_N$ then the estimators of the coefficients simplify to

$$\hat{c}_l = \frac{1}{N} \sum_{j=1}^N \xi_j \phi_l(x_j) \quad (18.5)$$

Note that if we use the Cauchy–Binet formula we have

$$\hat{c}_l = \frac{\sum_{1 \leq i_1 < ... < i_m \leq N} \Delta_l(\xi, i_1, ..., i_m) \Delta(i_1, ..., i_m)}{\sum_{1 \leq i < ... < i_m \leq N} \Delta^2(i_1, ..., i_m)} \quad (18.6)$$

where

$$\Delta(i_1, ..., i_m) = \det \|\phi_k(x_{i_j})\|_{k,j=1}^m \quad \text{and}$$

$$\Delta_l(\xi, i_1, ..., i_m) = \det \|\phi(x_{i_j}), ..., \phi_{l-1}(x_{i_j}), \xi(x_{i_j}), \phi_{l+1}(x_{i_j}), ..., \phi_m(x_{i_j})\|_{j=1}^m.$$

Considering the equality of the corresponding products of determinants when indices are interchanged we have

$$\hat{c}_l = \sum_{i_1,...,i_m} \Delta_l(\xi, i_1, ..., i_m) \Delta(i_1, ..., i_m) \Big/ \sum_{i_1,...,i_m} \Delta^2(i_1, ..., i_m). \quad (18.7)$$

If we introduce a probability distribution of sets of points $x_{i_1}, ..., x_{i_m}$ by the density

$$p(i_1, ..., i_m) = \Delta^2(i_1, ..., i_m) \Big/ \sum_{i_1,...,i_m} \Delta^2(i_1, ..., i_m) \quad (18.8)$$

then the random variable χ_l defined by

$$\chi_l(i_1, ..., i_m) = \Delta_l(\xi, i_1, ..., i_m)/\Delta(i_1, ..., i_m) \quad (18.9)$$

has expectation (conditionally on ξ fixed) equal to the value of the least squares estimator $\hat{c}_l$ determined by formula (18.4).

At the same time according to Cramer's rule the values $\chi_l(i_1, ..., i_m)$ are the solutions of the system of equations

$$\sum_{l=1}^{m} \chi_l(i_1, ..., i_m) \phi_l(x_{i_j}) = \xi_{i_j}, \quad j = i, ..., m. \quad (18.10)$$

The corresponding determinant differs from 0 with probability 1.

Thus we see that the distribution (18.8) is a Δ^2-distribution with respect to the uniform measure on $x_1, ..., x_N$. The correctness of the formula (18.3) for the case of D-regular $\{\phi_l\}$ arises from here. In the present notation this formulae can be rewritten in the following way:

$$Cov(\chi_l, \chi_r) = \begin{cases} \frac{1}{N} \sum_{j=1}^{N} \left(f(x_j) - \sum_{i=1}^{m} \langle f, \phi_i \rangle \phi_i(x_j) \right)^2 + \sigma^2, & l = r \\ 0, & l \neq r \end{cases} \quad (18.11)$$

where $\langle f, \phi_l \rangle = \frac{1}{N} \sum_{j=1}^{N} f(x_j) \phi_l(x_j)$.

Next we investigate two possibilities of sampling. Each of them leads to a different value of the final variance.

1. Randomization associated to an active experiment (independent replications are possible): A replication $x_{i_1}, ..., x_{i_m}$ is chosen at random according to the distribution (18.8) and independent values ξ_{i_j} are measured (observed) at each point. If we choose a new replication $x_{i'_1}, ..., x_{i'_m}$ then the observed values $\xi'_{i'_j}$ do not depend on the ξ_{i_j} obtained earlier.

2. Randomization is carried out after that we have got all N values of the function f (bootstrap, passive experiment).

In the first case, when we choose N_1 independent replications $x_{i_1}^{(k)},...,x_{i_m}^{(k)}$, $k=1,...,N_1$, we obtain for the average $\frac{1}{N_1}\sum_{k=1}^{N_1}\chi_l^{(k)}$ of estimators for the lth regression coefficient c_l

$$Var\left(\frac{1}{N_1}\sum_{k=1}^{N_1}\chi_l^{(k)}\right) = \frac{1}{N_1}(S^2+\sigma^2),$$

where

$$S^2 = \frac{1}{N}\sum_{i=1}^{N}\left(f(x_i) - \sum_{l=1}^{m}\langle f,\phi_l\rangle\phi_l(x_i)\right)^2$$

is the systematic error in the approximation of f by $\phi_1,...,\phi_m$.

If, moreover, N_2 repeated independent measurements are possible for each replication $x_{i_1}^{(k)},...,x_{i_m}^{(k)}$ then we have

$$Var\left(\frac{1}{N_1 N_2}\sum_{k=1}^{N_1 N_2}\chi_l^{(k)}\right) = \frac{1}{N_1}\left(S^2+\frac{\sigma^2}{N_2}\right). \tag{18.12}$$

In this case it is possible to separate the systematic component S^2 and the random σ^2 component of the variance when we repeat the experiment in the same m points. This procedure was described earlier by Ermakov (1975). As we can see much more experiments may be needed.

In the second case it is assumed that the experiment is carried out in N points (these are supposed to be different). The only aim of our procedure is to separate the systematic and the random components of the variance (bootstrap procedures).

As the result of the experiment we have got the observations $\xi_1,...,\xi_N$ at $x_1,...,x_N$. $\{\phi_l\}$ is assumed to be a D-regular orthonormal system of functions and $(i_1,...,i_m)$ and $(i'_1,...,i'_m)$ denote two independent samples according to the distribution (18.8).

Lemma 18.2.1 *For the estimators χ_l and χ'_l of the lth regression coefficients based on the replications $(i_1,...,i_m)$ and $(i'_1,...,i'_m)$, respectively, which are defined by (2.6) we have*

$$Cov(\chi_l,\chi'_l) = \frac{\sigma^2}{N}. \tag{18.13}$$

PROOF. Note that the overall covariance can be decomposed into the covariance of the conditional expectation and the expectation of the conditional covariance according to

$$Cov(\chi_l,\chi'_l) = Cov(\mathbf{E}(\chi_l|\varepsilon_1,...,\varepsilon_N),\mathbf{E}(\chi'_l|\varepsilon_1,...,\varepsilon_N)) + \mathbf{E}Cov(\chi_l,\chi'_l|\varepsilon_1,...,\varepsilon_N).$$

Moreover, $\mathbf{E}(\chi_l|\varepsilon_1,...,\varepsilon_N) = \mathbf{E}(\chi'_l|\varepsilon_1,...,\varepsilon_N) = \frac{1}{N}\sum_{i=1}^{N}\xi_j$ and the estimators χ_l and χ'_l are conditionally independent under fixed $\varepsilon_1,...,\varepsilon_N$. Thus the second term in the sum equals 0. For the first we have $Var(\frac{1}{N}\sum_{i=1}^{N}\xi_i) = \sigma^2/N$ which completes the proof. ∎

Now, let the repeated sampling of N_1 independent replications be carried out under the fixed observations $\xi_1,...,\xi_N$. Then

$$Var\left(\frac{1}{N_1}\sum_{k=1}^{N_1}\chi_l^{(k)}\right) = \frac{1}{N_1}Var(\chi_l) + \left(1 - \frac{1}{N_1}\right)Cov(\chi_l,\chi'_l) \qquad (18.14)$$

Here, the estimator $\chi_l^{(k)}$ corresponds to the kth independent replication.
The results can be summarized in the following theorem.

Theorem 18.2.1 *The variance of average of the l-th regression coefficient estimators (18.9) that correspond to N_l independent replicas in a passive experiment is expressed by*

$$Var\left(\frac{1}{N_1}\sum_{k=1}^{N_1}\chi_l^{(k)}\right) = \frac{1}{N_1}S^2 + \frac{N+N_1-1}{NN_1}\sigma^2.$$

PROOF. According to (18.14) we have

$$Var\left(\frac{1}{N_1}\sum_{k=1}^{N_1}\chi_l^{(k)}\right) = \frac{1}{N_1}(S^2 + \sigma^2) + \left(1 - \frac{1}{N_1}\right)\frac{\sigma^2}{N}$$

as was to be proved. ∎

These results enable us to use the randomization rather flexibly as we separate the errors of different kinds in the least squares method.

Note As was shown in Ermakov(1975) a randomized procedure for $N > M > m$ can be constructed. Namely, let as before $x_1,...,x_N$ be points where the experiment is carried out. Let $x_{i_1}^{(k)},...,x_{i_M}^{(k)}$ be the kth replication distributed according to the density

$$p(i_1,...,i_M) = \frac{\det\|[\phi_i,\phi_k]_M\|(i_1,...,i_M)}{\sum_{j_1=1}^{N}\cdots\sum_{j_M=1}^{N}\det\|[\phi_i,\phi_k]_M\|(j_1,...,j_M)} \qquad (18.15)$$

where $[\phi_i,\phi_k]_M = [\phi_i,\phi_k]_M(i_1,...,i_M) = \frac{1}{M}\sum_{r=1}^{M}\phi_i(x_{i_r})\phi_k(x_{i_r})$ and the list of indices in brackets denotes the replication that we use to calculate the least squares estimators

$$\chi_l^{(M)} = \frac{\det\|[\phi_1,\phi_k]_M,...,[\phi_{l-1},\phi_k]_M,[\xi,\phi_k]_M,...,[\phi_m,\phi_k]_M\|(i_1,...,i_M)}{\det\|[\phi_i,\phi_k]_M\|(i_1,...,i_M)}.$$

Then $\mathbf{E}(\chi_l^{(M)}|\xi) = \chi_l^{(N)}$ where $\chi_l^{(N)}$ is the least squares estimator of the coefficient c_l based on all N points. Unfortunately, we did not succeed, yet, in getting an explicit expression similar to (2.8), as we cannot get rid of the denominator in this expression.

The only setting where we are able to calculate $Var\chi_l^{(M)}$ is the case of piecewise constant functions ϕ_l (wavelets, histograms). However, the system $\{\phi_l\}$ is not D-regular in this case [Ermakov (1989)].

18.3 Δ^2-Distribution in Experimental Design

As an example let us consider the situation of polynomial regression on the unit interval [0,1]

$$\phi_l(x) = x^{l-1}, \qquad l = 1, ..., m. \tag{18.16}$$

With respect to the Lebesgue measure the density of the Δ^2-distribution is

$$p(x_1, ..., x_m) = \mathbf{c} \prod_{i<j} (x_i - x_j)^2, \tag{18.17}$$

i.e., up to a constant $\mathbf{c}$ it is equal to the squared Vandermonde determinant. The maximum of this determinant is known [Schur (1918)] and this allows for using the rejection method for simulating the density. By sampling realizations $x_1^{(k)}, ..., x_m^{(k)}$ and putting the received numbers in increasing order $x_{(1)}^{(k)}, ..., x_{(m)}^{(k)}$, where $x_{(1)}^{(k)} \leq x_{(2)}^{(k)} \leq ... \leq x_{(m)}^{(k)}$ and $\{x_{(1)}^{(k)}, ..., x_{(m)}^{(k)}\} = \{x_1^{(k)}, ..., x_m^{(k)}\}$, we can notice that these numbers are classified near some average values. It is interesting to study the behavior of these means $\overline{x_{(i)}} = \frac{1}{N} \sum_{k=1}^{N} x_{(i)}^{(k)}$ and, in particular, to compare these averages to the nodes of optimal experimental designs.

It is known that one of the widely used optimality criteria, the D-criterion, leads to experimental designs where the global maximum in (18.17) is attained. Consequently, the nodes of this design are the roots of the polynomial $x(1-x)P'_{m-1}(x)$, where P'_n is the derivative of the Legendre polynomial P_m of degree m.

On the other side, the so-called unbiased optimal designs (designs that minimize systematic error) are located at the roots of the Legendre polynomial [Ermakov and Sedunov (1979)].

Finally, we consider some results for the simulation of the Δ^2-distribution [Schwabe (1992)]. In some cases exact calculations are possible for the expectations of the ordered samples $x_{(1)}, ..., x_{(m)}$.

In order to get an impression of the shape of the randomization measure associated with the Δ^2-distribution for fitting polynomials (18.16) we generated random numbers by using the von Neumann rejection method. For $m = 2$, i.e.

linear regression, we generated $N = 10,000$ samples (x_1, x_2), while $N = 1,000$ were chosen for $m = 3$ and 5, and $N = 200$ for $m = 10$. The means $\overline{x_{(1)}}, ..., \overline{x_{(m)}}$ have been calculated from the ordered samples to get an estimate for their corresponding expectations $\mathbf{E}x_{(1)}, ..., \mathbf{E}x_{(m)}$. For lower degrees up to $m = 5$ exact results were obtained for the expectations.

Despite the apparent closeness in the simulation we can notice that the means are a little bit more scattered than the roots of P_n and that they are shifted towards the corresponding nodes of the D-optimal design, which are also listed in Table 18.1. This observation led us to investigate the distribution of the smallest component in the sample more closely.

Table 18.1

$m = 2$				$m = 3$			
P_2	$\overline{x_{(l)}}$	$\mathbf{E}x_{(l)}$	D-opt.	P_3	$\overline{x_{(l)}}$	$\mathbf{E}x_{(l)}$	D-opt.
0.2113	0.2005	0.2000	0.0000	0.1127	0.0988	0.1000	0.0000
0.7887	0.7980	0.8000	1.0000	0.5000	0.4986	0.5000	0.5000
				0.8873	0.9008	0.9000	1.0000

$m = 5$				$m = 10$		
P_5	$\overline{x_{(l)}}$	$\mathbf{E}x_{(l)}$	D-opt.	P_{10}	$\overline{x_{(l)}}$	D-opt.
0.0469	0.0376	0.0385	0.0000	0.0130	0.0107	0.0000
0.2308	0.2116	0.2178	0.1727	0.0675	0.0621	0.0402
0.5000	0.4965	0.5000	0.5000	0.1603	0.1536	0.1306
0.7692	0.7816	0.7822	0.8273	0.2833	0.2828	0.2610
0.9531	0.9615	0.9615	1.0000	0.4256	0.4231	0.4174
				0.5744	0.5838	0.5826
				0.7167	0.7274	0.7390
				0.8397	0.8512	0.8694
				0.9325	0.9431	0.9598
				0.9870	0.9896	1.0000

Lemma 18.3.1 *For the Δ^2-distribution defined in (18.17) the smallest component $x_{(1)} = \min(x_1, ..., x_m)$ has distribution function $F(x) = 1 - (1-x)^{m^2}$, density $f(x) = m^2(1-x)^{m(m-1)}$ and expectation $\mathbf{E}x_{(1)} = \frac{1}{m^2+1}$.*

The expectations $\mathbf{E}x_{(1)}$ listed in Table 18.1 and 18.2 are by a multiplicative factor smaller than the corresponding smallest roots $z_1^{(m)}$ of the Legendre polynomials P_m. In particular, for large degree m we have $z_1^{(m)} \sim \frac{1}{4}j_1^2 m^{-2} \approx 1.45 m^{-2}$ where j_1 is the smallest positive root of the Bessel function of the first kind of order zero [Szegö (1959)] and, thus, $\mathbf{E}x_{(1)} \sim m^{-2} \approx 0.69 z_1^{(m)}$, i.e. $\mathbf{E}x_{(1)}$ decreases with the same speed as $z_1^{(m)}$ to zero, which is the corresponding smallest node of the D-optimal design.

Table 18.2

m	6	7	8	9	10
$z_1^{(m)}$	0.0338	0.0254	0.0199	0.0159	0.0130
$\mathbf{E}x_{(1)}$	0.0270	0.0200	0.0154	0.0122	0.0099

More detailed investigation of the one- and two-dimensional marginal distribution of the order statistics $(x_{(1)}, ..., x_{(n)})$ can be found in Schwabe (1992).

The results obtained in the present section give rise to the impression that the randomized design associated with the Δ^2-distribution proposed by Ermakov (1975) is close to both the roots of P_m and to the nodes of the D-optimal design which are known to be of great use in approximation theory and experimental design, respectively.

Acknowledgments. This work was supported by the RFBR grant 96-01-00015G.

References

1. Ermakov, S. M. (1963). Interpolation on Random Points, *Journal of Comput. Math. and Math. Physics*, **3**, N 1, 550–554 (in Russian).

2. Ermakov, S. M. (1975). *Die Monte Carlo Methode und Verwandte Fragen*, Berlin: Deutscher Verlag der Wissenschaften; and München: Oldenbourg.

3. Ermakov, S. M. (1989). Random interpolation in the theory of experimental design, *Computational Statistics and Data Analysis*, **8**, 75–80.

4. Ermakov, S. M. (1993). On metric theorems in the interpolation theory, *Vestnik St Petersburg State Univ., ser. math., mech., astron.*, **1**, 26–29 (in Russian).

5. Ermakov, S. M. and Kurotschka, V. G. (1995). Zufällige Replikationen von Regressionsexperimenten in der Modellbildung, *Preprint Nr. A-31/95*, Freie Universität, Berlin, Fachbereich Mathematik.

6. Ermakov, S. M. and Sedunov, E. V. (1979). On unbiased designs for regression experiments in a finite-dimensional function space, *Vestnik Leningrad State Univ., ser. math., mech., astron.*, **7**, 11–21 (in Russian).

7. Ermakov, S. M. and Zolotukhin, V. G. (1960). Polynomial approximations and Monte Carlo method, *Journal Probability Theory Applic.*, **5**, N 4, 473–476 (in Russian).

8. Schur, J. (1918). Über die Verteilung der Wurzeln bei gewissen algebraischen Gleichungen mit ganzzahligen Koeffizienten, *Math. Z.*, **1**, 373–385.

9. Schwabe, R. (1992). On randomized designs for regression, In: *Model-Oriented Data Analysis. A Survey of Recent Methods* (Eds., V. Fedorov, W. G. Müller and I. N. Vuchkov), pp. 53–66. Heidelberg: Physica.

10. Szegö, G. (1959). *Orthogonal polynomials*, revised ed. American Math. Soc. Coll. Pub., **23**.

19

Nonstationary Generalized Automata with Periodically Variable Parameters and Their Optimization

A. Yu. Ponomareva and M. K. Tchirkov

St. Petersburg State University, St. Petersburg, Russia

Abstract: In this paper, principal conceptions of mathematical theory and optimization problems for generalized (defined under any associative body) periodically nonstationary finite automata with partially defined initial conditions and with periodically variation of all their structural parameters are formulated. Some special matrix methods and algorithms to construct reduced and minimal forms for these automata are theoretically proved and developed.

Keywords and phrases: Generalized automaton, periodically variable parameters, families of basic matrices, states minimization, reduced and minimal forms, matrix method, algorithm of optimization.

19.1 Introduction

The theory of mathematical models of finite automata have been actively developed over several decades. This is due to the theoretical interest to new problems of discrete mathematics and also due to the important fact that finite automata of different kinds are convenient for mathematical models describing a certain class of systems, devices, processes, and phenomena.

In the last years different kinds of generalized automata over certain algebraic systems are an important part of these models. In addition, one of the most important in mathematical theory of models of finite-automaton type turns out to be the problem of search for optimal forms of their representation under different criterions of optimality. One of the most actual classic optimization problems for such models is a problem of minimization for the number of system's states and the construction of their reduced (minimal) forms.

However it necessary to emphasize that presently there are more or less effective optimizing methods for completely definite finite-automaton models with constant structure such as stationary generalized finite automata given over noncommutative (in the general case) fields (or associative bodies) [Tchirkov and Shestakov (1990–1992)]. In the cases that a finite-automaton model is partially defined or has a variable structure the mathematical problems of its optimization have been studied only for several partial cases [see, for example, the book of Tchirkov (1983) and the papers of Miroshnichenko and Tchirkov (1990, 1994)]. At the same time for many problems (such as control, pattern recognition, artificial intelligence, the processing of large information arrays, coding and decoding of discrete messages, designing devices of automatics, telemechanics, and computers) under conditions of inaccuracy or uncertainty of information on processes being investigated or in the case of the other alternative variants and also in the case that the time dependence takes place it is required to solve pure mathematical problems of optimization of nondeterministic partially defined generalized mathematical models of finite-automaton type (including models having variable structure) given over different algebraic systems and the development of the corresponding methods and algorithms.

The prime object of this work is to develop on the basis of ideas and methods of the previous works [Tchirkov and Shestakov (1990–1992)] the analogous approach to problems involving theoretical justification and development of matrix optimization methods for a more general finite automaton model of a new type. This model is a generalized nonstationary finite automaton which is given over arbitrary associative body and have partially definite initial state weights and periodically variable structure of all parameters (input and output alphabets, state alphabet, weights of transitions and final state weights). In the very partial case a simular problem can be reduced to the minimization of a stationary automaton with much more number of states [Tchirkov, and Miroshnichenko (1990, 1994)]. In the general case it is required to develop the special methods [Ponomareva and Tchirkov (1998)].

19.2 Base Definitions and Problem Setting

Let an arbitrary (noncommutative in the general case) field $\mathcal{F}$ be given (it is also called an associative body). We shall say that a *generalized finite automaton with periodically variable parameters over the field* $\mathcal{F}$ is the following system

$$\mathcal{A}_{gv} = \langle X^{(\tau)}, A^{(\tau)}, Y^{(\tau)}, \tilde{\mathbf{r}}, \{\mathbf{R}^{(\tau)}(s,l)\}, \mathbf{q}^{(\tau)}, \tau = \overline{1,T} \rangle, \tag{19.1}$$

where T is a repeating period of automaton parameters; τ is a number of tact in the period; $X^{(\tau)}, Y^{(\tau)}$ are alphabets of input and output symbols admissible in the τ-th tact of the period, $|X^{(\tau)}| = n_\tau$, $|Y^{(\tau)}| = k_\tau$; $A^{(\tau)}$ is an alphabet of automaton states in the τ-th tact of the period, $|A^{(\tau)}| = m_\tau$; $\tilde{\mathbf{r}}$ is a matrix of dimension $q \times m_T$ with elements from $\mathcal{F}$, whose lines form the basis of a space $\mathcal{L}(\tilde{\mathbf{r}})$ of the admissible at the initial moment $t = 0$ initial vectors of statements' weights; $\{\mathbf{R}^{(\tau)}(s,l)\}$, $x_s \in X^{(\tau)}, y_l \in Y^{(\tau)}$ are a set $n_\tau k_\tau$ of rectangular $(m_{\tau-1} \times m_\tau)$-matrices of transition's weights with elements from $\mathcal{F}$; $\mathbf{q}^{(\tau)}$ is a column-vector with m_τ elements from $\mathcal{F}$ which gives final states's weights in the τ-th tact of the period.

At the initial moment $t = 0$ (in zero tact) the parameters of the automaton $\mathcal{A}_{gv}$ are given by the alphabet of state $A^{(T)}$ and by any initial state's weight vector $\mathbf{r} \in \mathcal{L}(\tilde{\mathbf{r}})$. For an arbitrary current moment (tact) $t = 1, 2, \ldots$ the parameters of the automaton $\mathcal{A}_{gv}$ are defined by the number of tact in the period $\tau = \tau(t) = (t-1)(\text{mod}T) + 1$.

We shall say that any pair of words (w, v), $w = x_{s_1} x_{s_2} \ldots x_{s_d}$, $v = y_{l_1} y_{l_2} \ldots y_{l_d}$ of the same length *admissible* for the automaton $\mathcal{A}_{gv}$ if for all $t = \overline{1, d}$ relations $x_{s_t} \in X^{(\tau)}$, $y_{l_t} \in Y^{(\tau)}$ are satisfied. Denote by Z_{adm} a set of all pairs of words (w, v) of any length $d = 0, 1, 2, \ldots$ which are admissible for the automaton $\mathcal{A}_{gv}$. Then a set $\tilde{\Phi} = \{\Phi_r | \mathbf{r} \in \mathcal{L}(\tilde{\mathbf{r}})\}$ of generalized maps $\Phi_r : Z_{\text{adm}} \to \mathcal{F}$ to be induced by the automaton $\mathcal{A}_{gv}$ is defined by the following expression

$$\Phi_r(w,v) = \mathbf{r} \prod_{t=1}^{d} \mathbf{R}^{(\tau(t))}(s_t, l_t) \mathbf{q}^{(\tau(d))},$$

where $(w,v) \in Z_{\text{adm}}$ and $\tau(d) = (d-1)(\text{mod}T) + 1$, $\mathbf{q}^{(0)} = \mathbf{q}^{(T)}$.

A problem of constructing a minimal form of the automaton $\mathcal{A}_{gv}$ (denoted by Min$\mathcal{A}_{gv}$) is as follows. In the set of generalized finite automata with periodically variable parameters having the same $T, X^{(\tau)}, Y^{(\tau)}$ and inducing the same set of generalized maps as the automaton $\mathcal{A}_{gv}$ it is required to find an automaton which has the least possible number of states in each tact of the period.

19.3 The Basic Matrices of the Automaton $\mathcal{A}_{gv}$

Consider an arbitrary tact of period τ and all possible instants (current tacts) having values $d_1 = lT + \tau$, $l = 0, 1, \ldots$. Denote by $\mathcal{L}(\mathcal{A}_{gv}^{(\tau)} \mathbf{q}^{(\tau)})$ a vector space generated by a column-vector $\mathbf{q}^{(\tau)}$ and by all column-vectors of the form

$$\mathbf{h}_q(w_2, v_2) = \prod_{t=d_1+1}^{d} \mathbf{R}^{(\tau(t))}(s_t, l_t) \mathbf{q}^{(\tau(d))}, \quad (w_1 w_2, v_1 v_2) \in Z_{\text{adm}}, \quad (19.2)$$

(where $w_1 w_2 = w$, $v_1 v_2 = v$). Let any matrix $\mathbf{H}_q^{(\tau)}$ of size $(m_\tau \times \eta_\tau)$, where $\eta_\tau = \dim \mathcal{L}(\mathcal{A}_{gv}^{(\tau)} \mathbf{q}^{(\tau)})$, such that its columns form the basis in the space $\mathcal{L}(\mathcal{A}_{gv}^{(\tau)} \mathbf{q}^{(\tau)})$, be called a *right-hand basic matrix* of the automaton $\mathcal{A}_{gv}$ in the τ-th tact of the period. We shall also say that a collection of matrices $\mathbf{H}_q^{(\tau)}$, $\tau = \overline{1,T}$ is a *family of the right-hand basic matrices* of the automaton $\mathcal{A}_{gv}$.

Analogously, denote by $\mathcal{L}(\widetilde{\mathbf{r}}\mathcal{A}_{gv}^{(\tau)})$ a vector space which for a given $\tau \in \overline{1, T-1}$ is generated by all vector-lines of the form

$$\mathbf{h}_r(w_1, v_1) = \mathbf{r} \prod_{t=1}^{d_1} \mathbf{R}^{(\tau(t))}(s_t, l_t), \quad \mathbf{r} \in \mathcal{L}(\widetilde{\mathbf{r}}), \ (w_1, v_1) \in Z_{\text{adm}}, \qquad (19.3)$$

and for a given $\tau = T$ by all vectors of form (19.3) and by the lines of matrix the $\widetilde{\mathbf{r}}$. Let us introduce the following notation. Any matrix $\mathbf{H}_r^{(\tau)}$ of size $(\xi_\tau \times m_\tau)$, where $\xi_\tau = \dim \mathcal{L}(\widetilde{\mathbf{r}}\mathcal{A}_{gv}^{(\tau)})$, such that its lines form the basis in space $\mathcal{L}(\widetilde{\mathbf{r}}\mathcal{A}_{gv}^{(\tau)})$, is called a *left-hand basic matrix* of the automaton $\mathcal{A}_{gv}$ in the τ-th tact of the period. A collection of matrices $\mathbf{H}_r^{(\tau)}$, $\tau = \overline{1,T}$ is called a *family of the left-hand basic matrices* of the automaton $\mathcal{A}_{gv}$.

According to the definitions of the right-hand $\mathbf{H}_q^{(\tau)}$ and the left-hand $\mathbf{H}_r^{(\tau)}$ basic matrices they have an infinite set of forms of representation, which are the following

$$\widetilde{\mathbf{H}}_q^{(\tau)} = \mathbf{H}_q^{(\tau)} \alpha, \quad \widetilde{\mathbf{H}}_r^{(\tau)} = \beta \mathbf{H}_r^{(\tau)}, \qquad (19.4)$$

where α, β are arbitrary nondegenerate square matrices of sizes $(\eta_\tau \times \eta_\tau)$ and $(\xi_\tau \times \xi_\tau)$ with elements from $\mathcal{F}$. Separating out any η_τ linearly independent lines $\mathbf{h}_{i_\nu}, \nu = \overline{1, \eta_\tau}$ in the matrix $\mathbf{H}_q^{(\tau)}$ and any ξ_τ linearly independent columns in the matrix $\mathbf{H}_r^{(\tau)}$, forming matrices $\alpha^{-1} = (\mathbf{h}_{i_\nu})_{\nu=\overline{1,\eta_\tau}}$, $\beta^{-1} = (\mathbf{h}^{j_\sigma})_{\sigma=\overline{1,\xi_\tau}}$, and finding their inverse matrices α, β we can find by transformation (19.4) *normalized* forms of representation for basic matrices $\widetilde{\mathbf{H}}_q^{(\tau)}$, $\widetilde{\mathbf{H}}_r^{(\tau)}$ such that in the matrix $\widetilde{\mathbf{H}}_q^{(\tau)}$ the lines with numbers $i_\nu, \nu = \overline{1, \eta_\tau}$ have the form $\widetilde{\mathbf{h}}_{i_\nu} = \mathbf{h}(\nu) = (0, ..., 0, 1, 0, ...0)$, where the element "1" is in the ν-th position while in the matrix $\widetilde{\mathbf{H}}_r^{(\tau)}$ the columns with numbers $j_\sigma, \sigma = \overline{1, \xi_\tau}$ take the form $\widetilde{\mathbf{h}}^{j_\sigma} = \mathbf{h}^{TR}(\sigma)$.

For the basic matrices being in the normalized form $\widetilde{\mathbf{H}}_q^{(\tau)}$, $\widetilde{\mathbf{H}}_r^{(\tau)}$ from relations $\widetilde{\mathbf{H}}_q^{(\tau)I} \widetilde{\mathbf{H}}_q^{(\tau)} = \mathbf{I}(\eta_\tau)$, $\widetilde{\mathbf{H}}_r^{(\tau)} \widetilde{\mathbf{H}}_r^{(\tau)I} = \mathbf{I}(\xi_\tau)$, where $\mathbf{I}(\eta_\tau)$, $\mathbf{I}(\xi_\tau)$ are the unit matrices of sizes $(\eta_\tau \times \eta_\tau)$, $(\xi_\tau \times \xi_\tau)$, we can find the corresponding *pseudoinverse* matrices $\widetilde{\mathbf{H}}_q^{(\tau)I}$, $\widetilde{\mathbf{H}}_r^{(\tau)I}$ by means of the following trivial method: matrix $\widetilde{\mathbf{H}}_q^{(\tau)I}$ is an $(\eta_\tau \times m_\tau)$-matrix such that all its columns with numbers $i \neq i_\nu$, $\nu = \overline{1, \eta_\tau}$ consist of zeros and the columns with numbers i_ν, $\nu = \overline{1, \eta_\tau}$ take the form $\mathbf{h}^{TR}(\nu)$, $\nu = \overline{1, \eta_\tau}$, and matrix $\widetilde{\mathbf{H}}_r^{(\tau)I}$ is an $(m_\tau \times \xi_\tau)$-matrix such that all its lines with numbers $j \neq j_\sigma$, $\sigma = \overline{1, \xi_\tau}$ consist of zeros and the lines with numbers j_σ $\sigma = \overline{1, \xi_\tau}$ take the form $\mathbf{h}(\sigma)$, $\sigma = \overline{1, \xi_\tau}$.

19.4 Algorithms for Construction of the Families of Basic Matrices

For construction of the family of the right-hand basic matrices $\mathbf{H}_q^{(\tau)}$, $\tau = \overline{1,T}$ the following recursive procedure can be used:

1. The zero approximation: all matrices $\mathbf{H}_q^{(\tau)}$, $\tau = \overline{1,T}$ are "empty" matrices containing no one column. The first approximation: each matrix $\mathbf{H}_q^{(\tau)}$ is a matrix of one column $\mathbf{H}_q^{(\tau)} = (\mathbf{q}^{(\tau)})$, $\tau = \overline{1,T}$.

2. If the $(\nu - 1)$-th and the ν-th approximations of the family of matrices $\mathbf{H}_q^{(\tau)}$, $\tau = \overline{1,T}$ are constructed and if in the ν-th approximation of the matrix $\mathbf{H}_q^{(\tau)}$ some new (in comparison with its $(\nu - 1)$-th approximation) basic column-vectors $\mathbf{h}_1^{(\tau)}, ... \mathbf{h}_{i_\nu}^{(\tau)}$ are added, then the $(\nu + 1)$-th approximation of the matrix $\mathbf{H}_q^{(\tau-1)}$ (here for $\tau = 1$ $\mathbf{H}_q^{(0)} = \mathbf{H}_q^{(T)}$) is obtained by adding to its ν-th approximation columns such that the columns of $(\nu + 1)$-th approximation of the matrix $\mathbf{H}_q^{(\tau-1)}$ form a basis in the space generated by columns of its ν-th approximation and by all the possible column-vectors of the form

$$\mathbf{h}^{(\tau-1)}(s,l) = \mathbf{R}^{(\tau)}(s,l)\mathbf{h}_j^{(\tau)}, \quad j = \overline{1, i_\nu}, \ x_s \in X^{(\tau)}, \ y_l \in Y^{(\tau)}. \quad (19.5)$$

The fulfillment of this rule sequentially for all $\tau = \overline{1,T}$ leads to the construction of the $(\nu + 1)$-th approximation of the family of the right-hand basic matrices.

3. The construction of the family of the right-hand basic matrices is ended if the $(\nu+1)$-th approximation obtained by the procedure of item 2 coincides with the ν-th approximation.

The analogous procedure for construction of the family of the left-hand basic matrices $\mathbf{H}_r^{(\tau)}, \tau = \overline{1,T}$, is as follows:

1. The zero approximation: all $\mathbf{H}_r^{(\tau)}$, $\tau = \overline{1,T}$, are "empty" matrices. The first approximation: all $\mathbf{H}_r^{(\tau)}$, $\tau = \overline{1, T-1}$, are "empty" matrices, and $\mathbf{H}_r^{(T)} = \tilde{\mathbf{r}}$.

2. If the $(\nu - 1)$-th and ν-th approximations of the family $\mathbf{H}_r^{(\tau)}$, $\tau = \overline{1,T}$ are constructed and in the ν-th approximation of the matrix $\mathbf{H}_q^{(\tau)}$ (in comparison with its $(\nu-1)$-th approximation) some new basic line-vectors $\mathbf{r}_1^{(\tau)}, ..., \mathbf{r}_{i_\nu}^{(\tau)}$ are added, then the $(\nu + 1)$-th approximation of the matrix $\mathbf{H}_r^{(\tau+1)}$ (here for $\tau = T$ $\mathbf{H}_r^{(T+1)} = \mathbf{H}_r^{(1)}$) is obtained by adding to its ν-th

approximation lines such that the lines of the $(\nu+1)$-th approximation of the matrix $\mathbf{H}_r^{(\tau+1)}$ form a basis in the space, generated by the lines of its ν-th approximation and by all the possible line-vectors of the form

$$\mathbf{r}^{(\tau+1)}(s,l) = \mathbf{r}_j^{(\tau)} \mathbf{R}^{(\tau)}(s,l), \quad j=\overline{1,i_\nu}, \ x_s \in X^{(\tau)}, \ y_l \in Y^{(\tau)}.$$

3. The construction of the family of the left-hand basic matrices comes to an end if the $(\nu+1)$-th approximation obtained by the procedure of item 2 coincides with the ν-th approximation.

19.5 Two Properties of Basic Matrices

It can be easily shown that for arbitrary left-hand and right-hand basic matrices of the automaton $\mathcal{A}_{gv}$ the following statements hold.

Lemma 19.5.1 *Let $\mathcal{A}_{gv}$ be a generalized automaton with periodically variable parameters (19.1), $\mathcal{L}(\mathcal{A}_{gv}^{(\tau)}\mathbf{q}^{(\tau)})$ be a vector space generated by a vector-column $\mathbf{q}^{(\tau)}$ and by all the vector-columns of form (19.2), $\widetilde{\mathbf{H}}_q^{(\tau)}$ be a right-hand basic matrix (in the normalized form) of the automaton $\mathcal{A}_{gv}$ in the τ-th tact of the period, and $\widetilde{\mathbf{H}}_q^{(\tau)I}$ be its pseudoinverse matrix. Then for all $\mathbf{q} \in \mathcal{L}(\mathcal{A}_{gv}^{(\tau)}\mathbf{q}^{(\tau)})$ the following equality*

$$\widetilde{\mathbf{H}}_q^{(\tau)} \widetilde{\mathbf{H}}_q^{(\tau)I} \mathbf{q} = \mathbf{q}. \tag{19.6}$$

is satisfied.

PROOF. Let $\widetilde{\mathbf{H}}_q^{(\tau)} = (\widetilde{\mathbf{h}}^j)_{j=\overline{1,\eta_\tau}}$, such that by definition of a right-hand basic matrix its vector-columns $\widetilde{\mathbf{h}}^j = (\widetilde{\mathbf{h}}_1^j, \ldots, \widetilde{\mathbf{h}}_{m_\tau}^j)^{TR}, \ j=\overline{1,\eta_\tau}$ make up a basis in the vector space $\mathcal{L}(\mathcal{A}_{gv}^{(\tau)}\mathbf{q}^{(\tau)})$. Whence it follows that for any vector-column $\mathbf{q} \in \mathcal{L}(\mathcal{A}_{gv}^{(\tau)}\mathbf{q}^{(\tau)})$ there exists a vector-column of coefficients $\alpha = (\alpha_1, \ldots, \alpha_{\eta_\tau})^{TR}$, $\alpha_i \in \mathcal{F}$, such that $\mathbf{q} = \widetilde{\mathbf{H}}_q^{(\tau)}\alpha$. In this case, taking into account the properties of pseudoinverse matrices,

$$\widetilde{\mathbf{H}}_q^{(\tau)} \widetilde{\mathbf{H}}_q^{(\tau)I} \mathbf{q} = \widetilde{\mathbf{H}}_q^{(\tau)} \widetilde{\mathbf{H}}_q^{(\tau)I} \widetilde{\mathbf{H}}_q^{(\tau)} \alpha = \widetilde{\mathbf{H}}_q^{(\tau)} \mathbf{I}(\eta_\tau) \alpha = \widetilde{\mathbf{H}}_q^{(\tau)} \alpha = \mathbf{q},$$

which proves this lemma. ∎

Lemma 19.5.2 *Let $\mathcal{A}_{gv}$ be a generalized automaton with periodically variable parameters (19.1), $\mathcal{L}(\widetilde{\mathbf{r}}\mathcal{A}_{gv}^{(\tau)})$ be a vector space generated for $\tau \in \overline{1,T-1}$ by all vectors of form (19.3), and for $\tau = T$ by all the lines of the matrix $\widetilde{\mathbf{r}}$ and by* all vectors of form (19.3), $\widetilde{\mathbf{H}}_r^{(\tau)}$ be a left-hand basic matrix (in the normalized form) of the automaton $\mathcal{A}_{gv}$ in the τ-th tact of the period, and $\widetilde{\mathbf{H}}_r^{(\tau)I}$ be its pseudoinverse matrix. Then for all $\mathbf{r} \in \mathcal{L}(\widetilde{\mathbf{r}}\mathcal{A}_{gv}^{(\tau)})$ the following equality holds:

$$\mathbf{r}\widetilde{\mathbf{H}}_r^{(\tau)I}\mathbf{H}_r^{(\tau)} = \mathbf{r} \; . \tag{19.7}$$

PROOF. Let $\widetilde{\mathbf{H}}_r^{(\tau)} = (\widetilde{\mathbf{h}}_i)_{i=\overline{1,\xi_\tau}}$, such that by definition of a left-hand basic matrix its vector-lines $\mathbf{h}_i = (\widetilde{h}_{i1},\ldots,\widetilde{h}_{im_\tau})$, $i = \overline{1,\xi_\tau}$ form a basis in the vector space $\mathcal{L}(\widetilde{\mathbf{r}}\mathcal{A}_{gv}^{(\tau)})$. Then for any vector-line $\mathbf{r} \in \mathcal{L}(\widetilde{\mathbf{r}}\mathcal{A}_{gv}^{(\tau)})$ there exists a vector-line of coefficients $\beta = (\beta_1,\ldots,\beta_{\xi_\tau})$, $\beta_j \in \mathcal{F}$ such that $\mathbf{r} = \beta\widetilde{\mathbf{H}}_r^{(\tau)}$. In this case, taking into account the properties of pseudoinverse matrices, we have

$$\mathbf{r}\widetilde{\mathbf{H}}_r^{(\tau)I}\widetilde{\mathbf{H}}_r^{(\tau)} = \beta\widetilde{\mathbf{H}}_r^{(\tau)}\widetilde{\mathbf{H}}_r^{(\tau)I}\widetilde{\mathbf{H}}_r^{(\tau)} = \beta\mathbf{I}(\xi_\tau)\widetilde{\mathbf{H}}_r^{(\tau)} = \beta\widetilde{\mathbf{H}}_r^{(\tau)} = \mathbf{r} \; ,$$

which proves the lemma. ∎

19.6 Reduced and Minimal Forms of the Automaton

Let us say that automaton (19.1) *is the left-hand reduced one (or it is of the left-hand reduced form)* if a family of its left-hand basic matrices in the normalized form is $\widetilde{\mathbf{H}}_r^{(\tau)} = \mathbf{I}(m_\tau)$, $\tau = \overline{1,T}$, i.e., consists of unit matrices only. Respectively, we shall say that automaton (19.1) *is the right-hand reduced one (or it is of the right-hand reduced form)* if a family of its right-hand matrices in the normalized form takes the form $\widetilde{\mathbf{H}}_q^{(\tau)} = \mathbf{I}(m_\tau)$, $\tau = \overline{1,T}$, i.e., consists of unit matrices only.

Assume that over the field $\mathcal{F}$ two generalized finite automata with periodically variable parameters

$$\mathcal{A}_{yv} = \langle X^{(\tau)}, A^{(\tau)}, Y^{(\tau)}, \widetilde{\mathbf{r}}_A, \{\mathbf{R}_A^{(\tau)}(s,l)\}, \mathbf{q}_A^{(\tau)}, \tau = \overline{1,T} \rangle \; , \tag{19.8}$$

$$\mathcal{B}_{gv} = \langle X^{(\tau)}, B^{(\tau)}, Y^{(\tau)}, \widetilde{\mathbf{r}}_B, \{\mathbf{R}_B^{(\tau)}(s,l)\}, \mathbf{q}_B^{(\tau)}, \tau = \overline{1,T} \rangle \; , \tag{19.9}$$

are given. Let these automata have the same repeating period T, the input $X^{(\tau)}$ and output $Y^{(\tau)}$ alphabets coinciding in each tact and, consequently, the same set Z_{adm} of admissible pairs of input and output words. Automata (19.8) and (19.9) are said to be *equivalent* (denotation $\mathcal{A}_{gv} \sim \mathcal{B}_{gv}$) if the sets $\widetilde{\Phi}_A$ and $\widetilde{\Phi}_B$ of generalized maps, induced by them, coincide: $\widetilde{\Phi}_A = \widetilde{\Phi}_B$, i.e. if for any $\mathbf{r}_A \in \widetilde{\mathbf{r}}_A$ there exists $\mathbf{r}_B \in \widetilde{\mathbf{r}}_B$ (and, conversely, for any $\mathbf{r}_B \in \widetilde{\mathbf{r}}_B$ there exists $\mathbf{r}_A \in \widetilde{\mathbf{r}}_A$) such that automata $\mathcal{A}_{gv}$ and $\mathcal{B}_{gv}$ induce one and the same generalized mapping.

We shall say that automaton (19.8) *is in the minimal form* if automaton (19.9) equivalent to it and such that

$$m_\tau^{(B)} = |B^{(\tau)}| \leq m_\tau^{(A)} = |A^{(\tau)}|, \quad \tau = \overline{1,T}, \quad \sum_{\tau=1}^{T} m_\tau^{(B)} < \sum_{\tau=1}^{T} m_\tau^{(A)}. \quad (19.10)$$

does not exist.

A *minimal form* of automaton (19.8) is any equivalent to it automaton (19.9) being in the minimal form.

19.7 Theorems on Reduced Forms

Let us prove two statements which are important for solving the optimization problems for automata of the type under consideration.

Theorem 19.7.1 *Let $\mathcal{A}_{gv}$ be a generalized finite automaton (19.8) with periodically variable parameters given over a field $\mathcal{F}$ and $\widetilde{\mathbf{H}}_r^{(\tau)}(\mathcal{A}_{gv})$, $\tau = \overline{1,T}$ be the family of its left-hand basic matrices reduced to the normalized form, and $\widetilde{\mathbf{H}}_r^{(\tau)I}(\mathcal{A}_{gv})$, $\tau = \overline{1,T}$, be a collection of their pseudoinverse matrices. Let $\mathcal{B}_{gv}$ be a generalized finite automaton (19.9) with periodically variable parameters which is given over the field $\mathcal{F}$ and obtained from the automaton $\mathcal{A}_{gv}$ by the following transformation*

$$\widetilde{\mathbf{r}}_B = \widetilde{\mathbf{r}}_A \widetilde{\mathbf{H}}_r^{(T)I}(\mathcal{A}_{gv}), \quad \mathbf{R}_B^{(\tau)}(s,l) = \widetilde{\mathbf{H}}_r^{(\tau-1)}(\mathcal{A}_{gv}) \mathbf{R}_A^{(\tau)}(s,l) \widetilde{\mathbf{H}}_r^{(\tau)I}(\mathcal{A}_{gv}),$$

$$\mathbf{q}_B^{(\tau)} = \widetilde{\mathbf{H}}_r^{(\tau)}(\mathcal{A}_{gv}) \mathbf{q}_A^{(\tau)} \quad (19.11)$$

for all $x_s \in X^{(\tau)}$, $y_l \in Y^{(\tau)}$, $\tau = \overline{1,T}$, such that for $\tau - 1 = 0$ it must be assumed that $\widetilde{\mathbf{H}}_r^{(0)}(\mathcal{A}_{gv}) = \widetilde{\mathbf{H}}_r^{(T)}(\mathcal{A}_{gv})$. Then $\mathcal{B}_{gv} \sim \mathcal{A}_{gv}$ and automaton $\mathcal{B}_{gv}$ is the left-hand reduced one.

PROOF. Let us prove firstly that $\mathcal{B}_{gv} \sim \mathcal{A}_{gv}$. By definition, automata (19.8) and (19.9) have one and the same set Z_{adm} of admissible pairs of words (w,v), $w = x_{s_1} x_{s_2} ... x_{s_d}$, $v = y_{l_1} y_{l_2} ... y_{l_d}$, $d = 0, 1, 2, ...$. Put $w = w_1 w_2$, $v = v_1 v_2$, $|w_1| = |v_1| = d_1$, $|w_2| = |v_2| = d_2$, $d = d_1 + d_2$. Since for all vector-lines $\mathbf{r}_A \in \widetilde{\mathbf{r}}_A$ and

$$\mathbf{h}_r^{(A)}(w_1, v_1) = \mathbf{r}_A \prod_{t=1}^{d_1} \mathbf{R}_A^{(\tau(t))}(s_t, l_t),$$

$$\mathbf{r}_A \in \widetilde{\mathbf{r}}_A, \quad d_1 = \overline{1,d}, \quad (w_1, v_1) \in Z_{\text{adm}},$$

the following relations hold $\mathbf{r}_A \in \mathcal{L}(\tilde{\mathbf{r}}_A \mathcal{A}_{gv}^{(T)})$, $\mathbf{h}_r^{(A)}(w_1, v_1) \in \mathcal{L}(\tilde{\mathbf{r}}_A \mathcal{A}_{gv}^{(\tau(d_1))})$, then by (19.7) of Lemma 19.5.2 and transformations (19.11), for automata $\mathcal{A}_{gv}$ and $\mathcal{B}_{gv}$ we have

$$\mathbf{h}_r^{(A)}(w, v)$$
$$= \mathbf{r}_A \tilde{\mathbf{H}}_r^{(T)I}(\mathcal{A}_{gv}) \tilde{\mathbf{H}}_r^{(T)}(\mathcal{A}_{gv}) \prod_{t=1}^{d} \mathbf{R}_A^{(\tau(t))}(s_t, l_t) \tilde{\mathbf{H}}_r^{(\tau(t))I}(\mathcal{A}_{gv}) \tilde{\mathbf{H}}_r^{(\tau(t))}(\mathcal{A}_{gv})$$
$$= \mathbf{r}_A \tilde{\mathbf{H}}_r^{(T)I}(\mathcal{A}_{gv}) \prod_{t=1}^{d} \tilde{\mathbf{H}}_r^{(\tau(t)-1)}(\mathcal{A}_{gv}) \mathbf{R}_A^{(\tau(t))}(s_t, l_t) \tilde{\mathbf{H}}_r^{(\tau(t))I}(\mathcal{A}_{gv}) \tilde{\mathbf{H}}_r^{(\tau(d))}(\mathcal{A}_{gv})$$
$$= \mathbf{r}_B \prod_{t=1}^{d} \mathbf{R}_B^{(\tau(t))}(s_t, l_t) \tilde{\mathbf{H}}_r^{(\tau(d))}(\mathcal{A}_{gv}) = \mathbf{h}_r^{(B)}(w, v) \tilde{\mathbf{H}}_r^{(\tau(d))}(\mathcal{A}_{gv}) . \quad (19.12)$$

But in this case for a generalized mapping induced by the automaton $\mathcal{A}_{gv}$ [taking into account (19.11) and (19.12)] the following equalities hold

$$\Phi_A(w, v) = \mathbf{h}_r^{(A)}(w, v) \mathbf{q}_A^{(\tau(d))}$$
$$= \mathbf{h}_r^{(B)}(w, v) \tilde{\mathbf{H}}_r^{(\tau(d))}(\mathcal{A}_{gv}) \mathbf{q}_A^{(\tau(d))}$$
$$= \mathbf{h}_r^{(B)}(w, v) \mathbf{q}_B^{(\tau(d))} = \Phi_B(w, v) ,$$

i.e., $\tilde{\Phi}_A \subseteq \tilde{\Phi}_B$. Conversely, let $\mathbf{r}_B$ be an arbitrary vector $\mathbf{r}_B \in \tilde{\mathbf{r}}_B$. Then [taking into account (19.11) and (19.12)] for generalized mapping induced by automaton $\mathcal{B}_{gv}$ we have:

$$\Phi_B(w, v) = \mathbf{h}_r^{(B)}(w, v) \mathbf{q}_B^{(\tau(d))}$$
$$= \mathbf{h}_r^{(B)}(w, v) \tilde{\mathbf{H}}_r^{(\tau(d))}(\mathcal{A}_{gv}) \mathbf{q}_A^{(\tau(d))}$$
$$= \mathbf{h}_r^{(A)}(w, v) \mathbf{q}_A^{(\tau(d))} = \Phi_A(w, v) ,$$

i.e., $\tilde{\Phi}_B(w, v) \subseteq \tilde{\Phi}_A(w, v)$ and, consequently, $\tilde{\Phi}_A = \tilde{\Phi}_B$, $\mathcal{A}_{gv} \sim \mathcal{B}_{gv}$.

Let us show now that the automaton $\mathcal{B}_{gv}$ is the left-hand reduced one, i.e. that for its left-hand basic matrices in the normalized form for any $\tau = \overline{1, T}$ an equality $\tilde{\mathbf{H}}_r^{(\tau)}(\mathcal{B}_{gv}) = \mathbf{I}(m_\tau^{(B)})$ is satisfied. From (19.12) for automata $\mathcal{A}_{gv}$ and $\mathcal{B}_{gv}$ for any $(w, v) \in Z_{\text{adm}}$ such that $|w| = |v| = d = lT + \tau$, $l = 0, 1, ...$, it follows that

$$\mathbf{h}_r^{(A)}(w, v) \tilde{\mathbf{H}}_r^{(\tau)I}(\mathcal{A}_{gv}) = \mathbf{h}_r^{(B)}(w, v) \tilde{\mathbf{H}}_r^{(\tau)}(\mathcal{A}_{gv}) \tilde{\mathbf{H}}_r^{(\tau)I}(\mathcal{A}_{gv})$$
$$= \mathbf{h}_r^{(B)}(w, v) \mathbf{I}(\text{rank} \tilde{\mathbf{H}}_r^{(\tau)}(\mathcal{A}_{gv}))$$
$$= \mathbf{h}_r^{(B)}(w, v) .$$

This relation is true for any vectors $\mathbf{h}_r^{(A)}(w, v) \in \mathcal{L}(\tilde{\mathbf{r}}_A \mathcal{A}_{gv}^{(\tau)})$, $\mathbf{h}_r^{(B)}(w, v) \in \mathcal{L}(\tilde{\mathbf{r}}_B \mathcal{B}_{gv}^{(\tau)})$, $(w, v) \in Z_{\text{adm}}$, $d = lT + \tau$, $l = 0, 1, ...$, and, consequently, for vectors forming the basis. Whence it follows directly that

$$\widetilde{\mathbf{H}}_r^{(\tau)}(\mathcal{A}_{gv})\widetilde{\mathbf{H}}_r^{(\tau)I}(\mathcal{A}_{gv}) = \mathbf{I}(\operatorname{rank}\widetilde{\mathbf{H}}_r^{(\tau)}(\mathcal{A}_{gv}))$$
$$= \widetilde{\mathbf{H}}_r^{(\tau)}(\mathcal{B}_{gv}) = \mathbf{I}(m_\tau^{(B)}),$$

which proves this theorem. ∎

Theorem 19.7.2 *Let the conditions of Theorem 19.7.1 be satisfied. Then if the automaton $\mathcal{A}_{gv}$ is the right-hand reduced one, then the $\mathcal{B}_{gv}$ is also the right-hand reduced one.*

PROOF. Let automaton (19.8) in Theorem 19.7.1 be the right-hand reduced one and, consequently, the family of its right-hand matrices $\mathbf{H}_q^{(\tau)}(\mathcal{A}_{gv})$, $\tau = \overline{1,T}$ contains only nondegenerate square matrices of order $\operatorname{rank}\mathbf{H}_q^{(\tau)}(\mathcal{A}_{gv}) = m_\tau^{(A)}$, $\tau = \overline{1,T}$, and in the normalized form it is as follows: $\widetilde{\mathbf{H}}_q^{(\tau)}(\mathcal{A}_{gv}) = \mathbf{I}(m_\tau^{(A)})$, $\tau = \overline{1,T}$.

Consider an arbitrary tact of a period τ. For the automaton $\mathcal{B}_{gv}$ by (19.11) vector-columns of form (19.2) can be found:

$$\mathbf{h}_q^{(B)}(w_2, v_2)$$
$$= \prod_{t=d_1+1}^{d} \mathbf{R}_B^{(\tau(t))}(s_t, l_t)\mathbf{q}_B^{(\tau(d))}$$
$$= \prod_{t=d_1+1}^{d} \widetilde{\mathbf{H}}_r^{(\tau(t-1))}(\mathcal{A}_{gv})\mathbf{R}_A^{(\tau(t))}(s_t, l_t)\widetilde{\mathbf{H}}_r^{(\tau(t))I}(\mathcal{A}_{gv})\widetilde{\mathbf{H}}_r^{(\tau(d))}(\mathcal{A}_{gv})\mathbf{q}_A^{(\tau(d))}$$
$$= \widetilde{\mathbf{H}}_r^{(\tau(d_1))}(\mathcal{A}_{gv}) \prod_{t=d_1+1}^{d} \mathbf{R}_A^{(\tau(t))}(s_t, l_t)\widetilde{\mathbf{H}}_r^{(\tau(t))I}(\mathcal{A}_{gv})\widetilde{\mathbf{H}}_r^{(\tau(t))}(\mathcal{A}_{gv})\mathbf{q}_A^{(\tau(d))}.$$
(19.13)

For any $d' = 0, 1, ..., d_2$ according to expression (19.3) and to the algorithm of construction of basic matrices (see Section 19.4) any line of the matrix $\widetilde{\mathbf{H}}_r^{(\tau(d_1))}(\mathcal{A}_{gv}) \times \prod_{t=d_1+1}^{d_1+d'} \mathbf{R}_A^{(\tau(t))}(s_t, l_t)$ belongs to the space $\mathcal{L}(\widetilde{\mathbf{r}}_A \mathcal{A}_{gv}^{(\tau(d_1+d'))})$ and expression (19.7) of Lemma 19.5.2 is true for it, i.e. the following equality

$$\widetilde{\mathbf{H}}_r^{(\tau(d_1))}(\mathcal{A}_{gv}) \prod_{t=d_1+1}^{d_1+d'} \mathbf{R}_A^{(\tau(t))}(s_t, l_t)\widetilde{\mathbf{H}}_r^{(\tau(d_1+d'))I}(\mathcal{A}_{gv})\widetilde{\mathbf{H}}_r^{(\tau(d_1+d'))}(\mathcal{A}_{gv})$$
$$= \widetilde{\mathbf{H}}_r^{(\tau(d_1))}(\mathcal{A}_{gv}) \prod_{t=d_1+1}^{d_1+d'} \mathbf{R}_A^{(\tau(t))}(s_t, l_t) \qquad (19.14)$$

is satisfied. In this case, applying formula (19.14) sequentially for $d' = 0, 1, 2, ..., d_2$, from relations (19.13) and (19.2) for all vector-columns of form (19.2) of the automaton $\mathcal{B}_{gv}$ the following condition holds

$$\mathbf{h}_q^{(B)}(w_2, v_2) = \widetilde{\mathbf{H}}_r^{(\tau(d_1))}(\mathcal{A}_{gv}) \prod_{t=d_1+1}^{d} \mathbf{R}_A^{(\tau(t))}(s_t, l_t) \mathbf{q}_A^{(\tau(d))}$$

$$= \widetilde{\mathbf{H}}_r^{(\tau(d_1))}(\mathcal{A}_{gv}) \mathbf{h}_q^{(A)}(w_2, v_2) \in \mathcal{L}(\mathcal{B}_{gv}^{(\tau)} \mathbf{q}_B^{(\tau)})$$

and, consequently, the right-hand basic matrix $\mathbf{H}_q^{(\tau)}(\mathcal{B}_{gv})$ of the automaton $\mathcal{B}_{gv}$ can be formed of a system of linearly independent columns of the matrix $\widetilde{\mathbf{H}}_r^{(\tau)}(\mathcal{A}_{gv}) \mathbf{H}_q^{(\tau)}(\mathcal{A}_{gv})$. Whence it follows that $\mathrm{rank} \mathbf{H}_q^{(\tau)}(\mathcal{B}_{gv}) = \mathrm{rank}(\widetilde{\mathbf{H}}_r^{(\tau)}(\mathcal{A}_{gv})$ $\times \mathbf{H}_q^{(\tau)}(\mathcal{A}_{gv}))$. Since by assumption the matrix $\mathbf{H}_q^{(\tau)}(\mathcal{A}_{gv})$ is a nondegenerate square matrix of order $m_\tau^{(A)}$ and the matrix $\widetilde{\mathbf{H}}_r^{(\tau)}(\mathcal{A}_{gv})$ has the size $(\xi_\tau^{(A)} \times m_\tau^{(A)})$, where $\xi_\tau^{(A)} = \mathrm{rank} \widetilde{\mathbf{H}}_r^{(\tau)}(\mathcal{A}_{gv}) \leq m_\tau^{(A)}$, by Sylvestr's inequality we have

$$\xi_\tau^{(A)} \leq \mathrm{rank} \mathbf{H}_q^{(\tau)}(\mathcal{B}_{gv}) \leq \min(\xi_\tau^{(A)}, m_\tau^{(A)}) = \xi_\tau^{(A)},$$

i.e., $\mathrm{rank} \mathbf{H}_q^{(\tau)}(\mathcal{B}_{gv}) = \xi_\tau^{(A)}$. According to transformation (19.11), the size of the matrix $\mathbf{R}_B^{(\tau)}(s, l)$ is $(m_{\tau-1}^{(B)} \times m_\tau^{(B)})$, where $m_\tau^{(B)} = \xi_\tau^{(A)}$, and, consequently, the matrix $\mathbf{H}_q^{(\tau)}(\mathcal{B}_{gv})$ of size $m_\tau^{(B)} \times \mathrm{rank} \mathbf{H}_q^{(\tau)}(\mathcal{B}_{gv})$ is a nondegenerate square matrix of order $m_\tau^{(B)}$, which in the normalized form is of the shape $\widetilde{\mathbf{H}}_q^{(\tau)}(\mathcal{B}_{gv}) = \mathbf{I}(m_\tau^{(B)})$. Since this conclusion is true for each tact of the period $\tau = \overline{1, T}$, we see that the automaton $\mathcal{B}_{gv}$ is also the right-hand reduced automaton. ∎

Theorem 19.7.3 *Let $\mathcal{A}_{gv}$ be a generalized finite automaton (19.8) with periodically variable parameters given over the field $\mathcal{F}$, and $\widetilde{\mathbf{H}}_q^{(\tau)}(\mathcal{A}_{gv})$, $\tau = \overline{1, T}$ is a family of its right-hand basic matrices reduced to the normalized form and $\widetilde{\mathbf{H}}_q^{(\tau)I}(\mathcal{A}_{gv})$, $\tau = \overline{1, T}$ is a collection of their pseudoinverse matrices. Let $\mathcal{B}_{gv}$ be a generalized finite automaton (19.9) with periodically variable parameters which is given over the field $\mathcal{F}$ and obtained from the automaton $\mathcal{A}_{gv}$ by means of the following transformation*

$$\widetilde{\mathbf{r}}_B = \widetilde{\mathbf{r}}_A \widetilde{\mathbf{H}}_q^{(T)}(\mathcal{A}_{gv}), \quad \mathbf{R}_B^{(\tau)}(s, l) = \widetilde{\mathbf{H}}_q^{(\tau-1)I}(\mathcal{A}_{gv}) \mathbf{R}_A^{(\tau)}(s, l) \widetilde{\mathbf{H}}_q^{(\tau)}(\mathcal{A}_{gv}),$$

$$\mathbf{q}_B^{(\tau)} = \widetilde{\mathbf{H}}_q^{(\tau)I}(\mathcal{A}_{gv}) \mathbf{q}_A^{(\tau)}, \qquad (19.15)$$

for all $x_s \in X^{(\tau)}$, $y_l \in Y^{(\tau)}$, $\tau = \overline{1, T}$, where for $\tau - 1 = 0$ one must assume that $\widetilde{\mathbf{H}}_q^{(0)}(\mathcal{A}_{gv}) = \widetilde{\mathbf{H}}_q^{(T)}(\mathcal{A}_{gv})$. Then $\mathcal{B}_{gv} \sim \mathcal{A}_{gv}$ and the automaton $\mathcal{B}_{gv}$ is the right-hand reduced one.

PROOF. Let us prove that $\mathcal{B}_{gv} \sim \mathcal{A}_{gv}$. Automata (19.8) and (19.9) have, by definition, one and the same set Z_{adm} of admissible pairs of words (w, v), $w = x_{s_1} x_{s_2} ... x_{s_d}$, $v = y_{l_1} y_{l_2} ... y_{l_d}$, $d = 0, 1, 2, ...$. Put $w = w_1 w_2$, $v = v_1 v_2$,

$|w_1| = |v_1| = d_1$, $|w_2| = |v_2| = d_2$, $d = d_1 + d_2$. Since for all vector-columns $\mathbf{q}_A^{(\tau)}$, $\tau = \overline{1,T}$ and

$$\mathbf{h}_q^{(A)}(w_2, v_2) = \prod_{t=d_1+1}^{d} \mathbf{R}_A^{(\tau(t))}(s_t, l_t) \mathbf{q}_A^{(\tau(d))}, \quad d_1 = \overline{0, d-1}, \quad (w,v) \in Z_{\text{adm}},$$

the following relations $\mathbf{q}_A^{(\tau)} \in \mathcal{L}(\mathcal{A}_{gv}^{(\tau)} \mathbf{q}_A^{(\tau)})$, $\mathbf{h}_q^{(A)}(w_2, v_2) \in \mathcal{L}(\mathcal{A}_{gv}^{(\tau(d_1))} \mathbf{q}_A^{(\tau(d_1))})$ be satisfied, then by (19.6) of Lemma 19.5.1 and transformation (19.15) for automata $\mathcal{A}_{gv}$ and $\mathcal{B}_{gv}$ the following relation holds:

$$\begin{aligned}
\mathbf{h}_q^{(A)}&(w_2, v_2) \\
&= \prod_{t=d_1+1}^{d} \widetilde{\mathbf{H}}_r^{(\tau(t)-1)}(\mathcal{A}_{gv}) \widetilde{\mathbf{H}}_q^{((\tau(t)-1)I}(\mathcal{A}_{gv}) \mathbf{R}_A^{(\tau(t))}(s_t, l_t) \\
&\quad \times \widetilde{\mathbf{H}}_q^{(\tau(d))}(\mathcal{A}_{gv}) \widetilde{\mathbf{H}}_q^{(\tau(d))I}(\mathcal{A}_{gv}) \mathbf{q}_A^{(\tau(d))} \\
&= \widetilde{\mathbf{H}}_q^{(\tau(d_1))}(\mathcal{A}_{gv}) \prod_{t=d_1+1}^{d} \widetilde{\mathbf{H}}_q^{(\tau(t)-1)I}(\mathcal{A}_{gv}) \mathbf{R}_A^{(\tau(t))}(s_t, l_t) \\
&\quad \times \widetilde{\mathbf{H}}_q^{(\tau(t))}(\mathcal{A}_{gv}) \widetilde{\mathbf{H}}_q^{(\tau(d))I}(\mathcal{A}_{gv}) \mathbf{q}_A^{(\tau(d))} \\
&= \widetilde{\mathbf{H}}_q^{(\tau(d_1))}(\mathcal{A}_{gv}) \prod_{t=d_1+1}^{d} \mathbf{R}_B^{(\tau(t))}(s_t, l_t) \mathbf{q}_B^{(\tau(d))} \\
&= \widetilde{\mathbf{H}}_q^{(\tau(d_1))}(\mathcal{A}_{gv}) \mathbf{h}_q^{(B)}(w_2, v_2) .
\end{aligned} \quad (19.16)$$

But in this case for a generalized mapping being induced by the automaton $\mathcal{A}_{gv}$ by (19.15) the following equalities

$$\begin{aligned}
\Phi_A(w,v) &= \mathbf{r}_A \mathbf{h}_q^{(A)}(w,v) \\
&= \mathbf{r}_A \widetilde{\mathbf{H}}_q^{(T)}(\mathcal{A}_{gv}) \mathbf{h}_q^{(B)}(w,v) \\
&= \mathbf{r}_B \mathbf{h}_q^{(B)}(w,v) = \Phi_B(w,v) ,
\end{aligned}$$

are true, i.e., $\widetilde{\Phi}_A \subseteq \widetilde{\Phi}_B$.

Conversely, let $\mathbf{r}_B$ be an arbitrary vector $\mathbf{r}_B \in \widetilde{\mathbf{r}}_B$. Then, according to (19.15) and (19.16) for the generalized mapping, induced by the automaton $\mathcal{B}_{gv}$, the following equalities

$$\begin{aligned}
\Phi_B(w,v) &= \mathbf{r}_B \mathbf{h}_q^{(B)}(w,v) = \mathbf{r}_A \widetilde{\mathbf{H}}_q^{(T)}(\mathcal{A}_{gv}) \mathbf{h}_q^{(B)}(w,v) \\
&= \mathbf{r}_A \mathbf{h}_q^{(A)}(w,v) = \Phi_A(w,v) ,
\end{aligned}$$

are true, i.e., $\widetilde{\Phi}_B(w,v) \subseteq \widetilde{\Phi}_A(w,v)$ and consequently $\widetilde{\Phi}_A = \widetilde{\Phi}_B$, $\mathcal{A}_{gv} \sim \mathcal{B}_{gv}$.

Let us show now that the automaton $\mathcal{B}_{gv}$ is the right-hand reduced one, i.e. that its right-hand basic matrices in the normalized form for any $\tau = \overline{1,T}$

Nonstationary Generalized Automata

satisfy an equality $\widetilde{\mathbf{H}}_q^{(\tau)}(\mathcal{B}_{gv}) = \mathbf{I}(m_\tau^{(B)})$. By (19.16) for automata $\mathcal{A}_{gv}$ and $\mathcal{B}_{gv}$ for any $(w, v) \in Z_{\text{adm}}$ such that $w = w_1 w_2$, $v = v_1 v_2$, $|w_1| = |v_1| = d_1 = lT + \tau$, $l = 0, 1, \ldots$, it follows that

$$\widetilde{\mathbf{H}}_q^{(\tau)I}(\mathcal{A}_{gv}) \mathbf{h}_q^{(A)}(w_2, v_2) = \widetilde{\mathbf{H}}_q^{(\tau)I}(\mathcal{A}_{gv}) \widetilde{\mathbf{H}}_q^{(\tau)}(\mathcal{A}_{gv}) \mathbf{h}_q^{(B)}(w_2, v_2)$$
$$= \mathbf{I}(\text{rank}\widetilde{\mathbf{H}}_q^{(\tau)}(\mathcal{A}_{gv})) \mathbf{h}_q^{(B)}(w_2, v_2) = \mathbf{h}_q^{(B)}(w_2, v_2) \ .$$

The last relation is true for arbitrary vectors $\mathbf{h}_q^{(A)}(w_2, v_2) \in \mathcal{L}(\mathcal{A}_{gv}^{(\tau)} \mathbf{q}_A^{(\tau)})$, $\mathbf{h}_q^{(B)}(w_2, v_2) \in \mathcal{L}(\mathcal{B}_{gv}^{(\tau)} \mathbf{q}_B^{(\tau)})$, $(w_1 w_2, v_1 v_2) \in Z_{\text{adm}}$, $d_1 = lT + \tau$, $l = 0, 1, \ldots$, and consequently for those forming the basis. Whence it follows directly that

$$\widetilde{\mathbf{H}}_q^{(\tau)I}(\mathcal{A}_{gv}) \widetilde{\mathbf{H}}_q^{(\tau)}(\mathcal{A}_{gv}) = \mathbf{I}(\text{rank}\widetilde{\mathbf{H}}_q^{(\tau)}(\mathcal{A}_{gv})) \widetilde{\mathbf{H}}_q^{(\tau)}(\mathcal{B}_{gv}) = \mathbf{I}(m_\tau^{(B)}) \ ,$$

The proof is completed. ∎

Theorem 19.7.4 *Let the conditions of Theorem 19.7.3 be satisfied. If the automaton $\mathcal{A}_{gv}$ is the left-hand reduced one, then the automaton $\mathcal{B}_{gv}$ is also the left-hand reduced one.*

PROOF. Let automaton (19.8) in Theorem 19.7.3 be the left-hand reduced one and consequently a family of its left-hand matrices $\mathbf{H}_r^{(\tau)}(\mathcal{A}_{gv})$, $\tau = \overline{1, T}$ contains nondegenerate square matrices of order $\text{rank}\mathbf{H}_r^{(\tau)}(\mathcal{A}_{gv}) = m_\tau^{(A)}$, $\tau = \overline{1, T}$ only and in the normalized form it is as follows: $\widetilde{\mathbf{H}}_r^{(\tau)}(\mathcal{A}_{gv}) = \mathbf{I}(m_\tau^{(A)})$, $\tau = \overline{1, T}$.

Consider an arbitrary tact of the period and, using expression (19.15), for the automaton $\mathcal{B}_{gv}$ find its vector-lines of form (19.3):

$$\mathbf{h}_r^{(B)}(w_1, v_1) = \mathbf{r}_B \prod_{t=1}^{d_1} \mathbf{R}_B^{(\tau(t))}(s_t, l_t)$$

$$= \mathbf{r}_A \widetilde{\mathbf{H}}_q^{(T)}(\mathcal{A}_{gv}) \prod_{t=1}^{d_1} \widetilde{\mathbf{H}}_q^{(\tau(t)-1)I}(\mathcal{A}_{gv}) \mathbf{R}_A^{(\tau(t))}(s_t, l_t) \widetilde{\mathbf{H}}_q^{(\tau(t))}(\mathcal{A}_{gv})$$

$$= \mathbf{r}_A \prod_{t=1}^{d_1} \widetilde{\mathbf{H}}_q^{(\tau(t)-1)}(\mathcal{A}_{gv}) \widetilde{\mathbf{H}}_q^{(\tau(t)-1)I}(\mathcal{A}_{gv})$$

$$\times \mathbf{R}_A^{(\tau(t))}(s_t, l_t) \widetilde{\mathbf{H}}_q^{(\tau(d_1))}(\mathcal{A}_{gv}) \ . \quad (19.17)$$

According to (19.2) and algorithm for construction of the right-hand basic matrices (see Section 19.4), any column of a matrix $\prod_{t=d'+1}^{d_1} \mathbf{R}_A^{(\tau(t))}(s_t, l_t) \widetilde{\mathbf{H}}_q^{(\tau(d_1))}(\mathcal{A}_{gv})$ belongs to a space $\mathcal{L}(\mathcal{A}_{gv}^{(\tau(d'))} \mathbf{q}_A^{(\tau(d'))})$ for any $d' = 0, 1, \ldots, d_1$ and satisfies (19.6) of Lemma 19.5.1, i.e., the following equality

$$\widetilde{\mathbf{H}}_q^{(\tau(d'))}(\mathcal{A}_{gv}) \widetilde{\mathbf{H}}_q^{(\tau(d'))I}(\mathcal{A}_{gv}) \prod_{t=d'+1}^{d_1} \mathbf{R}_A^{(\tau(t))}(s_t, l_t) \widetilde{\mathbf{H}}_q^{(\tau(d_1))}(\mathcal{A}_{gv})$$

$$= \prod_{t=d'+1}^{d_1} \mathbf{R}_A^{(\tau(t))}(s_t, l_t) \widetilde{\mathbf{H}}_q^{(\tau(d_1))}(\mathcal{A}_{gv}) \quad (19.18)$$

is true. In this case, using sequentially formula (19.18) for $d' = d_1, d_1 - 1, ..., 0$, from relation (19.17) and taking into account (19.3) we can find that for all vector-lines of form (19.3) the automaton $\mathcal{B}_{gv}$ satisfies the following condition

$$\begin{aligned}\mathbf{h}_r^{(B)}(w_1, v_1) &= \mathbf{r}_A \prod_{t=1}^{d_1} \mathbf{R}_A^{(\tau(t))}(s_t, l_t) \widetilde{\mathbf{H}}_q^{(\tau(d_1))}(\mathcal{A}_{gv}) \\ &= \mathbf{h}_r^{(A)}(w_1, v_1) \widetilde{\mathbf{H}}_q^{(\tau(d_1))}(\mathcal{A}_{gv}) \in \mathcal{L}(\widetilde{\mathbf{r}}_B \mathcal{B}_{gv}^{(\tau)})\end{aligned}$$

and consequently the left-hand basic matrix $\mathbf{H}_r^{(\tau)}(\mathcal{B}_{gv})$ of the automaton $\mathcal{B}_{gv}$ can be formed by the system of linearly independent lines of the matrix $\mathbf{H}_r^{(\tau)}(\mathcal{A}_{gv}) \times \widetilde{\mathbf{H}}_q^{(\tau)}(\mathcal{A}_{gv})$. Whence it follows that

$$\text{rank} \mathbf{H}_r^{(\tau)}(\mathcal{B}_{gv}) = \text{rank}(\mathbf{H}_r^{(\tau)}(\mathcal{A}_{gv}) \widetilde{\mathbf{H}}_q^{(\tau)}(\mathcal{A}_{gv})).$$

Since by assumption matrix $\mathbf{H}_r^{(\tau)}(\mathcal{A}_{gv})$ is a nondegenerate square matrix of order $m_\tau^{(A)}$ and the matrix $\widetilde{\mathbf{H}}_q^{(\tau)}(\mathcal{A}_{gv})$ is a matrix of size $(m_\tau^{(A)} \times \eta_\tau^{(A)})$, where $\eta_\tau^{(A)} = \text{rank} \widetilde{\mathbf{H}}_q^{(\tau)}(\mathcal{A}_{gv}) \leq m_\tau^{(A)}$, we see that according to Sylvestr's inequality the following inequality holds

$$\eta_\tau^{(A)} \leq \text{rank} \mathbf{H}_r^{(\tau)}(\mathcal{B}_{gv}) \leq \min(m_\tau^{(A)}, \eta_\tau^{(A)}) = \eta_\tau^{(A)},$$

i.e., $\text{rank} \mathbf{H}_r^{(\tau)}(\mathcal{B}_{gv}) = \eta_\tau^{(A)}$. According to transformation (19.15), a size of the matrix $\mathbf{R}_B^{(\tau)}(s, l)$ is $(m_{\tau-1}^{(B)} \times m_\tau^{(B)})$, where $m_\tau^{(B)} = \eta_\tau^{(A)}$. Consequently, the matrix $\mathbf{H}_r^{(\tau)}(\mathcal{B}_{gv})$ of the size $(\text{rank}\mathbf{H}_r^{(\tau)}(\mathcal{B}_{gv}) \times m_\tau^{(B)})$ is a nondegenerate square matrix of order $m_\tau^{(B)}$, which in the normalized form can be written as follows $\widetilde{\mathbf{H}}_r^{(\tau)}(\mathcal{B}_{gv}) = \mathbf{I}(m_\tau^{(B)})$. Since such conclusion is true for each tact of the period $\tau = \overline{1, T}$, we see that the automaton $\mathcal{B}_{gv}$ is the left-hand reduced one. ∎

The methods for construction of the automaton $\mathcal{B}_{gv}$ given in Theorems 19.7.1 and 19.7.3, together with the algorithms for finding the families of basic matrices of automata, considered previously, make it possible to find an effective method of constructing for each given generalized finite automaton $\mathcal{A}_{gv}$ with periodically variable parameters the left-hand or the right-hand reduced automaton $\mathcal{B}_{gv}$ equivalent to it. In addition to this, Theorems 19.7.2 and 19.7.4 show that the sequential application of constructions of Theorems 19.7.1 and 19.7.3 makes it possible to construct the automaton such that it is equivalent to automaton $\mathcal{A}_{gv}$ and it is simultaneously a right-hand and a left-hand reduced one.

19.8 Theorems on Minimal Forms

Let us pass to the verification of construction of the minimal form for the automaton $\mathcal{A}_{gv}$.

Theorem 19.8.1 *A generalized finite automaton (19.8) with periodically variable parameters given over the field $\mathcal{F}$ is in the minimal form if and only if it is the left-hand and right-hand reduced one.*

PROOF. The necessity follows directly from Theorems 19.7.1 and 19.7.3.

Let us prove the sufficiency. Suppose that an automaton $\mathcal{A}_{gv}$ [see (19.8)] is the left-hand and the right-hand reduced one, i.e., the family of its left-hand and right-hand basic matrices has the following properties: $\text{rank}\mathbf{H}_r^{(\tau)}(\mathcal{A}_{gv}^{(\tau)}) = m_\tau^{(A)}$, $\text{rank}\mathbf{H}_q^{(\tau)}(\mathcal{A}_{gv}^{(\tau)}) = m_\tau^{(A)}$, $\tau = \overline{1,T}$. Show that the automaton $\mathcal{A}_{gv}$ is in the minimal form, i.e. it has not been found the automaton $\mathcal{B}_{gv}$ which is equivalent to it [see (19.9)] and satisfies condition (19.10). Assume the converse. Automaton (19.9) is equivalent to the automaton $\mathcal{A}_{gv}$, condition (19.10) are satisfied for it, and τ is a tact of the period such that $m_\tau^{(B)} < m_\tau^{(A)}$. Consider $(w,v) \in Z_{\text{adm}}$ such that

$$w = w_1 w_2, \quad v = v_1 v_2, \quad |w_1| = |v_1| = d_1, \quad |w_2| = |v_2| = d_2,$$

$$d_1 = lT + \tau, \; l = 0, 1, 2, \ldots . \tag{19.19}$$

If the automata $\mathcal{A}_{gv}$ and $\mathcal{B}_{gv}$ are equivalent, then according to the definition of a generalized mapping being induced by these automata for suitable $\mathbf{r}_A \in \tilde{\mathbf{r}}_A$, $\mathbf{r}_B \in \tilde{\mathbf{r}}_B$ the following equalities

$$\begin{aligned}
\Phi_A(w,v) &= \left[\mathbf{r}_A \prod_{t=1}^{d_1} \mathbf{R}_A^{(\tau(t))}(s_t, l_t)\right] \left[\prod_{t=d_1+1}^{d} \mathbf{R}_A^{(\tau(t))}(s_t, l_t) \mathbf{q}_A^{(\tau(d))}\right] \\
&= \left[\mathbf{r}_B \prod_{t=1}^{d_1} \mathbf{R}_B^{(\tau(t))}(s_t, l_t)\right] \left[\prod_{t=d_1+1}^{d} \mathbf{R}_B^{(\tau(t))}(s_t, l_t) \mathbf{q}_B^{(\tau(d))}\right] \\
&= \Phi_B(w,v)
\end{aligned}$$

are true. Taking into account (19.2) and (19.3), it means that in the tact τ of the period

$$\mathbf{h}_r^{(A)}(w_1, v_1) \mathbf{h}_q^{(A)}(w_2, v_2) = \mathbf{h}_r^{(B)}(w_1, v_1) \mathbf{h}_q^{(B)}(w_2, v_2) \tag{19.20}$$

for any $(w,v) \in Z_{\text{adm}}$, satisfying condition (19.19).

Let us choose $(w_1^{(j)}, v_1^{(j)})$, $j = \overline{1, m_\tau^{(A)}}$, such that vector-lines $\mathbf{h}_r^{(A)}(w_1^{(j)}, v_1^{(j)})$ form a basis in a space $\mathcal{L}(\tilde{\mathbf{r}}_A \mathcal{A}_{gv}^{(\tau)})$, and choose $(w_2^{(j)}, v_2^{(j)})$, $j = \overline{1, m_\tau^{(A)}}$, such that vector-lines $\mathbf{h}_q^{(A)}(w_2^{(j)}, v_2^{(j)})$ form a basis in a space $\mathcal{L}(\mathcal{A}_{gv}^{(\tau)} \tilde{\mathbf{q}}_A^{(\tau)})$. Then from (19.20) it follows that

$$\mathbf{H}_r^{(\tau)}(\mathcal{A}_{gv}) \mathbf{H}_q^{(\tau)}(\mathcal{A}_{gv}) = \mathbf{H}_B^{(\tau)} \mathbf{Q}_B^{(\tau)} , \qquad (19.21)$$

where the lines of the matrix $\mathbf{H}_B^{(\tau)}$ are vectors $\mathbf{h}_r^{(B)}(w_1^{(j)}, v_1^{(j)})$, $j = \overline{1, m_\tau^{(A)}}$, and columns of the matrix $\mathbf{Q}_B^{(\tau)}$ are vectors $\mathbf{h}_q^{(B)}(w_2^{(j)}, v_2^{(j)})$, $j = \overline{1, m_\tau^{(A)}}$. Thus, the matrix $\mathbf{H}_B^{(\tau)}$ is of size $(m_\tau^{(A)} \times m_\tau^{(B)})$, and the matrix $\mathbf{Q}_B^{(\tau)}$ is of size $(m_\tau^{(B)} \times m_\tau^{(A)})$. From (19.21) and the supposition $m_\tau^{(B)} < m_\tau^{(A)}$ it follows that

$$\text{rank}(\mathbf{H}_r^{(\tau)}(\mathcal{A}_{gv}) \mathbf{H}_q^{(\tau)}(\mathcal{A}_{gv})) = m_\tau^{(A)} = \text{rank}(\mathbf{H}_B^{(\tau)} \mathbf{Q}_B^{(\tau)}) \leq m_\tau^{(B)} < m_\tau^{(A)} ,$$

This contradiction shows that our supposition on existence of the automaton $\mathcal{B}_{gv}$, such that it is equivalent to the automaton $\mathcal{A}_{gv}$, satisfies condition (19.10) and in the tact τ has $m_\tau^{(B)} < m_\tau^{(A)}$ is false and such automaton $\mathcal{B}_{gv}$ does not exist. Therefore, the automaton $\mathcal{A}_{gv}$ is in the minimal form. ∎

Theorems 19.7.1–19.7.4 and 19.8.1 result in the following theorem.

Theorem 19.8.2 *Let $\mathcal{A}_{gv}$ be a generalized finite automaton (19.8) with periodically variable parameters, given over the field $\mathcal{F}$, and $\tilde{\mathbf{H}}_r^{(\tau)}$, $\tau = \overline{1, T}$, be a family of its left-hand basic matrices reduced to the normalized form, and $\tilde{\mathbf{H}}_r^{(\tau)I}$, $\tau = \overline{1, T}$, be a set of their pseudoinverse matrices. Let $\mathcal{B}_{gv}$ be the automaton (19.9) obtained from the automaton $\mathcal{A}_{gv}$ by transformation (19.11) and $\tilde{\mathbf{H}}_q^{(\tau)}$, $\tau = \overline{1, T}$, be a family of its right-hand basic matrices reduced to the normalized form, and $\tilde{\mathbf{H}}_q^{(\tau)I}$, $\tau = \overline{1, T}$, be a set of their pseudoinverse matrices. Then if the following automaton*

$$\mathcal{D}_{gv} = \langle X^{(\tau)}, D^{(\tau)}, Y^{(\tau)}, \tilde{\mathbf{r}}_D, \{\mathbf{R}_D^{(\tau)}(s, l)\}, \mathbf{q}_D^{(\tau)}, \tau = \overline{1, T} \rangle \qquad (19.22)$$

is formed from the automaton $\mathcal{B}_{gv}$ by the following transformation [being analogous to transformation (19.15)]

$$\begin{aligned}
\tilde{\mathbf{r}}_D &= \tilde{\mathbf{r}}_B \tilde{\mathbf{H}}_q^{(T)}, \quad \mathbf{R}_D^{(\tau)}(s, l) = \tilde{\mathbf{H}}_q^{(\tau-1)I} \mathbf{R}_B^{(\tau)}(s, l) \tilde{\mathbf{H}}_q^{(\tau)}, \\
\mathbf{q}_D^{(\tau)} &= \tilde{\mathbf{H}}_q^{(\tau)I} \mathbf{q}_B^{(\tau)},
\end{aligned} \qquad (19.23)$$

for all $x_s \in X^{(\tau)}$, $y_l \in Y^{(\tau)}$, $\tau = \overline{1, T}$, where for $\tau - 1 = 0$ it must be assumed that $\tilde{\mathbf{H}}_q^{(0)I} = \tilde{\mathbf{H}}_q^{(T)I}$, then the automaton $\mathcal{D}_{gv}$ is the minimal form of the automaton $\mathcal{A}_{gv}$, $\mathcal{D}_{gv} = \text{Min } \mathcal{A}_{gv}$.

The same proposition also is true in the case that another order of transformations takes place and we have the transformation of the automaton $\mathcal{A}_{gv}$ into the automaton $\mathcal{B}_{gv}$ by formulas of type (19.15), and then the transformation of $\mathcal{B}_{gv}$ into $\mathcal{D}_{gv}$ by formulas of type (19.11).

19.9 The Algorithm of Optimization

Let a generalized automaton $\mathcal{A}_{gv}$ with periodically variable parameters (19.8) be given. Taking into account the above results the problem of constructing its minimal form can be solved by the following procedure for which fulfilling it is necessary:

1. To find a family of the left-hand basic matrices of the automaton $\mathcal{A}_{gv}$: $\mathbf{H}_r^{(\tau)}$, $\tau = \overline{1,T}$, to reduce them to the normalized form $\widetilde{\mathbf{H}}_r^{(\tau)}$, $\tau = \overline{1,T}$, to find pseudoinverse matrices $\widetilde{\mathbf{H}}_r^{(\tau)I}$, $\tau = \overline{1,T}$.

2. To transform the automaton $\mathcal{A}_{gv}$ into $\mathcal{B}_{gv}$ by formulas (19.11).

3. To find a family of the right-hand basic matrices of the automaton $\mathcal{B}_{gv}$: $\mathbf{H}_q^{(\tau)}$, $\tau = \overline{1,T}$, to reduce them to the normalized form $\widetilde{\mathbf{H}}_q^{(\tau)}$, $\tau = \overline{1,T}$, and to find pseudoinverse matrices $\widetilde{\mathbf{H}}_q^{(\tau)I}$, $\tau = \overline{1,T}$.

4. To transform the automaton $\mathcal{B}_{gv}$ into $\mathcal{D}_{gv}$ by formulas (19.23).

The result is $\mathcal{D}_{gv} = \text{Min } \mathcal{A}_{gv}$.

Note that the orders of transformations in the algorithm can be changed, i.e., at first we can make a transformation by the family of the right-hand basic matrices, and then by the family of the left-hand basic matrices. The result will be a minimal form of the automaton $\mathcal{A}_{gv}$ also.

19.10 Example

Let $\mathcal{A}_{gv}$ be a given over the field of real numbers generalized finite automaton with periodically variable parameters

$$\mathcal{A}_{gv} = \langle X^{(\tau)}, A^{(\tau)}, Y^{(\tau)}, \widetilde{\mathbf{r}}_A, \{\mathbf{R}_A^{(\tau)}(x_s, y_l)\}, \mathbf{q}_A^{(\tau)}, \tau = \overline{1,T} \rangle ,$$

where $T = 2$, $X^{(1)} = \{x_1, x_2\}$, $X^{(2)} = \{x_2, x_3\}$, $Y^{(1)} = \{y_1, y_2\}$, $Y^{(2)} = \{y_2, y_3\}$, $A^{(1)} = \{a_1, a_2, a_3, a_4\}$, $A^{(2)} = \{a_1, a_2, a_3, a_4, a_5\}$,

$$\tilde{\mathbf{r}}_A = \begin{pmatrix} 0,5 & 0,4 & 0,4 & 0,2 & 0,7 \\ 0,3 & 0,3 & 0,3 & 0,2 & 0,5 \end{pmatrix}, \quad \mathbf{q}_A^{(1)} = \begin{pmatrix} 0,6 \\ 0,5 \\ 0,1 \\ 0,5 \end{pmatrix}, \quad \mathbf{q}_A^{(2)} = \begin{pmatrix} 0,1 \\ 0,1 \\ 0,5 \\ 0,5 \\ 0,5 \end{pmatrix},$$

$$\mathbf{R}_A^{(1)}(x_1, y_1) = \begin{pmatrix} 0,2 & 0,1 & 0,05 & 0,04 \\ 0,59 & 0,49 & 0,39 & 0,45 \\ 0,11 & 0,21 & 0,21 & 0,29 \\ 0,05 & 0 & 0,05 & 0,04 \\ 0,05 & 0,05 & 0,05 & 0,05 \end{pmatrix},$$

$$\mathbf{R}_A^{(1)}(x_1, y_2) = \begin{pmatrix} 0,4 & 0,05 & 0,1 & 0,18 \\ 0,03 & 0,2 & 0,04 & 0,19 \\ 0,52 & 0,15 & 0,26 & 0,1 \\ 0,8 & 0,05 & 0,5 & 0,42 \\ 0,1 & 0,25 & 0,3 & 0,2 \end{pmatrix},$$

$$\mathbf{R}_A^{(1)}(x_2, y_1) = \begin{pmatrix} 0,2 & 0,1 & 0,1 & 0,13 \\ 0,54 & 0,36 & 0,38 & 0,39 \\ 0,11 & 0,59 & 0,12 & 0,42 \\ 0,1 & 0,25 & 0,2 & 0,36 \\ 0,3 & 0,2 & 0,1 & 0,05 \end{pmatrix},$$

$$\mathbf{R}_A^{(1)}(x_2, y_2) = \begin{pmatrix} 0 & 0,1 & 0,15 & 0,23 \\ 0,18 & 0,38 & 0,39 & 0,55 \\ 0,82 & 0,22 & 0,21 & 0,01 \\ 0,4 & 0,15 & 0,05 & 0 \\ 0,1 & 0,1 & 0,05 & 0,05 \end{pmatrix},$$

$$\mathbf{R}_A^{(2)}(x_2, y_2) = \begin{pmatrix} 0,17 & 0,65 & 0,35 & 0,12 & 0,37 \\ 0,02 & 0,35 & 0,3 & 0,07 & 0,07 \\ 0,05 & 0,05 & 0,1 & 0,05 & 0,04 \\ 0,05 & 0,5 & 0,25 & 0,05 & 0,2 \end{pmatrix},$$

$$\mathbf{R}_A^{(2)}(x_2, y_3) = \begin{pmatrix} 0,47 & 0,61 & 0,09 & 0,16 & 0,75 \\ 0,02 & 0,21 & 0,19 & 0,01 & 0,01 \\ 0,05 & 0,1 & 0 & 0,05 & 0,08 \\ 0,05 & 0,4 & 0,1 & 0,15 & 0,3 \end{pmatrix},$$

$$\mathbf{R}_A^{(2)}(x_3, y_2) = \begin{pmatrix} 0,95 & 0,45 & 0,07 & 0,12 & 0,95 \\ 0,25 & 0,1 & 0,02 & 0,07 & 0,63 \\ 0,15 & 0,05 & 0 & 0,05 & 0,7 \\ 0,5 & 0,25 & 0,05 & 0,05 & 0,1 \end{pmatrix},$$

$$\mathbf{R}_A^{(2)}(x_3, y_3) = \begin{pmatrix} 0,28 & 0,9 & 0,34 & 0,5 & 0,88 \\ 0,18 & 0,2 & 0,24 & 0,05 & 0,24 \\ 0,05 & 0,1 & 0,2 & 0,05 & 0,11 \\ 0,2 & 0,5 & 0,1 & 0,25 & 0,5 \end{pmatrix}.$$

It is necessary to find $\mathcal{D}_{gv} = \mathrm{Min}\,\mathcal{A}_{gv}$.

According to algorithm of Section 19.4 we form a family of the left-hand basic matrices of automaton $\mathbf{H}_r^{(\tau)}$, $\tau = 1, 2$. In result we obtain (in the normalized form)

$$\widetilde{\mathbf{H}}_r^{(1)} = \begin{pmatrix} 1 & 0 & 0 & -0,4 \\ 0 & 1 & 0 & 0,6 \\ 0 & 0 & 1 & 1 \end{pmatrix}, \quad \widetilde{\mathbf{H}}_r^{(2)} = \begin{pmatrix} 1 & 0 & 0 & 0 & 1 \\ 0 & 1 & 0 & 0 & 0,2 \\ 0 & 0 & 1 & 0 & -0,2 \\ 0 & 0 & 0 & 1 & 1 \end{pmatrix}.$$

$$\widetilde{\mathbf{H}}_r^{(1)'} = \begin{pmatrix} 1 & 0 & 0 \\ 0 & 1 & 0 \\ 0 & 0 & 1 \\ 0 & 0 & 0 \end{pmatrix}, \quad \widetilde{\mathbf{H}}_r^{(2)'} = \begin{pmatrix} 1 & 0 & 0 & 0 \\ 0 & 1 & 0 & 0 \\ 0 & 0 & 1 & 0 \\ 0 & 0 & 0 & 1 \\ 0 & 0 & 0 & 0 \end{pmatrix}. \quad (19.24)$$

In accordance with formulas (19.11), using the found left-hand basic matrices (19.24), from the automaton $\mathcal{A}_{gv}$ we now construct a new automaton

$$\mathcal{B}_{gv} = \langle X^{(\tau)}, B^{(\tau)}, Y^{(\tau)}, \widetilde{\mathbf{r}}_B, \{\mathbf{R}_B^{(\tau)}(x_s, y_l)\}, \mathbf{q}_B^{(\tau)}, \tau = 1, 2 \rangle,$$

$$B^{(1)} = \{b_1, b_2, b_3\}, \quad B^{(2)} = \{b_1, b_2, b_3, b_4\},$$

$$\widetilde{\mathbf{r}}_B = \begin{pmatrix} 0,5 & 0,4 & 0,4 & 0,2 \\ 0,3 & 0,3 & 0,3 & 0,2 \end{pmatrix}, \quad \mathbf{q}_B^{(1)} = \begin{pmatrix} 0,4 \\ 0,8 \\ 0,6 \end{pmatrix}, \quad \mathbf{q}_B^{(2)} = \begin{pmatrix} 0,6 \\ 0,2 \\ 0,4 \\ 1,0 \end{pmatrix},$$

$$\mathbf{R}_B^{(1)}(x_1, y_1) = \begin{pmatrix} 0,25 & 0,15 & 0,1 \\ 0,6 & 0,5 & 0,4 \\ 0,1 & 0,2 & 0,2 \\ 0,1 & 0,05 & 0,1 \end{pmatrix}, \quad \mathbf{R}_B^{(1)}(x_1, y_2) = \begin{pmatrix} 0,5 & 0,3 & 0,4 \\ 0,05 & 0,25 & 0,1 \\ 0,5 & 0,1 & 0,2 \\ 0,9 & 0,3 & 0,8 \end{pmatrix},$$

$$\mathbf{R}_B^{(1)}(x_2, y_1) = \begin{pmatrix} 0,5 & 0,3 & 0,2 \\ 0,6 & 0,4 & 0,4 \\ 0,05 & 0,55 & 0,1 \\ 0,4 & 0,45 & 0,3 \end{pmatrix}, \quad \mathbf{R}_B^{(1)}(x_2, y_2) = \begin{pmatrix} 0,1 & 0,2 & 0,2 \\ 0,2 & 0,4 & 0,4 \\ 0,8 & 0,2 & 0,2 \\ 0,5 & 0,25 & 0,1 \end{pmatrix},$$

$$\mathbf{R}_B^{(2)}(x_2, y_2) = \begin{pmatrix} 0,15 & 0,45 & 0,25 & 0,1 \\ 0,05 & 0,65 & 0,45 & 0,1 \\ 0,1 & 0,55 & 0,35 & 0,1 \end{pmatrix},$$

$$\mathbf{R}_B^{(2)}(x_2, y_3) = \begin{pmatrix} 0,45 & 0,45 & 0,05 & 0,1 \\ 0,05 & 0,45 & 0,25 & 0,1 \\ 0,1 & 0,5 & 0,1 & 0,2 \end{pmatrix},$$

$$\mathbf{R}_B^{(2)}(x_3, y_2) = \begin{pmatrix} 0,75 & 0,35 & 0,05 & 0,1 \\ 0,55 & 0,25 & 0,05 & 0,1 \\ 0,65 & 0,3 & 0,05 & 0,1 \end{pmatrix},$$

$$\mathbf{R}_B^{(2)}(x_3, y_3) = \begin{pmatrix} 0,2 & 0,7 & 0,3 & 0,4 \\ 0,3 & 0,5 & 0,3 & 0,2 \\ 0,25 & 0,6 & 0,3 & 0,3 \end{pmatrix}.$$

Let us minimize the constructed $\mathcal{B}_{gv}$, using its right-hand basic matrices $\mathbf{H}_q^{(1)}$, $\mathbf{H}_q^{(2)}$. According to algorithm of Section 19.4 we find the following family of the right-hand basis matrices (in the normalized form) and the collection of their pseudoinverse matrices

$$\widetilde{\mathbf{H}}_q^{(1)} = \begin{pmatrix} 1 & 0 \\ 0 & 1 \\ 0,5 & 0,5 \end{pmatrix}, \quad \widetilde{\mathbf{H}}_q^{(2)} = \begin{pmatrix} 1 & 0 & 0 \\ 0 & 1 & 0 \\ 0 & 0 & 1 \\ 1,5 & -0,5 & 0,5 \end{pmatrix},$$

$$\widetilde{\mathbf{H}}_q^{(1)I} = \begin{pmatrix} 1 & 0 & 0 \\ 0 & 1 & 0 \end{pmatrix}, \quad \widetilde{\mathbf{H}}_q^{(2)I} = \begin{pmatrix} 1 & 0 & 0 & 0 \\ 0 & 1 & 0 & 0 \\ 0 & 0 & 1 & 0 \end{pmatrix}. \tag{19.25}$$

According to (19.23), we use the found matrices (19.25) and from $\mathcal{B}_{gv}$ construct the following new automaton:

$$\mathcal{D}_{gv} = \langle X^{(\tau)}, D^{(\tau)}, Y^{(\tau)}, \widetilde{\mathbf{r}}_D, \{\mathbf{R}_D^{(\tau)}(x_s, y_l)\}, \mathbf{q}_D^{(\tau)}, \tau = 1, 2 \rangle,$$

$$D^{(1)} = \{d_1, d_2\}, \quad D^{(2)} = \{d_1, d_2, d_3\},$$

$$\widetilde{\mathbf{r}}_D = \begin{pmatrix} 0,8 & 0,3 & 0,5 \\ 0,6 & 0,2 & 0,4 \end{pmatrix}, \quad \mathbf{q}_D^{(1)} = \begin{pmatrix} 0,4 \\ 0,8 \end{pmatrix}, \quad \mathbf{q}_D^{(2)} = \begin{pmatrix} 0,6 \\ 0,2 \\ 0,4 \end{pmatrix},$$

$$\mathbf{R}_D^{(1)}(x_1, y_1) = \begin{pmatrix} 0,3 & 0,2 \\ 0,8 & 0,7 \\ 0,2 & 0,3 \end{pmatrix}, \quad \mathbf{R}_D^{(1)}(x_1, y_2) = \begin{pmatrix} 0,7 & 0,5 \\ 0,1 & 0,3 \\ 0,6 & 0,2 \end{pmatrix},$$

$$\mathbf{R}_D^{(1)}(x_2, y_1) = \begin{pmatrix} 0,6 & 0,4 \\ 0,8 & 0,6 \\ 0,1 & 0,6 \end{pmatrix}, \quad \mathbf{R}_D^{(1)}(x_2, y_2) = \begin{pmatrix} 0,2 & 0,3 \\ 0,4 & 0,6 \\ 0,9 & 0,3 \end{pmatrix},$$

$$\mathbf{R}_D^{(2)}(x_2, y_2) = \begin{pmatrix} 0,3 & 0,4 & 0,3 \\ 0,2 & 0,6 & 0,5 \end{pmatrix}, \quad \mathbf{R}_D^{(2)}(x_2, y_3) = \begin{pmatrix} 0,6 & 0,4 & 0,1 \\ 0,2 & 0,4 & 0,3 \end{pmatrix},$$

$$\mathbf{R}_D^{(2)}(x_3, y_2) = \begin{pmatrix} 0,9 & 0,3 & 0,1 \\ 0,7 & 0,2 & 0,1 \end{pmatrix}, \quad \mathbf{R}_D^{(2)}(x_3, y_3) = \begin{pmatrix} 0,8 & 0,5 & 0,5 \\ 0,6 & 0,4 & 0,4 \end{pmatrix}.$$

The construction of automaton $\mathcal{D}_{gv}$ concludes the solving of the minimization problem for the automaton $\mathcal{A}_{gv}$ since by Theorem 19.8.2 we have $\mathcal{D}_{gv} = \text{Min } \mathcal{A}_{gv}$.

Acknowledgments. This work is supported by a grant from the RFFR N 98-01-01008.

References

1. Tchirkov, M. K. (1983). *Partial Automata*, Leningrad University Press (in Russian).

2. Tchirkov, M. K. (1990). On Matrix Methods for Optimization of Generalized Automata, In *Discrete Systems and Their Software*, Leningrad University Press, pp. 3–18 (in Russian).

3. Tchirkov, M. K. (1991). On Matrix Methods for Optimization of Generalized Automata. *Annales Universitatis Scientiarum Budapestinensis de Rolando Eotvos nominatae. Sectio Computatorica*, Tomus 11, 175–191.

4. Shestakov, A. A. and Tchirkov, M. K. (1992). *Generalized Finite Automata: Behavioral Equivalence and Problems of Optimization*, Apatity: Russian academy of Science, Kolsky Science Centre Press (in Russian).

5. Miroshnichenko, I. D. and Tchirkov, M. K. (1990). Minimization of Finite Automata with Periodically Varying Structure, In *Optimization Problems of Discrete Systems*, Leningrad University Press, pp. 7–21 (in Russian).

6. Miroshnichenko, I. D. and Tchirkov, M. K. (1994). On Reduced Forms of Generalized Automata with a Periodically Variable Structure, In *International Workshop on Mathematical Methods and Tools in Computer Simulation. May 24–28, 1994, St. Petersburg*. St. Petersburg University: Preprint MM-94-01, p. 99–101.

7. Ponomareva, A. Yu. and Tchirkov, M. K. (1998). On a Special Method for States Minimization of Generalized Automata with Periodically Varying Parameters, In *Proceedings of the 3rd St. Petersburg Workshop on Simulation, June 28 – July 3, 1998*. St. Petersburg University Press, pp. 444–448.

20

Power of Some Asymptotic Tests for Maximum Entropy

M. Salicrú, S. Vives and J. Ocaña

Barcelona University, Barcelona, Spain

Abstract: In this chapter we propose some alternative improvements to the asymptotic chi-square distribution of the sample $(h-\phi)$-entropies under the hypothesis of maximum entropy. The usual asymptotic approach is based only on the consideration of the second-order Taylor expansion. We suggest some alternative approaches, based on including known information about moments, and the corresponding tests for the null hypothesis of maximum entropy. Formulae to compute the moments of $(h-\phi)$-entropies are also derived. A simulation study for the power of these tests indicates that, in all cases, the new testing procedures improve the results obtained with the rough asymptotic test.

Keywords and phrases: Entropy measures, asymptotic distribution, second-order expansion, moments, Monte Carlo simulation

20.1 Introduction

The concept of diversity appears in general associated with the idea of variability between the individuals of a population, in terms of the number of classes and the probability of each one of them. In correspondence with the heterogeneity of fields in which diversity measures have been applied, a large number of indexes have appeared in the literature [Shannon (1948), Havrda-Charvat (1967), Renyi (1961), Varma (1966), Arimoto (1971), Sharma and Taneja (1975) and Ferrari (1980), among others]. As a generalisation of these indexes, Salicrú *et al.* (1993) define the functional $(h-\phi)$-entropies as the family of measures of the form:

$$H_h^\phi(P) = h\left(\sum_{i=1}^M \phi(p_i)\right)$$

where $P = (p_1,\ldots,p_m)$ is a probability vector associated with a multinomial distribution and h and ϕ are functions twice times differentiable with continuity. When $\hat{P} = (\hat{p}_1,\ldots,\hat{p}_m)$ is an asymptotically normal estimation of P for a sample of size n, Salicrú et al. (1993) and Pardo et al. (1997), derive the asymptotic distribution of the sample functional $(h-\phi)$-entropies:

$$H_h^\phi(\hat{P}) = h\left(\sum_{i=1}^M \phi(\hat{p}_i)\right).$$

Under the hypothesis of maximum diversity,

$$H_0: H_h^\phi(P) = H_h^\phi(P)_{max}$$

where $H_h^\phi(P)_{max} = h(M\phi(M^{-1}))$ is the entropy associated with assigning probability M^{-1} to all classes, it can be stated that,

$$\frac{2nM\left[H_h^\phi(\hat{P}) - H_h^\phi(P)_{max}\right]}{c} \xrightarrow{L} \chi^2_{M-1}, \tag{20.1}$$

with

$$c = h'\left(M\phi\left(\frac{1}{M}\right)\right)\phi''\left(\frac{1}{M}\right).$$

More generally, for parametric families with probability density function $f(x,\Theta)$, we have the following general result under the hypothesis of maximum diversity:

$$2n\left[H_h^\phi(f(x,\hat{\Theta})) - H_h^\phi(f(x,\Theta))\right] \xrightarrow{L} \sum \beta_i \chi^2_{1,i},$$

that is, this difference statistic asymptotically approaches a linear combination of χ^2 distributions, each with one degree of freedom [Pardo et al. (1997)].

Since we are dealing with the asymptotical distribution, here we attempt to improve the efficiency of the associated tests by including known information about moments. In this way, we can evaluate the expectation and variance of the general statistic $H_h^\phi(\hat{P})$ (asymptotic formulae in Section 20.6). More concise (and exact) formulae for the first two moments of some particular diversity indexes also exist. Section 20.5 states the case of the Havrda-Charvat index but the methodology is also directly applicable to Renyi's entropy.

On the other hand, and for the statistic associated with asymptotical tests, the parameter alpha is determined by the one that the asymptotical distribution is closest to the exact distribution.

20.2 Taylor Series Approximations

Briefly, result (20.1) arises from considering the Taylor expansion of the function $H_h^\phi(\hat{P})$ at point P. Therefore:

$$H_h^\phi(\hat{P}) = H_h^\phi(P) + T^t(\hat{P} - P) + \frac{1}{2}(\hat{P} - P)^t C (\hat{P} - P) + R_n,$$

where

$$T^t = (t_1, \ldots, t_M) = \left(\frac{\partial H_h^\phi(P)}{\partial p_1}, \ldots, \frac{\partial H_h^\phi(P)}{\partial p_M} \right)$$

$$C = (c_{ij}) = \left(\frac{\partial^2 H_h^\phi(P)}{\partial p_i \partial p_j} \right) \quad i, j = 1, \ldots, M.,$$

and R_n is the remainder corresponding to the second-order polynomial. When the first-order term, $T^t(\hat{P} - P)$ is equal to zero, which occurs at least for the maximum entropy associated with $P = (M^{-1}, \ldots, M^{-1})$, we have:

$$\frac{2nM \left[H_h^\phi(\hat{P}) - H_h^\phi(P) \right]}{h'\left(M\phi\left(\frac{1}{M}\right)\right) \phi''\left(\frac{1}{M}\right)} \approx n^{\frac{1}{2}}(\hat{P} - P)^t C\, n^{\frac{1}{2}}(\hat{P} - P), \qquad (20.2)$$

with $n^{\frac{1}{2}}(\hat{P} - P) \xrightarrow{L} N(0, \Sigma)$, $\Sigma = (\sigma_{ij})_{i,j=1,\ldots,M}$ and $\sigma_{ij} = \delta_{ij} p_i - p_i p_j$ where δ_{ij} stands for Dirac's delta, ($\delta_{ij} = 0$ if $i = j$ and 1 otherwise). Using known results on the distribution of quadratic forms, we can show that:

$$n^{\frac{1}{2}}(\hat{P} - P)^t C\, n^{\frac{1}{2}}(\hat{P} - P) \approx \sum \beta_i \chi_{1,i}^2,$$

where the β_i are the eigenvalues of $C\Sigma$ and $\chi_{1,i}^2$ are independent random variables distributed according to a chi-square distribution with one degree of freedom. In the present case, Salicrú et al. (1993) show that the distribution of $\sum \beta_i \chi_{1,i}^2$ could be approximated by a chi-square distribution with $M - 1$ degrees of freedom.

Looking at the problem in this way, the distribution attributed to the statistic $2n \left[H_h^\phi(\hat{P}) - H_h^\phi(P) \right]$ is based only on the consideration of the second-order Taylor expansion term of the function $H_h^\phi(P)$. It is possible to improve this approximation using constants that collect some information about the true moments of the statistic. This is the point of view of the alternatives discussed in the next paragraphs.

Alternative 1

Consider the aproximation:

$$2n\left[H_h^\phi(\hat{P}) - H_h^\phi(P)\right] \approx a_1 \chi_\nu^2,$$

where a_1 and ν are obtained by considering the method of moments and the rank of the quadratic form as a whole. The rationale of this approach lies in the possibility of adjusting for the true expectation of $2n\left[H_h^\phi(\hat{P}) - H_h^\phi(P)\right]$, so a_1 and ν are obtained by solving these equations:

$$\left.\begin{array}{l} E\left(2n\left[H_h^\phi(\hat{P}) - H_h^\phi(P)\right]\right)_{H_0} = 3Da_1\nu \\ \operatorname{rank} C\,\Sigma = 3D\nu \end{array}\right\},$$

which, in our case, leads to $\nu = 3DM - 1$ and

$$a_1 = 3D(M-1)^{-1}\left[2nE\left(H_h^\phi(\hat{P})\right)_{H_0} - 2nh\left(M\phi(M^{-1})\right)\right]. \qquad (20.3)$$

Alternative 2

Consider the approximation

$$2n\left[H_h^\phi(\hat{P}) - H_h^\phi(P)\right] \approx a_2 \chi_\nu^2 + b_2.$$

In a parallel way to the previous case, the distribution can be adjusted for the true mean and variance of $2n\left[H_h^\phi(\hat{P}) - H_h^\phi(P)\right]$. Then, a_2, b_2 and ν, are obtained as a result of solving these equations:

$$\left.\begin{array}{l} E\left(2n\left[H_h^\phi(\hat{P}) - H_h^\phi(P)\right]\right)_{H_0} = 3Da_2\nu + b_2 \\ VAR\left(2n\left[H_h^\phi(\hat{P}) - H_h^\phi(P)\right]\right)_{H_0} = 3D2a_2^2\nu \\ \operatorname{rank} C\,\Sigma = 3D\nu \end{array}\right\}.$$

Therefore,

$$\left.\begin{array}{l} \nu = 3DM - 1 \\ a_2 = 3D - \left[\dfrac{VAR\left(2nH_h^\phi(\hat{P})\right)_{H_0}}{2(M-1)}\right]^{\frac{1}{2}} \\ b_2 = 3DE\left(2n\left[H_h^\phi(\hat{P}) - H_h^\phi(P)\right]\right)_{H_0} - a_2(M-1) \end{array}\right\}.$$

The alternatives discussed in the preceding paragraphs were suggested by similar approximations described in Satterthwaite (1946), Kotz et al. (1967) and Rao and Scott (1981). As Sections 20.5 and 20.6 indicate, it is possible to compute the expectation and variance for a wide range of families of diversity indexes.

20.3 A Simulation Study

In order to study the power of the tests based on alternatives 1 and 2, we have performed a simulation study for the following entropy indexes and parameters:

Havrda-Charvat entropy ($\alpha = 2, 3, 5$),

$$H^{HC}_\alpha(P) = \frac{1}{\alpha - 1}\left(1 - \sum_{i=1}^M p^\alpha_i\right).$$

Shannon's entropy:

$$H^S(P) = -\sum_{i=1}^M p_i \log p_i.$$

Renyi's entropy ($\alpha = 2, 3, 5$)

$$H^R_\alpha(P) = \frac{1}{1-\alpha} \log \sum_{i=1}^M p^\alpha_i.$$

For these cases we have carried out a comparative simulation study to evaluate the efficiency of the chi-square tests based on the following statistics:

rough asymptotic test:

$$E_1 = \frac{2n\left[H^\phi_h(\hat{P}) - h(M\phi(M^{-1}))\right]}{c} \approx \chi^2_{M-1} \qquad (20.4)$$

test based on alternative 1:

$$E_2 = \frac{2n\left[H^\phi_h(\hat{P}) - h(M\phi(M^{-1}))\right]}{a_1} \approx \chi^2_{M-1} \qquad (20.5)$$

test based on alternative 2:

$$E_3 = \frac{2n\left[H^\phi_h(\hat{P}) - h(M\phi(M^{-1}))\right] - b_2}{a_2} \approx \chi^2_{M-1} \qquad (20.6)$$

to test the hypothesis:

$$H_0 : H^\phi_h(P) = H^\phi_h(P)_{max}$$
$$H_1 : H^\phi_h(P) < H^\phi_h(P)_{max} \left(H^\phi_h(P)_{max} = h(M\phi(M^{-1}))\right).$$

For the shake of the comparison, we have also included three more tests. The first one is the classical and less specific chi-square test for adherence to equiprobability:

$$E_4 = 3Dn \sum_{i=3D1}^{M} \frac{(\hat{p}_i - M^{-1})^2}{M^{-1}} \approx \chi^2_{M-1}. \qquad (20.7)$$

The second one is the likelihood ratio test (with Wilks approximation)

$$E_5 = 3D2n \left[\sum_{i=3D1}^{M} \hat{p}_i \log \hat{p}_i + \log M \right], \qquad (20.8)$$

and the third is based on a Monte Carlo estimation of the *true* null distribution of the statistic:

$$E_6 = 3DH_h^\phi(\hat{P}). \qquad (20.9)$$

In order to specify the test based on (20.9), under given fixed conditions (entropy index, n and M), 50,000 $\hat{P}$ vectors were generated as $\hat{P} = 3D\frac{X}{n}$, for X multinomially distributed with n and $P = 3D\left(M^{-1}, \ldots, M^{-1}\right)$ parameters. For each of these $\hat{P}$ vectors, the entropy index was calculated. These samples values were ordered and the upper 5% quantile, $C_{0.05}$, was determined. This value establishes the limit of the acceptance region.

Note the correspondence between the chi-square test for adherence to equiprobability and rough asymptotic test for Havrda-Charvat entropy (α=3D2) (20.4) [Nayak (1985)] and the likelihood ratio test (Wilks approximation) with rough asymptotic test for Shannon's entropy (20.4).

We used Monte Carlo simulation to estimate the power of preceding tests. These simulation studies were performed under all the following possible combinations of P, M and n parameters:

Probability vector P: Six P configurations were considered. These included the null hypothesis of maximum entropy, $P_0 = 3D\left(M^{-1}, \ldots, M^{-1}\right)$, and five intermediate = points between P_0 and a point corresponding to minimum entropy. In this study we have used a characterisation of the multinomial parametrical space. This characterization is based on the choice of 4 pattern vectors and the consideration of intermediate vectors, according to the Euclidean distance criteria stated in Vives *et al.* (1996).

Number of classes: M=3D10 and M=3D20.

Multinomial sample size: n=3D40 and n=3D100.

For each one of the preceding experimental conditions 50,000 vectors of relative frequencies, $\hat{P}$, were generated. From each of these vectors, the statistics

E_1, E_2, E_3, E_4, E_5 and E_6 were calculated and the corresponding tests were evaluated. The power of each test was estimated as the percentage of times that the null hypothesis was rejected. The $\hat{P}$ vectors were generated as $\hat{P} = 3Dn^{-1}X$ for X multinomial. These multinomial vectors were generated by repeatedly producing binomial values, according to a well known conditional algorithm (Dagpunar, 1988). The binomial values were generated by means of an optimised inversion algorithm. Random number generation was based on the substract-with borrow generator described in Marsaglia and Zaman (1991). The generation algorithms were implemented in Delphi in the class library ObjGenRV. This software is an optimised version of the Borland Pascal object-oriented generators described in Ocaña (1995). The simulated conditions (entropy indexes, number of classes and multinomial sample sizes) correspond to common situations in several research areas. These include Genetics (linkage disequilibrium and genetic polymorphism measured by means of the Havdra-Charvat index of degree α=3D2, known as Nei's quadratic index), Human Population Biology (population genetic structure), and Genetic Epidemiology (association studies) among others.

The study of the asymptotic behaviour of the statistic associated to the Havrda-Charvat and Renyi entropies in relation to the recovered obtained for the different parameters alpha, was carried out, generating 50,000 values associated with a Multinomial distribution with M=3D10, 20, n=3D40, 100 and we evaluated the probability of acceptance of the equiprobability hypothesis for the different values of ponderation alpha ($\alpha = 3D0.1, \ldots, 7$) being the level of significance of the study ($\varepsilon = 3D0.05$)

20.4 Results and Discussion

Table 20.1 refers to the Havrda-Charvat (α=3D3) index. It displays the estimation of the acceptance probability for the null hypothesis of maximum entropy using tests based on statistics E_1 to E_6. E.V stands for *entropy value* and corresponds to the true value of the entropy index, $H_3^{HC}(P)$. The first row always corresponds to the case of sampling under the null hypothesis $P = 3D(M^{-1}, \ldots, M^{-1})$ while subsequent rows correspond to P vectors associated with various several conditions. Tests were always performed under a nominal 5% significance level, and thus, the first row should be always close to 95%. E_6 plays the role of a nearly "exact" test used as a reference for the asymptotic tests. In all cases, E_2 (alternative 1) partially corrects the deviations of the rough asymptotic test E_1, while E_3 (alternative 2) provides a much better correction, inducing a nearly exact test, comparable to E_6. E_4 (adherence to the equiprobability distribution) corresponds to a conservative test, and

E_5 yields similar results to asymptotic test E_1. These results are reproduced for the Havrda-Charvat($\alpha = 3D2, 5$) and for the other indexes (Shannon, and Renyi $\alpha = 3D2, 3, 5$), as shown in section (20.7).

Table 20.1: Havrda-Charvat ($\alpha = 3D3$) $M = 3D20$, $n = 3D40$, $\varepsilon = 3D0.05$.

E.V	E1	E2	E3	E4	E5	E6
0.49875	85.1	91.6	94.9	95.7	88.6	95.01
0.49815	61.3	71.2	81.2	91.2	65.4	82.10
0.49781	39.86	54.03	63.71	75.1	42.9	65.03
0.49765	33.0	47.19	57.15	70.21	34.76	58.64
0.4974	23.11	36.30	46.49	63.14	23.24	48.09
0.49721	17.17	27.8	39.7	59.57	17.08	40.12
0.49701	12.10	20.1	30.1	50.1	12.29	31.2

The results obtained in the study of the statistics associated to the Havrda-Charvat and Renyi entropies in relation to the parameter alpha (Table 20.2), they suggest that the conservative behaviour of the tests associated with the entropies of Renyi when increasing the parameter alpha, is because the asymptotic statistic is a growing function of alpha. In that sense, the parameter alpha that optimizes the recovery of the null hypothesis (it is closer to the significance level) for the entropy of Renyi is $\alpha = 3D1.2$. On the other hand, the concavity of the statistic to the Havrda-Charvat entropies leads to the good one in two points. The parameters $\alpha = 3D1.2$ and $\alpha = 3D2.3$ optimize the recovered of the null hypothesis for sample sizes risen in connection with the number of classes. For samples of reduced size, the parameters that optimize the recovery are obtained for $\alpha > 1.2$ and $\alpha < 2.3$. This result is a consequence of the negative structural bias that appears when considering the estimates of measures of diversity, and of the bias compensation that takes place when a considering conservative test.

As a final conclusion, the test based on statistic E_3 (20.6) should be recommended in practice. It is only slightly more difficult to compute it than the rough asymptotic test and it performs better in any circumstances. In fact its power is very close to that of the exact test based on the tabulation of the sampling distribution.

Table 20.2: Asymptotic behaviour of Havrda-Charvat and Renyi entropies versus parameter alpha $M = 3D20$, $n = 3D40$, $\varepsilon = 3D0.05$.

Alpha Values	Havrda-Charvat	Renyi	Alpha Values	Havrda-Charvat	Renyi
1.0	0.887	0.887	2.1	0.947	0.999
1.1	0.917	0.93	2.2	0.943	1
1.2	0.935	0.956	2.3	0.936	1
1.3	0.946	0.972	2.4	0.927	1
1.4	0.954	0.983	2.5	0.919	1
1.5	0.957	0.989	2.6	0.908	1
1.6	0.958	0.993	2.7	0.895	1
1.7	0.959	0.996	2.8	0.881	1
1.8	0.958	0.997	2.9	0.867	1
1.9	0.956	0.998	3.0	0.852	1
2.0	0.956	0.999	3.1	0.833	1

20.5 Expectation and Variance of the Havrda-Charvat Entropies

Let $H_\alpha(P)$ stand for the "population" Havrda-Charvat entropy of order α calculated over the probability vector $P = 3D(p_1, \ldots, p_M)$ and $H_\alpha(\hat{P})$ for the corresponding sample entropy calculated over a relative frequency vector $\hat{P} = 3D(\hat{p}_1, \ldots, \hat{p}_M)$ computed from a sample of size n. Then, we have:

Theorem 20.5.1

$$E\left[H_\alpha\hat{P} - H_\alpha(P)\right]$$
$$= 3D \frac{1}{\alpha - 1}\left[\sum_{i=3D1}^{M} p_i^\alpha - \sum_{i=3D1}^{M}\left(\sum_{j=3D1}^{\alpha} \frac{n(n-1)\ldots(n-j+1)p_i^j}{j!\, n^\alpha}\right.\right.$$
$$\left.\left. \times \left(\sum_{k=3D0}^{j} \binom{j}{k}(-1)^k(j-k)^\alpha\right)\right)\right] \qquad (20.10)$$

and

$$VAR\left[H_\alpha(\hat{P}) - H_\alpha(P)\right] = 3D\frac{A_1 + A_2 + A_3}{(n^{\alpha-1}(\alpha-1))^2},$$

where:

$$A_1 = 3D\ E\left(\sum_{i,j=3D1\ i\neq j}^{M} (n_i^\alpha n_j^\alpha)\right)$$

$$A_2 = 3D \ E\left(\sum_{i=3D1}^{M} n_i^{2\alpha}\right)$$

$$A_3 = 3D \ \left[E\left(\sum_{i=3D1}^{M} n_i^{\alpha}\right)\right]^2,$$

which reduce to

$$A_1 = 3D \sum_{i,j=3D1 \ i\neq=j}^{M} \left[\sum_{j_1}^{\alpha}\sum_{j_2}^{\alpha} \frac{n(n-1)\ldots(n-j_1-j_2+1)}{j_1!\,j_2!} p_1^{j_1} p_2^{j_2} \right.$$
$$\left. \times \sum_{k_1=3D0}^{j_1-1}\sum_{k_2=3D0}^{j_2-1} \binom{j_1}{k_1}(-1)^{k_1}(j_1-k_1)^{\alpha} \binom{j_2}{k_2}(-1)^{k_2}(j_2-k_2)^{\alpha}\right]$$

$$A_2 = 3D \sum_{i=3D1}^{M}\sum_{j=3D1}^{2\alpha} \frac{n(n-1)\ldots(n-j+1)p_i^j}{j!} \sum_{k=3D0}^{j-1}\binom{j}{k}(-1)^k(j-k)^{2\alpha}$$

and

$$A_3 = 3D \left(\sum_{i=3D1}^{M}\sum_{j=3D1}^{\alpha}\frac{n(n-1)\ldots(n-j+1)p_i^j}{j!}\right.$$
$$\left. \times \sum_{k=3D0}^{j-1}\binom{j}{k}(-1)^k(j-k)^{\alpha}\right)^2.$$

The *proof* of the preceding result lies in the following previous results.

The joint distribution of the random vector $(n_1,\ldots,n_M) = 3Dn(\hat{p}_1,\ldots,\hat{p}_M)$ is multinomial with parameters $(p_1,\ldots,p_M;n)$, factorial moment generating function: $g(t_1,\ldots,t_M) = 3DE\left(t_1^{n_1},\ldots,t_M^{n_M}\right) = 3D(t_1p_1+\ldots+t_Mp_M)^n$ and factorial moments

$$\mu_{[r_1,\ldots,r_M]}$$
$$= 3D \ E\left(n_1(n_1-1)\ldots(n_1-r_1+1)\ldots n_M(n_M-1)\ldots(n_M-r_M+1)\right)$$
$$= 3D \ \left.\frac{\partial^{\sum r_i} g(t_1,\ldots,t_M)}{\partial t_1^{r_1}\ldots \partial t_M^{r_M}}\right|_{t_1=3D\ldots,t_M=3D1}$$
$$= 3D \ n(n-1)\ldots(n-\sum_{i=3D1}^{M} r_i+1)p_1^{r_1}\ldots p_M^{r_M}.$$

Then, its moments are given by the following expression:

Lemma 20.5.1

$$E(n_1^{r_1} \ldots n_s^{r_s})$$
$$= 3D \left[\sum_{j_1=3D0}^{r_1} \ldots \sum_{j_s=3D0}^{r_s} \frac{E\left(n_1^{[j_1]} \ldots n_s^{[j_s]}\right)}{j_1! \ldots j_s!} \right.$$
$$\left. \times \left[\sum_{k_1=3D0}^{j_1} \ldots \sum_{k_s=3D0}^{j_s} \binom{j_1}{k_1}(-1)^{k_1}(j_1-k_1)^{r_1} \ldots \binom{j_2}{k_2}(-1)^{k_s}(j_s-k_s)^{r_s} \right] \right]$$
(20.11)

where $n^{[j]} = 3D\, n(n-1)\ldots(n-j+1)$.

The preceding Lemma (20.5.1) is a direct consequence of the relationship

$$X^s = 3D \sum_{i=3D1}^{s} \frac{X^{[i]}}{i!} \sum_{j=3D0}^{i} \binom{i}{j}(-1)^j (i-j)^s, \qquad (20.12)$$

which may be proved as follows:

For a given real function $g(x)$, for $x_0, \ldots, x_n \in \Re$ and for the family of functions of the form:

$$g_r[x_0, \ldots, x_r] = 3D \sum_{i=3D0}^{r} \frac{g(x_i)}{\prod_{j=3D0\, j\neq i}^{r}(x_i - x_j)},$$

for which:

$$g_r[x_0, \ldots, x_r] = 3D \frac{g_{r-1}[x_0, \ldots, x_{r-1}] - g_{r-1}[x_1, \ldots, x_r]}{x_0 - x_r},$$

it is possible to apply the Newton interpolation formula:

$$g(X) = 3D\ g(x_0) + \sum_{i=3D1}^{n} g_i[x_0, \ldots, x_i](x-x_0)\ldots(x-x_{i-1})$$
$$+ g_{n+1}[x, x_0, \ldots, x_n](x-x_0)\ldots(x-x_n) \qquad (20.13)$$

as a result of combining the following equalities:

$$\left. \begin{array}{l} = g_{n+1}[x, x_0, \ldots, x_n] = 3D \frac{g_n[x,x_0,\ldots,x_{n-1}]}{x-x_n} - \frac{g_n[x_0,\ldots,x_n]}{x-x_n} \\ = \frac{g_n[x,x_0,\ldots,x_{n-1}]}{x-x_n} = 3D \frac{g_{n-1}[x,x_0,\ldots,x_{n-2}]}{(x-x_n)(x-x_{n-1})} - = \frac{g_{n-1}[x_0,\ldots,x_{n-1}]}{(x-x_n)(x-x_{n-1})} \\ \vdots \\ \frac{g_1[x,x_0]}{(x-x_n)\ldots(x-x_1)} = 3D \frac{g(x)}{(x-x_n)\ldots(x-x_0)} - \frac{g(x_0)}{(x-x_n)\ldots(x-x_n)} \end{array} \right\}.$$

By particularising(20.13) to the polynomial $Q(x) = 3DX^s$ and replacing $x = 3D0, x_1 = 3D1, \ldots, x_s = 3Ds$ can we write(20.12), as

$$Q_r[x_0, \ldots, x_r] = 3D \sum_{i=3D0}^{r} \frac{i^s}{i!(r-i)!(-1)^{r-i}} = 3D \frac{1}{r!} \sum_{j=3D0}^{r} \binom{r}{j}(-1)^j(r-j)^s$$

and the remainder of the Newton interpolation formulae is zero, as $Q(x)$ and $Q(x_0) + \sum_{i=3D1}^{s} Q_i[x_0, \ldots, x_i] x^{[i]}$ are equal in $s+1$ points ($x_0 = 3D0, x_1 = 3D1, \ldots, x_s = 3Ds$). By particularising the expression Lemma(20.5.1) to the univariate case we have

$$E(n_i^r)$$
$$= 3D \left(\sum_{i=3D1}^{M} \sum_{j=3D1}^{\alpha} \frac{n(n-1)\ldots(n-j+1)p_i^j}{j!} \sum_{k=3D0}^{j-1} \binom{j}{k}(-1)^k(j-k)^\alpha \right)$$

and the bivariate case,

$$E(n_1^r n_2^s)$$
$$= 3D \sum_{j_1=3D1}^{r} \sum_{j_2=3D1}^{s} \left[\frac{E\left(n_1^{[j_1]}(n_2^{[j_2]}\right)}{j_1! j_2!} \right.$$
$$\times \left. \left[\sum_{k_1=3D0}^{j_1-1} \sum_{k_2=3D0}^{j_2-1} \binom{j_1}{k_1}(-1)^{k_1}(j_1-k_1)^r \binom{j_2}{k_2}(-1)^{k_2}(j_2-k_2)^s \right] \right]$$

from which we can compute the expectation and the variance, respectively.

20.6 Expectation and Variance of the Functional (h, ϕ)-Entropies

From previous results obtained in section 20.5 we have developed a generic expression for the expected value and the variance of the functional (h, ϕ)-entropies. Recall that

$$h\left(\sum_{i=3D1}^{M} \phi(\hat{p}_i)\right)$$
$$= 3Dh\left(\sum_{i=3D1}^{M} \phi(p_i)\right)$$
$$+ \sum_{k=3D1}^{\infty} \frac{1}{k!} \frac{1}{n^k} \left(\sum_{i_1 \ldots i_k=3D1}^{M} A_k(i_1, \ldots, i_k)(n_1 - np_1)^{r_1} \ldots (n_M - np_M)^{r_M} \right),$$

where

$$A_k(i_1,\ldots,i_k) = \frac{\partial^k h\left(\sum_{i=1}^M \phi(p_i)\right)}{\partial p_{i_1}\ldots \partial p_{i_k}} \quad (r_1+\ldots+r_k = k)$$

is the multiplicity corresponding to the terms $(\hat{p}_1 - p_1,\ldots,\hat{p}_M - p_M)$. Thus,

$$E\left(h\left(\sum_{i=1}^M \phi(\hat{p}_i)\right)\right)$$
$$= h\left(\sum_{i=1}^M \phi(p_i)\right)$$
$$+ \sum_{k=1}^{\infty} \frac{1}{k!} \frac{1}{n^k} \left(\sum_{i_1\ldots i_k=1}^M A_k(i_1,\ldots,i_k) B_k(r_1,\ldots,r_M)\right),$$

where

$$B_k(r_1,\ldots,r_M) = E\left[(n_1 - np_1)^{r_1}\ldots(n_k - np_k)^{r_k}\right]$$
$$\times \sum_{j_1=0}^{r_1} \sum_{j_2=0}^{r_2} \ldots \sum_{j_k=0}^{r_k} (-1)^{k-j_1\ldots-j_k} \binom{r_1}{j_1}\binom{r_2}{j_2}\ldots$$
$$\binom{r_k}{j_k} n^{k-j_1\ldots-j_k} p_1^{r_1-j_1}\ldots p_k^{r_k-j_k} E\left(n_1^{j_1}\ldots n_k^{j_k}\right)$$

and $E\left(n_1^{j_1}\ldots n_k^{j_k}\right)$ is the expected value obtained in section 20.5. In this context the variance of the functional is given by:

$$VAR\left(h\left(\sum_{i=1}^M = \phi(\hat{p}_i)\right)\right) = E\left(H_{h^2}^{\phi}(P)\right) - \left[E\left(H_h^{\phi}(P)\right)\right]^2,$$

where: $E\left(H_{h^2}^{\phi}(P)\right)$ $\left[E\left(H_h^{\phi}(P)\right)\right]^2$ are the expected values obtained from applying the previous expression for the functions h^2 and h respectively.

20.7 Tables

20.7.1 Tables of simulation results

E.V	E1	E2	E3	E4	E5	E6
0.95	95.7	95.7	94.9	95.7	88.6	95.1
0.948	91.2	91.2	91.8	91.2	65.4	92.1
0.9455	75.1	75.1	86.5	75.2	42.9	86.75
0.942	70.21	70.21	76.7	70.21	34.76	77.07
0.9375	63.14	63.14	61.8	63.14	23.24	62.08
0.932	59.57	59.57	57.75	59.57	17.08	57.8
0.924	50.1	50.1	17.31	50.1	12.29	17.49

Havrda-Charvat($\alpha = 3D2$) M $=3D$ 20, n $=3D$ 40 $\varepsilon = 3D0.05$

E.V	E1	E2	E3	E4	E5	E6
0.249998	37.7	88.75	94.89	95.7	88.6	95.0
0.249995	8.6	70.41	89.81	91.2	65.4	90.1
0.249994	6.72	46.92	71.49	75.1	42.9	72.0
0.249993	4.71	40.88	67.12	70.21	34.76	67.65
0.249991	2.5	31.83	59.65	63.14	23.24	60.23
0.24999	1.44	26.18	54.38	59.57	17.08	54.95
0.249989	0.8	20.86	48.73	50.1	12.29	49.32

Havrda-Charvat($\alpha = 3D5$) M $=3D$ 20, n $=3D$ 40 $\varepsilon = 3D0.05$

E.V	E1	E2	E3	E4	E5	E6
2.99573	88.6	95.63	95.02	95.7	88.6	95.0
2.91345	65.4	76.8	76.2	91.2	65.4	75.2
2.86492	42.9	60.29	59.76	75.1	42.9	58.4
2.84186	34.76	51.41	50.35	70.21	34.76	49.83
2.80298	23.24	38.8	38.1	63.14	23.24	37.46
2.77397	17.08	28.59	28.41	59.57	17.08	27.4
2.74255	12.29	20.41	20.12	50.1	12.29	19.14

Shannon M $=3D$ 20, n $=3D$ 40 $\varepsilon = 3D0.05$

E.V	E1	E2	E3	E4	E5	E6
2.99573	99.89	97.2	95.03	95.7	88.6	95.1
2.84731	97.71	82.41	81.0	91.2	65.4	80.8
2.77258	95.25	64.36	62.10	75.1	42.9	62.08
2.73985	93.43	57.21	54.44	70.21	34.76	54.41
2.68824	89.10	44.6	42.14	63.14	23.24	42.10
2.65242	79.2	35.8	33.6	59.57	17.08	33.6
2.61565	69.4	27.6	25.1	50.1	12.29	25.12

Renyi($\alpha = 2$) M = 20, n = 40 $\varepsilon = 0.05$

E.V	E1	E2	E3	E4	E5	E6
2.99573	100	98.07	95.06	95.7	88.6	95.01
2.79971	99.9	92.32	83.2	91.2	65.4	82.10
2.71592	99.84	80.25	66.14	75.1	42.9	65.03
2.68148	99.78	75.50	59.83	70.21	34.76	58.64
2.62954	99.68	66.86	49.35	63.14	23.24	48.09
2.5950	99.60	60.43	41.72	59.57	17.08	40.12
2.56067	99.46	52.92	33.92	50.1	12.29	31.2

Renyi($\alpha = 3$) M = 20, n = 40 $\varepsilon = 0.05$

E.V	E1	E2	E3	E4	E5	E6
2.99573	100	98.34	94.98	95.7	88.6	95.0
2.74484	100	96.84	89.96	91.2	65.4	90.1
2.66116	100	86.52	71.08	75.1	42.9	72.0
2.62869	100	83.83	66.68	70.21	34.76	67.65
2.58127	100	78.74	59.15	63.14	23.24	60.23
2.55052	100	74.79	53.87	59.57	17.08	54.95
2.52046	100	70.18	48.18	50.1	12.29	49.32

Renyi($\alpha = 5$) M = 20, n = 40 $\varepsilon = 0.05$

References

1. Arimoto, S. (1971). Information-theoretical considerations on estimation problems, *Information and Control*, **6**, 189–194.

2. Cavalli-Sforza, L. L., Menozzi, P., and Piazza, A. (1994). *The History and Geography of Human Genes*, Princeton, NJ: Princeton University Press.

3. Dagpunar, J. (1988). *Principles of Random Variate Generation*, Oxford Science Publication.

4. Ferreri, C. (1980). Hypoentropy and related heterogeneity, divergence and information measures, *Statistica*, **40**, 55–118.

5. Havrda, J. and Charvat, F. (1967). Quantification method in classification processes: Concept of structural α-entropy, *Kybernetika*, **3**, 30–35.

6. Kotz, S., Johnson, N. L. and Boyd, D. W. (1967). Series representation of distribution of quadratic forms in normal variables. I Central Case, *A.M.S*, 823–837.

7. Marsaglia, G. and Zaman, A. (1991). A new class of random number generators, *Annals of Applied Probability*, **1**, 462–480.

8. Ocaña, J. (1985). *Bibliotecas orientadas a objetos de generadores de distribuciones*, Departament d'Estadística, Universitat de Barcelona.

9. Pardo, L., Morales, D., Salicrú, M. and Menéndez, M. L. (1997). Large sample behaviour of entropy measures when the parameters are estimated, *Communications in Statistics—Theory and Methods*, **26 (2)**, 483–501.

10. Rao, J. N. K. and Scott A. J. (1981). The analysis of categorical data from complex sample surveys: chi squared tests for goodness of fit and independence in two way tables, *Journal of the American Statistical Association*, **76**, 221–230.

11. Renyi, A. (1961). On measures of entropy and information, In *Proceedings of the 4th Berkeley Symposium on Mathematical Statistics and Probability*, Volume **1**, pp. 547–561, Berkeley, CA: University of California Press.

12. Roychoadhury, A. K. and Nei, M. (1988). *Human polymorphic genes. World distribution*, Oxford University Press.

13. Salicrú, M. and Calvo, M. (1988). Medidas de incertidumbre asociadas a J-Divergencias, *Trabajos de Estadística*, **3 (2)**, 121–140.

14. Salicrú, M., Menéndez, M. L., Morales, D. and Pardo, L. (1993). Asymptotic distribution of $(h - \phi)$-entropies, *Communications in Statistics—Theory and Methods*, **22 (7)**, 2015–2031.

15. Satterthwaite, F. E. (1946). An approximate distribution of quadratic forms, *Annals of Mathematical Statistics*, **34**, 1582–1584.

16. Shannon, C. E. (1948). A mathematical teory of communications, *Bell. System Tech. J*, **27**, 379–423.

17. Sharma, B. D. and Taneja, I. J. (1975). Entropy of type (α, β) and other generalised measures in information theory, *Metrika,* **22,** 205–215.

18. Varma, R. S. (1966). Generalisations of Renyi's entropy of order α, *Journal of Mathematical Science,* **1,** 34–48.

19. Vives, S., Ocaña, J. and Salicrú, M. (1996). Simulation of the asymptotic behaviour of Havrda-Charvat entropies $(\alpha = 3D3)$.*Proceedings in Computational Statistics (Short communications),* pp. 237–238, University of Barcelona Press.

20. Vives, S., Salicrú, M. and Ocaña, J. (1997). Moments of the statistics associated at (h, ϕ)-entropies, *Technical Reports* **255** 1–10, University Of Barcelona Press.

21

Partially Inversion of Functions for Statistical Modelling of Regulatory Systems

A. G. Bart, N. P. Alexeyeff (Klochkova) and N. Botchkina
St. Petersburg State University, St. Petersburg, Russia

Abstract: In this paper inversion of functions is considered as one of the applications of the reflection principle which is the basis for self-regulation in complex systems. A set of partially inverse functions for a real-valued measurable function is described parametrically. The generalised binomial distributions are defined using the symmetry of fiducial distributions which is a stochastic form of this reflection principle. These methods are illustrated by examples from neurophysiology and sales marketing.

Keywords and phrases: Partially inverse functions, fiducial distributions, branching processes, compound Poisson distribution, neuronal trees, crucial points, sales marketing

Introduction

In 1930 Fisher suggested a new method of randomized parameter estimation using fiducial distributions defined by

$$F^*(\theta \mid x) = 1 - F(x \mid \theta) \qquad (21.1)$$

with natural conditions for the family of distribution functions $F(x \mid \theta)$ with θ as a parameter. For many years this idea was related to the famous Berens and Fisher problem and to other similar statistical problems. A theory was created to generalise this idea using a group-theoretical approach [Klimov (1973)]. But Fisher's idea may be considered as a method of describing regulation in complicated systems whose major control mechanism is aggregation of elements. Recently such problems have become more important, for example, membrane regulation in biosystems, the organisation of social groups, and the interaction between the introduction of a new product and market laws themselves.

Let us consider a stochastic process $x = \xi(\theta, \omega)$ and construct another process $\theta = \xi^-(x, \omega)$ such that the trajectories of these processes with fixed ω are mutually inverse. Then the fiducial identity [Fisher (1930)] relates the one-dimensional distributions of these processes. The difficulty of this approach is that inversion for measurable functions must be properly defined.

In papers by Bart (1987) and Bart, Klochkova and Kozhanov (1993) the theory of generalized inversion for measurable functions was constructed. As a consequence of this definition the operation of *partial inversion* was also defined. The extreme inverse functions for real functions were investigated in detail. Following Bart (1987), in this paper, we describe parametrically the set of partially inverse functions for real measurable functions. Note that if a function of interest has singularities such as jumps or nonmonotonicity it can be recovered from the double inverse functions up to a set of measure zero [Bart, Klochkova and Kozhanov (1993)]. Therefore the trajectories of stochastic processes are a natural domain for partial inversion.

In this paper we discuss possible applications of the reflection principle to practical problems using the technique of partial inversion. Reconciliation of the rhythmic and spatial structure of biosystems in terms of fiducial distributions (with neuronal trees as an example) will be discussed in Section 21.3.

In particular partial inversion of binomial processes leads to generalized binomial distributions described in Section 21.2. Using these distributions stable regulation (i.e. thresholds and the scale of immunity) is defined within a statistical model of a given biosystem.

In Section 21.4 we analyse sales marketing dynamics and consequences of control (by advertisement) in this regulatory system.

Some proofs regarded by the authors as technical are carried out in the Appendix.

21.1 A Method for Partial Inversion of Functions

21.1.1 The parametrical description of the partial inverse functions

In this section we suggest the generalization of the inversion operation for functions with jumps and nonmonotonicity by means of truncating the function's domain and expanding its image. The main problem of parametrical description of the possible functions inverse to a given one is to construct a common parametrization of all real measurable sets. We consider one example of such a parametrization.

Let $\mathcal{X}$ be a measurable real set and μ be a σ-finite measure equivalent to Lebesgue measure. The minimal interval R containing the set $\mathcal{X}$ we call the *scope* of the set $\mathcal{X}$, i.e. $R = R(\mathcal{X}) = (\inf \mathcal{X}, \sup \mathcal{X}) \cup \mathcal{X}$. Now we consider three cases for the measure of the scope.

If $0 < \mu(R) < \infty$ we can construct a bijection $\alpha \in [0,1] \longrightarrow z_\alpha \in \bar{R}$ ($\bar{R}$ is the closure of R) defined implicitly by

$$\alpha = \mu(x \leq z_\alpha | R) = \mu\{(x \leq z_\alpha) \cap R\}/\mu(R), \tag{21.2}$$

and we define the point $x_\alpha \in \overline{\mathcal{X}}$ by $x_\alpha = \sup\{(x\,;\, x \leq z_\alpha) \cap \mathcal{X}\}$.

If $\mu(R) = \infty$ we truncate the scope R changing it into $R^\alpha = I^\alpha \cap R$ where

$$I^\alpha =: (-\infty; 1/(1-\alpha)\,];\quad [-1/\alpha; 1/(1-\alpha)];\quad [-1/\alpha, \infty)$$

respectively where R is infinite in the positive direction only, in both directions or in the negative direction only.

Because of the σ-finiteness of the measure μ there exists a sequence of sets R_n of finite-measure such that $R_n \uparrow R$.

For parametrization of this kind of set it is sufficient to change (21.2) in the previous construction to the limit case:

$$\alpha = \lim_{n\to\infty} \mu(x \leq z_\alpha | R_n \cap R^\alpha).$$

Finally for the case $\mu(R) = 0$ and $\mathcal{X}$ is not empty then the pre-image consists of one point only which is the image of every $\alpha \in [0,1]$.

Below we call μ the form of parametrization and α the partiality parameter. Note that we can consider the dual parametrization of elements of the set $\mathcal{X}$ which associates to every $\beta \in [0,1]$ the element $x^\beta \in \mathcal{X}$ in the following way

$$\beta = \mu(x \geq z^\beta | R), \qquad x^\beta = \inf\{(x\,;\, x \geq z^\beta) \cap \mathcal{X}\}.$$

We call this parametrization of $\mathcal{X}$ the β-parametrization and the previous one the α-parametrization. Consider a measurable function $f(x)$ mapping $\mathcal{X}$ onto $\mathcal{Y}$. For every $y - f(x)$ we define its pre-images:

the left pre-image: $\overleftarrow{\mathcal{X}}_y = \{x : f(x) \geq y\}$,

the right pre-image: $\overrightarrow{\mathcal{X}}_y = \{x : f(x) \leq y\}$,

the real pre-image: $\tilde{\mathcal{X}}_y = \overrightarrow{\mathcal{X}}_y \cap \overleftarrow{\mathcal{X}}_y$

and the complete pre-image:

$$\mathcal{X}_y = \{x :\ \forall \delta > 0,\ X_\delta \cap \overrightarrow{\mathcal{X}}_y \neq \emptyset,\ X_\delta \cap \overleftarrow{\mathcal{X}}_y \neq \emptyset\},$$

where $X_\delta = (x - \delta, x + \delta)$.

In contrast to the real pre-image the complete pre-image also contains the jump points of the function.

Now if we apply the parametrization method described above to the sets $\mathcal{X}_y$ then for every $\alpha \in [0,1]$ the function $f_\alpha^-(y) = x_\alpha$ is partially inverse to $f(x)$.

Note that at the points where the original function is strictly monotonic and continuous, all its partially inverse functions have the same values. The partially inverse functions with parameters $\alpha = 0;1$ or $\beta = 0;1$ become the extreme inverse functions considered in Bart, Klochkova and Kozhanov (1993).

21.1.2 Double inversion

We denote by $f_{\gamma\delta}^{-2}(x) = (f_\gamma^-)_\delta^-$ — the double partial inversion of the function f with parameters γ and δ.

We shall use here *the constant indicator* $\rho(x)$ which equals 1 if $x \in int X$ and $\Delta > 0$ such that for every δ, $(0 < \delta < \Delta)$, $\mu(R(f(X_\delta))) = 0$ where $X_\delta = (x - \delta, x + \delta)$ and 0 otherwise. $\rho^-(y) = \rho_\alpha^-(y)$ means the constant indicator for the partially inverse function f_α^-.

Theorem 21.1.1 *1) If $\rho(x) = 1$ then for every $x_1 \neq x_2, x_i \in X_\delta, i = 1,2$ (from the definition) and for every partially inverse function $f_\alpha^-(y)$ at least one of x_i does not belong to $f_\alpha^-(\mathcal{Y})$.*

2) If $\rho_\alpha^-(y) = 1$ then there exists a neighbourhood of y Y_δ such that for every $y_1 \neq y_2, y_i \in Y_\delta, i = 1,2$ at least one of them belongs to $\mathcal{Y}^- \setminus \mathcal{Y}$.

3) If the complete prototype $\mathcal{X}_y$ is not empty and $x \in \mathcal{X}_y$ then

a) for every partially inverse function $\rho(x)\rho^-(y) = 0$;

b) if at the point x $f(x)$ has a singularity then $\rho(x) + \rho^-(y) > 0$.

The original function can be recovered in terms of constant indicators and double partially inverse functions. Denote the double inverse function to f with α-parametrization for the first inversion and with right inversion for the second as f_α^{-2}; and the double inverse function to f with β-parametrization for the first inversion and with left inversion for the second as f_β^{-2}; ρ_α and ρ_β are the corresponding constant indicators.

Corollary 21.1.1 *Let $A(x)$ and $B(x)$ be subsets of $[0,1]$ such that for the function $f : \mathcal{X} \longrightarrow \mathbf{R}^1$ for every $\alpha \in A(x)$ and $\beta \in B(x)$ functions f_α^{-2} and f_β^{-2} are defined. Then*

1) For every $x \in \mathcal{X}$

$$A(x) \neq \emptyset, \ B(x) \neq \emptyset.$$

2) If there exist $\alpha \in A(x)$ and $\beta \in B(x)$ such that $\rho_\alpha(x)\rho_\beta(x) = 0$ then for every point of continuity of the function f the following holds:

$$f(x) = f_\alpha^{-2}(x) + \rho_\alpha(x)(f_\beta^{-2} - f_\alpha^{-2}) = f_\beta^{-2} + \rho_\beta(f_\alpha^{-2} - f_\beta^{-2}).$$

Partially Inversion of Functions for Statistical Modelling 359

3) *If for every* $\alpha \in A(x)$ *and* $\beta \in B(x)$ $\rho_\alpha(x)\rho_\beta(x) = 1$ *then* $f(x)$ *is constant in a neighbourhood of the point* x *and*

$$f(x) = f_\alpha^{-2}(x) = f_\beta^{-2}(x).$$

The first statement follows immediately from the parameter definition. For the second item we note that the only possible change to the function f after double partial inversion is to convert parts of the function into a constant function. If the function $f(x)$ is continuous and, for example, $\rho_\beta(x) = 0$, then x belongs to the narrowing region and $f_\beta^{-2} = f(x)$.

Note also that at the jump points this equality may not hold [Bart, Klochkova and Kozhanov (1993)].

Finally the last statement follows from the theorem and from the fact that orientations of parametrizations after repeated inversion are reciprocal.

21.2 Generalized Binomial Distributions

Consider the standard binomial scheme. Let a single event ω be an infinite sequence of 0's and 1's, let p be a fixed probability of success, $k = \xi(n, \omega \mid p) = \xi(n, \omega)$ be the number of 1's amongst the first n elements. Given ω and p, the partially inverse functions to $\xi(n, \omega)$ are $\xi_\alpha^-(k) = \xi_\alpha^-(k) = \lfloor (1-\alpha)\xi_0^-(k) + \alpha\xi_1^-(k) \rfloor$, where

$$\xi_0^-(k) = \min\{n; \xi(n) \geq k\}, \quad \xi_1^-(k) = \max\{n; \xi(n) \leq k\},$$

$\alpha \in [0, 1]$, $\lfloor x \rfloor$ is the integer part of x and $\lceil x \rceil$ is the least integer not less than x. Letting ω vary we have obtained two random processes with mutually inverse trajectories. If k is fixed, then we have the following equality between distributions $\xi_\alpha^-(k) = \xi_0^-(k) + \xi_\alpha^-(0)$, where $\xi_\alpha^-(0) = \lfloor \alpha\xi_1^-(0) \rfloor$ are generalized geometrical distributions $\beta_-^*(j+1 \mid 1, p, \alpha) = p_j q^{\alpha_j}$, where $\alpha_j = \lceil \frac{j}{\alpha} \rceil$, $q_j = q^{\alpha_j \mid 1-\alpha_j}$, $p_j + q_j = 1$, $j = 1, 2, \ldots$. The generalized negative binomial distribution (that is, its scale modification) $\beta_-^*(k+j \mid k, p, \alpha)$ is defined as the convolution of k independent generalized geometrically distributed random variables.

Generalized beta and binomial distributions can be defined using the well-known fiducial identities for the ordinary binomial distributions:

$$\beta_+^{*<}(k \mid n, p, \alpha) = 1 - \beta_-^{*\leq}(n \mid k, p, \alpha), \quad \beta^{*<}(p \mid k, j, \alpha) = \beta_-^{*<}(k+j \mid k, p, \alpha),$$

where $n = k + j$ and the signs $<, \leq$ correspond to the cumulative distribution functions with this sign, e.g. $F^{\leq}(x) = P\{\xi \leq x\}$.

Remark 21.2.1 Introduction of rational $\alpha = \frac{s}{m}$ leads to clusterization the aggregate of Bernoulli trials in groups of m trials where each of them consists of s inhomogeneous subgroups. If $s = 1$ then subgroups become homogeneous and the generalized binomial distributions are the ordinary simple ones with probability of success $1 - q^m$. This property is crucial for analysis of regulatory systems since it determines their rhythmic balance.

Thus the interval $[0, 1]$ of parameter q values can be divided into three parts. The first part is the region of stable regulation where q can be expressed as $q = q_0^m$, the second is $q > q_0$ where rhythmic balance fails and the introduced generalizations are necessary, and the third region where though the representation $q = q_0^m$ is formally possible nevertheless $m > m_1$ — the largest number for which the system can really be stable $q_1 = q_0^{m_1}$ and q_0 are the thresholds of the system's stable regulation. In biological applications of papers [Bart (1987), Bart, Klochkova and Kozhanov (1993) and Bart and Ivanov (1995)] they are called the immune thresholds and m is called the immune scale.

Since grouping methods are of importance in statistics, the introduced generalizations can be used as a new technique. We estimate parameters of generalized geometrical distribution using a stratified sample method. It is based on the following decomposition of the random variable into the independent summands given rational α.

Lemma 21.2.1 $\xi_{\frac{s}{m}}(1, p) = s\xi_0^-(1 - q^m) + \eta_s$ *(in law)* .

PROOF. Denote $j = st + i$, $i = 0, 1, ..., (s-1)$ and a random variable η_s with distribution $P(\eta_s = i) = \frac{p_i q^{\alpha i}}{1 - q^m}$. Then the generating function of $\xi_{\frac{s}{m}}(1, p)$

$$G_0(\nu) = \frac{1}{1 - \nu^s q^m} \sum_{i=0}^{s-1} \nu^i p_i q_{i-1} ... q_0$$

can be represented as a product of two generating functions: one of the random variable η_s and another of a geometrically distributed random variable multiplied by s. This corresponds to the fact that the original random variable is a sum of these two independent random variables. ∎

Let this distribution sample be presented with the frequencies N_j for the value j, $\sum_{j=0}^{\infty} N_j = N$. Put $j = st + i$, $i = 0, 1, ..., (s-1)$, $n_{ti} = \frac{N_{st+i}}{N}$, $n_{\cdot i} = \sum_{t=0}^{\infty} n_{ti}$.

Theorem 21.2.1 *Let s be known, then:*

1) If m is known, then the maximum likelihood estimate (MLE) for q is $\hat{q} = (\frac{\bar{x}^*}{1+\bar{x}^*})^m$, where $\bar{x}^* = \frac{1}{N}\sum_{j=0}^{\infty} \lfloor \frac{j}{s} \rfloor N_j$.

2) If m is unknown, then the MLE for q_i is

$$\hat{q}_i = 1 - \frac{n_{\cdot i}}{1 + \bar{x}^* - n_{\cdot 0} - ... - n_{\cdot i-1}}.$$

PROOF. 1) It is sufficient to consider a sample for the values y of the geometrically distributed random variable $\xi_0^-(1 \mid 1 - q^m)$. The sample mean is

$$\bar{y} = \sum_{t=0}^{\infty} t n_t. = \frac{1}{N} \sum_{t=0}^{\infty} t \sum_{i=0}^{s-1} N_{st+i} = \bar{x}^*. \qquad (21.3)$$

Therefore the maximum likelihood estimation for geometrical distribution determines the pointed result.

2) Estimation by distribution of η_s.

Find the logarithm of the likelihood function

$$\ln L(x \mid q_0, \ldots, q_{s-1}) = N \left[\sum_{i=0}^{s-1} n_{.i} \ln(1 - q^m) + \sum_{i=1}^{s-1} n_{.i} \sum_{r=0}^{i-1} \ln q_r - \ln(1 - q^m) \right].$$

From the previous item we have $\bar{x}^* = \frac{\hat{q}^m}{1 - \hat{q}^m}$.

So the equations for the maximum likelihood method are

$$-\frac{n_{.i}}{1 - \hat{q}_i} + \sum_{l=i+1}^{s-1} \frac{n_{.l}}{\hat{q}_i} - \frac{\bar{x}^*}{\hat{q}_i} = 0, \quad i = 0, 1, \ldots, (s-2)$$

$$-\frac{n_{.(s-1)}}{1 - \hat{q}_{s-1}} + \frac{\bar{x}^*}{\hat{q}_{s-1}} = 0.$$

∎

The structure of generalized positive binomial distributions is connected with the Galton-Watson branching processes with s particles. This is conveniently expressed by their generating functions. Let the generating function of the generalized positive binomial distributions $\beta_+^*(k \mid n, p, \frac{s}{m})$ be denoted by $f_n(\lambda)$.

The following recurrent formula defines $s - 1$ more distributions by its generating function.

$$f_n^{(i)}(\lambda) = \lambda f_{(n-1)}(\lambda) p_i + \sum_{t=i+1}^{n+i-1} \lambda q_i \ldots q_{t-1} p_t f_{n-t+i-1}(\lambda) + q_i \ldots q_{n+i-1}, \qquad (21.4)$$

$i = 0, \ldots, s - 1$.

Theorem 21.2.2 1. $f_n(\lambda) = f_n^{(0)}(\lambda)$.

2. $F_n^{(i)}(\lambda) = [\lambda P + Q]^n I_s$, where the vector $I_s = [1 \ldots 1]^T$, the vector function $F_n(\lambda) = [f_n^{(0)}(\lambda), \ldots, f_n^{(s-1)}(\lambda)]^T$ and matrices:

$$P = \begin{pmatrix} p_0 & 0 & \ldots & 0 \\ p_1 & 0 & \ldots & 0 \\ \vdots & \vdots & \ddots & \vdots \\ p_{s-1} & 0 & \ldots & 0 \end{pmatrix}, \quad Q = \begin{pmatrix} 0 & q_0 & \ldots & 0 \\ \vdots & \vdots & \ddots & \vdots \\ 0 & 0 & \ldots & q_{s-2} \\ q_{s-1} & 0 & \ldots & 0 \end{pmatrix}.$$

3. The generating function of generalized negative binomial distributions (scale modification)

$$G_0^k(\nu) = ((E - \nu Q)^{-1} P e_1)^k, \tag{21.5}$$

where E is a unit matrix of order s, $\quad e_1 = [1, 0, \ldots, 0]^T$ is the first orth.

21.3 Applications of Fiducial Distributions to Neurophysiology

This technique is convenient for observations with not only instrumental noise (often normally distributed), but also with fluctuations of the major regulatory processes themselves. This so-called structural noise usually has a nonlinear influence. Consider, as an example, neuronal interaction, which is one of the major problems of neurophysiology. Here, we see increasing complexity of noises at the higher interaction levels. The reflection principle stated above provides a description of the common regulatory mechanism at all levels [Bart (1987)].

The lowest level of interactions is the level having only one place of immediate signal transmission via a mediator (synapse). At this level the real-life observations of a postsynaptic potential (PSP) amplitude (according to the recent PSP-quantum theory) have the following structure: $\zeta = Q\xi + \xi_N$, where ξ stands for a PSP quantum number — binomial random variable, Q is the size of a quantum and ξ_N is normally distributed instrumental noise. A technique for dealing with this noise using standard parameter estimation and applying statistical simulation methods can be found in Bart, Klochkova and Kozhanov (1993). The regulatory aspect appears in the fiduciality of distributions of the number of quanta already thrown and the number of quanta yet to be thrown. For interneuronal junctions with registered inhomogenous contribution of many synapses, the same form of observations and their interpretation can be used if we replace the binomial distributions by the generalized binomial distributions. Here, the partiality parameter describes the structural noise and has an obvious morphological interpretation [Bart (1987)].

Investigation of neuronal networks is more involved. Origin data are three-dimensional reconstructions of neuronal trees or their two-dimensional projections with synapses, branch points and other morphological information marked. The synapses are allocated in clusters in the space of neuronal tree. So we apply a compound Poisson scheme for fitting their spatial allocation. If we divide the space into single blocks (cubes or squares), we can study relation (1) between distributions of the number of synapses in a single block and the distance between successive synapses. The number of synapses in a single block has the following structure: $\zeta = \xi_1 + \xi_2 + ... + \xi_\nu$, where ξ_i are binomial (possibly generalized binomial) and independent and ν is a Poisson random variable. The fiducial distribution for compound Poisson distribution is the distribution a random variable ζ^- fitting an equation $\xi^- = \zeta^-\nu$, where ξ^- is a negative binomial random variable, and ν is independent Poisson noise. It can be interpreted as showing that the expectation time of synapse measured with a Poisson distributed time unit has a negative binomial distribution.

PC programmes were created for dealing with neuronal trees, including suitable data transformation, synapse cloud extraction, parameter estimation for the compound Poisson and the negative binomial distributions, and several methods of multidimensional statistics.

Having applied these methods to 10 reconstructions of neuronal trees we found that the estimated parameters of fiducial distributions of corresponding variants did not contradict this model.

21.4 Advertisement for Sales Marketing

The other example of applying this method we consider here is a regulatory system for a product's sales marketing in order to find the region of stable regulation mentioned in Remark 21.2.1. Real data is considered only for showing the methodology.

21.4.1 Sanogenesis (compensation) curve

In this case we shall apply our partial inversion method to the function $S(t) = e^{-\eta t}\cos\tau t$, $t > 0$. In medical applications it is known as the sanogenesis curve [Bart, Bondarenko and Boiko (1980)]. The zero-points of this function and its derivatives $S^{(n)}(t) = Ae^{-\eta t}\cos(\tau t - n\varphi)$, $A = (-1)^n(\sigma^2 + \tau^2)^{n/2}$, $\mathrm{tg}\,\varphi = \eta/\tau$, are of the type $(2\pi j + n\varphi)/\tau$. They determine the points T_L^j on decreasing and T_P^j on increasing for $n = 0$, and also the latent points (the maximum points) T_l^j for $n = 1$. The latent points form an arithmetic progression with difference $d = T_l^{j+1} - T_l^j = 2\pi/\tau$. The values of the function $S(t)$ at these points form a geometric progression with ratio $q = S(T_l^{j+1})/S(T_l^j)$. The jumps between the maximal values of $S(t)$ can be described in terms of the geometrical distribution $S(T_l^j) - S(T_l^{j+1}) = c(q^j - q^{j+1})$, where $c = e^{\eta\varphi/\tau}\cos\varphi$.

We define the right inverse function as a supremum of the corresponding right pre-image. Denote the double inverse function of $S(t)$ with partiality parameter γ of the first inverse and the second right inverse as $H_\gamma(t) = S_{\gamma 1}^{-2}$.

Graphs of this function with $\gamma = 1$ and $\gamma = 0.5$ are shown in the (Fig. 21.1). Double inversion extracts new special points which are projections of max-

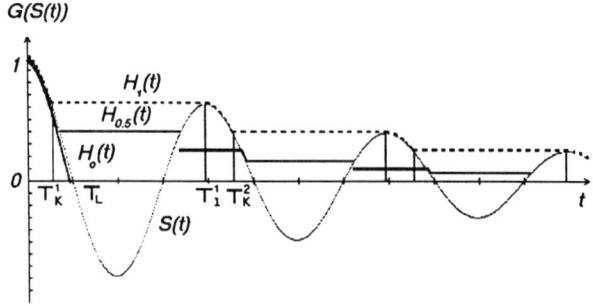

Figure 21.1: Graphs of the function $H_\gamma(t) = S_{\gamma 1}^{-2}$ with $\gamma = 1$ and $\gamma = 0.5$.

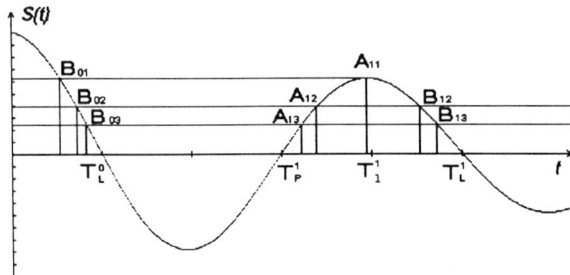

Figure 21.2: Crucial points of sanogenesis curve.

points of $S(t)$ at its previous decreasing branches T_k^{ij} and at its previous increasing branches T_r^{ij}. We shall call them crucial points.

Let us note that the sub-indices of the points in our notation are not numbers (and super-indexes are). They distinguish only the time order of points.

$$S(T_k^{0j}) = S(T_k^{1j}) = \ldots = S(T_k^{j-1,j}) = S(T_r^{1j}) = \ldots = S(T_r^{j-1,j}) = cq^j,$$

and

$$T_k^{0j} \in [0, T_L^0], \quad T_k^{ij} \in [T_l^i, T_L^i], \quad T_r^{ij} \in [T_P^i, T_l^i], \tag{21.6}$$

where $j \geq 1$ $0 \leq i < j$.

Points B_{ij} and A_{ij} on the graph of $S(t)$ at (Fig. 21.2) have coordinates $(T_k^{ij}, S(T_k^{ij}))$ and $(T_r^{ij}, S(T_r^{ij}))$ respectively. Denote $T_k^j = T_k^{j-1,j}$.

It is obvious that the crucial points also form an arithmetic progression: $T_k^{i+1,j+1} = T_k^{ij} + d$, $T_k^{ij} = T_k^{0,j-i} + id$. For example, the following fact follows from (21.6) and (Fig. 21.2):

$$T_l^i + d < T_k^{ij} + d < T_L^i + d$$

if and only if $T_l^{i+1} < T_k^{ij} + d < T_L^{i+1}$. Besides,

$$\begin{aligned} S(T_k^{ij} + d) &= e^{-\eta(T_k^{ij} + 2\pi/\tau)} \cos\tau(T_k^{ij} + 2\pi/\tau) \\ &= e^{-\eta T_k^{ij}} \cos(\tau T_k^{ij}) e^{-\eta 2\pi/\tau} = cq^{j+1}. \end{aligned}$$

Denote $\Delta_i = T_r^{1,i+1} - T_k^{0,i+1}$, $\delta_i = T_k^{1,i+1} - T_r^{1,i+1}$, $D_i = \Delta_i + \delta_{i+1}$ $d_i = \Delta_i + \delta_i$, $i \geq 0$. In this notation it is easy to see that for every $j \geq 0$

$$T_r^{j+1,j+i+1} - T_k^{j,j+i+1} = \Delta_i \qquad T_k^{j+1,j+i+1} - T_r^{j+1,j+i+1} = \delta_i. \qquad (21.7)$$

Besides $\Delta_i > d/2$, $D_i > d$, $\delta_i < d/2$, $d_i < d$.

It is essential to investigate the properties of $H_\gamma(t)$ as a function of γ. In particular the pre-image's scope given the value $S(t) = cq^m$ is $R_m = T_l^m - T_k^{0m} = \Delta_0 + \sum_{j=1}^{m-1} d_j$.

The following values of the partiality parameter

$$a_{s,m} = \left(\Delta_{m-s} + \sum_{j=m-s+1}^{m-1} d_j\right) \bigg/ \left(\Delta_0 + \sum_{j=1}^{m-1} d_j\right) \qquad (21.8)$$

$$b_{s,m} = \left(\sum_{j=m-s}^{m-1} d_j\right) \bigg/ \left(\Delta_0 + \sum_{j=1}^{m-1} d_j\right), \qquad (21.9)$$

determine the pre-image's points corresponding to the crucial points T_r^{sm} and T_k^{sm} respectively.

The most essential property of the function $H_\gamma(t)$ is that for some rational $0 < \gamma = s/m < 1$ the values of the function $H_\gamma(t)$ at the jump points can be described as a generalized geometrical distribution with some parameter α.

If in this case α coincides with γ we say that there is *accordance* between generalized double inversion of function $S(t)$ and generalized geometrical distribution. The conditions of such accordance for $\gamma = \frac{1}{m}$ are stated in the following theorem.

Theorem 21.4.1

1. Let R_M be the scope of the pre-image for $S(T_l^M)$, $M = rm + l$, $l \neq 0$. If $l > m/2$ then the following condition holds

$$\Delta_{M-r-1} > d\frac{2l-1}{2(m-1)}, \qquad (21.10)$$

$$b_{t,M} < 1/m < a_{t+1,M}, \qquad (21.11)$$

where $t = \lfloor M/m \rfloor$. In case $l < m/2$ the inequality (21.11) holds independently of the condition (21.10).

2. Let the partiality parameter $\gamma = 1/m$, $m > 2$. Define $q_1 = cq^\mu$, $q_0 = cq$,

$$\mu = \sup\{m : \Delta_{m-2} > (2m-3)d/(2m-2)\}. \tag{21.12}$$

Then $1 < \mu < \infty$ and in $[q_1, q_0]$ the function $H_\gamma(t)$ is in accordance with the generalized geometrical distribution.

PROOF. 1. Denote $D_i^k = \sum_{j=i}^k d_j$. Firstly, we show that $b_{t,M} < 1/m$, where $t = \lfloor M/m \rfloor$.

This inequality is equivalent to $mD_{m-t}^{M-1} < D_0^{M-1}$ or $(m-1)D_{m-t}^{M-1} < D_0^{M-t-1}$, which is true because of (21.7) and because for $M = rm + l$ and $l \neq 0$ the inequality $mt < M + 1/2$ holds.

Indeed,

$$(m-1)D_{M-t}^{M-1} < (m-1)td < (M-t-1/2)\,d < D_0^{M-t-1}, \tag{21.13}$$

as: $D_0^{M-t-1} = \Delta_0 + \delta_1 + \cdots + \Delta_{M-t-2} + \delta_{M-t-1} + \Delta_{M-t-1} = D_0 + \cdots + D_{M-t-2} + \Delta_{M-t-1} > (M-t-1)d + d/2$.

Consider now the inequality

$$1/m < a_{t+1,M}, \tag{21.14}$$

which is equivalent to the inequality

$$m\Delta_{M-t-1} + mD_{M-t}^{M-1} > D_0^{m-1}. \tag{21.15}$$

Denote $Q_1 = (m-1)td + (m-1)\Delta_{M-t-1}$ $Q_2 = (M - t - 1/2)\,d$. It's clear that $Q_1 > Q_2$ as inequality $\Delta_{M-t-1} > d\frac{M-mt-1/2}{m-1}$ is equivalent to (21.10) with $M = rm + l$, $l \neq 0$. Therefore with (21.7) we have

$$(m-1)\Delta_{M-t-1} + (m-1)D_{M-t}^{M-1} > Q_1 > Q_2 > D_0^{M-t-2} + \delta_{M-t-1},$$

which is equivalent to (21.15), i.e. (21.14) holds.

2. Use the inequality (21.10) with $M = rm + l$.

Firstly, we show that it is enough to consider only case $l = m - 1$. If (21.10) does not hold for $l = m - 2$, then it does not hold for $l = m - 1$ since

$$\begin{aligned}
\Delta_{r(m-1)+m-2} &< \Delta_{r(m-1)+m-3} \\
&< (2m-5)\,d/(2m-2) \\
&< (2m-3)\,d/(2m-2) \stackrel{def}{=} A_m.
\end{aligned} \tag{21.16}$$

So in case $l = m - 1$ the inequality (21.10) is equivalent to

$$\Delta_{r(m-1)+m-2} > A_m. \tag{21.17}$$

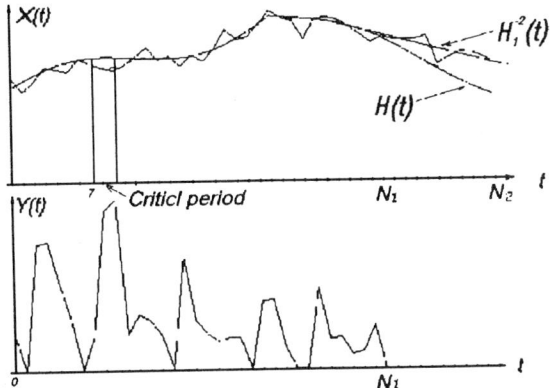

Figure 21.3: Dynamics of advertisement ($Y(t)$) and sales percentage ($X(t)$).

Since the sequence Δ_j decreases we consider $r = 0$ and $\Delta_{m-2} > A_m$ to obtain the largest m for which the inequality is true.

It's clear that for $m = 2$ the inequality is true. Under $m \to \infty$ we have $\Delta_{m-2} \searrow d/2$, $A_m \nearrow d$. So a finite μ exists such that (21.12) holds for which

$$b_{0,\mu-1} < 1/\mu < a_{1,\mu-1}. \tag{21.18}$$

■

21.4.2 Example

Consider two time series: X_t, $t = 1, \ldots, T_x$, — the product's sales percentage, and y_t, $t = 1, \ldots, T_y$ — the product's advertisement cost.

We propose the following dynamic model:

$$H(t) = \gamma_0 e^{-\beta_0|t-T_0|} + \gamma e^{-\beta|t-T|} \cos\alpha(t-T). \tag{21.19}$$

Parameters are estimated by the least squares method independently of the scale parameter γ_0:

$$\beta_0 = -\ln \sum_{i=0}^{n-1} x_t x_{t+1} / \sum_{i=0}^{n-1} x_t^2, \quad \cos\hat{\alpha} = \frac{1}{2\sqrt{ab}}, \quad \hat{\beta} = \frac{1}{2}\ln\frac{\beta}{\alpha},$$

where

$$a = \frac{\sum u_i \sum u_i v_i - \sum v_i \sum u_i^2}{(\sum u_i)^2 - n \sum u_i^2}, \quad b = \frac{\sum v_i - n\hat{a}}{\sum v_i},$$

$u_t = z(t+2)/z(t)$, $\quad v(t) = z(t+1)/z(t)$, $\quad z_t = x_t - H(t|\gamma = 0)$. For the product under consideration (Fig. 21.3) parameters were estimated on the base of the first 33 points:

$$\hat{T}_0 = 24, \ \hat{\alpha} = \frac{\pi}{13} = 0.242, \ \hat{\beta} = 0.007, \ T = 5, \ \hat{\gamma} = 0.104.$$

The region of stable regulation is shown at the picture (Fig. 21.3). It corresponds to the $2.269 < t < 4.271$ which is the region between the crucial points T_k^{01} and T_k^{04} since the largest m allowing the accordance of different inversing types of function $S(t) = e^{-\beta t}\cos\alpha t$ is 4.

The forecasting of sales percentage after the 33rd point was performed based on the estimated parameters.

Data of sales percentage shows that the optimal control based on the function H_{11}^{-2} ($\gamma = 1$) determines the real direction of developing. Such optimality, from our point of view, can be explained by the great advertisement taken place at the crucial time region which is the region of stable regulations. This conclusion is supported by some analogous examples.

Acknowledgments. Financial support was provided by the Russian Fundamental Research (Grant 99-04-49696).

Appendix

PROOF OF THEOREM 21.1.1. 1) Let f_α^- be partially inverse to f and let $x_i \in X_\delta$ be such that $y_i \in Y_\alpha$ exist where: $x_i = f^-(y_i)$, $i = 1, 2$. Then $\rho(x) = 1$ implies $f(x_1) = f(x_2) = ff^-(y_1) = ff^-(y_2)$, implies by the definition of generalized inversion [Bart, Klochkova and Kozhanov (1993)] ($O_1 : ff^-(y) = y$) we have $y_1 = y_2$ and therefore $x_1 = x_2$.

2) The proof is similar to the previous one using the other definition [Bart, Klochkova and Kozhanov (1993)] ($O_3 : f^-f(x) = x$).

The third statement follows from the two previous because at a jump point of f $\rho(x) = 0$ and for any $y \in \mathcal{Y}^- \setminus \mathcal{Y}$ $\rho^-(y) = 1$. ∎

PROOF OF THEOREM 21.2.2. We begin by proving the following lemma which introduces a shorter notation.

$$\beta_-^*(n, k) = \beta_-^*(n - k \mid 1, p, \alpha) = P(\xi_\alpha^-(0) + \cdots + \xi_\alpha^-(0) = n - k) \quad (21.20)$$

Note that $\beta_+^*(0, n) = q^{\alpha n}$, $\beta_+^*(k, n) = \beta_+^*(k \mid n)..$

Lemma 21.4.1 *For every s, $1 \le s \le k - 1$*

$$\beta_\pm^*(n, k) = \sum_{t=0}^{n-k} \beta_-^*(t + s, s)\beta_\pm^*(k - s, n - t - s).$$

Partially Inversion of Functions for Statistical Modelling

PROOF. The recurrent formula for the generalized negative binomial distribution is obvious after dividing up the k summands in (21.20) into 2 groups of s and $k-s$ summands and applying the formula of composite probability.

For the generalized positive binomial distribution it is sufficient to use the recurrent formula for the corresponding negative binomial distribution we have just proved:

$$\beta_{\pm}^{*>}(n,k) = \sum_{t=0}^{n-k} \beta_{-}^{*}(t+s,s)\beta_{\pm}^{*>}(k-s, n-t-s) + \beta_{-}^{*>}(n-k+s,s)$$

and then apply the fiducial identity which is the definition of considering the distribution

$$\beta_{+}^{*<}(k,n) = P(\xi_{\alpha}^{+}(n) < k) = P(\xi_{\alpha}^{-}(k) > n) = \beta_{-}^{*>}(n,k).$$

∎

1) Applying Lemma 21.4.1 for the generalized positive binomial distribution with $s=1$ we have

$$\begin{aligned}
f_n(\lambda) &= \sum_{k=0}^{n} \lambda^k \beta_+^*(k \mid n,p,\alpha) \\
&= \sum_{k=0}^{n} \sum_{t=0}^{n-k} \beta_-^*(t+1,1)\beta_+^*(k-1, n-t-1) \\
&= \sum_{k=1}^{n} \lambda \sum_{t=0}^{n-k} \lambda^{k-1} \beta_-^*(t+1,1)\beta_+^*(k-1,n-t-1) + q^{\alpha n}
\end{aligned}$$

changing the order of summation and considering the generating function

$$= \sum_{t=0}^{n-1} \lambda f_{n-t-1}(\lambda) \beta_-^*(t+1,1) + q^{\alpha n};$$

Comparing with (21.4) we have the statement.

2) Induction by n.
The start of the induction was proved in the previous item.

$$\begin{aligned}
f_{n+1}^i(\lambda) &= \lambda f_n^i(\lambda) p_i + \sum_{t=i+1}^{n+i} \lambda q_i \cdots q_{t-1} p_t f_{n-t+i}(\lambda) + q_{i+1} \cdots q_{n+i} \\
&= \lambda p_i f_n(\lambda) + q_i(\lambda f_{n-1}^i(\lambda) p_{i+1} \\
&\quad + \sum_{t=i+2}^{n+i} \lambda q_i \cdots q_{t-1} p_t f_{n-t+i}(\lambda) + q_{i+1} \cdots q_{n+i})
\end{aligned}$$

$$= \lambda p_i f_n^0(\lambda) + f_{n+1}^0(\lambda)$$
$$= \lambda f_n^0(\lambda) p_0 + q_0 (\sum_{t=1}^{n} \lambda q_1 \cdots q_{t-1} p_t f_{n-t}(\lambda) + q_1 \cdots q_n$$
$$= \lambda p_0 f_n^0(\lambda) + q_0 f_n^1(\lambda).$$

Therefore $F_{n+1}(\lambda) = [\lambda P + Q] F_n(\lambda) = [\lambda P + Q]^n I_s$, $F_0(\lambda) = I_s$.

3) By direct calculation of determinant and algebraic adjuncts of matrix $[E - \nu Q]$ we have the inverse matrix.

$$det[E - \nu Q] = 1 - \nu^s q_0 \cdots q_{s-1} = 1 - \nu^s q^m,$$

$$[E - \nu Q]^{-1} = \begin{pmatrix} 1 & \nu q_0 & \nu^2 q_0 q_1 & \cdots & \nu^{s-1} q_0 q_1 \cdots q_{s-2} \\ \nu^{s-1} q_1 \cdots q_{s-1} & 1 & \nu q_1 & \cdots & \nu^{s-2} q_1 q_2 \cdots q_{s-2} \\ \nu^{s-2} q_2 \cdots q_{s-1} & \nu^{s-1} q_0 \cdots q_{s-1} & 1 & \cdots & \nu^{s-3} q_2 q_3 \cdots q_{s-2} \\ \vdots & \vdots & \vdots & \ddots & \vdots \\ \nu q_{s-1} & \nu^2 q_0 q_{s-1} & \nu^3 q_0 q_1 q_{s-1} & \cdots & 1 \end{pmatrix}.$$

The t-th component of the vector $G(\nu) = (E - \nu Q)^{-1} P e_t$ where e_t is the t orth, is of the type

$$G_t(\nu) = \frac{1}{1 - \nu^s q^m} (\nu^{s-t} q_t \cdots q_{s-1} \sum_{i=0}^{t-1} \nu^i p_i q_{i-1} \cdots q_0 + p_t + \sum_{i=t+1}^{s-1} \nu^{i-t} p^i q_{i-1} \cdots q_t).$$

Substituting $t = 0$ we have our result. ■

References

1. Fisher, R. A. (1930). Inverse probability, *Proceedings of the Royal Cambridge Society*, **26**, 528–535.

2. Klimov, G. P. (1973). *Invariant Inferences in the Statistics*, 186p, Moscow: MSU (in Russian).

3. Bart, A. G. (1987). Integrity and the Control over the Biological Systems, In *Biometrical aspects of the organism integrity Study*, pp.141–151, Moscow: Nauka (in Russian).

4. Bart, A. G. and Klochkova, N. and Kozhanov, V. M. (1993). The Universal Scheme of Regulations in Biosystems for the Analysis of Neuron Junctions as an Example, In *Model-Oriented Data Analysis* (Ed. W. G. Muller, H. P. Wynn and A. A. Zhigljavsky), pp. 167–177, Heidelberg: Physica-Verlag.

5. Bart, A. G., Klochkova, N., Kozhanov, V. M. and Chmykhova, N. M. (1997). The Study of Interneuronal Interaction Mechanisms on the Basis of the Reflections Principle, *Evolutionary Biochemistry and Physiology*, **33**, No **4,5**, pp.462–474 (in Russian).

6. Bart, A. G., Dityatev, A. E. and Kozhanov, V. M. (1988). Quantal Analysis of the Postsynaptic Potentials in the Interneuronal Synapses: Recovery of a Signal from the Noise, *Neurophysiology* **20**, pp. 479–487, Kiev (in Russian).

7. Bart, A. G. and Ivanov, S. L. (1995). Distributional Analysis of Gingival Indices, *Oral Care Statistical Studies (selected papers)*, (Ed., A. A.Zhigljavsky), pp. 167–176, University of St. Petersburg.

8. Bart, A. G., Bondarenko, B. B. and Boiko, B. I. (1980). Mathematical analyisis of the chronic glomerulonephritis, *Glomerulonephritis*, (Ed., S. I. Riabov), pp. 213–225, Leningrad: Medicine (in Russian).

22

Simple Efficient Estimation for Three-Parameter Lognormal Distributions with Applications to Emissions Data and State Traffic Rate Data

N. Balakrishnan and Jun Wang

McMaster University, Hamilton, ON, Canada

Abstract: We propose some simple efficient estimators for the three-parameter lognormal distribution. In addition to their computational superiority, these estimators are more efficient (in terms of mean square errors) than the modified moment estimators proposed by Cohen and Whitten and are also more efficient than the local maximum likelihood estimators in many cases. Several illustrative examples are finally presented, some dealing with the analysis of emissions data.

Keywords and phrases: Lognormal distributions, estimation, bias, mean square errors

22.1 Introduction

The lognormal distribution has had a long and rich history; see, for example, Aitchison and Brown (1957), Crow and Shimizu (1988), and Chapter 14 of Johnson, Kotz and Balakrishnan (1994). The differences between the normal and the lognormal distributions and their respective roles in statistical applications have been highlighted by Kotz (1973). In spite of all these developments and discussions, the estimation problem for the three-parameter lognormal distribution has not been resolved satisfactorily. As has been noted by Cohen, Whitten and Ding (1985), the maximum likelihood method of estimation can often lead to inadmissible estimates and the local maximum likelihood estimators (also discussed by these authors) need not always exist. On the other hand, the moment estimators are easy to determine and implement, but they fail to make full use of the available sample information and consequently possess variances which are often unduly large.

For this reason, Cohen, Whitten and Ding (1985) suggested modified moment estimators for this case and displayed their efficiency (as compared to the maximum likelihood estimators and the local maximum likelihood estimators) through extensive Monte Carlo simulations. This modified method of moment estimation has also been adopted successfully for a number of other three- parameter distributions such as gamma, Weibull and inverse Gaussian; see, for example, the papers by Cohen and Whitten (1982, 1985, 1986) and Cohen, Whitten and Ding (1984), and the books by Cohen and Whitten (1988) and Balakrishnan and Cohen (1991).

The two-parameter (location and scale, with fixed shape parameter) lognormal distribution has also been studied and applied quite extensively. For example, Gupta, McDonald and Galarneau (1974) presented tables of single and product moments of order statistics (for sample sizes up to 20) from the standard form of this distribution. Life test sampling plans based on this distribution have been discussed earlier by Gupta (1962). Gibbons and McDonald (1975), by making use of the tables prepared by Gupta, McDonald and Galarneau (1974), discussed the exact best linear unbiased estimation of the location and scale parameters of this distribution and illustrated them with emissions data. Selection procedures based on this distribution have been discussed by McDonald (1979) who then used them to analyse state traffic rates. For the emissions data, McDonald, Vance and Gibbons (1995) discussed some discrimination methods between the lognormal and Weibull distributions.

In this paper, we propose some simple efficient estimators which are functions of order statistics for the three-parameter lognormal distribution. In addition to their computational superiority, these estimators are more efficient (in terms of mean square errors) than the modified moment estimators proposed by Cohen, Whitten and Ding (1985) and are also more efficient than the local maximum likelihood estimators in many cases. In particular, when the shape parameter σ of the lognormal distribution is close to 0 (i.e, when the underlying lognormal distribution is very close to normal), the proposed estimators significantly outperform both the modified moment estimators and the local maximum likelihood estimators. Finally, we present several examples in order to illustrate the method of estimation presented in this paper; some of these examples deal with the analysis of emissions data.

22.2 Explicit Estimators

Let Y be a $lognormal(\gamma, \beta, \sigma)$ random variable, with location (threshold) parameter γ, scale parameter β and the shape parameter σ. Its probability density function is [see Johnson, Kotz and Balakrishnan (1994, pp. 208)]

$$f(y, \gamma, \beta, \sigma) = \frac{exp\{-[ln(y-\gamma) - ln\beta]^2/(2\sigma^2)\}}{\sqrt{2\pi}\sigma(y-\gamma)}, \quad y > \gamma, \ \beta > 0, \ \sigma > 0. \tag{22.1}$$

Sometimes, the parameter $\omega = exp(\sigma^2)$ is used as the shape parameter instead of σ.

Then, we propose the following simple explicit estimators (called NEW) for the three parameters of the distribution in (22.1):

$$\hat{\gamma} = y_{(1)} - [0.27 + 0.0078 \ min\{100, n\}](y_{(5)} - y_{(1)}), \tag{22.2}$$

$$\hat{\sigma} = \frac{2}{p}\sqrt{|ln\frac{m_p}{m_q^2}|}, \tag{22.3}$$

$$\hat{\beta} = \frac{\bar{y} - \hat{\gamma}}{exp[\hat{\sigma}^2/2]}, \tag{22.4}$$

where $y_{(1)}, y_{(2)}, ..., y_{(n)}$ are the order statistics obtained from the given sample of size n from three-parameter lognormal distribution in (22.1), and

$$m_a = \frac{1}{n}\sum_{i=1}^{n}(y_{(i)} - \hat{\gamma})^a, \quad a > 0,$$

$$t = \frac{2}{3}\sqrt{|ln\frac{m_3}{m_{1.5}^2}|}, \quad s = min\{\frac{1}{10}exp[18/(10t)], 12\},$$

$$p = max\{s, 1/2\}, \quad q = p/2.$$

In fact, these NEW estimators can be thought of as coming from an infinite family of estimators with different coefficients for the first several order statistics and combinations of exponents. The choice made above appears (from our study) to be the one having an overall good performance (in terms of both bias and mean square errors).

22.3 Simulation Results

In this section, we present some simulation results concerning the bias, variances and mean square errors of these estimators. We also make comparisons with the corresponding quantities for Cohen, Whitten and Ding's (1985) modified method-of-moments estimators (MME's) as well as the local maximum likelihood estimators (LMLE's) given by Cohen, Whitten and Ding (1985).

In the following table, we have presented the simulation results on the bias and mean square errors of the NEW estimators for various choices of the shape parameter ω and sample size $n = 100$ determined from 10,000 Monte Carlo runs using IMSL FORTRAN program. For the purpose of comparison, we have also included the corresponding bias and mean square errors of the modified moment estimators and the local maximum likelihood estimators taken from the tables of Cohen, Whitten and Ding (1985).

It is quite clear from this table that the new estimators are on the whole superior in all the cases considered as compared to the MME's both in terms of both bias and mean square error. Also, in most cases, these estimators turn out to be better than the LMLE's. In particular, we observe from the table that when ω is close to 1 (i.e., when σ is close to 0), both the modified method-of-moments estimators as well as the local maximum likelihood estimators of γ and β have very large mean square error values as compared to the new estimators. This suggests that in cases where ω is close to 1, these two estimators should not be used. Realize that in this situation the lognormal distribution is very close to the normal distribution [see Johnson, Kotz and Balakrishnan (1994, pp. 212-216)] so that the discrimination between the two distributions as well as efficient estimation of the underlying parameters becomes a very critical problem.

		Mean Square Errors of the Estimators					
		ω					
PARAMETER	ESTIMATOR	1.03	1.1	1.25	1.5	2.0	3.0
γ	NEW	0.311	0.122	0.036	0.014	0.007	0.004
	MME	6.036	0.885	0.052	0.018	0.008	0.004
	LMLE	12.751	1.031	0.034	0.014	0.006	0.002
σ	NEW	0.010	0.010	0.010	0.010	0.011	0.012
	MME	0.010	0.010	0.010	0.011	0.016	0.024
	LMLE	0.008	0.008	0.009	0.010	0.012	0.013
β	NEW	0.310	0.121	0.041	0.022	0.018	0.024
	MME	6.078	0.909	0.064	0.033	0.028	0.040
	LMLE	12.779	1.043	0.043	0.027	0.018	0.018
		Bias of the Estimators					
γ	NEW	0.550	0.331	0.157	0.061	0.003	-0.019
	MME	-0.996	-0.260	-0.064	-0.031	-0.024	-0.031
	LMLE	-0.700	-0.116	-0.014	0.007	0.016	0.013
σ	NEW	0.084	0.042	0.028	0.009	-0.017	-0.034
	MME	0.006	-0.019	-0.021	-0.025	-0.042	-0.094
	LMLE	0.010	0.003	0.005	0.012	0.019	0.021
β	NEW	-0.549	-0.326	-0.152	-0.052	0.008	0.051
	MME	1.001	0.269	0.079	0.048	0.062	0.107
	LMLE	0.699	0.115	0.014	-0.012	-0.019	-0.027

22.4 Illustrative Examples

Example 1: Cohen, Whitten and Ding (1985) discussed the following data representing the maximum flood level in millions of cubic feet per second for the Susquehanna River at Harrisburg, Pennsylvania, over 20 four-year periods from 1890 to 1969:

0.654 0.613 0.315 0.449 0.297 0.402 0.379 0.423 0.379 0.3235

0.269 0.740 0.418 0.412 0.494 0.416 0.338 0.392 0.484 0.265

For this data, by assuming the three-parameter lognormal distribution in (22.1), we find the NEW estimates of the parameters (from Eqs. (22.2)-(22.4)) to be

$$\hat{\gamma} = 0.240, \quad \hat{\sigma} = 0.569, \quad \hat{\beta} = 0.156$$

with corresponding estimates for the population mean, variance and the third standard moment as 0.423, 0.013 and 2.093, respectively.

It needs to be pointed out here that Cohen, Whitten and Ding (1985) have determined the MME's of location parameter, shape parameter, scale parameter, population mean, variance and the third standard moment as:

0.171, 0.470, 0.225, 0.422, 0.016, 1.614,

respectively. Similarly, they have also reported the corresponding LMLE's as

0.185, 0.507, 0.210, 0.424, 0.017, 1.783.

Example 2: The following data consisting of fatigue life in hours of ten bearings of a certain type was used by Cohen, Whitten and Ding (1985). They are arranged in increasing order of magnitude:

152.7 170.2 172.5 173.3 193.0 204.7 216.5 234.9 262.6 422.6

For this data, using NEW estimation method, we obtain the estimates

$$\hat{\gamma} = 138.676, \quad \hat{\sigma} = 0.805, \quad \hat{\beta} = 59.147$$

with corresponding estimates for the population mean, variance and the third standard moment as 220.481, 6109.169 and 3.739.

For the same data, Cohen, Whitten and Ding (1985) have reported the MME's of the location parameter, shape parameter, scale parameter, population mean, variance and the third standard moment as

$$132.38, \ 0.764, \ 65.85, \ 220.547, \ 6161.55, \ 3.377,$$

respectively. Similarly, they have reported the corresponding moment estimates as
$$80.97, \ 0.5524, \ 121.620, \ 220.487, \ 6150.440, \ 1.864.$$

Example 3: The following data, taken from Gibbons and McDonald (1975), present the automotive emmision in grams/mile. Once again, they are presented in increasing order of magnitude:

0.39 0.42 0.42 0.43 0.44 0.46 0.47 0.49 0.51 0.53 0.55 0.56 0.62 0.85

In this case, the NEW estimates of the three parameters turn out to be

$$\hat{\gamma} = 0.371, \quad \hat{\sigma} = 0.732, \quad \hat{\beta} = 0.106$$

with corresponding estimates for the population mean, variance and the third standard moment as 0.510, 0.014 and 3.122, respectively.

It should be noted that Gibbons and McDonald (1975) assumed that the shape parameter of the underlying lognormal distribution $\sigma = 1$ and then obtained the corresponding best linear unbiased estimates (BLUE's) of the location and scale parameters to be 0.373 and 0.096, using which we get their estimates of the population mean, variance and the third standard moment to be 0.531, 0.043, 6.185.

It is of interest to mention here that the Kolmorov-Smirnov distance between the fitted two-parameter lognormal distribution function and the empirical distribution function in this case turns out to be 0.10961; on the other hand, the Kolmogorov-Smirnov distance between the three-parameter lognormal distribution function and the empirical distribution function turns out to be 0.07838. The plots of these three are presented in Figure 22.1, where the dashed line corresponds to the two-parameter model while the solid line corresponds to the three-parameter model. Observe that the two-parameter lognormal model assumed by Gibbons and McDonald (1975) fits the lower order statistics very well while having a large departure in fitting the larger order statistics; but, the three-parameter model fits the data very well overall.

Example 4: The following data, taken from McDonald (1979), reports the motor vehicle traffic fatalities per year per 100,000,000 vehicle miles for the

Three-Parameter Lognormal Distributions

year 1976 for 49 states in USA (excluding Hawaii):

    4.0 4.1 3.7 3.1 3.7 2.2 3.2 2.1 3.0 3.4 4.1 3.0 3.4 4.0 3.2 3.3
    4.4 2.8 2.6 2.8 3.3 3.0 4.9 3.8 4.6 3.3 4.5 2.9 2.1 5.3 3.4 3.9
    3.3 2.8 3.6 3.8 2.8 2.1 3.7 4.3 3.2 3.7 3.1 3.4 2.8 3.3 4.1 3.2 6.4

If we assume the three-parameter lognormal distribution in (22.1) for this data, we obtain the NEW estimates for the three parameters to be

$$\hat{\gamma} = 1.774, \quad \hat{\sigma} = 0.385 \quad \hat{\beta} = 1.588$$

with corresponding estimates for the population mean, variance and the third standard moment as 3.484, 0.467 and 1.263, respectively.

In order to verify the suitability of the assumed three-parameter lognormal distribution for this data, we constructed a Q-Q plot (by assuming the shape parameter $\sigma = 0.385$) and then calculated the correlation coefficient between the observed quantiles and the corresponding population quantiles (of the estimated model) to be 0.984. The Q-Q plot for this example is presented in Figure 22.2. Through simulation, we then found that the p-value for this correlation statistic is 0.323 which clearly indicates that the three-parameter lognormal distribution is quite suitable for the data at hand.

Example 5: As in the previous example, the following data from McDonald (1979) represent motor vehicle traffic fatalities per year per 100,000,000 vehicle miles for the year 1970 for 49 states in USA (excluding Hawaii):

    6.4 6.3 5.4 4.2 5.2 2.7 5.1 4.3 5.2 6.2 6.9 4.2 4.8 4.9 4.9 5.4
    7.1 4.5 3.8 3.5 4.1 4.4 7.7 5.7 6.5 4.3 7.4 4.4 3.2 7.6 4.5 6.0
    4.5 4.6 4.7 5.1 4.0 3.0 6.2 5.1 6.6 5.2 5.5 4.6 4.3 4.2 6.6 4.6 6.4

In this case, the NEW estimates turn out to be

$$\hat{\gamma} = 1.983, \quad \hat{\sigma} = 0.203, \quad \hat{\beta} = 3.096$$

with corresponding estimates of the population mean, variance and the third standard moment as 5.143, 0.421 and 0.625.

By constructing a Q-Q plot as in the last example (presented as Figure 22.3), we found the correlation coefficient from the plot to be 0.990 with the corresponding p-value to be 0.551. This clealy reveals that the three-parameter lognormal distribution is a good model for the data at hand.

22.5 Concluding Remarks

The simple estimators (which are functions of order statistics) proposed in this paper are more efficient that the modified moment estimators (in terms of mean square errors). They are also observed to be more efficient than the local maximum likelihood estimators in many instances. Also, in cases where the shape parameter σ of the lognormal distribution is close to 0 (with the corresponding lognormal distribution being close to the normal), these estimators significantly outperform both the modified moment estimators and the local maximum likelihood estimators. Furthermore, the estimators proposed in this paper (being explicit estimators) are also simple to implement. Based on all these considerations, we recommend the usage of these estimators for the estimation of the location, scale and shape parameters of lognormal distributions.

References

1. Aitchison, J. and Brown, J. A. C. (1957). *The Lognormal Distribution*, Cambridge University Press, Cambridge.

2. Balakrishnan, N. and Cohen, A. C. (1991). *Order Statistics and Inference: Estimation Methods*, Academic Press, San Diego.

3. Cohen, A. C. and Whitten, B. J. (1982). Modified moment and maximum likelihood estimators for parameters of the three-parameter gamma distribution, *Communications in Statistics - Simulation and Computation*, **11**, 197–216.

4. Cohen, A. C. and Whitten, B. J. (1985). Modified moment estimation for the three-parameter inverse Gaussian distribution, *Journal of Quality Technology*, **17**, 147–154.

5. Cohen, A. C. and Whitten, B. J. (1986). Modified moment estimation for the three-parameter gamma distribution, *Journal of Quality Technology*, **18**, 53–62.

6. Cohen, A. C. and Whitten, B. J. (1988). *Parameter Estimation in Reliability and Life Span Models*, Marcel Dekker, New York.

7. Cohen, A. C., Whitten, B. J. and Ding, Y. (1984). Modified moment estimation for the three-parameter Weibull distribution, *Journal of Quality Technology*, **16**, 159–167.

8. Cohen, A. C., Whitten, B. J. and Ding, Y. (1985). Modified moment estimation for the three-parameter lognormal distribution, *Journal of Quality Technology*, **17**, 92–99.

9. Crow, E. L. and Shimizu, K. (Eds.) (1988). *Lognormal Distributions: Theory and Applications*, Marcel Dekker, New York.

10. Gibbons, D. I. and McDonald, G. C. (1975). Small-sample estimation for the lognormal distribution with unit shape parameter, *IEEE Transactions on Reliability*, **R-24**, 290–295.

11. Gupta, S. S. (1962). Life test sampling plans for normal and lognormal distributions, *Technometrics*, **4**, 151–175.

12. Gupta, S. S., McDonald, G. C. and Galarneau, D. I. (1974). Moments, product moments and percentage points of the order statistics from the lognormal distribution for samples of size twenty and less, *Sankhyā, Series B*, **36**, 230–260.

13. Johnson, N. L., Kotz, S. and Balakrishnan, N. (1994). *Continuous Univariate Distributions, Vol. 1*, John Wiley & Sons, New York.

14. Kotz, S. (1973). Normality vs. lognormality with applications, *Communications in Statistics*, **1**, 113–132.

15. McDonald, G. C. (1979). Nonparametric selection procedures applied to state traffic rates, *Technometrics*, **21**, 515–523.

16. McDonald, G. C., Vance, L. C. and Gibbons, D. I. (1995). Some tests for discriminating lognormal and Weibull distributions - An application to emissions data, In *Recent Advances in Life-Testing and Reliability* (Ed., N. Balakrishnan), pp. 475–490, CRC Press, Boca Raton.

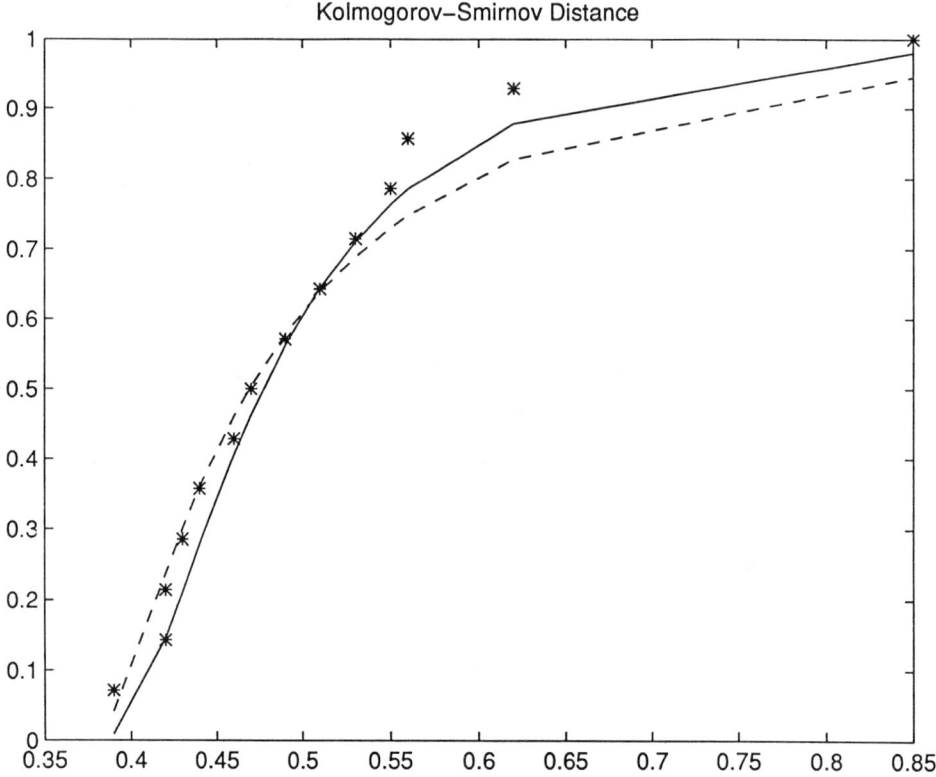

Figure 22.1: Empirical, two-parameter lognormal, and three-parameter lognormal distribution functions for the data in Example 3.

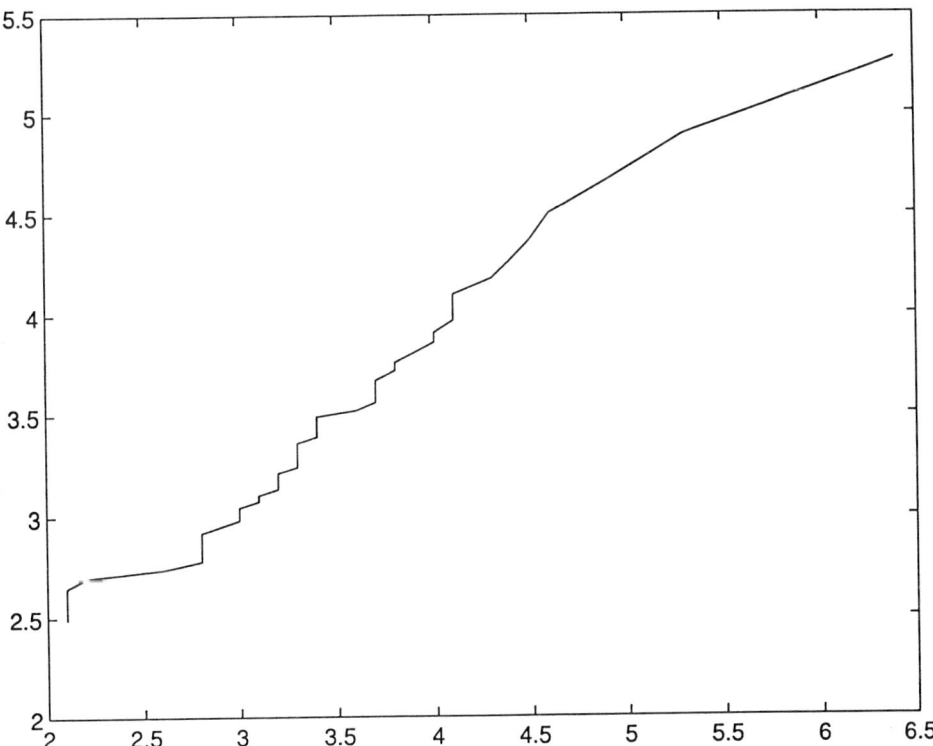

Figure 22.2: Q-Q Plot for Example 4.

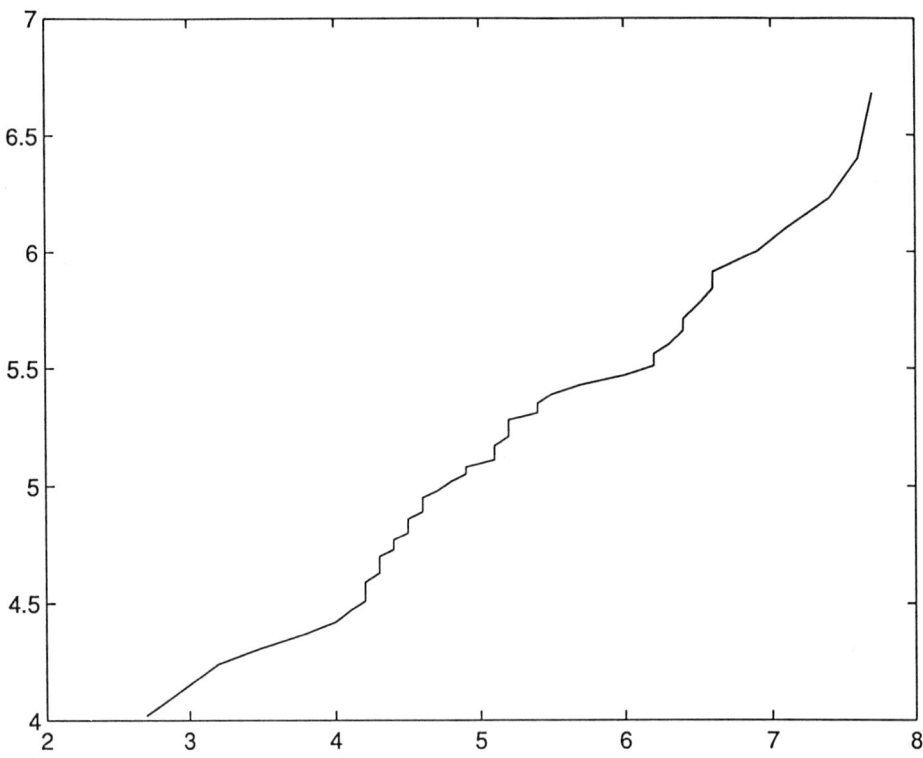

Figure 22.3: Q-Q Plot for Example 5.

Subject Index

Acyclic fork-join queueing networks, 63
Algorithm of optimization, 315
Algorithms, 165
Alias sets, 165
Ancillary statistics, 293
Approximate confidence interval, 207, 245
Approximate theory, 189
Asymptotic distribution, 33

Best linear unbiased estimators, 293
Bias, 373
Block designs, 165
Bootstrap, 305
Branching processes, 355

Change of variables, 197
Coefficients of skewness and kurtosis, 207, 245
Complexity, 47
Compound Poisson distribution, 355
Conditional inference, 293
Confidence intervals, 293
Confounded interactions, 165
Convex optimization problems, 197
Copolymerization, 153
Crucial points, 355

Dependent tests, 47
D-optimal designs, 189
Double moments, 207, 245
Double-exponential model, 135
Doubly truncated distribution, 207, 245
Dynamic state equation, 63

Edgeworth approximation, 207, 245
Efficiency, 135
Entropy measures, 337

Error-in-variables, 153
Estimation, 373
Estimator by absorption, 3
Estimator by collision, 3
Exact moments, 207, 245
Expected total sample size, 283
Experimental design, 153, 305
Extrapolation, 117

Families of basic matrices, 315
Fiducial distributions, 355
Fractional factorials, 165
Functional relations, 153

Generalized automaton, 315
Generalized M-estimation, 117

Hajek bound, 153

Incomplete quadratic models, 189
Indifference-zone, 283
Integration, 47
Invariance generating transformations, 197

Lack of fit, 117
Least squares estimator, 305
Legendre polynomials, 117
Linear regression, 189, 305
Locally optimal design, 135
Logistic model, 135
Lognormal distributions, 373
L-optimality, 197

Matrix method, 315
(max,+)-algebra, 63
Mean square errors, 373
Measure-valued random variables, 29
Moments, 337

Monte Carlo, 47
Monte Carlo estimators, 17
Monte Carlo method, 3
Monte Carlo simulation, 337
Multi-level method, 47
MV-optimality, 197

Navier-Stokes equation, 17
Neumann boundary value problem, 17
Neumann-Ulam scheme, 3
Neuronal trees, 355
Non-linear functionals, 29

Optimal design, 117
Optimal design of experiments, 197
Order statistics, 207, 245, 293
Orthogonal regression, 153

Pareto distribution, 207, 293
Partially inverse functions, 355
Periodically variable parameters, 315
Pivotal quantity, 207, 245,
Polynomial regression, 117, 197
Power function distribution, 245
Progressive censoring, 293

Quadruple moments, 207, 245

Randomization, 305
Ranking absolute means, 283
Recurrence relations, 207, 245
Reduced and minimal forms, 315
Resolution numbers, 165
Robustness, 135

Sales marketing, 355
Second-order expansion, 337
Service cycle time, 63
Single moments, 207, 245
Standardized criteria, 135
States minimization, 315
Stochastic dynamic systems, 63
Stratification technique, 29
Symmetric designs, 197
Symmetry, 197
System of nonlinear equations, 3

Tolerance intervals, 293
Total least squares, 153
Triple moments, 207, 245
Two-stage procedure, 283
Type-II right censoring, 293

Vector spaces, 165

Wavelets, 117
Weighted least squares, 117